Studies in Universal Logic

This series is devoted to the universal approach to logic and the development of a general theory of logics. It covers topics such as global set-ups for fundamental theorems of logic and frameworks for the study of logics, in particular logical matrices, Kripke structures, combination of logics, categorical logic, abstract proof theory, consequence operators, and algebraic logic. It includes also books with historical and philosophical discussions about the nature and scope of logic. Three types of books will appear in the series: graduate textbooks, research monographs, and volumes with contributed papers.

Jean-Yves Béziau · Dale Jacquette

Editors

Around and Beyond the Square of Opposition

 Birkhäuser

Editors
Jean-Yves Béziau
Institute of Philosophy
Federal University of Rio de Janeiro
Rio de Janeiro, RJ, Brazil

Dale Jacquette
Institute of Philosophy
University of Bern
Bern, Switzerland

ISBN 978-3-0348-0378-6   ISBN 978-3-0348-0379-3 (eBook)
DOI 10.1007/978-3-0348-0379-3
Springer Basel Heidelberg New York Dordrecht London

Library of Congress Control Number: 2012938884

Mathematics Subject Classification: 03-XX, 03Axx, 03Bxx

Printed on acid-free paper

Springer is part of Springer Science+Business Media (www.springer.com)

Preface

The theory of inferences and oppositions among categorical propositions, based on Aristotelian term logic, is pictured in a striking square diagram. The graphic representation of contradictories, contraries, subcontraries and subalterns intended as a foundation for syllogistic logic can be understood and applied in many different ways with interesting implications for various disciplines, notably including epistemology, linguistics, mathematics, psychology. The square can also be generalized in other two-dimensional and multi-dimensional graphic depictions of logical and other relations, extending in breath and depth the original Aristotelian theory. The square of opposition is accordingly a very attractive theme which has persisted down through the centuries with no signs of disappearing or even diminishing in fascination. For the last 10 years, there has been a new growing interest for the square due to new discoveries and challenging interpretations. This book presents a collection of previously unpublished papers by well-regarded specialists on the theory and interpretation of the concept and application of the square of opposition from all over the world. We thank all the authors who have contributed a paper to this book, and the referees who have analyzed, commented on, and made invaluable recommendations for improving the essays.

Rio de Janeiro, Brazil   Jean-Yves Béziau
Bern, Switzerland   Dale Jacquette

Contents

Contributors

Jean-Pascal Alcantara Centre Georges Chevrier-UMR 5605, Université de Bourgogne, Dijon, France

Jean-Yves Béziau CNPq – Brazilian Research Council, UFRJ – University of Brazil, Rio de Janeiro, Brazil

Ferdinando Cavaliere Cesenatico, FC, Italy

Saloua Chatti Department of Philosophy, Faculté des Sciences Humaines et Sociales, University of Tunis, Tunis, Tunisia

Ka-fat Chow The Hong Kong Polytechnic University, Hong Kong, China

Manuel Correia Facultad de Filosofía, Pontificia Universidad Católica de Chile, Santiago de Chile, Chile

Duilio D'Alfonso University of Calabria, Arcavacata di Rende, CS, Italy

C. de Ronde Departamento de Filosofía "Dr. A Korn", Universidad de Buenos Aires-CONICET, Buenos Aires, Argentina; Center Leo Apostel and Foundations of the Exact Sciences, Vrije Universiteit Brussel, Brussels, Belgium

Lorenz Demey Center for Logic and Analytical Philosophy, Institute of Philosophy, KU Leuven – University of Leuven, Leuven, Belgium

Jean-Pierre Desclés LaLIC (Langues, Logiques, Informatique et Cognition), Université de Paris Sorbonne, Paris, France

G. Domenech Instituto de Astronomía y Física del Espacio, Buenos Aires, Argentina

Antonino Drago University of Pisa, Pisa, Italy

H. Freytes Universita degli Studi di Cagliari, Cagliari, Italy; Instituto Argentino de Matemática, Buenos Aires, Argentina

Stamatios Gerogiorgakis University of Erfurt, Erfurt, Germany

Raffaela Giovagnoli Pontifical Lateran University, Vatican City, Italy

Dale Jacquette University of Bern, Bern, Switzerland

John T. Kearns Department of Philosophy and Center for Cognitive Science, University at Buffalo, the State University of New York, Buffalo, NY, USA

Srećko Kovač Institute of Philosophy, Zagreb, Croatia

Wolfgang Lenzen Dept. of Philosophy, University of Osnabrueck, Osnabrueck, Germany

Thierry Libert Département de mathématiques, Université Libre de Bruxelles (U.L.B.), Brussels, Belgium

Baptiste Mélès Université Blaise Pascal, Clermont-Ferrand, France

Anca Pascu LaLIC, Université de Bretagne Occidentale Brest, Brest Cedex 03, France

Stephen Read Department of Philosophy, University of St Andrews, St Andrews, Scotland, UK

Fabien Schang LHSP Henri Poincaré (UMR7117), Université de Lorraine, Nancy, France

Pieter A.M. Seuren Max Planck Institute for Psycholinguistics, Nijmegen, The Netherlands

Hartley Slater University of Western Australia, Crawley, WA, Australia

Mark Weinstein Department of Educational Foundations, Montclair State University, Upper Montclair, NJ, USA

Part I
Historical and Critical Aspects of the Square

The New Rising of the Square of Opposition

Jean-Yves Béziau

Abstract In this paper I relate the story about the new rising of the square of opposition: how I got in touch with it and started to develop new ideas and to organize world congresses on the topic with subsequent publications. My first contact with the square was in connection with Slater's criticisms of paraconsistent logic. Then by looking for an intuitive basis for paraconsistent negation, I was led to reconstruct S5 as a paraconsistent logic considering ¬□ as a paraconsistent negation. Making the connection between ¬□ and the O-corner of the square of opposition, I developed a paraconsistent star and hexagon of opposition and then a polyhedron of opposition, as a general framework to understand relations between modalities en negations. I also proposed the generalization of the theory of oppositions to polytomy. After having developed all this work I have begun to promote interdisciplinary world events on the square of opposition.

Keywords Square of opposition · Paraconsistent logic · Negation · Modal logic

Mathematics Subject Classification Primary 03B53 · Secondary 03B45 · 03B20 · 03B22

> *One can never tell when a old, discarded theory will suddenly burst into renewed life.*
>
> Brendan Larvor [35]

1 The Square Adventure

In June 2007 I organized the first world congress on the square of opposition in Montreux, Switzerland. In June 2010 was organized a second edition of this event in Corsica and presently a third edition is on the way for June 2012 in Beirut, Lebanon. After the first event was published a special issue of the journal *Logica Universalis* dedicated to the square of about 200 pages with 13 papers [15] and the book *The Square of Opposition— A General Framework for Cognition* of about 500 pages with 18 papers [19]. Most of the papers in the present book *Around and Beyond the Square of Opposition* are related to the talks presented at the second congress in Corsica and a double special issue of *Logica Universalis* [18] of about 250 pages with 10 papers has also been released following this event. Thus about 1500 pages have been recently published based on the square of opposition, a very productive tool.

J.-Y. Béziau, D. Jacquette (eds.), *Around and Beyond the Square of Opposition*, 3–19
Studies in Universal Logic, DOI 10.1007/978-3-0348-0379-3_1, © Springer Basel 2012

The aim of this paper is to tell how all this square animation did arise. In this paper I will speak about my first encounter with this creature and how our relation has developed changing the shape of both of us. There are mainly four stages in this square adventure: Los Angeles, CA, 1995; Fortaleza, Brazil, 1997; Stanford, CA, 2001; and Neuchâtel, Switzerland, 2003.

2 The Slater Affair (Los Angeles 1995)

In 1995 I was spending some time in Los Angeles, CA, as a Fulbright scholar at the Department of Mathematics of UCLA (invited by Herb Enderton). I was 30 years old and had never yet met the square face to face. Like any logician I had heard about it, but didn't know exactly what it was. For me it was connected with Aristotelian logic and syllogistic, something out-of-date, of historical interest, if any, similar to astrology.

I was then asked to write a *Mathematical Review* of a paper by Hartley Slater entitled "Paraconsistent logics?" [49] challenging the very existence of paraconsistent logics, a bunch of organisms I had been working on since a couple of years. I was quite interested to know up to which point I had been wasting time studying non-existent beings.

The weapon used by Slater to annihilate the paraconsistent ghosts was related to the square of opposition, a good reason to study this magic stick. In his paper the notions of contradictories and subcontraries, essential features of the square, are used to discuss the question whether a paraconsistent negation is a negation or not. Slater recalls that Priest and Routley did argue at some point that da Costa's negation is not a negation because it is a subcontrary functor but not a contradictory functor: in the paraconsistent logic C1 of da Costa, p and $\sim p$ cannot be false together, but sometimes can be true together, this is exactly the definition of subcontrariety.

An essential feature of a paraconsistent negation $\sim$ is to allow the rejection of the *ex-contradictio sequitur quod libet*: $p, \sim p \not\models q$. If we interpret this semantically, this means that there is a model (valuation, world) in which both p and $\sim p$ are true and in which q is false, for one p and for one q. This means therefore that $\sim$ is not a contradictory forming functor, but a subcontrary forming functor. So what applies to da Costa's paraconsistent negation seems to apply to any paraconsistent negation. That is on this basis that Slater argues that there is no paraconsistent negation at all, using the very idea of Priest and Routley that a negation should be a contradictory forming functor.

How can then the two pseudo-Australians argue that their paraconsistent negations are negations but not the Brazilian ones? That sounds quite impossible, unless there is a play on words, that's what Slater emphasizes. Priest has developed a paraconsistent logic LP with three-values, the difference with Łukasiewicz's logic is that the third value is considered as designated, that is the reason why in LP, we have $p, \sim p \not\models q$, since p and $\sim p$ can both have this third value. Now how can Priest argue that the paraconsistent negation of LP is a contradictory forming functor? It is because he is not calling the third value "true", despite the fact that it is designated. In my paper "Paraconsistent logic!" I comment the situation as follows:

> Priest's conjuring trick is the following: on the one hand he takes truth to be only 1 in order to say that his negation is a contradictory forming relation, and on the other hand he takes truth to be $\frac{1}{2}$ and 1 to define LP as a paraconsistent logic. However it is reasonable to demand to someone to keep his notion of truth constant, whatever it is. ([13], p. 21)

The subtitle of the quoted paper is "A reply to Slater". In this paper I argue that a contradictory forming functor is necessarily a classical negation, with a proof of a general theorem sustaining this fact. This entails that: (1) a paraconsistent negation cannot be a contradictory functor but not that: (2) a paraconsistent negation is not a negation—unless we admit that a negation has to be a classical negation. For me Slater is wrong arguing that:

> If we called what is now 'red', 'blue', and vice versa, would that show that pillar boxes are blue, and the sea is red? Surely the facts wouldn't change, only the mode of expression of them. Likewise if we called 'subcontraries', 'contradictories', would that show that 'it's not red' and 'it's not blue' were contradictories? Surely the same points hold. And that point shows that there is no 'paraconsistent' logic. ([49], p. 451)

A paraconsistency negation like the one of C1 is not definable in classical logic, so in this logic there is a new operator, it is not just a question of shifting names:

> It is important to note, against Slater, that the paraconsistent negations in the logics of da Costa and Priest (...) cannot be defined in classical logic. Thus paraconsistent logic is not a result of a verbal confusion similar to the one according to which we will exchange "point" for "line" in geometry, but rather the shift of meaning of "negation" in paraconsistent logic is comparable to the shift of meaning of "line" in non-Euclidean geometry. ([4], this idea was further developed in [8])

Nevertheless one should have good reasons to sustain that such new operators are indeed negations, one cannot define a paraconsistent negation only negatively by the rejection of the *ex-contradictio*. This is what I have explained at length in my papers "What is paraconsistent logic?" [5] showing also that it was not easy to find a bunch of positive criteria for positively define a paraconsistent negation. The square can help to picture a positive idea of paraconsistent negation: the very notion of subcontrariety can be used to support the existence of paraconsistent negations rather than their non-existence. I was led to this conclusion not by a mystical contemplation of the square, nor by a philological study of this legendary figure, but indirectly by trying to find a meaningful semantics for paraconsistent logics.

3 S5 Is a Paraconsistent Logic (Fortaleza/Rio de Janeiro 1997)

Paraconsistent logic was my point of departure for exploring the logic world. I did a Master thesis in mathematical logic on da Costa's paraconsistent logic C1 at the university of Paris 7 under the supervision of Daniel Andler in 1990 (see [3]). As I have explained elsewhere [14] my interest for paraconsistent logic was based on questions regarding the foundations of logic rather than by a childish attraction to paradoxes. If the principle of (non)contradiction is not a fundamental principle, what are the fundamental principles of logic, if any? How is it possible to reason without the principle of contradiction? Those were the questions I was exploring.

Investigating these questions I did study in details various paraconsistent logics and I noticed that generally they were *ad hoc* artificial constructions or/and had some problematic features. I was wondering if there was any intuitive idea corresponding to the notion of paraconsistent negation nicely supported by, or expressed through, a mathematical framework. In August 1997 I was taking part to the 4th WOLLIC (Workshop on Logic, Language, Information and Computation) in Fortaleza, Brazil and I was discussing the question of intuitive basis for paraconsistent negation with my colleague Arthur de Vallauris Buchsbaum. Influenced by *The Kybalion* [34], he was telling me that something can be considered true from a certain point of view and false from another viewpoint.

I was sympathetic to this idea having worked on the philosophy of quantum mechanics during my *Maîtrise de Philosophie* at the Sorbonne under the supervision of Bernard d'Espagnat (see [2]). I had been in particular studying the work of Heisenberg, Bohr and Bohm. From a certain viewpoint an object o is a particle, from another viewpoint it is a wave. Since a particle is something that cannot be a wave, this means that from a viewpoint it is true that o is a particle and from another viewpoint it is true that o is not a particle. Using classing negation this means that from this other viewpoint it is false that o is a particle. Bohr developed *complementarity* to explain this paradoxical situation, saying that there is no contradiction here since these two viewpoints are distinct visions that are not conflicting since the two contradictory features do not manifest in the same experiment. But what is then the reality beyond these two viewpoints? The Copenhagen interpretation eludes the problem, rejecting the idea of an objective reality beyond experiments. This can lead to a relativism not in the sense of Einstein, but in the sense of post-modernism, as caricatured by Sokal in his hoax quoting in fact Bohr: "A complete elucidation of one and the same object may require diverse points of view which defy a unique description" and Heisenberg: "As a final consequence, the natural laws formulated mathematically in quantum theory no longer deal with the elementary particles themselves but with our knowledge of them. Nor is it any longer possible to ask whether or not these particles exist in space and time objectively" ([52], p. 218).

David Bohm, like d'Espagnat, had a different philosophical idea, more Kantian, in the sense that for him these contradictory viewpoints are the expression of the limitations of our thought to capture the unthinkable noumenal reality. Bohm uses metaphors to explain paradoxes of quantum physics, for example he explains inseparability by the metaphor of an aquarium in a room with a fish filmed by two perpendicular cameras. In another room someone may see the two movies on two different screens trying to understand the strange interaction between the two fishes without knowing that it is in fact the same fish (see [24]).

Discussing with Arthur, I was disagreeing with him saying that everything can be considered as true form a certain viewpoint and false from another viewpoint. This seems to me too easy, too trivial. I had then the following idea: the negation of p is false if and only if p is considered true from all viewpoints. If p can be considered as true from a viewpoint and false from another viewpoint, then $\sim p$ can be considered also as true. What does this mean exactly, true from which viewpoint? This can be clarified using a Kripkean semantics. Viewpoints are considered as worlds and we say:

$\sim p$ is false in a world w iff p is true in all worlds accessible to w.

To simplify I decided to consider a universal accessibility relation, then we have:

$\sim p$ is false in a world w iff p is true in all worlds.

If p is not true in all worlds, this trivially means that there is a world w in which p is false, but this also means that $\sim p$ is true in all the worlds and we may have a world w in which p is true, so, in this world, p and $\sim p$ are both true.

Back to Rio de Janeiro where I was living I started to work the details of this semantics. I decided to call Z the logic with such a paraconsistent negation by reference to LeibniZ, because it was connected to the semantics of possible worlds. I developed this paper in view of the projected conference in July 1998 in Torun commemorating the 50 years of Jaśkowski's paper on paraconsistent logic [32]. Jaśkowski's logic is called *discussive*

Fig. 1 Paraconsistent
negation as a modality

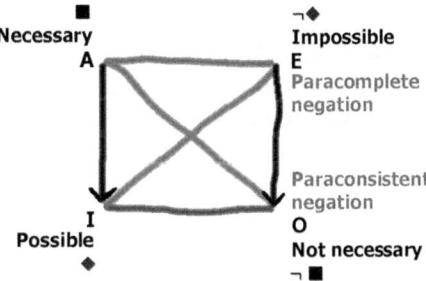

logic because the motivation was to formalize the logic of people discussing together. Different people may have different ideas, according to someone the proposition p may be true, according to someone else the proposition p may be false. The worlds of the logic Z can be considered as different opinions people may have. According to Z if everybody agrees that the proposition p is true, then everybody has to agree that the proposition $\sim p$ is false, but if someone thinks that the proposition p is false, then the proposition $\sim p$ is true for everybody and someone may also think that the proposition p is true, then for him p and $\sim p$ are both true. So my paper was called "The paraconsistent logic Z—a possible solution to Jaśkowski's problem" and presented at the Jaśkowski's commemorative conference. For a reason I ignored the paper was not subsequently published in the double issue of *Logic and Logical Philosophy* related with the conference, but much later in a issue of this same journal [12] and is having a certain success (cf. [38, 43–45]).

The main result of this paper is to axiomatize the logic Z which has, as primary connectives, classical conjunction, disjunction and implication and the negation $\sim p$ defined according to the above possible worlds semantical condition. In this paper I give a Hilbert's style axiomatization for Z, proving the completeness theorem generalizing the method of completeness for modal logic. I also show that it is possible to translate Z into S5, by interpreting $\sim p$ as $\neg\Box p$. This is easily shown by checking that:

$$\neg\Box p \text{ is false in a world } w \text{ iff } p \text{ is true in all worlds.}$$

I wrote this paper in December 1997 and then discussed these questions with Claudio Pizzi, who was visiting Rio at this time. The results of this discussion are described as follows in my paper "Adventures in the paraconsistent jungle":

> I discussed my discovery with the Italian logician Claudio Pizzi, who has a little castle in Copacabana and uses to come there frequently. Pizzi told me two important things: one right and one wrong. The wrong was that the operator $\neg\Box$ corresponds to contingency, the right was that S5 and Z are equivalent since it is possible in S5 to define classical negation and necessity with $\neg\Box$, conjunction and implication. ([14], p. 71)

This discussion will lead to my paper "S5 is a paraconsistent logic and so is first-order logic" [9], where I show how possible worlds semantics can be used to develop paraconsistent logics. Despite the fact that Z is the same logic as S5, it is a quite interesting construction, since it gives a very different picture of the situation: we have a logic with conjunction, disjunction, implication and a paraconsistent negation as the only primitive connectives. And after discussing with Pizzi I was on the way to develop a connection between the modal square and an intuitive interpretation of paraconsistent negation, with the picture shown in Fig. 1. The modality $\neg\Diamond$, i.e. impossible, is considered as a paracomplete negation, whose intuitionistic case is a particular case. This is connected with

Gödel's result: it is possible to translate intuitionistic logic into the modal logic S4 [26]. I presented a talk on this subject in March 1998 at the PUC-Rio before presenting it in Poland in July 1998 at the Jaśkowski's event. The following year I was again in Poland taking part to the 11th International Congress of Logic, Methodology and Philosophy of Science (Kraków, August 20–26, 1999), where I had the opportunity to meet Slater in flesh and blood, being convinced of his existence, but this didn't convince me of the non-existence of paraconsistent negations.

4 The O-Corner (Stanford 2001)

In 2000 and 2001 I was a visiting scholar at the CSLI/EPGY at Stanford University invited by Patrick Suppes. During this time I had the opportunity to discuss the question of $\neg\Box$ and the square with various people at Stanford and also circulating around the world.

At Stanford University I discussed the question of $\neg\Box$ with Johan van Benthem during Spring 2001, who oriented me, as Pizzi did it before, to contingency in particular the work of Lloyd Humberstone from Melbourne. So at the beginning of May 2001 I wrote to Humberstone. Discussing with him I was led to completely clarify the relation between $\neg\Box$ and contingency. Here is the e-mail exchange I had with Humberstone (May 8, 2001):

> JYB: I have been interested recently on the logic of contingency and Johan van Benthem told me that perhaps you can give me some information about the subject. I was led to the study of the operator of contingency defined standardly as $\neg\Box$, because it has some interesting properties from the point of view of paraconsistent logic. It is in fact dual of the intuitionistic negation considered as $\Box\neg$.
> LH: Now in fact as far as I know, nobody has ever proposed that contingency is definable as $\neg\Box$, since $\neg\Box$ is consistent with (indeed follows from) being impossible $\Box\neg$ and so is not a notion of contingency. The standard definition is rather: $\neg\Box \wedge \neg\Box\neg$ or alternatively $\diamond \wedge \diamond\neg$.

So $\neg\Box$ was not contingency. What it was, what was a good name for it, other than the pure negative name "not necessary"? This question would be clarified a couple of months later when still at Stanford I attended a talk by Seuren. Anyway at this time I just had learned more about contingency, had a look at Humberstone's work, who has also been working on non-contingency [29]—the contingent theme being a long time Australian topic, Routley was one of the first to work on the subject [48]. At this time I remembered the work of Blanché about the hexagon [20] that I read several years ago before. I was thus facing the picture shown in Fig. 2. This picture replaces the wrong picture that can be found in many different papers or books, and that I had in mind (see Fig. 3). The mistake has been cleared up in the 1950s by Robert Blanché reconstructing the square within two triangles forming an hexagon (see [20–22]). Jean-Louis Gardies, a follower of Blanché, explains that during the history of logic since Aristotle there was a confusion about modalities, people having in mind rather a triangle but trying to express the relation between the modalities with a square similar to the square of quantifiers (see [25]). Blanché was nevertheless able to keep the parallel between quantifiers and modalities by elaborating a hexagon of modalities which also clarifies some confusions about the theory of quantifiers (making the difference between "at least one" and "some but not all"). The triangle of modalities is shown in Fig. 4. According to this triangle, contingent is the same as possible and means not necessary, where "not" is interpreted as a contrary negation: "it is necessary that god exists" and "it is contingent that god exists" cannot be true together, but they can be false together, in this case "it is impossible that god exists" is true.

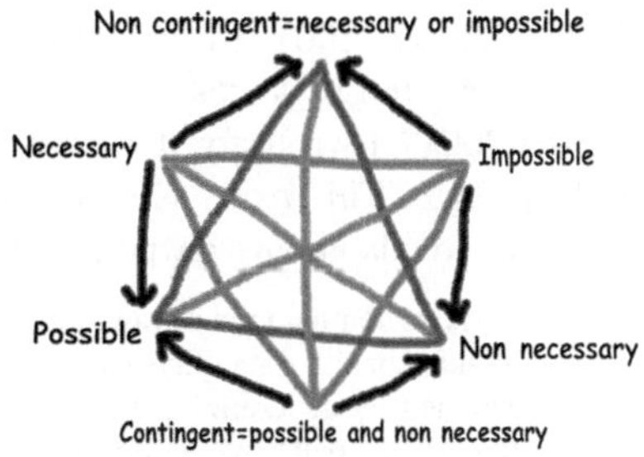

Fig. 2 Hexagon of modalities

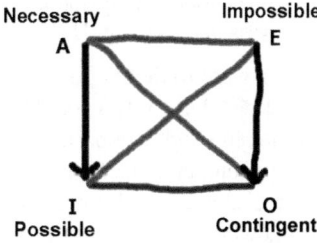

Fig. 3 Wrong square of modalities

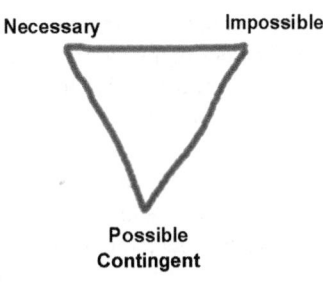

Fig. 4 Triangle of contrary modalities

Fig. 5 Triangle of subcontrary modalities

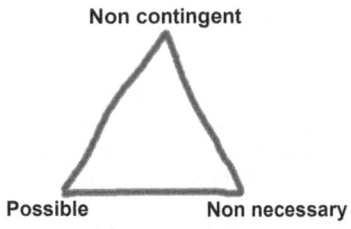

But "not necessary" as $\neg\Box$ appears as one of the vertices of the dual triangle, the triangle of subcontrariety (Fig. 5), where possible here has the standard meaning given by the modal square, not excluding necessity. My idea was to go on studying $\neg\Box$, a primitive and fundamental connective of the paraconsistent logic Z. The Australians, like Routley and Humberstone, have been working on modal logic based on contingency as primitive, but have they been working with $\neg\Box$ as primitive? Humberstone wrote me: "I don't know of any work specifically on taking $\neg\Box$ as a primitive" (e-mail 18.05.2001). Bob Meyer that I met at the Annual Meeting of the Society for Exact Philosophy (Montreal, May 11–14, 2001) also told me that he had never heard about this or/and the fact that $\neg\Box$ could be considered as a paraconsistent negation.

June 25–28, 2001, I organized, jointly with Darko Sarenac a workshop on paraconsistent logic in Las Vegas, part of the International Conference on Artificial Intelligence (IC-AI'2001). I wrote for this event the paper "The logic of confusion" and presented it in Las Vegas. This paper is a generalization of the techniques used in my Z paper and I decided to talk about viewpoints rather than possible worlds, here is a summarize of it:

> The logic of confusion is a way to handle together incompatible "viewpoints". These viewpoints can be information data, physical experiments, sets of opinions or believes. Logics of confusion are obtained by generalizing Jaśkowski-type semantics and combining it with many-valued semantics. In this paper we will present a general way to handle together incompatible viewpoints which promotes objectivism via paraconsistent logic. This general framework will be called logic of confusion. The reason for the name is that we want to put (*fusion*) together (*con*) different viewpoints. ([7], p. 821)

I then went to the Annual Meeting of the Australasian Association of Philosophy in Tasmania (Hobart, July 1–6, 2001) and to Melbourne for a one day workshop (July 12, 2001) at the philosophy department with Graham Priest, Ross Brady, Otávio Bueno, and where I met Humberstone. I gave in Melbourne a talk about modality and paraconsistency based on the square which I presented few days later in Perth at the University of Western Australia where I was visiting Slater.

Back to Stanford via South-Africa and Brazil presenting talks about this theme, I attended in the Fall 2001 an interesting lecture by Pieter Seuren from Max Planck Institute for Psycholinguistics (Language and Cognition Group), Nijmegen, The Netherlands. It was a talk entitled "Aristotelian predicate calculus restored" presented at CSLI CogLunch on 25 October 2001.

This was the first time I heard about the non-lexicalization of the O-corner. "O-corner" is the traditional name for the corner where $\neg\Box$ is positioned in the square of modality. As A, E, I, this is an artificial name, all being mnemotechnic abbreviations of some Latin expressions describing the corners of the square of quantification. Larry Horn, author of the famous book *A Natural History of Negation* [28], has shown that there was no known languages in the world where there is a primitive word for the O-corner of the quantificational square. Horn's analysis is summarized as follows by Hoeksema in a paper dedicated to JFAK for his 50th birthday:

> Striking is the observation in [Horn 1989, p. 259] that natural languages systematically refuse to lexicalize the O-quantifier, here identified with "not all". There are no known cases of natural languages with determiners like "nall"; meaning "not all". Even in cases that look very promising (like Old English, which has an item nalles, derived from alles, "all"; by adding the negative prefix ne- the same that is used in words like never, naught, nor, neither), we end up empty-handed. Nalles does not actually mean "not all" or "not everything", but "not at all" [Horn 1989, p. 261]. Jespersen [1917] suggested that natural language quantifiers form a Triangle, rather than a Square. ([27], p. 2)

Based on the non-lexicalization of the O-corner, Seuren was arguing, following the Danish linguist Jespersen, that the O-corner didn't make sense, that the square was wrong. I was no really convinced by such linguistical argument telling him that the equation

$$\text{exists} = \text{has a natural name}$$

seems wrong to me, and that a concept may exist even if he has no name in natural language, like many mathematical concepts, for example the number 0 (funny because of the similarity with the "O" of the O-corner which can be seen as a zero corner). Mathematics is an intelligent way to develop new concepts. The square is a simple mathematical construction and the O-corner makes sense within this mathematical construction. But of course it is always interesting to make the bridge between mathematics and reality, to turn mathematics reality. For me the fact that paraconsistent negation was exactly at the location of the O-corner in the modal square was interesting: these two attacked concepts could mutually be strengthened by sitting together.

At this time I was also working on a four-valued modal logic influenced by Łukasiewicz who in his last paper [36] presents a four-valued system of modal logic starting with the modal square of opposition. Łukasiewicz's logic has several defects, I started to rethink the problem with the same starting point: the square of opposition. I present this work at the Annual Meeting of the Australasian Logic Society, November 28–30, 2001, in Wellington, New Zealand. I have developed this work along the years and it was finally published in 2011 [16]. Before that I wrote another paper entitled "Paraconsistent logic from a modal viewpoint" [11] mixing the idea of this four modal logic and the square vision of paraconsistent logic, contents of which was presented at a paraconsistent workshop organized by João Marcos at the 14th European Summer School in Logic, Language and Information (ESSLLI), August 5–16, 2002 in Trento, Italy (João Marcos has been extending my work in different directions, see [39] and [40]).

5 Polyhedron and Polytomy of Oppositions (Neuchâchel 2003)

I settled down in August 2002 in my country of origin, Switzerland, where I would be a SNF (Swiss National Science Foundation) Professor for six years at the University of Neuchâchel. This is during this period that I fully expanded the square activities:

- raising the ideas of polyhedron of oppositions and polytomy of oppositions (2002–2003),
- directing Alessio Moretti's PhD thesis "The geometry of logical opposition" (2004–2008),
- organizing the first world congress on the square of opposition in Montreux in June 2007.

Arriving in Neuchâchel, I finished to elaborate the ideas that will lead to the paper "New light on the square of oppositions and its nameless corner" [10], published in *Logical Investigations* in 2003, the logic journal of the Russian Academy of Sciences where had been previously published in 2002 my paper on S5 as a paraconsistent logic.

In this paper I systematically developed the parallel between the three oppositions of the square (contradiction, contrariety, subcontrariety) and three kinds of negations (classical, paracomplete and paraconsistent) and I argued that subalternation is not really an opposition (since the square is about three oppositions, I decided to use the expression "square of oppositions", rather than the standard "square of opposition").

For this reason the square diagram seemed to me quite artificial and that was one more reason for me to emphasize the hexagon. I started to use colors for a better visual representation, deciding to have contradiction in red, contrariety in blue, subcontrariety in green, and choosing black for subalternation (see Fig. 6).

Considering subalternation as secondary I was most of the time rather picturing the Star of David (Fig. 7).

In the Fall 2002 I gave a series of lectures in Italy presenting a talk entitled "The square of oppositions, modal logic and paraconsistent logic":

Fig. 6 Colored hexagon of
oppositions

Fig. 7 Star of David of
oppositions

Fig. 8 Octagon of
oppositions

Fig. 9 Paraconsistent star of
oppositions

Fig. 10 Paracomplete star of
oppositions

- on November 12, 2002 in Cagliari, invited by Francesco Paoli,
- on November 15, 2002 in Napoli, invited by Nicola Grana,
- on November 20, 2002 in Siena, invited by Claudio Pizzi, with whom I had discussed these questions in Rio in 1999.

Fig. 11 Linking the stars

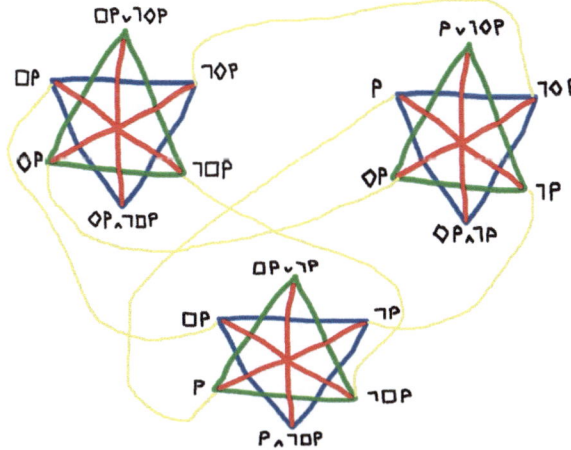

Fig. 12 Stellar dodecahedron
of oppositions

Back to Switzerland, during the winter 2002–2003, I started to conceive a polyhedron of oppositions. I was led to this polyhedron by trying to systematically picture the relations between the 4 modalities and p and $\neg p$. It is nice to say that the three notions of opposition of the square correspond to three notions of negation, but in the modal square there is not the proposition p itself that can be confronted to these three negations. The simplest idea is to picture the octagon (Fig. 8), where $\neg p$ also appears. But I was dissatisfied with this picture mainly for esthetic reasons.

I thought that stars were more beautiful and had the idea to draw the two stars shown in Figs. 9 and 10. I saw then that a way to put together the three stars according to the links—as shown in Fig. 11— was to construct a polyhedron. My intuitive idea was that this polyhedron was the stellar dodecahedron shown in Fig. 12.

I was still in touch with Humberstone and at this time he sent me a preliminary draft of [30] where two other hexagons corresponding to these two stars are presented. The difference with my work is that no paraconsistent and/or paracomplete interpretations are provided and one cannot find here the idea to put the three hexagons together building a three-dimensional object (influenced by my work Humberstone wrote later on [31]).

Fig. 13 Double square of oppositions

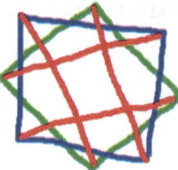

Fig. 14 Double square in Hagia Sofia

Not a specialist of polyhedra, I was looking on the web and I found a webpage on this topic produced by Hans Smessaert, a linguist from the Catholic University of Leuven, Belgium. I e-mailed him, explaining my discovery, he kindly replied to me, describing a fourth hexagon. At the same time I discussed this question with my friend Alessio Moretti who independently arrived at the same conclusion. And both Hans and Alessio were led to consider a tetradecahedron rather than a stellar dodecahedron in particular because they were figuring subalternations. An improvement of this polyhedron was constructed later on by Régis Pellissier (see details of the story in [42] and see the references of the work of Smessaert, Pellissier and Luzeaux in the bibliography [37, 46, 47, 50, 51]).

I had met Alessio at the ESSLLI 2000 in Birmingham where he was attending my tutorial on paraconsistent logic and we met again at the ESSLLI in Trento in 2002. He was quite interested by the square and I proposed him to supervise a PhD on this theme at the University of Neuchâchel. The title of his PhD is: "The geometry of logical opposition" (see [42]), but he generally prefers to use the expression "Theory of n-opposition" (see e.g. [41]) to name the work he is developing around the square (an ambiguous terminology, as pointed out by Alexandre Costa-Leite, since this theory is not about multiplication of oppositions). This is related with an idea I had at the same time I was developing the polyhedron of oppositions early in 2003. My idea was to generalize the hexagon of Blanché into the double square shown in Fig. 13.

Later on in 2003 I would have the pleasure to see plenty of double squares when visiting Turkey. I was taking part to the 21st World Congress of Philosophy at the Congress center in Istanbul (August 10–17, 2003) and the floor was covered by such squares and I also saw such squares visiting the famous temple Hagia Sophia. A picture I took is shown in Figure 14.

Later on Alessio told me that it is better to figure the idea beyond this double square by a bi-simplex, since we have then the same distances between the vertices of each "square". The generalization from the hexagon to a double square or a bi-simpex is quite natural

Fig. 15 Truth-valued square
of contrariety

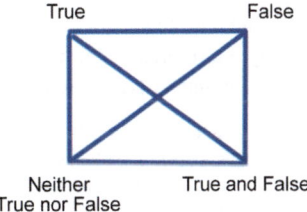

from a mathematical viewpoint, it is like going from trivalent logic to quadrivalent logic. And we are facing the same problem: does this make sense? Can we find interesting philosophical motivations and applications of these generalizations? In his logic book Kant argues that only dichotomy is *a priori*, polytomy is *a posteriori* (see [33]). But Blanché's masterpiece *Structures intellectuelles. Essai sur l'organisation systématique des concepts* gives us good reasons to think that trichotomy can be put on the *a priori* side: a ternary structure is ruling conceptualization (see [23]). However it is difficult to find some systematic n-tomic ($n > 3$) structures and we can agree with Kant that at some point polytomy is rather empirical, maybe though the case of quadritomy can reasonably put on the *a priori* side, considering for example the square of contrariety (Fig. 15) corresponding to the famous four-valued logic of Dunn-Belnap [1].

Having developed these new ideas about the square, I gave in 2003 a series of lectures on this topic around the world:

- March 6, in Brussels (Belgium),
- March 10, in Lausanne (Switzerland),
- April 11, in Rio de Janeiro,
- April 24, in Florianópolis (Brazil),
- May 17, in Vancouver (at the 31st Annual Meeting of the Society for Exact Philosophy),
- May 29, in Moscow at the 4th Smirnov's Readings,
- August 10, in Oviedo (Spain) at the 12th International Congress of Logic, Methodology and Philosophy of Science,
- September 9, at the Second Principia International Symposium, Santa Catarina, Brazil,
- September 16, at the Academy of Science of Buenos Aires, Argentina,
- October 15, in Geneva,
- October 18, in Amsterdam.

6 Squaring the World (2007–2011)

In 2003, 2004 and 2005 I was working intensively to develop universal logic. After having organized an international workshop on universal logic In Neuchâtel in October 6–8, 2003, I started the preparation of a big world congress and school on universal logic in Montreux in 2005 (UNILOG'05) and then launched the journal *Logica Universalis* and the book series *Studies in Universal Logic* with Birkhäuser. Universal logic is a general theory of logical systems trying in particular to study concepts that are beyond specific systems of logic and that can be therefore applied to a huge variety of logics (see my paper

[6] that can be seen as a complement to the present one). The square of opposition is typically a universal tool, as explained in the preface of a special issue of *Logica Universalis* on the square I edited with Gillman Payette:

> The square of opposition is something which is very much in the spirit of universal logic. It is a general framework than can be used to appraise existing logics and develop new ones (…) It can be applied to logical notions such as: quantifiers, negation and all kinds of modalities (ontic, deontic, alethic, epsitemic).
>
> The square of opposition helps to turn logic more universal. The square can not only be used as a tool for a systematic study of classes of logics and for developing bridges between logics but also as a central logical key to semiotics, ethnology, linguistics, psychology, artificial intelligence, computer science. ([15], p. 1)

UNILOG'05 was a great success with more than 200 participants from about 50 different countries. Everybody enjoyed very much the atmosphere of Montreux and the friendly hotel Helvétie in which the event took place. I decided therefore to organize the first world congress on the square of opposition in the same location. However the event was projected in a different way, a small interdisciplinary 3 day congress: the event took place June 1–3, 2007, in Montreux. We had 10 invited speakers, among them: Larry Horn, Pascal Engel, Alessio Moretti, Pieter Seuren, Hans Smessaert, Terence Parsons, Jan Woleński, and we had plenty of good contributors (about 60 people from all around the world). I wanted to really develop interdisciplinarity, gathering people from philosophy, linguistics, mathematics, computer science (the basic square), but also form psychology, ethnology and art. Since we were in Montreux, jazz was natural. We had a jazz show with compositions based on the square. This was organized by my friend Michael Frauchiger, from the Lauener Foundation in Bern.[1] We also presented a movie we had produced especially for the event, a remake of the Biblical story of Salomé, the square being used to articulate the relationships between the four main characters: Herod, Herodias, Salomé, Saint John the Baptist.

---

[1] Henri Lauener was a Swiss philosopher who liked very much music, after his death a philosophy prize was created and during the award ceremony there is always a band playing music.

The trailer of the movie and extract of the square jazz show can be seen in the DVD which was produced after the event by Catherine Chantilly, included in the book *The Square of Opposition—A General Framework for Cognition* [19] that I have edited with Gillman Payette after the event.

At the final round square table of the first congress in Montreux it was discussed if there would be future editions of the event, where and when. Among the participants there was a colleague, Pierre Simonnet, working at the University Pascal Paoli in Corté, Corsica, who suggested to organize the event there. I liked the idea since I had been living in Corsica as a child and had wonderful time. I visited Pierre in Corsica in March 2008 and again in October 2009 to prepare the event. The event took place in Corsica in June 17–20, 2010, with the same format as in Montreux. Among the participants, there were Damian Niwinski, Stephen Read, Dale Jacquette and Pierre Cartier—the famous Bourbachic mathematician, and many old and new contributors. At the end of this second event we discussed the organization of the third world congress on the square of opposition, and it seems that this series of events is now launched for many years. The third square event is scheduled in June 2012 at the American University of Beirut and the fourth square congress will probably happen in Munich in 2014.

Working on the organization of these events, I have also been continuously working on the square, always having new ideas, applying the square to various fields. Among many talks on the square around the world I presented "Extensions of the square of opposition and their applications" at ECAP08 (6th European Conference on Computing and Philosophy, June 16–18, 2008, Montpellier, France), "The logical geometry of economy" at a congress on Epistemology of Economical Sciences (October 1–2, 2009, Buenos Aires, Argentina). I also applied the square to semiotics (talk at Dany Jaspers's CRISSP seminar in Brussels, October 28, 2010) and music (talk at the seminar *mamuphi*—mathematics, music and philosophy—organized by François Nicolas at the ENS Ulm in Paris, November 5, 2011).

I see the square as a way to go beyond dichotomy. Dichotomy is a thinking methodology, promoted by the Pythagoreans, which was essential for the development of science. But dichotomy is limited and many dichotomic pairs are quite artificial. Blanché has rightly pointed out that trichotomy is more interesting and this is the basis of the hexagon of oppositions. In the hexagon, trichotomy appears at two levels: there are the two trichotomic triangles, but there are also the three notions of oppositions from the square of oppositions: contrariety which is the basis of the blue triangle, subcontrariety, which is the basis of the dual green triangle and contradiction linking these two trichotomic triangles. Blanché's hexagon is a really powerful tool based on trichotomy (see [17]).

References

1. Belnap, N.D.: A useful four-valued logic. In: Dunn, M. (ed.) Modern Uses of Multiple-Valued Logic, pp. 8–37. Reidel, Boston (1977)
2. Beziau, J.-Y.: L'holomouvement selon David Bohm. MD Thesis, Department of Philosophy, University Panthéon-Sorbonne, Paris (1986)
3. Beziau, J.-Y.: La logique paraconsistance C1 de Newton da Costa. MD Thesis, Department of Mathematics, University Denis Diderot, Paris (1990)
4. Beziau, J.-Y.: Review of [49]. Mathematical Rev., 96e03035

5. Beziau, J.-Y.: What is paraconsistent logic? In: Batens, D., et al. (eds.) Frontiers of Paraconsistent Logic, pp. 95–111. Research Studies, Baldock (2000)
6. Beziau, J.-Y.: From paraconsistent to universal logic. Sorites **12**, 5–32 (2001)
7. Beziau, J.-Y.: The logic of confusion. In: Arabnia, H.R. (ed.) Proceedings of the International Conference of Artificial Intelligence IC-AI'2002, pp. 821–826. CSREA, Las Vegas (2001)
8. Beziau, J.-Y.: Are paraconsistent negations negations? In: Carnielli, W., et al. (eds.) Paraconsistency: The Logical Way to the Inconsistent, pp. 465–486. Dekker, New York (2002)
9. Beziau, J.-Y.: S5 is a paraconsistent logic and so is first-order classical logic. Log. Investig. **9**, 301–309 (2002)
10. Beziau, J.-Y.: New light on the square of oppositions and its nameless corner. Log. Investig. **10**, 218–232 (2003)
11. Beziau, J.-Y.: Paraconsistent logic from a modal viewpoint. J. Appl. Log. **3**, 7–14 (2005)
12. Beziau, J.-Y.: The paraconsistent logic Z—a possible solution to Jaskowski's problem. Logic Log. Philos. **15**, 99–111 (2006)
13. Beziau, J.-Y.: Paraconsistent logic! (A reply to Salter). Sorites **17**, 17–26 (2006)
14. Beziau, J.-Y.: Adventures in the paraconsistent jungle. In: Handbook of Paraconsistency, pp. 63–80. King's College, London (2007)
15. Beziau, J.-Y., Payette, G. (eds.): Special Issue on the Square of Opposition. Logica Univers. **2**(1) (2008)
16. Beziau, J.-Y.: A new four-valued approach to modal logic. Log. Anal. **54**, 18–33 (2011)
17. Beziau, J.-Y.: The power of the hexagon. In: [18]
18. Beziau, J.-Y. (ed.): Special Double Issue on the Hexagon of Opposition. Logica Univers. **6**(1–2) (2012)
19. Beziau, J.-Y., Payette, G. (eds.): The Square of Opposition—A General Framework for Cognition. Peter Lang, Bern (2012)
20. Blanché, R.: Sur l'opposition des concepts. Theoria **19**, 89–130 (1953)
21. Blanché, R.: Opposition et négation. Rev. Philos. **167**, 187–216 (1957)
22. Blanché, R.: Sur la structuration du tableau des connectifs interpropositionnels binaires. J. Symb. Log. **22**, 17–18 (1957)
23. Blanché, R.: Structures intellectuelles. Essai sur l'organisation systématique des concepts. Vrin, Paris (1966)
24. Bohm, D.: Wholeness and the Implicate Order. Routlege, London (1980)
25. Gardies, J.-L.: Essai sur la logique des modalités. PUF, Paris (1979)
26. Gödel, K.: Zum intuitionisticischen Aussagenkalkül. Akad. Wiss. Wien, Math.-Nat. Wiss. Kl. **64**, 65–66 (1932)
27. Hoeksema, J.: Blocking effects and polarity sensitivity. In: Essays Dedicated to Johan van Benthem on the Occasion of His 50th Birthday. Amsterdam University Press, Amsterdam (1999)
28. Horn, L.R.: A Natural History of Negation. University Chicago Press, Chicago (1989)
29. Humberstone, L.: The logic of non-contingency. Notre Dame J. Form. Log. **36**, 214–229 (1995)
30. Humberstone, L.: Modality. In: Jackson, F.C., Smith, M. (eds.) The Oxford Handbook of Contemporary Philosophy. Handbook of Analytical Philosophy, pp. 534–614. Oxford University Press, Oxford (2005)
31. Humberstone, L.: Contrariety and subcontrariety: the anatomy of negation (with special reference to an example of J.-Y. Béziau). Theoria **71**, 241–262 (2005)
32. Jaśkowski, S.: Rachunek zdan dla systemów dedukcyjnych sprzecznych. Stud. Soc. Sci. Torun., Sect. A **I**(5), 55–77 (1948)
33. Kant, I.: Logik (1800)
34. The Kybalion: A Study of the Hermetic Philosophy of Ancient Egypt and Greece. Yogi Publication Society (1908)
35. Larvor, B.: Triangular logic. The Philosophers' Magazine **44**, 83–88 (2009)
36. Łukasiewicz, J.: A system of modal logic. J. Comput. Syst. **1**, 111–149 (1953)
37. Luzeaux, D., Sallantin, J., Dartnell, C.: Logical extensions of Aristotle's square. Logica Univers. **2**, 167–187 (2008)
38. Omori, H., Wagarai, T.: On Beziau's logic Z. Logic Log. Philos. **18**, 305–320 (2008)
39. Marcos, J.: Nearly every normal modal logic is paranormal. Log. Anal. **48**, 279–300 (2005)

40. Marcos, J.: Modality and paraconsistency. In: Bilkova, M., Behounek, L. (eds.) The Logica Yearbook 2004, pp. 213–222. Filosofia, Prague (2005)
41. Moretti, A.: Geometry of modalities? Yes: through n-opposition theory. In: Beziau, J.-Y., Costa Leite, A., Facchini, A. (eds.) Aspects of Universal Logic. Travaux de logique, vol. 17, pp. 102–145. Université de Neuchâtel, Neuchâtel (2004)
42. Moretti, A.: The geometry of logical opposition. PhD Thesis, University of Neuchâtel (2009)
43. Mruczek-Nasieniewska, K., Nasieniewski, M.: Paraconsistent logics obtained by J.-Y. Beziau's method by means of some non-normal modal logics. Bull. Sect. Log. **37**, 185–196 (2008)
44. Mruczek-Nasieniewska, K., Nasieniewski, M.: Beziau's logics obtained by means of quasi-regular logics. Bull. Sect. Log. **38**, 189–203 (2009)
45. Osorio, M., Carballido, J.L., Zepeda, C.: Revisiting Z (to appear)
46. Pellissier, R.: "Setting" n-opposition. Logica Univers. **2**, 235–263 (2008)
47. Pellissier, R.: 2-Opposition and the topological hexagon. In: [19]
48. Montgomery, H., Routley, R.: Contingency and non-contingency bases for normal modal logics. Log. Anal. **9**, 341–344 (1966)
49. Slater, B.H.: Paraconsistent logics. J. Philos. Log. **24**, 451–454 (1995)
50. Smessaert, H.: On the 3D visualisation of logical relations. Logica Univers. **3**, 303–332 (2009)
51. Smessaert, H.: The classical Aristotelian hexagon versus the modern duality hexagon. In: [18]
52. Sokal, A.: Transgressing the boundaries: towards a transformative hermeneutics of quantum gravity. Social Text **46–47**, 217–252 (1996)

J.-Y. Béziau (✉)
CNPq – Brazilian Research Council, UFRJ – University of Brazil, Rio de Janeiro, Brazil
e-mail: jyb@ifcs.ufrj.br

Logical Oppositions in Arabic Logic:
Avicenna and Averroes

Saloua Chatti

Abstract In this paper, I examine Avicenna's and Averroes' theories of opposition and compare them with Aristotle's. I will show that although they are close to Aristotle in many aspects, their analysis of logical oppositions differs from Aristotle's by its semantic character, and their conceptions of opposition are different from each other and from Aristotle's conception. Following Al Fārābī, they distinguish between propositions by means of what they call their "matter" modalities, which are determined by the meanings of the propositions. This consideration gives rise to a precise distribution of truth-values for each kind of proposition, and leads in turn to the definitions of the logical oppositions. Avicenna admits the four traditional oppositions, while Averroes, who seems closer to Aristotle and especially to Al Fārābī, does not mention subalternation, but admits sub-contrariety. Nevertheless, we can find that Averroes defends what Parsons calls SQUARE and [SQUARE], because he holds **E** and **I**-conversions and the truth conditions he admits are just those that make all the relations of the square valid, while Avicenna defends SQUARE and [SQUARE] only for the waṣfī reading of assertoric propositions. They also give a special attention to the indefinite which in Averroes' view is ambiguous, while Avicenna treats it as a particular. Some points of their analysis prefigure the medieval concepts and distinctions, but their opinion about existential import is not as clear as the medieval one and does not really escape the modern criticisms.

Keywords Logical oppositions · Matter necessity · Possibility and impossibility · Existential import · Indefinites · waṣfī vs ḏātī readings of propositions

Mathematics Subject Classification Primary 03A05 · Secondary 03B10

1 Introduction

In this paper, I will examine Avicenna's and Averroes' views on the specific topic of the so-called logical oppositions in order to compare between them both on the one hand and between them and Aristotle on the other hand. As is well known, Aristotle identifies opposition with incompatibility, since the only logical relations that he holds to be oppositions are contradiction and contrariety. But can one say the same thing about Avicenna and Averroes? Are their respective accounts of the notion of opposition distinct from Aristotle's one? Are they distinct from each other? How are logical oppositions characterized in their respective systems? And how is the notion of opposition itself viewed in these systems? In order to answer these questions, I will start by analyzing Avicenna's system then

J.-Y. Béziau, D. Jacquette (eds.), *Around and Beyond the Square of Opposition*, 21–40
Studies in Universal Logic, DOI 10.1007/978-3-0348-0379-3_2, © Springer Basel 2012

I will turn to Averroes' one and finally, I will compare between them both and Aristotle in order to determine the way they define this notion. This comparative analysis shows that the notion of opposition is different in the three systems, and gives rise to three quite distinct figures.

2 Avicenna's Analysis of the Logical Oppositions

The logical oppositions are analyzed by Avicenna (980–1037) in his book entitled *al-shifā'* (*The Cure*), more precisely in *al-Maqūlāt* (*Categories*) and *al-'Ibāra* (*Peri Hermeneias*). In *al-'Ibārā*, he defends his views on oppositions and presents what we would consider as a square of oppositions since he analyzes the four traditional relations which are: contradiction, contrariety, subcontrariety and subalternation. But before turning to his analysis of these relations let us first see how he defines the notion of opposition itself. This definition is presented in *al-Maqūlāt* (*Categories*) where Avicenna considers many types of oppositions which could concern the terms or the propositions. He defines opposition in general by saying that "The opposites do not conjoin in the same subject by any aspect in any time" [12, p. 241].[1] The opposites could be terms or propositions expressed by sentences, and opposition itself could be expressed by means of negation or by other means; as examples, he gives the following pairs: "horse/non-horse" and "even/odd". Avicenna follows here Aristotle's *Categories* (10) when he distinguishes between oppositions by virtue of correlation such as "double" and "half", of possession and privation, of contrariety such as "sick" and "healthy" and finally the opposition of truth-values. This last one affects propositions and is expressed by means of a negation; thus singular propositions such as "Zayd is a man" and "Zayd is not a man" are opposed because they do not have the same truth-value. The real opposition between propositions is, then, the opposition between truth-values. This opposition in truth-values is specifically made by the negation since Avicenna, like Aristotle, thinks that sentences that contain opposite predicates are not contradictories when the subject does not exist, because in that case, they are both false. Avicenna gives the following example: "Zayd who does not exist [*al ma'dūm*] is seeing" does not contradict "Zayd who does not exist is blind", but it does contradict "Zayd who does not exist is not seeing" [12, p. 259], because this last sentence is true. In the same vein, the sentence "Stones are sick" does not contradict "Stones are healthy" but it does contradict "Stones are not sick" [12, p. 258], because the first two sentences are false while the last one is true. But the notion of opposition is more general than contradiction, contrariety or correlation. It can be seen as a genus that includes several species [12, p. 245].

How can we apply this to the different kinds of propositions that are expressed by singular, quantified or non quantified sentences? To answer this question let us see what Avicenna says in *al-'Ibāra*. In this book, he classifies propositions into three kinds: (1) Singulars, (2) Indefinites (i.e. not quantified) and (3) Quantified, i.e. Universal and Particular propositions. We have to notice here that Avicenna uses a specific term to indicate the presence of a quantification. The quantifier is expressed by the word '*sūr*' [14, p. 52] and

---

[1] Wilfrid Hodges in [19, p. 13] translates this passage in the following way: "We say: opposing pairs are those which don't combine in a single subject from a single aspect at a single time together."

the quantified propositions by '*musawwara*'.[2] This word '*sūr*' is also used by Al Fārābī in *al-Qawl fi al-'Ibārā*, for instance. Al Fārābī explicitly mentions Alexander of Aphrodisias and other commentators of Aristotle in many of his writings, which could explain the closeness of his terminology to the Greek commentators' one.[3] He defines the quantifier in the following way: "It is the word that indicates that the judgment made by the predicate is about part of the subject or about the total subject" [1, p. 118] he adds that there are four quantifiers, which are "All, None (lā' wāḥid), Some and Not all" [1, p. 118] and are used in the four kinds of propositions. Avicenna mentions the same classification in *al-'Ibāra* [14, p. 54], but Averroes seems to be clearer on that point since he talks about only two quantifiers by saying: "I mean by quantifier the words «all» and «some»" [9, p. 91]. This generic word '*sūr*' is not, however, exactly equivalent to the modern quantifier, for (1) in Avicenna's and Al Farabi's account, such words may be mixed with negation, the separation from negation occurs only in Averroes' account, (2) the particular quantifier does not stress specifically on existence, as is the case with the modern existential quantifier. But this grouping may be seen as prefiguring the medieval distinction between two separate kinds of terms: the *syncategorematic* and the *categorematic* terms. As it is expressed by Jean Buridan, this distinction occurs between terms that signify by themselves, and "may be subject or predicate *per se*" in propositions (the *categorematic* terms), and terms that do not signify in isolation but only in connection with other terms in the proposition (the *syncategorematic* terms, e.g. 'not', 'or' and the like). Terms like 'nobody', 'nothing', 'somewhere' are said to be 'mixed' [17, p. 96]. As noted by [24, section 4], the origin of the words '*syncategorematic*' and '*categorematic*' is grammatical and can be found in "Priscian's *Institutiones grammaticae* II, 15". The grammatical distinctions are also made by Arabic logicians, since they distinguish between the noun, the verb and the particle, which is defined exactly like the *syncategorematic* terms: the particle does not signify in isolation but only in connection with something else. But they do not provide a complete listing of *logical syncategorematic* terms.

Let us now turn to the logical oppositions as they are defined by Avicenna. We will focus here on his *al-'Ibāra*, since we find them stated there quite systematically, but we will mention also *al-Qiyās*. Not surprisingly, Avicenna considers the singular propositions as being contradictories, that is, they never share the same truth-value. Whenever a singular proposition is negated, it becomes false if the affirmative is true and vice versa.[4] However, with respect to other kinds of propositions, that is, the quantified and the non quantified propositions, Avicenna distinguishes between several possibilities which are treated in much detail. This treatment reveals many kinds of oppositions, which are all related in one way or another with the notion of truth-value.

Let us start by the quantified propositions. These are the particular and the universal propositions; they are explicitly quantified by adding in one case 'some' and in the other

---

[2] Avicenna uses also the expression 'maḍkūrat as sūr' and the word 'maḥsūra' to designate the quantified propositions. As to Al Fārābī, he uses the expression 'ḍawāt al aswār' (i.e. those that contain quantifiers) [2, p. 121].

[3] I thank one anonymous referee who drew my attention to the fact that the word '*sūr*' is "heir to the notion of «*prosdiorismos*» in the Greek commentators of Aristotle".

[4] However in [14, p. 70], Avicenna says that the singular propositions which are in the future are not necessarily true or false and seems to agree with Aristotle in his treatment of the problem of the "future contingents", although he gives more details on the possible propositions.

case 'all'. When we add negations, the universal negative is expressed by "No A is B" and the particular negative by "Not all A are B". As in the Aristotelian tradition, Avicenna considers that the universal negative (**E**) and the particular affirmative (**I**) on the one side, and the universal affirmative (**A**) and the particular negative (**O**) on the other are contradictories, that is, they never share the same truth-value. But what is added is the subdivision of these quantified propositions into three kinds which are: Necessary, Impossible and Possible. Necessary, Possible and Impossible must be understood here in terms of the relation between the subject and the predicate. The necessity, possibility or impossibility is internal and is not expressed by a specific word. As an example of a necessary proposition, Avicenna gives the following: "Every man is an animal", this proposition is necessary because of the fact that being an animal is an essential attribute of men. The example corresponding to an impossible proposition is the following: "No man is a stone" [14] which expresses the fact that "stone" cannot be a feature of the subject "man". This sentence expresses an impossible proposition because Avicenna defines material impossibility in this way: it is "what is permanent and whose affirmation is necessarily false" [14, p. 47].[5] This means that in an impossible proposition, the predicate is never adequate for the subject, which makes the affirmative proposition always false (and consequently the negative one always true). In a possible proposition, the predicate does not express an essential attribute of the subject, but could be predicated of it; the example is a sentence where the subject is "man" and the predicate "writer". When we add the universal quantifier, we obtain the following proposition "All men are writers" which is false as well as the corresponding universal negative which is "No man is a writer" [14, p. 46]. These modalities are called "matter"[6] modalities because they are related to the essences of the objects concerned and express material necessity or impossibility or possibility. When the inherence of the predicate into the subject is permanent, the proposition is necessary, when it is not, the proposition is possible, when the predicate is never convenient for the subject, the proposition is impossible. Necessity and impossibility are related to the notion of permanence, while in the notion of possibility, there is no permanence. These modalities must, however, be distinguished from the explicit (verbal) modalities which are expressed by specific words such as "necessary" (= *wājib*), "possible" (= *mumkin*) and "impossible" (= *mumtana'*). Avicenna makes a clear distinction between the two kinds of modalities by saying that a sentence which contains an explicit modality could be false as is the case with the following example "All men are necessarily writers" [14, p. 112], while a sentence with a matter modality is never false when it is affirmative and necessary, for instance. The falsity of this sentence is explained by the fact that it has "a modality which disagrees with its matter" [14, p. 112]. "Matter" modalities are considered in the general theory of (categorical) syllogisms while explicit modalities are studied in the theory of

---

[5]The Arabic sentence is the following: "… yadūmu wa-yajibu kadhibu ījābihi…yusammā māddat al- imtinā'."

[6]We find the expression *"matter modalities"* in Al Fārābī's text too and the word *matter* seems to have a long history starting from Aristotle and his Greek commentators. The *matter* (*hūlē*) in Aristotle is opposed to the *tropos*, which has a rather vague sense and could mean the *form* of the proposition (see [18, p. 298]). In Ammonius' text, it is related to modalities since he says in his *De Interpretatione*: "These relations <between subject and predicate> they call the matter of propositions and they say they are necessary, impossible or possible" (cited in [8, p. 233]). The necessity, possibility or impossibility are, according to this author, "due to the very nature of the objects" [8, p. 233]. In Avicenna's text, the modal sense is clear as we have seen.

modal syllogisms. This notion of "matter" modalities could be related to the medieval notions "*materia necessaria*", "*materia contingenti*", and "*materia impossibili*", which were discussed by many authors "in early medieval logic and <were> dealt with in mid-thirteenth-century books", for instance, in Thomas Aquinas' writings "who wrote that universal propositions are false and particular propositions are true in contingent matter (*In Perihem.*I. 13, 168)" [21, Sect. 3]. Similar distinctions are made by William of Sherwood, according to the same author [21]. This distribution of truth-values is exactly the same in Avicenna's (and Al Fārābī's) analysis, as will appear in the following list:

A necessary: True   E necessary: False
A impossible: False   E impossible: True
A possible: False   E possible: False

I necessary: True   O necessary: False
I impossible: False   O impossible: True
I possible: True   O possible: True

These truth conditions follow directly from what Avicenna says about the truth-values of the considered propositions [14, p. 47]. It shows that the contradictory propositions which never share the same truth-value in any matter are, when they are quantified, **A** and **O** on the one hand and **E** and **I** on the other. We have thus six contradictory propositions which are: (1) Necessary **A** and Necessary **O**, (2) Impossible **A** and Impossible **O**, (3) Possible **A** and Possible **O**, (4) Necessary **E** and Necessary **I**, (5) Impossible **E** and Impossible **I**, and (6) Possible **E** and Possible **I**.[7]

All these oppositions are contradictions since in all these pairs, only one proposition is true, the other being false. Avicenna explains this very precisely by saying exactly which one is true and which one is false in all the cases. They are then totally opposed *whatever matter they may have* as he notes in the chapter devoted to the analysis of the contradictory propositions [14, pp. 66–75]. This means that contradiction is the strongest kind of opposition.

The second kind of opposition is contrariety. The contrary propositions are as in the Aristotelian tradition the two universal propositions, that is, **A** and **E**. As we can see, they do not share the same truth-value when they are Necessary and when they are Impossible since Necessary **A**: "Every man is an animal" is true and opposed to Necessary **E**: "No man is an animal" which is false. Impossible **A**: "Every man is a stone" is false and opposed to Impossible **E**: "No man is a stone" which is true. But these propositions do share the same truth-value when they are Possible, since Possible **A**: "Every man is a writer" and Possible **E**: "No man is a writer" are both false. Contrariety is, then, defined in the traditional way: it is the relation between propositions that are never true together but might be false together. But the cases of truth and falsity are determined more precisely by taking into consideration the matter of the propositions. It is less strong than contradiction because it concerns only two propositions (**A** and **E**) opposed in two modes.

Then, he considers subcontrary propositions which are the two particular propositions. As the table shows, these do not share the same truth-value when they are Necessary and when they are Impossible; but they are both true when they are Possible. Thus, **I** necessary: "Some men are animals" is true and opposed to **O** Necessary: "Not all men are

---

[7]Unlike Aristotle, Avicenna does not use the word "contingency" in [14, p. 122].

animals" which is false, and **I** impossible: "Some men are stones" is false and opposed to **O** Impossible: "Not all men are stones" which is true. Possible particulars are exemplified by: "Some men are writers" and "Not all men are writers" which are both true. Subcontrariety is then an opposition which makes the propositions never false together but sometimes true together. This is even less strong than contrariety since we have two propositions opposed in two modes, but the propositions are true in the third mode, thus not opposed at all in that mode: this makes subcontrariety less strong than contrariety.

Finally, we have the subaltern propositions which are **A** and **I** on the one hand and **E** and **O** on the other, the word used for subalternation being "*Tadākhul*". These are opposed when they are possible as we can see in the following examples: **A** possible: "Every man is a writer" is false, while **I** possible: "Some men are writers" is true. The same holds with **E** possible: "No man is a writer", which is false, while **O** Possible: "Not all men are writers", is true. But they do share the same truth-value when they are Necessary (they are both true) and when they are Impossible (they are both false).

What is interesting here is the way Avicenna expresses this relation, since he says: "As to those that differ in quantity but not in quality, *let us call them* subalterns, we *find* that those which are affirmative are true in the Necessary, and that the negative subalterns are true in the Impossible, and both do not share the same truth-value in the Possible, but the particulars are true in that case, and examine that by yourself" [14, p. 48, my emphasis]. This shows that his characterization of this kind of opposition and the other ones follows from the observation of the distribution of the propositions' truth-values which in turn depends upon the senses of the propositions involved: this makes it even *more semantic* than in Aristotle's account. The semantic character is related both to the consideration of the meanings of the propositions and to the distribution of truth-values which is quite systematic in Avicenna's account, while it is not in Aristotle's and the traditional logicians' one, even if they define also the oppositions by considering the truth-values of the propositions. Avicenna, unlike Aristotle and the traditional logicians, does not say: since the contrary propositions are never true together *therefore*, when one of them is true, the other must be false; rather, he *finds that*, in consideration of the distribution of the truth-values of all kinds of propositions that *they are never true together*. He presents then a kind of truth table similar to the well known contemporary semantic method, though less achieved since only the cases of truth are considered, but not all the cases. We see then that his method is the ancestor of the semantic contemporary method of truth tables.

Regarding subalternation, it is not defined exactly as in Aristotle's text since Aristotle says: "For in demolishing or establishing a thing universally we also prove it in particular; for if it belongs to all, it belongs also to some, and if to none, not to some" [5, III, 6, 119a, 34–36]. Aristotle does not mention any differences in truth-values between the propositions, but rather relations of implication. Moreover, he does not consider subalternation as an opposition at all. But we can notice that, according to Avicenna, subalternation, even if characterized in a different way, is the less strong opposition since only two pairs of propositions in the Possible matter are opposed by their truth-values. We can notice also that the Arabic word used by Avicenna to express this relation, which is "*Tadākhul*" is *not* synonymous with the traditional word "Subalternation" since it does not have the same linguistic meaning. While "Subalternation" derives from the Latin words "*alter*" which means "other" and "*sub*" which means "under" and evokes the notion of dependence (upon the other) and thus implication, the Arabic word comes from the root "*dakhala*" which is a verb meaning "to enter", the other verb, which is closer to the word used, is "*tadākhala*" and means "to enter into each other". The ideas involved then are the ideas of inclusion and of the relation between the whole and the part: the part is included into the whole, therefore what is true of the whole is true of the part. It seems that Avicenna has chosen this word by himself without relying on a specific tradition, which appears clearly in the preceding quotation where he says: "*let us* call them [...]".[8]

The oppositions involve more or less differences in truth-values but the differences are either total or partial i.e. concern only some cases. The strongest opposition is contradiction since it involves all the pairs of propositions concerned and all the modes or matters but the other ones are different in degree so that we can say that subalternation is the less strong one, while contrariety, which involves one pair of propositions and two matters (the propositions being false in the third matter), and subcontrariety, which involves also one pair of propositions and two matters, are intermediates. As we can see, the oppositions between quantified propositions lead to a Square of oppositions in Avicenna's view since he admits the four oppositions.

Regarding the indefinites, Avicenna tends to defend an Aristotelian position according to which these propositions should be considered as particulars even though they do not contain explicitly any quantification. This opinion is expressed explicitly at page 51 of *al-'Ibāra* where he says "the indefinite has the force of the particular". But he spends much time and place to explain why this should be so. Being perfectly aware that this kind of propositions might be considered, in ordinary usage, as universal propositions, he tries to explain that one should avoid this kind of interpretation because it might be confusing and even misleading. For when we add a universal quantifier to an indefinite proposition, it might become false while it was true without quantification. For instance, when we say "White is necessarily white" this is true, but when a quantifier is added, we obtain the following sentence: "All what is described as white is necessarily white" [14, p. 52] which is false according to him, since what is white now might not be white later.

But what happens if we negate an indefinite proposition? Does it become false if its corresponding affirmative is true? According to Avicenna, the indefinite when negated, is *not* the contradictory of its corresponding affirmative. He says that at page 67 where he claims: "the indefinite has no contradictory" and also "the indefinites [...] are like the particulars, they should be said to be subcontraries" [14, p. 66, my translation] since they might be true together as witnessed by the two following sentences: "Men are beautiful", "Men are not beautiful" [14, p. 67]. But Avicenna gives examples of indefinite sentences which do not share the same truth-value, such as "Stones are sick" and "Stones are not sick" [12, p. 258], and he says explicitly that such sentences are contradictory, as appears in the following quotation: "Two contradictories are not false together [such as] when we say: 'Stones (or Zayd who does not exist) are seeing', 'Stones (or Zayd who does not exist) are not seeing' " [12, p. 259]. In other passages of *al-'Ibāra*, he defends the opinion that, although the indefinites seem contradictory in impossible and necessary matters, one should treat them in general without focusing on the different matters, since he says: "One must treat the indefinites as propositions, which is more general than the three matters, and not consider them matter by matter. So the indefinite in the necessary matter, in so far as it is an indefinite, is a particular judgment" [14, p. 69]. From these quotations, one can conclude the following: the indefinites are true in the possible, but they do not

---

[8] As far as I know, there is no mention of this word in Al Fārābī's texts; Al Fārābī does not mention nor include subalternation in his treatment of oppositions.

share the same truth-value in the impossible and the necessary. Their truth conditions are then the same as those of the two particulars. But this is not quite satisfying since if we consider the indefinite as a particular and only as a particular, first it should behave as such in all circumstances, which is not obvious nor warranted, secondly, the particular has a contradictory which is the universal negative and this does not fit with what Avicenna says about the fact that the indefinite has no contradictory and makes his opinion somewhat confused as we will show in the last part. But we could say the same thing about Aristotle himself who tends to consider the indefinite as a particular without treating it exactly as a particular.

The shape corresponding to this analysis of the opposition is then a square since the four oppositions are admitted and the indefinite is not characterized with enough preciseness. This square is the following:

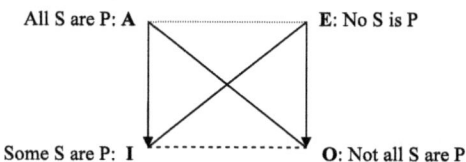

It seems then that Avicenna defends what Terence Parsons calls SQUARE, that is, all the relations of the square for the propositions he examined in that first treatise. But the question remains to determine whether or not he defends SQUARE, as well as [SQUARE] i.e. the relations of the square plus **E** and **I** conversions, for all the readings of the propositions that he talks about in *al-Qiyās*. We have also to examine his treatment of the question of existential import. We will return to both topics in the last section.

3 Averroes' Views on the Oppositions

Regarding Averroes (1126–1198), things are different because his aim in writing his treatises was to comment on Aristotle's logical writings. These treatises are the following: (1) *Talkhīṣ Kitāb al-Maqūlāt* [Paraphrase of the *Categories*], (2) *Talkhīṣ Kitāb al-'Ibāra* [Paraphrase of the *Peri Hermeneias*], (3) *Talkhīṣ Kitāb al Analitīqa al Awwil* (or *al-Qiyās*) [Paraphrase of the *Prior Analytics*], (4) *Talkhīṣ Kitāb al Analitīqa at thānī* (or *al-Burhān*) [Paraphrase of the *Posterior Analytics*), (5) *Talkhīṣ Kitāb al-Jadal* [Paraphrase of *The Topics*], (6) *Talkhīṣ Kitāb al-Moughālaṭa* [Paraphrase of the *Sophistical Refutations*]. They have been grouped and edited recently by Gérard Jehamy under the title *Talkhīṣ Manṭiq Aristū* (Lebanon, 1982). We will also mention the Cairo editions of these different treatises.

As we can see from the very beginning, Averroes follows Aristotle's text faithfully. But this does not mean that his opinions are exactly similar to Aristotle's as we will see in the following. For it happens to him to diverge from Aristotle's text even if his aim is to explain it. For instance, he writes sometimes "He (i.e. Aristotle) says... and we say..." [9, p. 92, p. 137, etc.] which shows that when commenting the text, he gives his opinion on it.

Averroes says that there are exactly six oppositions between the different kinds of propositions. These kinds are the following: (1) The singular propositions, (2) The indefinite propositions, (3) The quantified propositions which are the particular and the universal propositions. The first opposition is the one between the two singular propositions which is, as in Avicenna and Aristotle's views, a contradiction without any doubt. The second is the opposition between the two indefinites or non quantified propositions: these are in Averroes' view either contraries when they are meant to be universal propositions, or subcontraries if they are meant to be particular propositions. The indefinites could be true in possible "matters", in which case they are particular and would be subcontraries, but if in that same matter they are interpreted as universal, then they are contraries [9, pp. 92–93]. The example he gives to illustrate this fact is "Men are white" and "Men are not white" both propositions are true if they are interpreted as particular, but if they are interpreted as universal, both are false. Their relation depends then upon how one interprets the possible indefinites, since when the matters of the indefinite propositions are impossible or necessary, they never share the same truth-value, whether they are universal or particular. For instance "Men are animals" and "Men are not animals" differ in their truth-values: the first is true, the second is false, whether they are particular or universal, and we can say the same thing about "Men are stones" which is false and "Men are not stones" which is true. This means that, according to Averroes, the indefinites are ambiguous because in the vernacular language both interpretations are admissible. He does not share Aristotle's opinion, nor Avicenna's and Al Fārābī's [1, p. 122] one, because all of them treat the indefinite as a particular, which makes the two indefinites be subcontraries. However, this position is not quite convincing for several reasons that we will examine in the last section.

The third is the opposition between quantified propositions. These are the following: the contradictories, the contraries and the subcontraries. If we consider that there are two pairs of contradictories, we have really six oppositions, which are: (**1**) *Singulars*, (**2**) *Indefinites*, (**3**) *Contradictories$_1$*, (**4**) *Contradictories$_2$*, (**5**) *Contraries*, and (**6**) *Subcontraries*.

The contradictories are those which never share the same truth-value such as the singular propositions. Other contradictories are the quantified propositions which never share the same truth value "in all matters" [9, p. 92]. These are **A** and **O** on the one side and **E** and **I** on the other. To illustrate this, he gives the following example: "All men are white" and "Not all men are white". Then we have the contraries which are the two universal propositions and do not share the same truth value "in the Necessary and the Impossible" [9, p. 92] but are both false when they are Possible. The examples given are the following: "Every man is white", "No man is white" which are both false. The subcontraries are the two particular propositions, which are never true or false together in the Impossible and the Necessary but are true together in the Possible.

However, like Aristotle (and Al Fārābī), Averroes does *not* mention the subalterns which he seems to ignore completely. This might be explained by the fact that he is commenting on the *Peri Hermeneias* in which we don't find any mention of the subalterns. Regarding the treatise corresponding to the *Topics*, which is *al-Jadal* [10, p. 558], Averroes does not say anything different from Aristotle for he just summarizes his ideas by saying that what is true of the whole (or is false of it) is also true (or false) of the part, without entering into more details. He does not even use the word "*Tadākhul*" which was used by Avicenna and so known by the Arabic logicians. This shows that he probably does not consider subalternation as an opposition or else that he did not want to adopt Avicenna's views but rather to return to Aristotle's ones (even if it is with some modifications) and probably to Al Fārābī's opinions, since Al Fārābī admits exactly the same kinds of oppositions: contradiction, contrariety and subcontrariety and differs from Averroes only in his treatment of the indefinites which, according to him, are subcontrary.

It seems then that he follows Al Fārābī's and Avicenna's views by adopting the same classification of matters, but he is closer to Aristotle and especially to Al Fārābī in his classification of the kinds of oppositions. He differs from Aristotle by considering sub-contrariety as a *real* opposition while it is only a *verbal* opposition in Aristotle's view and from Aristotle, Al Fārābī and Avicenna by considering explicitly the indefinites as ambiguous while in these authors' view they ought to be considered as particulars. The shape corresponding to this classification is, then, the following if we do not include the singulars:

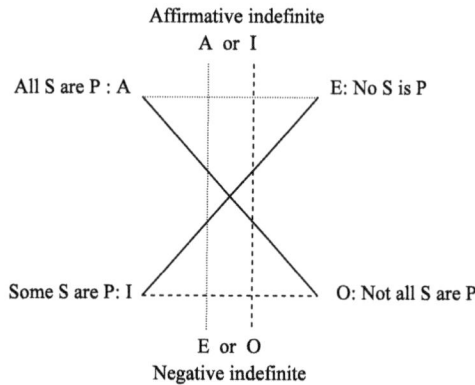

This shape is different from Aristotle's and Avicenna's ones. It shows that the indefinite might be included inside the square since it has the specificity of being ambiguous and could not be assimilated either to the particular or to the universal. If we add the singular propositions, we have an even more extended shape which will contain one more horizontal line similar to the diagonals. However, if 'or' is taken as an inclusive disjunction, the proposition 'A or I' is equivalent to 'I' and the proposition 'E or O' is equivalent to O,[9] when A and E have existential import, given that both I and O have existential import. This makes these two vertices superfluous, hence there is no hexagon since the ambiguous character of the indefinites disappears. However when A and E do not have existential import, while I and O do have it, there is no equivalence and the propositions 'A or I' and 'E or O' remain ambiguous and different respectively from I and O. The problem is then to determine whether A, E, I and O have existential import or not in Averroes' theory. It makes no doubt that existential import is attributed to A and to I as is the case with Aristotle and Al Fārābī, but what about E and O? According to many authors, Aristotle himself gives to both universals an existential import, but others such as Terence Parsons, say on the contrary that only affirmatives have existential import in Aristotle's theory, the two negatives being free of it, because O is expressed in the following way: "Not all S are P", which allows it not to have an import (see [23]). As to Averroes, he says that O might be expressed either by "Not all S are P" or by "Some S are not P" (the example given is: "Not all men are white", and "Some men are not white" [9, p. 92]) which suggests that, according to him, both formulations are equivalent since he adds "as to the particular negative, it is expressed in both ways" [9, p. 92]. But the second formulation means that O does have an existential import; if it

---

[9]This observation is due to Fabien Schang. I thank him for having pointed it out to me in an informal discussion.

is equivalent to the first one, then **O** has an import according to Averroes. As to Aristotle's text, if we follow Tricot, the French translator of *De Interpretatione*, then **O** has an import since he says in [6, note 1, p. 90]: "The following Aristotle's example: οὐ πας ανθρωπος λευκος is translated in Latin by *non omnis homo est halbus*, which is equivalent to *quidam homo non est albus*, which we have expressed in French by *quelque homme n'est pas blanc*". In the English translation, however, the same example is translated simply as: "Not every man is white" [3, p. 5]. And the ancient Arabic translation is also the same as the English one since Aristotle's example is expressed as: "Not all men are white" [7, p. 106]. This means that regarding Aristotle's text, the Latin translation is ambiguous, but the other ones corroborate Parsons' theory. As to Averroes, however, the text is clear: **O** could be expressed both ways, and this means that it can have an existential import. Regarding the singular proposition, Averroes defends Aristotle's position, saying that whenever the subject does not exist, the sentences "Socrates is sick" and "Socrates is healthy" are both false, while the negative sentence "Socrates is not sick" is true [9, p. 66]. So if we generalize this to the universal negative, we would say that it does not have an existential import in Averroes' account. But things are not so clear for Averroes seems to treat both universals in the same way in *al-Qiyās*, for instance, since he states that "the universal negative is the one where the predicate is negated from the whole of the subject as when we say 'no single man is a stone' " ([9, p. 138], [11, p. 62]), while "the universal affirmative is the one where the predicate is affirmed of the whole subject" ([11, p. 138], [9, p. 62]). This way of expressing things suggests that in the negative universal, the negation puts on the predicate. If this is so, then **E** has an import as well as **A**. Therefore if the disjunction is inclusive, '**E** or **O**' will be equivalent to **O**.

But if we consider 'or' as an exclusive disjunction i.e. '**A** or **I** but not both' and '**E** or **O** but not both', the indefinites remain also ambiguous and different from the quantified propositions. However, this creates other problems which we will consider in our last part. Anyway the figure is more complex then Aristotle's and Avicenna's ones since we could also add the singular propositions which are explicitly included by Averroes into the class of opposed propositions and construct a new kind of hexagon where the new line is horizontal. The problem is that Averroes did not specify precisely the logical relations between the singulars and the other types of propositions, so we could not really credit him with the discovery of this kind of hexagon. Furthermore, he did not consider subalternation as an opposition which makes his figure different from a closed hexagon.

But Averroes' position about the notion of opposition itself is different from both Avicenna's and Aristotle's ones. For in Aristotle's view, the notion of opposition is meant to be incompatibility since he admits only contradiction and contrariety while in Avicenna's view, the opposition is defined by the difference in truth-values but there are different oppositions which are more or less strong depending on the number of propositions involved. In Averroes' view subcontrariety is considered as an opposition but subalternation is not, which means that he considers opposition as a plural notion but it is limited to three main patterns: it is thus less extended than Avicenna's notion but more extended than Aristotle's. In the last part, we will try to characterize the differences and the affinities between these three theories.

4 Differences and Affinities Between the Three Views

According to Aristotle, only contradiction and contrariety are considered as oppositions and the shape he admits is not really a square but just a fragment of it as has been shown by Terence Parsons [23] and other people. This means that the notion of opposition according to him must be understood in the following way:

Two propositions are opposed to each other if and only if:

(1) They have the same subject and the same predicate but one of them is affirmative and the other is negative
(2) Either they never share the same truth-value or they are never true together

By admitting only contradictory propositions which never share the same truth-value, and contrary propositions which are never true together, Aristotle considers opposition as being incompatibility. This shows as Jean-Yves Béziau, for instance, has noted that Aristotle "defends an asymmetrical view, privileging the principle of contradiction over the principle of excluded middle" [15, p. 224] since contrariety respects the first principle but not the second (contrariety is true when the two propositions are false and false when they are true, which is not in accordance with the principle of excluded middle), while subcontrariety respects the second principle. This creates according to J.-Y. Béziau, some kind of "asymmetry" which is not legitimate if we consider that both principles have the same importance and are "dual". Moreover, we can demonstrate that the two principles are equivalent to each other by using De Morgan's laws and the law of double negation. If we consider this equivalence, J.-Y. Béziau is right in saying that "it makes no sense" [15, p. 224] to admit contrariety and reject subcontrariety at least in a bivalent system, which makes Aristotle's view incomplete or even somewhat incoherent.

Regarding Avicenna, as we have seen, the notion of opposition could be seen as the difference of truth-values. Whenever there is such a difference, there is some kind of opposition; the propositions might then be opposed to each other totally or partially and there is some kind of graduation in the oppositions as we have seen in the first section. The notion of opposition itself seems then to be plural unlike Aristotle's notion and it goes from the strongest kind to the less strong one, which is subalternation. We could say that in this theory the opposition is either total (complete) or partial. This shows that both principles are respected: there is accordingly no asymmetry in the theory. The oppositions are characterized semantically by the distribution of the truth-values in the table lines corresponding to each of them. In his distribution of truth-values, Avicenna makes contradiction correspond to classical negation, contrariety to incompatibility ($|$) (or to the negation of conjunction $\sim(\wedge)$) since the values he retains are the three lines where this relation is true, subcontrariety to inclusive disjunction ($\vee$), and subalternation to implication ($\supset$), for the values he retains are just those which make these operators true. The distribution that Avicenna presents corresponds to the truth cases of these different operators, which shows that his characterization of the oppositions does not differ from the classical one but is more precise in that it determines exactly the cases of truth and falsity of the propositions.

We could then define opposition by distinguishing between a complete opposition and partial oppositions in the following way:

(I) Two propositions are opposed completely if and only if:
  (1) They have the same subject and the same predicate but one of them is affirmative and the other is negative
  (2) They never share the same truth-value whatever matter they have
(II) Two propositions are partially opposed if and only if:
  (1) They have the same subject and the same predicate but one of them may deny the other
  (2) They do not share the same truth-value in one or two matters

This distinction between the two kinds of oppositions is justified by the fact that Avicenna considers contradiction as the most important opposition and that he says that the opposition is "a genus which could be divided into species" [12, p. 245]. The attention he pays to the matter of the propositions could be justified by his definition of logic which, in his view, is a very general study which analyzes not only the forms of the arguments but also the matter of the propositions involved in them. He compares the logician to an architect who must take care not only of the shapes of his buildings but also of the materials he is using in order to arrive to a good result. In the same way the logician must take care of the form and the matter in order for the argument to be conclusive, for if the form is good but not convenient for the matter the argument will not be conclusive [13, pp. 6–7].

Avicenna agrees with Aristotle in giving a great importance to the principle of contradiction since he devotes a whole chapter [14, chapter 10] to the notion of contradiction and defines opposition by means of it. In his view, contradiction is the most complete opposition because it respects the principle of contradiction. He agrees with him also by considering that the opposition between propositions is introduced (most of the times) by negation.

However, his position regarding the indefinites, which are considered as particulars and are said to have no contradictory, is not very convincing. First, there is no reason why one kind of propositions could not have a contradictory, secondly if the affirmative indefinite is particular, its negation would be universal, which contradicts the general opinion that the indefinites (affirmative or negative) must be seen as particulars.

Averroes seems to be closer to Aristotle and to Al Fārābī than Avicenna is, for as we have seen, he follows Aristotle and his text faithfully and agrees with him in many points, for instance, in not considering subalternation as an opposition. Furthermore, the figure that corresponds to what he says about oppositions is very close to the one that could represent Al Fārābī's opinion about oppositions, the only difference being related to the indefinite, which would not be included in Al Fārābī's figure if it were drawn. But the theory he defends about opposition is, in the final analysis, slightly different from Aristotle's theory. For he admits subcontrariety as an opposition, which distinguishes him from Aristotle and he uses the same way as Avicenna in classifying the propositions into Necessary, Possible and Impossible, which we do not find in Aristotle's texts. Moreover, he distributes the truth-values in the same way as Avicenna. Regarding the indefinites, his opinion is different from the other ones since he considers it explicitly as ambiguous and not only as a particular. His opinion seems to be plural but restricted to three main kinds of oppositions, which makes it less limited than Aristotle's notion but more limited than Avicenna's one. The reason for that may be that he is not convinced by Aristotle's claim that subcontrariety is a verbal opposition, although he does not comment explicitly on this claim. But he thinks like Aristotle, that opposition involves a difference in the quality of the propositions concerned: this is clear from the conditions he states himself in order for an opposition to hold. These conditions are the following: (1) The subject and the predicate must be the same in all aspects in both propositions, (2) There must be only one affirmation and one negation, (3) There is only one negation opposed to a single affirmation [9, p. 94]. We could, then, define his notion of opposition in the following way:

Two propositions are opposed to each other if and only if:

(1) They have the same subject and the same predicate but one of them is affirmative and the other is negative
(2) Either they never share the same truth-value or they are never true together or they are never false together

But his treatment of the indefinites is not very convincing even if it does not contain incoherencies. For in his view, the indefinite has either a subcontrary or a contrary proposition depending on what it says; he does not say what its contradictory is and seems to share Avicenna's opinion that it does not have any contradictory. But as we have already noted, (1) there is no reason why one kind of proposition should not have a contradictory, (2) if we consider the disjunction as inclusive, the indefinite is no more ambiguous; therefore we could try to save this ambiguous character by considering that 'or' is exclusive so that the affirmative indefinite would mean "**A** or **I** but not both" and the negative one "**E** or **O** but not both". This "solution", however, is not quite satisfying because the indefinites would be in this case equivalent and not contraries nor subcontraries as we can show by considering their formulas.

$\mathbf{A} \veebar \mathbf{I}$ is formalized by:

$$[(\exists x)Sx \wedge (x)(Sx \supset Px)] \veebar (\exists x)(Sx \wedge Px),$$

where **A** has existential import.

$\mathbf{E} \veebar \mathbf{O}$ is formalized by:

$$[(\exists x)Sx \wedge (x)(Sx \supset {\sim}Px)] \veebar {\sim}(x)(Sx \supset Px).$$

If we consider a universe containing only two elements, that is, $\{x_1, x_2\}$, then $\mathbf{A} \veebar \mathbf{I}$ would be rendered thus:

$$\{(Sx_1 \vee Sx_2) \wedge [(Sx_1 \supset Px_1) \wedge (Sx_2 \supset Px_2)]\} \veebar [(Sx_1 \wedge Px_1) \vee (Sx_2 \wedge Px_2)].$$

And $\mathbf{E} \veebar \mathbf{O}$ is rendered thus if **E** has existential import:

$$\{(Sx_1 \vee Sx_2) \wedge [(Sx_1 \supset {\sim}Px_1) \wedge (Sx_2 \supset {\sim}Px_2)]\} \veebar {\sim}[(Sx_1 \supset Px_1) \wedge (Sx_2 \supset Px_2)].$$

If we construct the truth table of $\mathbf{A} \veebar \mathbf{I}$, and that of $\mathbf{E} \veebar \mathbf{O}$[10] we find that both formulas are just equivalent since the exclusive disjunction in both cases is false in all the lines except lines 2 and 3 where they are both true. This result does not correspond to Averroes' opinion and is not intuitively satisfying. If on the other hand, we consider that **A** and **E** have no existential import and the indefinites are expressed by '$\mathbf{A} \vee \mathbf{I}$' and '$\mathbf{E} \vee \mathbf{O}$', then we find by constructing the truth tables that the propositions are subcontraries since the inclusive disjunction is valid but no other relation expressing the square oppositions is valid, for $\supset$ is false in lines $1 + 5 + 6 + 9$, ${\sim}(\wedge)$ is false in lines $2 + 3 + 13$ to 16,

---

[10]The reader can check the values of this relation by himself.

and $\underline{\vee}$ is false in lines $2 + 3 + 13$ to 16. Even if this interpretation is more satisfying intuitively than the preceding one, it does not correspond either to Averroes' text which says that the indefinites are either subcontrary or contrary. Here, they are only subcontrary. Furthermore, it is highly improbable that **A** does not have an import in Averroes' view, even if things are more ambiguous regarding **E**. Averroes' opinion seems then confused and not very convincing despite its plausibility.

What about the problem of existential import in Averroes' and Avicenna's theories? As we have seen earlier, their wording of **O** is indeed "Not all A are B", but Averroes equates this wording with "Some A are not B", which seems to give existential import to **O**. We can add that it happens also to Avicenna to express **O** as follows: "Some men are not writers" [14, p. 51], which raises also the problem of existential import for him; elsewhere he also says: "Not all men are writers, but rather some of them" [14, p. 54], which shows even more clearly that **O** has existential import, and seems close to the **Y** vertex defended by Blanché (see [16, p. 97]). So what Terence Parsons says about Aristotle, that is: "Aristotle's articulation of the **O** form is *not* the familiar 'Some S is not P' or one of its variants; it is rather 'Not every S is P'. With this wording Aristotle's doctrine automatically escapes the modern criticism" [23, section 2.2] does not seem to apply to our two Arabic logicians even if their wording is indeed 'Not all S are P'. According to Parsons, this wording of **O** solves all the problems about existential import. Parsons' argument is the following: if S is empty, **I** will be false, therefore, **E** will be true and then, "**O** must be true" because it is entailed by **E**; moreover, since **A** "has existential import", "if S is empty the **A** form must be false" [23, section 2.2]. This argument leads to the opinion that "*affirmatives* have existential import, and *negatives* do not" [23, section 2.2]. But (1) it is circular because it presupposes what has to be proved, that is, the validity of the relations of the square, (2) Parsons does not show how one can formalize the particular negative in a way that neutralizes its existential import. For if we formalize **A** by $(\exists x)Sx \wedge (x)(Sx \supset Px)$ [20, chapter 2, §26] and **O** by $\sim (x)(Sx \supset Px)$ (which corresponds literally to "Not all S are P") and construct a truth table by considering that there are only x_1 and x_2 in the universe, the following line of the table (where 1 means true, and 0 means false):

$$\{(Sx_1 \vee Sx_2) \wedge [(Sx_1 \supset Px_1) \wedge (Sx_2 \supset Px_2)]\} \; \underline{\vee} \sim [(Sx_1 \supset Px_1) \wedge (Sx_2 \supset Px_2)]$$
$$\quad 0 \; 0 \; 0 \quad 0 \quad 0 \; 1 \; 0 \quad 1 \quad 0 \; 1 \; 0 \qquad 0 \; 0 \quad 0 \; 1 \; 0 \quad 1 \quad 0 \; 1 \; 0$$

shows that there is no contradiction since, as we can see, *there is* a case of falsity under $\underline{\vee}$ which means that $\underline{\vee}$, which is the exclusive disjunction, is not valid. This shows that "Not all S are P" should be formalized in another way since, when it is expressed as above (i.e. by $\sim (x)(Sx \supset Px)$), it is exactly equivalent to $(\exists x)(Sx \wedge \sim Px)$ and does not neutralize the existential import of **O**. So Parsons' solution would be convincing only if one gives the right formalization of **O** when it has no import.[11] Besides that, it is not obvious that the Aristotelian reading of **E** makes it free of existential import. On the contrary, many authors assume that Aristotle, as well as most traditional logicians, regards **E** as having existential import. This is, for instance, what Mark McIntire says in the following: "Classical logicians typically presupposed that universal propositions do have existential import" [22]. And this is also the opinion reported by Michael Wreen who says: "The chief difference

---

[11] The right reading of **O** when it is *without* import, is given by Parsons in [24]. It is the following: "Either nothing is A, or something is A that is not B" [24, p. 6].

between classical (Aristotelian) logic and modern (Russellian) logic, it's often said, is a difference of existential import. (1) In classical logic, all categorical propositions ("All S is P"; "Some S is P"; and so on) have existential import; in modern logic, particular affirmative (PA) and particular negative (PN) propositions do while universal affirmative (UA) and universal negative (UN) do not, have existential import" [26, p. 59]. According to that opinion, there is no difference between the universal affirmative and the universal negative regarding existential import since he says "*all* categorical propositions...". This interpretation of **E**-propositions is corroborated by Aristotle's text itself which says: "By universal, I mean a statement that something belongs to all or none of something; by particular that it belongs to some or not to some or not to all..." [4, I, 1, 24a, 17–19] and where he talks about **E**-propositions in the same way as about **A**-propositions: they both concern the whole of a certain class, **A** affirms something of that whole and **E** denies something of that same whole. But if **E** has existential import in Aristotelian framework, then the solution given by Parsons is not really Aristotelian as he says, but corresponds only to the Medieval theories. As we have seen, **O** without import is expressed by: "Either nothing is A, or something is A that is not B", and this wording corresponds to Ockham's and Buridan's interpretation of **O** (see [24, p. 5]).

As to the other relations, subalternation between **A** and **I** holds when **A** has an existential import, but the one between **E** and **O** does *not* hold when **O** is expressed by $\sim(x)(Sx \supset Px)$ (i.e. when **O** has existential import), as is shown by this line of the table:

$$\left[(Sx_1 \supset \sim Px_1) \wedge (Sx_2 \supset \sim Px_2)\right] \supset \sim\left[(Sx_1 \supset Px_1) \wedge (Sx_2 \supset Px_2)\right]$$
$$0\ 1\ \ 1\ \ \ \ 1\ 0\ 1\ \ 1\ \ \ \ \ \ 0\ 0\ \ \ \ 0\ 1\ 0\ \ \ 1\ \ 0\ 1\ 0$$

If **E** has an existential import, subalternation holds indeed. Regarding subcontrariety, which is expressed by: $(\exists x)(Ax \wedge Bx) \vee \sim(x)(Ax \supset Bx)$ when both propositions have existential import, things are not much better since it does not hold in that case. This means that the Aristotelian as well as the traditional views are still confused and contain some incoherencies. These incoherencies are not avoided by Avicenna and Averroes, who do not say explicitly that **E** and **O** are free of existential import and seem to assume that they do have an import.

But the medieval formalization of **O**, given by Parsons in [24], which could be expressed in the modern symbolism by: "$\sim(\exists x)Sx \vee \sim(x)(Sx \supset Px)$" or equivalently "$\sim(\exists x)Sx \vee (\exists x)(Sx \wedge \sim Px)$, even if it is a good way to render **O** when it has *no* import and appears to solve all the relations of the square, could be also challenged. This formalization is equivalent by De Morgan's law to the following one:[12] $\sim[(\exists x)Sx \wedge (x)(Sx \supset Px)]$, which says simply: it is not the case that there are S and that all these S are P. It neutralizes the existential import of **O** since what it says is just $\sim$**A** when **A** has existential import. With this formalization, and if **A** has existential import, **E** has no import and **I** has existential import, the relations of the square, i.e. $\sim(A \wedge E)$, $A \underline{\vee} O$, $E \underline{\vee} I$, $A \supset I$, $E \supset O$, $O \vee I$ are all valid as the truth-tables show very clearly. To see this, let us take $A \underline{\vee} O$ and $O \vee I$ when **O** is formalized in that way, we have the following formulas where we consider a universe of only one element $\{x_1\}$ (but even with two elements, the relation is valid):

A : $(\exists x)Sx \wedge (x)(Sx \supset Px) = Sx_1 \wedge (Sx_1 \supset Px_1)$ (with one element)

O : $\sim[(\exists x)Sx \wedge (Sx \supset Px)] = \sim[Sx_1 \wedge (Sx_1 \supset Px_1)]$ (under the same conditions).

---

[12]This formalization of **O** comes from a suggestion made by Fabien Schang in an informal discussion. I thank him for fruitful discussion about this topic.

The exclusive disjunction is expressed by: $[Sx_1 \wedge (Sx_1 \supset Px_1)] \veebar \sim[Sx_1 \wedge (Sx_1 \supset Px_1)]$ and is without any doubt valid. If we formalize **I** by: $(\exists x)(Sx \wedge Px)$, subcontrariety is expressed by the following: $\sim[Sx_1 \wedge (Sx_1 \supset Px_1)] \vee (Sx_1 \wedge Px_1)$.[13] This formula is also valid. In the same way all the other relations of the square are valid, which means that these formalizations are the ones that show the validity of Parsons' and Buridan's solution. But we can show that this solution is not the only one that saves the relations of the square, since there are other alternatives that have the same effect, i.e. make all the relations of the square valid. These alternatives may be exhibited by a systematic examination[14] and they do not all require **O** to lack an import.

Now, as we have said earlier, this solution may be challenged, since the aforementioned formalization of **O** might seem to some people counter-intuitive. As a matter of fact, **O** may have an existential import in some cases. An example of such cases is mentioned by an anonymous referee who considers the following sentence: "Some politicians do not tell the truth". He rightly notes that if we translate **O** in the way Parsons and the medieval logicians translate it, this would lead to the following formulas: $(\exists x)Sx \supset \sim(x)(Sx \supset Px)$, which means: "If there are politicians, then not all of them tell the truth". This in turn leads to its equivalent formula by contraposition, that is: $(x)(Sx \supset Px) \supset \sim(\exists x)Sx$, which means: "If all the politicians tell the truth, then there are no politicians". This last formula seems very counter-intuitive, so the proposed formalization of **O** seems to be unacceptable, because of its undesirable consequences. We can answer by saying that this counter-intuitive character is related to the way the sentence is expressed, since that sentence has the following structure: "Some S are not P". And obviously, when one expresses **O** in that way one gives an import to it. But the formula we have given is not supposed to express **O** *with* import, on the contrary, it expresses an **O** which *does not have* an import. Our formula, as well as Terence Parsons' one is indeed a *reformulation* of **O** by "Not all S are P", which is supposed to account for an **O** *without* import. So the consequences seem in that case quite natural: they reflect the lack of import of that kind of negative particular. But this criticism shows above all that **O** should not be taken to always lack an import, and this is quite right, since in many cases, **O** does have an import. Anyway, this solution does not seem to be Averroes' and Avicenna's one, since what they say about the existential import of **O** is not as clear as what the medieval logicians say. Their wording of **O** does not mean that they do not give it an import.

What about their opinion about conversion? Let us start by Averroes, since his opinion is clearer than Avicenna's one. According to Averroes, **E**-conversion holds as well as **I**-conversion. **E**-conversion is stated very clearly in the following quotation: "As to the absolute universal premises, the negative converts in a way that preserves its quantity" ([11, p. 70], [9, p. 144]) and, **I**-conversion is also stated in the following: "As to the particular affirmative, I say also that it converts to a particular" ([9, p. 145], [11, p. 72]). So, if we consider that, despite the fact that Averroes does not talk about subalternation, he admits the same truth conditions for the different propositions as Avicenna, then we could say that he defends SQUARE as well as [SQUARE]. For these truth conditions state just the values that make true the implication between **A** and **I** on the one hand, and between **E** and **O** on the other. So we can say that subalternation holds indeed in his system even if it

---

[13] The reader may check the validity of this relation and all the others by constructing truth tables with the given formulas.

[14] This examination is made in another article written with Fabien Schang, which is under consideration.

is not really an opposition. Since he admits both conversions, we can say that he defends [SQUARE] too.

But things are more complex with Avicenna. We have seen that he holds SQUARE for the propositions he talks about in *al-Ibāra*. He also admits I-conversion for assertoric propositions. However, his opinion about E-conversion and about assertoric propositions in general is not the same as those defended by Aristotle and Averroes, since the analysis he presents in *al-Qiyās* distinguishes between two readings of this kind of propositions, and invalidates E-conversion in one of these two readings. To understand this, let us consider those readings. In [13], Avicenna defines the assertoric by saying that "it is more general than the necessary" [13, p. 28] because an assertoric sentence does not contain any modal word, so that what is important in that kind of sentences it does not require any necessity or non necessity in the relation between the subject and the predicate [13, p. 26]. Even if its matter is necessary, its necessity is different from perpetuity. For instance, the sentence "Every man is an animal" does not mean "Every man is perpetually an animal" (as when we say that "God is perpetually existent") but rather that "All men are animals as long as they exist" [*mā' dāma ḍātuhu wa jawharuhu mawjūdan*] [13, pp. 21–22]. In other absolute (or assertoric) sentences, we could have 'at some times' instead of 'as long as they exist' as in the following example: "All who wake sleep, (at some times)" [13, p. 23] (i.e. not necessarily 'as long as they exist'). This reading is called the ḍātī reading and is translated by Tony Street by the word 'substantial' [25, p. 551]. It is different from the waṣfī reading (translated as the "descriptional" reading [25, p. 551] which says the following: "All what is white is visible, as long as it is white" [13, p. 22] (one can also say: "while white" (see [25, p. 551]). Now in the ḍātī reading, the negative universal absolute does *not* convert, for if we take the following A-sentence: "All men are laughing (at some times)", which is an absolute one, and is true since it happens to everyone to laugh from time to time, we may have a corresponding absolute E-sentence, which would say: "No man is laughing, (at some times)", which means that no man could be said to laugh all the time, so that one can say that it happens to everyone not to laugh, at some times. This seems to be the way Avicenna understands this sentence since he says: "the predicate 'is laughing' can be negated from every man, at some times" [13, p. 82]. This sentence does not convert, because its conversion would lead to the following sentence: "No laughing thing is a man, (at some times)", which cannot be true, since "it is impossible to negate the predicate 'man' from what is laughing in effect" [13, p. 82] because a laughing thing cannot be said not to be a man. So for the ḍātī reading of absolute sentences, [SQUARE] does not hold, because conversion does not hold. Note that even SQUARE does not hold for this reading, since as we can see from the examples given (that is: "All men are laughing (at some times)" and "No man is laughing (at some times)") A and E may be true together, so they are no more contraries. But for the waṣfī reading, conversion may hold when we express E-sentences in the following way: "No As are Bs, while As". This may convert as: "No Bs are As, while Bs" [25, p. 551]. For instance: "Nothing that sleeps, wakes, while sleeping", which converts as "Nothing that wakes, sleeps while awake" [25, p. 551]. So, in this interpretation of absolute sentences [SQUARE] would hold as well as SQUARE, since when we say "All men are laughing, while men" and "No man is laughing, while being a man", both sentences are false together, and we can say that A and E, in this waṣfī reading, are contraries.

5 Conclusion

It follows from the above that Avicenna's and Averroes' treatment of the notion of opposition and of the opposed propositions are quite different from each other and different from Aristotle's treatment too. The notion of opposition appears to be stronger in Aristotle's view, it is considered as plural in Averroes' and Avicenna's views, but these authors differ in that Avicenna distinguishes between a strong opposition which is contradiction and other species which are less strong and seem to be partial oppositions while Averroes considers that the difference of quality is fundamental to define opposition and does not admit for this reason subalternation although he does admit subcontrariety unlike Aristotle. Their analysis seems also to prefigure the medieval distinctions and classifications and it is based on a method which we could characterize as semantic since it relies on a distribution of truth-values which follows itself from the meanings of the sentences. They seem to give an existential import to the quantified propositions, so their theories do not escape the modern criticisms. Nevertheless, each of them defends, in his own way, what Terence Parsons calls SQUARE and [SQUARE]. Avicenna's analysis is more subtle and complex than the two others, since it shows that for the substantive reading of the absolute propositions (the so-called ḏātī reading), neither SQUARE, nor [SQUARE] hold. But for the descriptive (or waṣfī) reading of the assertoric **E**-propositions, he defends both SQUARE and [SQUARE].

Acknowledgements   I would like to dedicate this article to my colleague Dr Hatem Zeghal, recently deceased, who gave me Avicenna's books and whose knowledge and advice were very precious to me. I am also grateful to my colleague Professor Mokdad Arfa for fruitful discussions, to Professor Jean-Yves Béziau for his precious help and advice, to Fabien Schang who read an earlier version of this paper and made very valuable remarks, to the anonymous referees for their useful and helpful comments, suggestions and criticisms, and to Amirouche Moktefi who procured me some useful references.

References

1. Al Fārābī: Al Qawl fi al-'Ibāra. Tekī dench Proh, M. (ed.). Al Mantiqiyāt lil Fārābī, vol. 1. Qom, Iran (1409 of Hegira, approximately 1988)
2. Al Fārābī: Charḥ al-'Ibāra. Tekī dench Proh, M. (ed.). Al Mantiqiyāt lil Fārābī, vol. 2. Qom, Iran (1409 of Hegira, approximately 1988)
3. Aristotle: De Interpretatione. Barnes, J. (ed.). The Complete Works of Aristotle, the Revised Oxford Translation. Bollingen Series LXXI* 2, vol. 1. Princeton
4. Aristotle: Prior Analytics. Barnes, J. (ed.). The Complete Works of Aristotle, the Revised Oxford Translation. Bollingen Series LXXI* 2, vol. 1. Princeton
5. Aristotle: Topics. Barnes, J. (ed.). The Complete Works of Aristotle, the Revised Oxford Translation. Bollingen Series LXXI* 2, vol. 1. Princeton
6. Aristotle: De l'interprétation. Tricot, J. (French trans.). Vrin, Paris (1969)
7. Aristotle: Kitāb al-'Ibāra. Badawī, A. (ed.). Mantiq Aristū, vol. 1. Beyrouth (1980)
8. Asad, Q.A.: The Jiha/tropos-mādda/Hulē in Arabic Logic and its significance for Avicenna's modals. In: Rahman, S., Street, T., Tahiri, H. (eds.) The Unity of Science in the Arabic Tradition. LEUS, vol. 11, pp. 229–253 (2008)
9. Averroes: Kitāb al-Maqūlāt, Kitāb al-'Ibāra, Kitāb al-Qiyās. Jehamy, G. (ed.). Talkhīṣ Mantiq Aristū, vol. 1. Beyrouth (1982)
10. Averroes: Kitāb al-Jadal. Jehamy, G. (ed.). Talkhīṣ Mantiq Aristū, vol. 2. Beyrouth (1982)
11. Averroes: Middle Commentary on Aristotle's Prior Analytics. Critical edition by Kassem, M., completed, revised and annotated by Butterworth, C.E. and Abd al-Magīd Harīdī, A. The General Egyptian Book Organization, Cairo (1983)

12. Avicenna: al- Shifā', al-Mantiq 2: al-Maqūlāt. Anawati, G., El Khodeiri, M., El-Ehwani, A.F., Zayed, S. (eds.), Madkour, I. (rev. and intr.). Cairo (1959)
13. Avicenna: al-Shifā', al-Mantiq 4: al-Qiyās. Zayed, S. (ed.), Madkour, I. (rev. and intr.). Cairo (1964)
14. Avicenna: al-Shifā', al-Mantiq 3: al-'Ibāra. El Khodeiri, M. (ed.), Madkour, I. (rev. and intr.). Cairo (1970)
15. Béziau, J.-Y.: New light on the square of oppositions and its nameless corner. Log. Investig. **10**, 218–233 (2003)
16. Blanché, R.: Sur l'opposition des concepts. Theoria **19**, 89–130 (1953)
17. Buridan, J., King, P.: Jean Buridan's Logic: The Treatise on Supposition, The Treatise on Consequences. Synthese Historical, vol. 27. Kluwer Academic, Dordrecht (1985)
18. Hasnawi, A.: Avicenna on the quantification of the predicate. In: Rahman, S., Street, T., Tahiri, H. (eds.) The Unity of Science in the Arabic Tradition. LEUS, vol. 11, pp. 295–328 (2008)
19. Hodges, W.: Ibn Sīnā and conflict in logic. http://wilfridhodges.co.uk/ (2010)
20. Kleene, S.C.: Mathematical Logic. Wiley, New York (1967). French translation: Logique mathématique. A. Colin, Paris (1971)
21. Knuutila, S.: Medieval theories of modalities. In: Zalta, E.N. (ed.) Stanford Encyclopedia of Philosophy, http://plato.stanford.edu/entries/modality-medieval (2008)
22. McIntire, M.: Categorical logic. http://www.markmcintire.com/phil/chapter3.html (2011)
23. Parsons, T.: The traditional square of opposition. In: Zalta, E.N. (ed.) Stanford Encyclopedia of Philosophy. http://plato.stanford.edu/entries/square/index.html (2006)
24. Parsons, T.: Things that are right with the traditional square of opposition. Logica Univers. **2**, 1 (2008)
25. Street, T.: Arabic logic. In: Gabbay, D., Woods, J. (eds.) Handbook of the History of Logic, vol. 1, Elsevier, Amsterdam (2004)
26. Wreen, M.: Existential import. Rev. Hispanoam. Filos. **16**, 47 (1984)

S. Chatti (✉)
Department of Philosophy, Faculté des Sciences Humaines et Sociales, University of Tunis, 94, Boulevard 9 avril 1938, Tunis, Tunisia
e-mail: salouachatti@yahoo.fr

Boethius on the Square of Opposition

Manuel Correia

Abstract This article intends to reconstruct the textual tradition of the square of oppositions from the earliest textual sources just as treated in Boethius' commentaries on
Aristotle's *De Interpretatione* and his treatises on syllogistic, *De syllogismo categorico*
and *Introductio ad syllogismos categoricos*. The research discovers two different tracks.
One way comes from Plato's *Sophist* and Aristotle's *De Interpretatione*, and the aim is
to distinguish contrariety from contradiction. The second influence also starts from Aristotle, but now in connection with his *Prior Analytics* and its commentaries and treatises
on categorical syllogistic, where the aim is to show the square as one of the three main
chapters of the complete theory of categorical logic. I suggest that this double ingredient
has accompanied the development of the square from the very original beginning of logic.

Keywords Categorical logic · Square of oppositions · Logical/material truths · Tradition
of commentaries and treatises on Aristotelian logic · Theory of oppositions

Mathematics Subject Classification Primary 03-02 · Secondary 03B65

Even though the earliest known source for the square of opposition is the treatise *Peri
Hermeneias* [6], which has been attributed to the Platonic philosopher Apuleius of
Madaura, Boethius is a primary source for the study of the square, since he says to be
following the earlier doctrine of some ancient Peripatetics, in particular Theophrastus,
who is also named in Apuleius' treatise.[1] Boethius does not presume to be an authority in logic, he rather assumes to be selecting and translating into Latin the most serious
Greek material of Peripatetic provenance he had at his disposal. That makes interesting
his logical writings, for he is an essential source of information in some specific areas of

---

[1]Boethius names Theophrastus in many places of his commentaries on Aristotle's *De Interpretatione*
[16]. Here (*in Int* 2, 12, 3ff.), he characterizes Theophrastus as a critical commentator of Aristotle, able to
go beyond Aristotle's teaching. In another passage, Boethius refers to Theophrastus as a *vir doctissimus*
(*in Int* 2, 389, 19), a phrase he uses only to refer to the Neoplatonist Porphyry (cf. *doctissimus vir*:
in Int 2, 15, p. 276). In his treatises on syllogistic [18, 19], Boethius refers to Theophrastus indirectly
when suggesting that he would be one of the ancient Greeks who "have bequeathed much to posterity in
their learned treatises" (translation by Thomsen Thörnqvist [18, p. 102]). But he also refers to his name in
connection with the five indirect syllogistic moods Theophrastus added to those ones Aristotle had found
in his *Prior Analytics* [9]. Apuleius in his *Peri Hermeneias* [6] names Theophrastus twice and both in a
more distant relation. One concerns the five indirect moods Theophrastus found in addition (cf. 193, 8).
The other relates to the case of a second *Darapti* in the third direct figure (cf. 189, 26).

J.-Y. Béziau, D. Jacquette (eds.), *Around and Beyond the Square of Opposition*, 41–52   41
Studies in Universal Logic, DOI 10.1007/978-3-0348-0379-3_3, © Springer Basel 2012

Aristotelian logic. This is the case with the square of Oppositions, which plays a central part in his work on categorical logic.[2] Indeed, the square is referred to in his (i) expositions of Aristotelian categorical syllogistic, [18, 19], and his (ii) two commentaries on Aristotle's *De Interpretatione* (= *De Int*), [16].[3] As a result, the square is examined from two different aspects:

(a) As a theory of negation. (Here, the square is related to the opposition of categorical propositions, which goes back to Plato's seminal discussion on negation in his dialog *Sophist*.)
(b) As a theory of logical consequence. (Here, the square is related to the syllogistic theory as one of its essential preambles.)

I suggest that (a) and (b) accompanied the commentaries and the treatises on the square from the original and first development of the Peripatetic tradition in logic. It is also likely that (a) is more related to the tradition of commentaries on *De Int* and (b) takes its essentials from the tradition on Aristotle's *Prior Analytics* [9].

1 The Square and the Tradition of *De Interpretatione*

In (a), the question is which categorical opposition is the strongest and divides more completely or exclusively truth from falsity. In his commentaries on *De Int* (= *in Int*), [16], Boethius suggests that Aristotle resolved this question when he went beyond Plato's *Sophist* [26] in distinguishing contrariety from contradiction. Indeed, Plato in this dialog did not distinguish one concept from another, when he defines negation of being as the *other* than being (*Sophist* 257c–258c).[4] According to Boethius, Aristotle's phrase "nor is there any correct name for it <for the indefinite name>",[5] and a similar one for the indefinite verb (*De Int* 16b13), are claims that the past was unable to explain satisfactorily the indefinite signification of name and verbs, negation, and not-being within the categorical proposition. Boethius' reference to some more ancient authors (*antiquiores*)[6] points out some ancient philosophers, in particular Plato, as those who fail in this attempt. Boethius' commentary insists that these more ancient authorities (*antiquiores*) did not make an explicit difference (*nuncupata differentia*) between being and non-being. Similarly, the commentary makes manifest that the imposition of the name 'indefinite verb',

---

[2]Boethius is also the author of one treatise on hypothetical or conditional logic. In his *De hypotheticis syllogismis* [17] Boethius gives an idea of the first Peripatetic developments and later systematization. Boethius and his sources understand categorical logic as the logic of single sentences that can be true or false, without any condition (cf. *in Int* 2, 186, 13ff.: Categoricas propositiones Graeci vocant, quae sine aliqua condicione positionis pronuntur (…); cf. also *in Int* 2, p. 112, 10–26), [16].

[3]Boethius wrote a minor and a major commentary on Aristotle's *De interpretatione* (= *in Int*), [16]. His treatises on syllogistic are also double (*De syllogismo categorico* and *Introductio ad syllogismos categoricos*: [18, 19]), but in this case their unity is problematic. See [17–19].

[4]As is generally accepted Aristotle bore in mind Plato's *Sophist* [26] when he addressed his criticism to the old-fashioned way of solving the problem of the existence of being and plurality (cf. *Metaphysics* N, 2, 1088b. 28–1089a 6, [7]). To this concern, cf. [5, 9, 12, 14, 20, 27].

[5]*De Int* 16a29: *ou men oude keitai ónoma*, [8, 11].

[6]*Antiquiores*: cf. *in Int*, p. 58, 18; *in Int*, p. 59, 18–26; *in Int* 2, p. 63, 5–14, [16].

made by Aristotle historically, is a remarkable fact, because a name for these expressions had not been imposed by the ancients (*ab antiquis*), despite the difference between this kind of verb and the simple and pure one.[7] Boethius suggests that these *antiquiores* were not Peripatetics but some more ancient authors than Aristotle, who are found in the late Parmenidean tradition, and Plato's discussion in *Sophist* 257c–258c [26] appears to be the immediate precedent of Aristotle's discussion on the nature of the indefinite terms and negation. Boethius' comments suggest that the lack of a specific denomination for expressions like 'not-man' and 'non-being' is due to the fact that nobody before Aristotle had become really cognizant of the nature of that kind of expressions (*et usque ad Aristotelem nullus noverat quid esset id quod non homo diceretur*). Here *noscere* must signify a real knowledge and it should imply Aristotle's disapproval of Plato's solution in *Sophist*. Thus, while it is unquestionable that Plato in the *Sophist* tries to determine the nature of not-being and even that he, as a product of this attempt, remains able to give a certain account of the nature of expressions like 'non-beautiful', 'non-being' and similar ones, Boethius suggests that Aristotle's attempt would be an accurate account of those expressions and that would be tested by the technical name that Aristotle gave to the difference (*differentia*) between simple terms and terms with a negative particle, which denotes the explicit difference (*nuncupata differentia*). Thus, in Boethius' commentary we find a history of negation, while Ammonius Hermeias,[8] the Alexandrian contemporary of Boethius, prefers a non-historical recount.

Non-being and the nature of negation have been also put at examination by Aristotle's *De Int* [8]. Here Aristotle would also criticize Plato's *Sophist* [26] because non-being is rather the *absence* of being and not *the other than* being.[9] The absence of being would be a

---

[7]*in Int* 18–26, p. 59: haec enim differentia quae est *non currit* et *non laborat*, quae a verbo puro et simplici distat, nulla apud antiquiores vocabulo nuncupata est differentia. differentiam autem vocavit id quod dicitur *non currit* et *non laborat* ab eo quod est *currit* et *laborat*. sed quoniam his nullum est ab antiquis nomen impositum, Aristoteles nomen ipse constituit dicens: *sit verbum infinitum*, [16].

[8]Ammonius and all the post-Ammonian commentators refer to certain *palaioteroi*, but they interpret their existence in the mythical way that Proclus' commentary on Plato's *Cratylus* suggested. Ammonius makes a passing allusion to these ancient authors at *in Int*, p. 41, 16–20, [4]. His allusion also supposes that they were more ancient than Aristotle, which is clear from his use of the adjective comparative *palaioteroi*. But Ammonius seems to have understood the existence of these ancient men in a non-historical sense, since when he comments on this point, he says that even though there is an obvious difference between a definite and an indefinite verb, this last name was disregarded (*pareóratai*: p. 52, 6) by those who have given the names, and some lines later he adds that this difference should have been established by the 'fathers of the names' (p. 52, 9). So it is evident that Ammonius' comments on this point lead to a non-historical interpretation, which was probably influenced by Proclus' interpretation of Plato's *Cratylus* (see *Cratylus* 389a 2 *et passim*).

[9]Berti [14, p. 122] remarks that Aristotle in *Metaphysics* Γ, 7, [7], when he explains and defends the validity of the principle of *tertium exclusum*, maintains in his argument that being and not-being are a contradictory pair (1012a. 5–8). Similarly, some lines later (1012a. 15–18), Aristotle will affirm that not-being is the denial of being. In this passage, as is known, Aristotle says that "when a man, on being asked whether a thing is white, says 'no', he has denied nothing except that it is; and its not being is a negation" (translation by Ross [7]). Tricot [10]—whose translation Berti follows—says "et le Non-être est une negation", which seems to be an assertion of equivalence, and too general. On the other hand, Berti [14] has also observed that Aristotle (*Met.* Γ, 2, 1004a 2; and I, 3, 1054a 30, [14]) includes 'the same' and 'the other' in his division of contraries, which implies that Aristotle thinks that they are opposite by contrariety and not by contradiction. Berti has also shown that Aristotle distinguishes clearly 'the other' from not-being, because 'the other' is one of the forms or species of multiple, and it is opposite to 'the same', which is a species of the one.

strict negation (= contradiction), whereas the other than being would be the accidental or general negation (= contrariness). This is the closest reference to the essence of negation, for Plato assures plurality by denying being (and so making a parricide over Parmenides, *Sophist* 241d3) while Aristotle would think that this is not the specific function of a denial, but to show the contradictory opposite of affirmation.

According to Aristotle's criticism of *Metaphysics* N, 2, [7] Plato's error is the same as that of the ancients (cf. the archaic solution: *Metaphysics* 1089a1–2), namely, to think that it was necessary to prove that not-being exists and has a certain reality, i.e. a certain being, in order to prove the existence of plural things.[10] But, according to Aristotle, it was not necessary, in order to derive the plurality of beings, to show that non-being is, since being and non-being are already multiple by themselves. In fact, as Aristotle implies in *Met.* N, 2, being and not-being have already many senses as there are categories, and in addition to this they are used to mean what is true and false, and finally to express what is actual and potential (cf. 1089a 26–31).

By means of this sort of history of logic, Boethius suggests that Aristotle's *De Int*, [8, 11], through his insistent saying that a single affirmation has only one single negation (cf. 17a33–34; 17b37; 18a8; 20b3–4), is calling attention to the fact that Aristotle has a formal concept of opposition and that Plato's theory of negations has been completed by Aristotle by distinguishing contrariety from contradiction.

This would explain why the Neoplatonic commentators embarks in a complete classification and calculus of the number of categorical oppositions,[11] since they would attempt to confirm the thesis that every affirmation has its own negation by means of that expedient. This classification and calculus would be implicit in Aristotle's saying that "there will not be any more opposition than these" (*De Int* 20a1: Ackrill translation [11]). From this argument, it follows that Aristotle wants to outline a square to display the different concepts of categorical opposition. The expedient is rather common in *De Int* [8, 11], for Aristotle adds another diagram (*De Int* 19b20–24) that aims at stating that simple negation follows from indefinite affirmation and indefinite affirmation follows from privative affirmation (i.e., not being X follows from being non-X and this follows from being un-X).[12] Actually, in the square of opposition, he would be purposely interested in representing diagonal relations where contradictory pairs are described, since Plato might only express contrary and subcontrary oppositions: the A/E and I/O relations of opposition.[13] Plato's advance in the *Sophist* [26] had given a first difference in opposite pairs of categorical propositions when he defined not being, i.e., the negation of being, as the *other than* being, for this allows him to define non being, negation, and falsity, but he did not recognize kinds of opposition. Aristotle, on his hand, defines negation as the absence of the thing that has been affirmed (cf. *Metaphysics* 1004a16, [10]) and finds a clear distinction between contrariety and contradiction, namely, contradiction never fails in dividing truth

---

[10]*Met.* 1089a. 4–6. "They <the Platonics> thought it necessary to prove that what is not is; for only in this way—from being and from something else—would it be possible for there to be many existing things" (translation by Annas [5, p. 119]).

[11]For example, Ammonius in his commentary on Aristotle's *De Int* argues so. Cf. *in Int*, p. 86, 26–p. 87, 7; p. 159, 24–29; p. 214, 6–24, [4].

[12]For the history of this diagram, see Correia [21, pp. 41–56].

[13]We follow later tradition of naming the universal affirmative proposition an A, the universal negation an E, the particular affirmation an I, and the particular negation an O.

and falsity, and so contradiction must be defined as the specific and strongest negation in categorical propositions. By contrast, contrary propositions can be false together.

This doctrine of negation has a Peripatetic provenance but a Neoplatonic consent and it is what Boethius introduces in Latin terms through his commentaries on *De Int* [15, 16, 18]. Aristotle's square of opposition would present contradiction as the strongest opposition between two opposite categorical propositions, since it divides entirely truth and falsity, whereas contrariety divides truth and falsity but not completely, because if the matter of the proposition is merely *possible* and one proposition is false (e.g., "Every man is just") the other is not necessarily true (i.e., "No man is just").[14] Subcontrariety is clearly the weakest opposition, for it does not even divide truth and falsity, although if one of two is false, the other must be true.[15] This is the reason why in Aristotle's *De Int* [8, 11] and, as I think, in the amplitude of the Aristotelian logic, the denial of a categorical proposition is its contradictory opposition; and so when we need to deny a categorical proposition, what we actually need is to find its contradiction.[16] Aristotle knows that while A and E propositions, which are contrary opposites, are also mutually exclusive (and so a *per absurdum* argument could rest on contrary propositions),[17] contrariety is only a secondary and accidental opposition.

2 The Square and the Tradition of *Prior Analytics*

In (b), the square appears to be related to the tradition of commentaries on Aristotle's *Prior Analytics* [9] and now Boethius' teaching seems to stem from Theophrastus' own *Prior Analytics* and perhaps own square. As told, Boethius develops this perspective in his syllogistic treatises [18, 19], which owe much to Alexander of Aphrodisias, Theophrastus

---

[14] As I will explain in detail in the second part of this paper, the most ancient Peripatetic tradition of commentaries think that Aristotle would determine the truth values of opposition by supposing that every proposition has a signification given by the matter of the proposition. The matter of the proposition is an ancient doctrine also accepted by Neoplatonic commentators of Aristotelian logic. Cf. Correia [22, p. 397, n. 22].

[15] Subcontrariety is presented by Aristotle in *De Int* 17b29–33, [8, 11].

[16] In other words, Ammonius Hermeias in his commentary on Aristotle's *De Int* [4] illustrates this with the position of the negative particle in the proposition that will be denied: "Now, that the negation arises when the affirmation takes on the negative particle, is clear. But where in the affirmation one must place it, in order to make the negation, and why this is so, we must specify. I say, therefore, that one must not join it to the subject, but to the predicate; first, because the predicate is more important, as has been said (70, 4f.), and prior to the subject, which is also why the whole sentence is called 'predicative' (so, if we want to destroy the affirmation and make a negation, we must not attach the negative particle, which is the cause of the destruction, to the less important of the parts, but to the more important, since in animals too, more than in other living things, the whole does not perish if just any part is destroyed, but only if one of the more important part <is destroyed>" (Blank's translation, [15]). This biological comparison can stem from Dionysius Thrax (*Scholia*, p. 516, 28–36), [31], where the name and the verb are compared to the brain and the heart in the human body, and the other parts of the phrase to such parts as a hand or a foot: a man can live without a hand, but not without brain and heart.

[17] In *De syllogismo categorico* 96, 6ff., Boethius defends this doctrine by saying "And we should not be disturbed because in some syllogisms the conclusion is opposed to the contrary proposition, while in others to the contradictory proposition. Since the error will be committed in both cases: either accepting both contrary propositions or accepting both contradictory propositions" [18].

and Eudemus, and also to the Neoplatonic Porphyry, who is said to have made a sort of appraisal of Theophrastus' critical review of Aristotle's logic.[18] It is accepted that Boethius found his Greek material for commenting on the square in this work in Porphyry's writings on Aristotle's logic.[19]

As a result, in every ancient Aristotelian commentator, either Peripatetic or Neoplatonic, the square is made to come from a division of the categorical proposition. The division can be traced as early as Alexander of Aphrodisias and Apuleius. It does not intend to divide propositions in species according to their properties (quality, quantity, etc.), but intends to divide the categorical propositions that do not have terms in common from those that have some or both terms in common either in the same order or in the converse one.[20]

Thus, a propositional pair can always be divided into whether (a) it does not have any terms in common or (b) it does have some terms in common. For there are some propositions that have no term in common with another, e.g. 'Every man is an animal' and 'Virtue is good'; but there are some others that certainly have some terms in common, e.g. 'Every man is wise' and 'No man is just'; or both terms in common, as in 'Every man is wise' and 'Every wise (being) is a man'. Accordingly, Boethius accepts that the pair having no term in common cannot be studied by logic, but the pair having either one or both terms in common can be studied by logic.[21]

The second level divides propositions into (b.1) both terms in common and (b.2) one term in common. And these propositions that have both terms in common are divided into (b.1.1) both common terms with the same order of terms (e.g., 'Every man is an animal'/'No man is an animal') or (b.1.2) both common terms with a change of order (e.g., 'Every man is an animal'/'Every animal is a man'). Finally, as to those propositions that (b.2) have only one term in common, Boethius finds three types: the common term is (b.2.1) subject in one proposition and predicate in the other (e.g., 'Every man is animal'/'Every Greek is a man'), or the common term is (b.2.2) the predicate in both, or (b.2.3) the subject in both.

---

[18]Boethius, *De syllogismo categorico* 101, 6–11: "And I took over these things from an introduction to the Categorical Syllogisms, by following Aristotle and borrowing certain things from Theophrastus and Porphyry, as much as the brevity of an introduction allowed me" (my translation from [18]).

[19]I would like to refer here to Shiel [28] and the special proof he gives to prove this sifting place occupied by Porphyry. See also Thomsen Thörnqvist [18, p. 141].

[20]This is the reason why the most likely author of this division is Theophrastus. Indeed, before Boethius, Alexander of Aphrodisias already assumed this division in his comment on Aristotle's *Prior Analytics* [1] and Apuleius [6] also gives textual evidence of being dependent of this division in his *Peri Hermeneias*.

[21]Indeed, the division is in Boethius' *De syllogismo categorico* 17, 10ff., [18], but it is neither in the commentaries on *De Int* [16] nor at any other ancient commentary on this treatise, but a similar division takes place in *Introductio ad syllogismos categoricos* 26, 7–8, [19]; in Apuleius' *Peri Hermeneias* 183, 9–19, [6]. Also "the procedure is closely paralleled", as Thomsen Thörnqvist [18, xxii] has recently corroborated it, in the commentaries on *Analytics* by Alexander (*in An Pr* 45, 10ff.), [1], Ammonius (*in An Pr* 35, 36ff.), [3], and Philoponus (*in An Pr* 40, 31ff.), [25]. This certainly proves that syllogistic treatises by Boethius belong to the tradition of commentaries on Aristotle's *Prior Analytics* [9], rather than to the commentaries on Aristotle's *De Int*. See also Lee [24, 65–74].

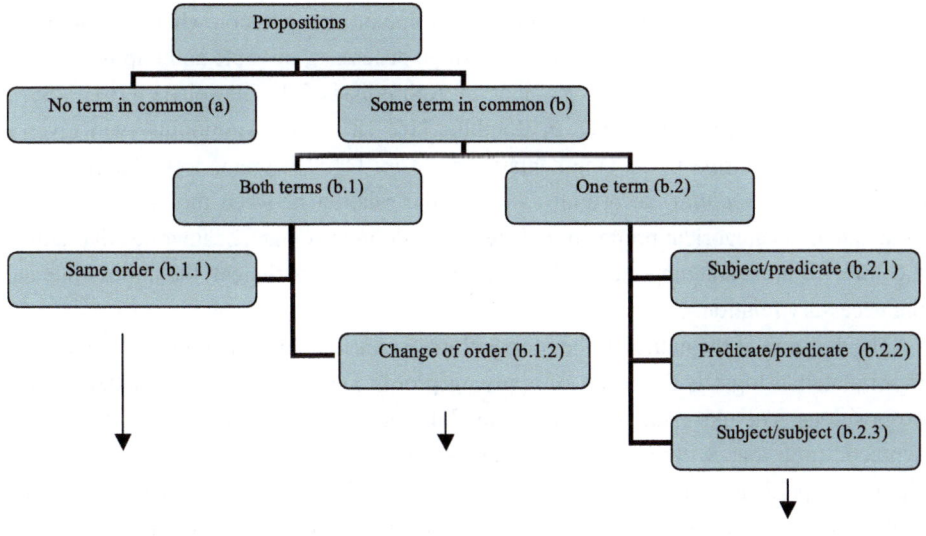

The square of Opposition   The theory of Conversion   The theory of Syllogism

This division gives theoretical unity to *De syllogismo categorico* (against McKinlay's conjecture that this treatise is rather an editorial phenomenon than a book designed by Boethius),[22] for the most important logical operations of the Aristotelian logic find their place in it. Indeed, (b.1.1) gives room to the doctrine of the square of Opposition, while (b.1.2) classifies the theory of conversion both simple and by contraposition. Finally, the theory of syllogism is under (b.2), where the three figures of the syllogism correspond to (b.2.1), (b.2.2) and (b.2.3). In this point, I differ from De Rijk [23, p. 18], since he thinks that (b.2) does not play any further role in *De syllogismo categorico* [18].[23]

According to this tradition, the square of Opposition can be identified with the logic of the propositions having both terms in common, the subject and the predicate, in the same order. When two categorical propositions are in this situation (e.g., 'Every man is just' and 'Some man is not just'), the differences of quality and quantity are of relevance to logical truth. This is certainly the origin of the two kinds of logical relations that commentators distinguished in their comments on the square: relations of opposition (when quality and quantity are important) and relations of the parts and the whole (when quantity is important).

Boethius goes to prove the logical relations between opposite propositions in the square. The methodology used by all of the commentators, either Peripatetic or Neoplatonic, is based on the matters of the proposition, a doctrine that can be represented in the following points: (i) every well-formed categorical proposition has its own matter of signification. (ii) The signification of a proposition is given by the matters of the proposition, i.e., the relation between the nature of the subject term and of the predicate term.

---

[22]Cf. Correia [22, p. 395, n. 14].

[23]Thomsen Thörnqvist ([18, p. 164] and [19, pp. 140–143]) and Correia [22, p. 394, n. 13] have been independently stated that this remark by De Rijk [23, p. 18] is not convincing. What really happens is that the account of pairs of propositions having one term in common anticipates the three figures of the categorical syllogism.

(iii) It is always possible to calculate the truth-value of a proposition when its matter has been identified. (iv) There are only three matters: necessary, possible and impossible.

This doctrine is known by Alexander of Aphrodisias [2],[24] Porphyry [16],[25] Syrianus,[26] and it is a commonplace in Boethius [16, 18, 19].[27] Ammonius even gives a report of this doctrine in his commentary on *De Int* [4].[28] It seems to be based on the doctrine of *predicabilia*, according to which the relation between the subject and the predicate in a categorical proposition depends on definite semantic categories that define the ambit of understanding the significations of terms: the contingent, the impossible and the necessary relation.

The doctrine of the matter of the propositions seems to demonstrate that a formal truth in Aristotelian logic is the result of applying proofs related to the three matters of the propositions. The universal necessity of this kind of proof seems to be supported by the syntactic truth that, within a categorical proposition, there are no more ways to relate a significant predicate term to a significant subject term than those given by the necessary, impossible and possible matters. This amounts to say that the subject whether is entirely contained in the predicate or entirely separated of he predicate or partially included en the predicated. Apuleius of Madaura, in his *Peri Hermeneias* [6] called the matter of the propositions *significationes* and when he opened to the question of how a logical truth can be based on significations that seems to be indefinite in number, he points out that these significations are not innumerable but perfectly countable, since they are no more

---

[24]Cf. Alexander *in Top*, p. 192, 7–8, [2]. Also Boethius *in Int* 21–24, p. 136. Also in *in An Pr* 84, 16–20, [16].

[25]Boethius [16] follows Porphyry (*in Int* 2, 2–6, pp. 279–283) in using the doctrine of the matter of propositions for proving results.

[26]Boethius names Syrianus on this concern in: *in Int* 2, p. 322, 29–p. 323, 13. There is no clarity concerning whether Syrianus is distinguishing matter from mode. The vagueness of the expression used by Boethius in his report of Syrianus, namely, *qualitates propositionum*, makes difficult to decide whether Syrianus was thinking of what Ammonius called later the matters of the proposition, or of the modal propositions. Zimmermann [32, p. lxxxix, n. 3] thinks that Syrianus refers here to modalities and not to matters. Lack of textual material here is decisive.

[27]E.g., *Introductio ad syllogismos categoricos* 29, 9ff. Also *in Int*, p. 137, 15–16; and *in Int* 2, p. 148, 3–155, 16, [16, 18].

[28]Ammonius Hermeias [3], *in Int*, p. 88, 12–23, gives an eloquent confirmation of the importance and interest of this doctrine for the development of ancient logic when he says: "I am talking about the relation according to which the predicate term either always holds of the subject term, as when we say the sun moves or man is an animal, or never holds <of it>, as when we say 'The sun stands still' or 'Man is winged', or sometimes holds and sometimes does not hold, as when we say Socrates walks or reads. Those who care about the technical treatment of these things call these relations the matters (*hulai*) of the propositions, and they say that one of them is necessary (*anagkaia*), another impossible (*adunatos*), and the third contingent (*endekhomenê*). The reason for these names is obvious, but they decided to call these relations 'matters' in the first place because they are seen together with the things which underlie (*hupokeimena*) the propositions and are not obtained from our thinking or predicating, but from the very nature of the things" (Blank's translation, [15]).

than five.[29] The doctrine of significations is the most likely antecedent of the doctrine of matters of propositions.

That the matter of the proposition is not the same as the three modalities of categorical propositions, distinguished by Aristotle in *De Int*, Chaps. 11 and 12 [4], is also affirmed by Ammonius, although the difference between one and another seems to have been confused sometimes.[30]

If this is so, ancient Aristotelian logic seems to keep away from both a formalistic dependence on rules and a psychological dependence on matters and significations, and this is why I would rather consider misleading the explanation of what is a logical truth for ancient logicians given recently by Barnes [13] in terms that "any logical form will determine a formal rule and any formal rule will conversely determine a logical form".[31]

It would be also an error to suppose that ancient Aristotelian logic accepted a sort of material logic, as Sullivan [6, pp. 71–75] advances in his appraisal of this special characteristic of Apuleian logic. He says that there are two logics in Apuleius, one material, the other formal.[32] But ancient Aristotelian logic is neither a sort of material logic nor a sort of formal-rule logic. Differently to both positions, ancient Peripatetic and Neoplatonic logicians are relating the syntactic concept of logical consequence to the semantic one. In this relation, formal rules are related to the matters of the proposition, and somehow to the capacity of the human mind of conceiving examples in the three different signification forms. There is no psychologism in this ancient doctrine, for the aim is to prove that an inference or conclusion is a logical truth when it is true in every matter of categorical predication. Boethius is eloquent in distinguishing material truth from logical truth when he defines the conversion of categorical propositions: "In general, by 'propositional conversion' I mean conversion without exception, that is, in every material instance" (p. 32, 9, Thomsen Thörnqvist's translation, [18]).

---

[29]In *Peri Hermeneias* VI, 4–6, p. 182, [6], Apuleius adds that the doctrine consists of verifying each proposition, with the help of the significations of terms (*significationes*), whether the truth and falsity keep identical once the proposition is converted. And he lists the significations for, he says, they are not innumerable (*innumerae*): the property, the genus, the difference, the essence and the accident. These are transformed into necessary, impossible and contingent matters already in Ammonius.

[30]*in Int*, p. 215, 9–16, [18]. It is worth noting that Philoponus' commentary on Aristotle's *Prior Analytics*, by following Ammonius, says that Aristotle established this distinction between matter (*hulai*) and mode (*tropos*) in *De Int*. Cf. Philoponus *in An Pr*, pp. 44–45, [25]: "It must be said that we consider the matters <of the propositions> in the very nature of the facts stated in the proposition (because it is necessary that what is predicated of the subject either always belongs to it, or never, or sometimes belongs and sometimes does not, and this is the reason why we call these 'matters': for without these the principle of understanding of propositions does not subsist either. On the other hand, we say that the modes are put in the very expression that we make and that they are put from outside to the terms that complete the proposition and maintain what is predicated."

[31]Barnes [13, p. 79] says that "An argument—I conjecture—concludes methodically provided that there is some *formal rule* which validates it. And there will be such a formal rule just in case the argument is valid in virtue of a logical form. (For any logical form will determine a formal rule and any formal rule will conversely determine a logical form.) Hence methodically concluding arguments are, in effect, formally valid arguments; and unmethodically concluding arguments are materially valid arguments."

[32]Sullivan [30, pp. 71–75].

If we leave aside the unusual formulae Boethius uses to describe the notion of the matters of the proposition,[33] and we follow Ammonius' easier formulation dividing the matter into three, the square could be reconstructed as follows:

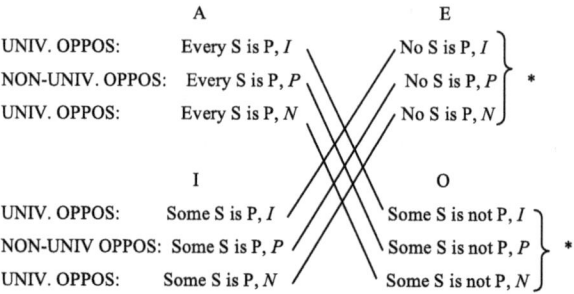

		A			E
UNIV. OPPOS:	Every S is P, I			No S is P, I
NON-UNIV. OPPOS:	Every S is P, P			No S is P, P
UNIV. OPPOS:	Every S is P, N			No S is P, N

*: These are valid in two material instances only

Where UNIVERSAL OPPOSITION = it does divide completely truth and falsity (as contradiction does) in any matter; and NON-UNIVERSAL OPPOSITION = it does not divide completely truth and falsity but only in some matters. And where *I, P, N* are the matters of the propositions and *I* = impossible, *P* = possible and *N* = necessary. Diagonal relations correspond to contradictory opposite pairs and they are UNIVERSAL OPPOSITIONS in any matter. The brackets indicates that since this relation of opposition, when every matter is considered, is valid only in two instances, the relation is not a logical truth.

Two final and last observations: (i) The accidental or possible matter is what generates the *rule* for contrary and subcontrary propositions. And that seems to be the reason why Aristotle and Boethius propose the squares' propositions with accidental matter, i.e., with 'white' said of 'man', as Aristotle, and with 'just' said of 'man', as Boethius. It is because both 'white' and 'just' are accidental or possible predicates of man. Ammonius Hermeias also uses an accidental matter: Every man walks. And (ii) within the square, the rule is not properly speaking a rule; it is just the description of the exception of a true contradiction (= UNIVERSAL OPPOSITION). Here, so to speak, the exception makes the rule. For example, if one says, "contrary propositions cannot be true together but they can be false together", what one is pointing out is that contrary propositions behave differently from contradictory propositions, which are mutually exclusive. In this square, contradiction, which is represented by diagonal propositions, is a perfect relation of opposition, because it divides completely truth and falsity. Accordingly, opposition and negation are primarily said of contradiction. Let us take I = The man is a stone; P = The man is a grammarian; N = The man is an animal. When the A is true in any or every matter, then its contradiction, the O, is false in any or every matter, and vice versa, when the first is false the other must be true. This is both the original result and the actual unique rule.

---

[33] For example, in the case of impossible matter, Boethius uses the formula: At si id de subiecto praedicetur quod vel numquam subiecto valeat convenire, ut lapis homini. ("But if <we take> this <matter> that never is true when predicated of the subject, as 'stone' of 'man', then (…)", cf. *Introductio ad syllogismos categoricos*, p. 53, 10–15.) Similar formulae are used to describe the notion off the other matters of proposition, [19].

3 Conclusions

In the third place and lastly, I summarize by concluding that: (i) The square of Opposition finds its roots in the dialectical discussions on negation, falsity and non-being in Plato's *Sophist* [26], and so the square finally stems from Plato's solution to the problem of how non being is, if one does neither accept that everything is true nor that everything is false. But Aristotle criticizes Plato's solution that non-being is the other than being, because one should start from the plurality of being as given and not to pursuit it as an aim. Also, because what is a non-being is not the other than being, but the absence of what has been affirmed; opposition in this way becomes a complex concept that must be further divided. This being the case, quality and quantity are the properties to be studied, and that explains why Aristotle divides opposition into contradiction, contrariety and subcontrariety. Thus, the square illustrates Aristotle's refutation of Platonic logic by showing that contradiction is the strongest opposition of those that can affect a simple or categorical affirmation; then, contrariety and finally subcontrariety.

On the other hand, (ii) the tradition of commentaries on *Prior Analytics* already assumes Aristotle's teaching on negation and takes the square as a description[34] of the logical inferences there are when propositions have the same subject—and predicate terms in the same order. There are many relations of inference and proofs become very important. The doctrine of the matter of the propositions supports these proofs and shows that there are logical or universal relations of opposition and material or non-universal relations of opposition. This doctrine also makes the general conviction that ancient Aristotelian logic (whether Peripatetic or Neoplatonic) aimed at organizing syntactical and semantic aspects of logical truth and it is neither a material logic nor a formal-rule one.

Acknowledgements This article was made possible thanks to Fondecyt N° 1095206. I am also grateful to Prof. Pierre Simonnet for his great help and Prof. Pieter Seuren for his comments on the oral version of this paper at the Second World Congress on the Square of Opposition, June 17–20, Corsica. I am also very grateful to the editor and the referees for their comments on my paper.

References

1. Alexander of Aphrodisias: Alexandri in Aristotelis Analyticorum Priorum Librum I Commentarium. Wallies, M. (ed.). Commentaria in Aristotelem Graeca, vol. 2.1. Berlin (1883)
2. Alexander of Aphrodisias: Alexandri Aphrodisiensis in Aristotelis Topicorum Libros Octo Commentaria. Wallies, M. (ed.). Commentaria in Aristotelem Graeca, vol. 2.2. Berlin (1891)
3. Ammonius: Ammonius in Aristotelis Analyticorum Priorum Librum I Commentarium. Wallies, M. (ed.). Commentaria in Aristotelem Graeca, vol. 4.6. Berlin (1890)
4. Ammonius: Ammonii In Aristotelis De Interpretatione Commentarius. Busse, A. (ed.) Commentaria in Aristotelem Graeca, vol. 4.6. Berlin (1895)
5. Annas, J.: Aristotle's Metaphysics Books M and N. Oxford (1976)
6. Apuleius: Apulei Opera Quae Supersunt, vol. 3. Apulei Platonici Madaurensis de Philosophia Libri, Liber PERI ERMENEIAΣ. Thomas, P. (ed.). Teubner, Leipzig (1908)

---

[34]Diagrams are common in Boethius' writings on Aristotelian logic. They are both *descriptions* and *tools* for the memory. For example, in *in Int* 2, p. 324, 15, [18], he calls a diagram a *subsidium memoriae* (a tool for the memory); also *in Int* 2, p. 152, 7–10, [19]. He uses *descriptio* in *De syllogismo categorico*, p. 21, 1, [18].

7. Aristotle: Aristotle's Metaphysics. A revised text with introduction and commentary by Ross, W.D. Oxford University, Oxford (1924)
8. Aristotle: Aristotelis Categoriae et Liber de Interpretatione. Minio-Paluello, L. (ed.). Oxford (1949)
9. Aristotle: Aristotle's Prior and Posterior Analytics. A revised text with introduction and commentary by Ross, W.D. Clarendon, Oxford (1949)
10. Aristotle: La Métaphysique. Translation with notes by Tricot, J., vols. i–ii. Paris (1962)
11. Aristotle: Aristotle's Categories and De Interpretatione. Translation with notes by Ackrill, J.L. Oxford (1963)
12. Aubenque, P.: Le Problème de l'être chez Aristotle, 5th edn. Paris (1983)
13. Barnes, J.: Logical form and logical matter. In: Alberti, A. (ed.) Logica, Mente e Persona. Studi sulla filosofia antica, p. 79. Florence (1990)
14. Berti, E.: Quelques remarques sur la conception Aristotélicienne du Non-être. Rev. Philos. Anc. I, 115–142 (1983)
15. Blank, D.: Ammonius on Aristotle on Interpretation 1–8. Translation with notes in: Sorabji, R. (ed.) Ancient Commentators of Aristotle. London (1996)
16. Boethius: Anicii Manlii Severini Boetii Commentarii in Librum Aristotelis PERI ERMHNEIAS, prima et secunda editio. Meiser, C. (ed.). Leipzig (1877–1880)
17. Boethius: De Hypotheticis Syllogismis. Obertello, L. (ed.). Paideia Editrice, Brescia (1969)
18. Boethius: Anicii Manlii Seuerini Boethii De syllogismo categorico. Thomsen Thörnqvist, C. (ed.). A critical edition with introduction, translation, notes and indexes. Studia Graeca et Latina Gothoburgensia, vol. LXVIII. Acta Universitatis Gothoburgensis, University of Gothenburg (2008)
19. Boethius: Anicii Manlii Severini Boethii Introductio ad syllogismos categoricos. Thomsen Thörnqvist, C. (ed.). A critical edition with introduction, commentary and indexes. Studia Graeca et Latina Gothoburgensia, vol. LXIX. Acta Universitatis Gothoburgensis, University of Gothenburg (2008)
20. Cherniss, H.: Aristotle's Criticism of Plato and the Academy, vol. I. Baltimore (1944)
21. Correia, M.: Es lo mismo ser no-justo que ser injusto? Aristóteles y sus comentaristas. Méthexis. Int. J. Anc. Philos. XIX, 41–56 (2006)
22. Correia, M.: The syllogistic theory of Boethius. Anc. Philos. 29, 391–405 (2009)
23. De Rijk, L.: On the chronology of Boethius' works on logic (I and II). Vivarium 2(1–2), 1–49 and 122–162 (1964)
24. Lee, T.-S.: Die griechische Tradition der aristotelischen Syllogistik in der Späntantike, eine Untersuchung über die Kommentare zu den Analytica Priora von Alexander Aphrodisiensis, Ammonius und Philoponus. Hypomnemata, vol. 79. Göttingen (1984)
25. Philoponus, J.: Ioannis Philoponi in Aristotelis Analytica Priora. Wallies, M. (ed.). Commentaria in Aristotelem Graeca, vol. 13.1–2. Berlin (1905)
26. Plato: Opera Omnia. Volume I: Euthyphro, Apologia, Crito, Phaedo, Cratylus, Theaetetus, Sophista, Politicus. Recognivit brevique adnotatione critica instruxerunt. Duke, E.A., Hicken, W.F., Nicoll, W.S.M., Robinson, D.B., Strachan, J.C.G. Clarendon, Oxford (1995)
27. Reale, G.: Aristotele. Metafisica, vols. i, ii, iii. Milan (1993)
28. Shiel, J.: Boethius' commentaries on Aristotle. In: [29], pp. 349–372. Text originally in: Medieval and Renaissance Studies, vol. 4, pp. 217–244 (1958)
29. Sorabji, R. (ed.): Aristotle Transformed. The Ancient Commentators and Their Influence. London (1990)
30. Sullivan, M.W.: Apuleian Logic. The Nature, Sources and Influence of Apuleius's Peri Hermeneias. Studies in Logic and the Foundations of Mathematics. North-Holland, Amsterdam (1967)
31. Thrax, D.: Scholia in Dionysii Thracis Artem Grammaticam. Hilgard, A. (ed.). Grammatici Graeci 1, 3. Teubner, Leipzig (1901)
32. Zimmermann, F.W.: Al-Farabi's Commentary and Short Treatise on Aristotle's de Interpretatione. Translation, introduction and notes. Oxford (1991)

M. Correia (✉)
Facultad de Filosofía, Pontificia Universidad Católica de Chile, Campus San Joaquín, Av. Vicuña Mackenna 4860, Macul, Santiago de Chile, Chile
e-mail: mcorreia@uc.cl
url: http://www.uc.cl/filosofia

Leibniz, Modal Logic and Possible World Semantics: The Apulean Square as a Procrustean Bed for His Modal Metaphysics

Jean-Pascal Alcantara

Abstract Even if Leibniz didn't have the opportunity to actually conceive an explicit modal logic system, remains the fact that he had worked out a modal metaphysics, of which the inaugural act, in his *Elementa juris naturalis* (c. 1671) was an obvious reference to the Apulean square of opposition. Later, scholars acknowledged in this passage probably one of the first sketch of deontic logic of norms. His modal metaphysics rather deals with the so-called alethic modalities, sometimes expounded through a language such as R.M. Adams wondered whether Leibniz could be "a sort of grandfather of possible worlds semantics for modal logic". In the following study, the Apulean square is used as a hermeneutic tool: however, is the square really well-fitted to express some basic implications of the modal metaphysics? The most remarkable point is that Leibniz was well aware of the **K**-distributive axiom $\Box(p \supset q) \supset (\Box p \supset \Box q)$ common to the main modal systems today. This awareness dissuaded him to trust the too easy solution of the necessitarianism issue grounded on the well-known distinction coming from Boethius between a "*necessitas consequentiae*" and a "*necessitas consequentis*".

Thus Leibniz must consider a proof-theory solution style (for a contingent proposition, the reducibility of the predicate to its subject cannot be achieved). A new square could be making up, however implying a difficulty: if the modal metaphysics cannot admit the logical convertibility of the contingent propositions of which the *dictum* of one is exactly the negation of the *dictum* of the other, known since Aristotle. This is the reason for what the Apulean square could represent a Procrustean bed for Leibniz' modal metaphysics. At the end, a new modality square may be drawn, according to which each modality in the corner is expressed by quantifications on possible worlds. Whether possible worlds semantics could supersedes Leibniz' own explanation of contingency thanks to this new square of modalities, i.e. a possible worlds square of modalities, will be the topic of a second part of this study.

Keywords Deontic logic · **K**-Distributive axiom of modal logic systems · Blanché's hexagon · Necessitarianism · Proof theory · Modal metaphysics · Algebra of concepts · Convertibility of the contingent propositions · Possible worlds semantics · Boethius · Calvin · Leibniz

Mathematics Subject Classification 03A10

As Leibniz had not himself developed a logical modal calculus, given that he did not even devoted time for constituting a propositional calculus (but after all, he was still a logician

J.-Y. Béziau, D. Jacquette (eds.), *Around and Beyond the Square of Opposition*, 53–71
Studies in Universal Logic, DOI 10.1007/978-3-0348-0379-3_4, © Springer Basel 2012

of terms, so inclined to an intensional treatment of concepts) however he obviously had done more than merely to sketch some logical calculus. Particularly Wolfgang Lenzen demonstrated that Leibniz's strict implication calculus, inferred from algebra of concepts isomorphic to the one then built by Boole, amounted to either of the weakest Lewis's systems, S_2 or S_3 (which cannot encapsulate the necessitation rule: $p \supset \Box p$) ([20], [22]: 38–39).

Most certainly Leibniz remains the master-builder of a modal metaphysics of which the inaugural act, since the *Elementa juris naturalis* (1671), was the position of some interdefinissability relations which are obtained inside the alethic modalities between the deontic categories, then recognized by G.H. Von Wright and G. Kalinowski at the foundation of deontic logic ([13], [14, VI, 1: 465–480], [29]). Here the explicit recourse to the Apulean square grounded the first elements of logic of norms.

Further, as it was often argued, Leibniz could gather all the necessary and sufficient tools which would make him the founder of a possible worlds semantics before her time and more accurately, the one fitted for S_5, which is the more powerful of the classic systems of modal logic. At the beginning of his synthetic great book *Leibniz. Determinist, Theist, Idealist*, Robert Merrihew Adams wondered if Leibniz would not be "a sort of grandfather of possible worlds semantics for modal logic" before he appreciably withdrawed his hypothesis at the end of an analysis of the diverse sides of Leibniz's modal metaphysics ([1]: 9).

Though as still the author of differential analysis, Leibniz preferred to explain contingency—that is, the main theological attribute of the created world—as the result of a divine calculus of a kind of utility function, as to say (representing the best series of the compatible *possibilia* together), inside the frame of a proof theory.

Although he appealed to the venerable theological as well as modal distinction, inherited from Boethius, between a necessity of the consequence and a necessity of the consequent, nowadays we may uphold the thesis according to which Leibniz was fully aware that this distinction runned up against an equivalence of the entailment:

$$\Box(p \supset q) \supset (\Box p \supset \Box q)$$

where we identify the basis axiom of **K** modal system.

The so-called analytical truth theory, determining that a predicate must be contained into the subject of any true proposition, supposes in the case of the contingent truths, that the algorithm of the resolution to an identity goes on indefinitely (process compared by Leibniz himself to the infinite development of an irrational number by the mean of continuous fractions), whereas the demonstration of necessary truths always may be achieved in a finite number of steps.

Both modalities of possibility and contingency depend on the excess of demonstrability with respect to the effectibility of a computation. Thus it will be the touchstone of the length of the conceptual analysis which completes the discernment of modalities. However isn't there something unusual, in spite of the attractive power of this theory on the modern logicians as Tarski, noticeably highlighted thanks to the notion of ω-consistence ([28]: 288)—we intend to deal with this issue in the future second part of this paper—?

On the other hand, some commentators, Benson Mates or Hidé Ishiguro, didn't believe that the criterion of the infinite complexity concerning the analysis of the contingent propositions was so trusty in order to switch a modality into another. The scope of the above criterion was even not appraised worthy to be considered by Bertrand Russell along his chapter dealing with the contingent propositions ([25]: 28–34).

Then the way of possible worlds semantics seemed to provide a more suitable expression of the modal notions in the still relatively recent English-speaking interpretations of Leibniz's metaphysics.

Before to draw the new issues arising from the recourse of those semantical approaches, I intend in this paper to consider some of the following points (however unfortunately keeping for another occasion the points 4 and 5):

1. Does the Apulean square, which explicitly appears in the first writings about law, constitute a suitable hermeneutical tool to track the elaboration of the Leibnizian modal metaphysics?
2. Was Leibniz aware that the compatibilist solution, provided by some religious thinkers as Thomas of Aquino, Calvin or Pascal, to the problem of the coherency of a contingent creation by a necessary being, could not itself agree with the distributive axiom of modal logic?
3. If the answer would affirmative, we could be recognize under this trouble a motivation to enquire for another solution, which would rebuild the modal though, thus the Apulean square, from a proof-theory.
4. According to the possible worlds semantics, we have to leave the syntactical method, where all true proposition is derivable inside the considered system, for a semantical conception, the truth of a proposition holding with respect to a model.
5. So a new generation of issues is brought to light, particularly this one: may the S_5 system, involving the axiom $\sim \Box p \supset \Box \sim \Box p$, agree with the Leibnizian thesis (following from the point 3) by virtue of it could be the case for a proposition to be undemonstrable without being demonstratibly undemonstrable?

1 Is the Apulean Square Compatible with Leibniz's Modal Metaphysics?

The founders of modern juridical logic, i.e. G.H. Von Wright, had never missed to recall that they had had a precursor. As a young lawyer, Leibniz indeed had conceived a logic system of norms, connecting the deontic categories of obligatory (*debitum*), allowed (*licitum*), prohibited (*illicitum*) and facultative (*indifferentum*) with the Aristotelian modalities marshalled by Apuleius in his square of oppositions.

Modalities: deontic	alethic	*potest* + negation	epistemic
Justum, Licitum (allowed)	possible	*potest fieri a quodam viro bono*	That which is understandable
Injustum, Illicitum (prohibited)	impossible	*non potest fieri a nullo* (*seu non quodam*) *viro bono*	That which is not understandable
Aequum, Debitum (obligatory)	necessary	*non potest non fieri a* (*omni, seu non quodam non*) *viro bono*	That which the opposite is not understandable
Indebitum, Omissibile (facultative)	contingent	*potest non fieri a* (*quodam non*) *viro bono*	That which the opposite is understandable

On this table which can be read in *Elementa juris naturalis* (1671), we see that Leibniz exhibited a semantic analysis of the deontic modalities from the notion of possibility diversely combined with the negation.

Upholding that one might in safety "transposes" (*transferre*) from the alethic modalities to the so-called "*juris modalia*" all the complications, transpositions and oppositions noticeable in Aristotle's syllogistic and his followers, Leibniz revealed an isomorphism on the ground which he might have drawn the deontic square:[1]

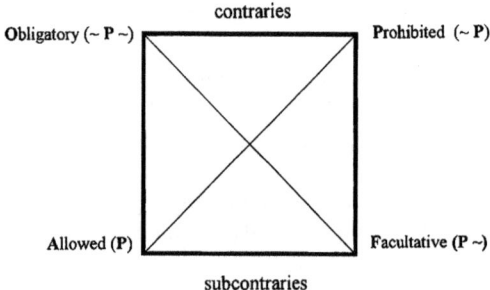

In a paper entitled "Leibniz on alethic and deontic modal logic", W. Lenzen asserted that this square becomes analogous to the one of alethic modalities interpreted in words of possible worlds, if we mean "world" by *casus*:

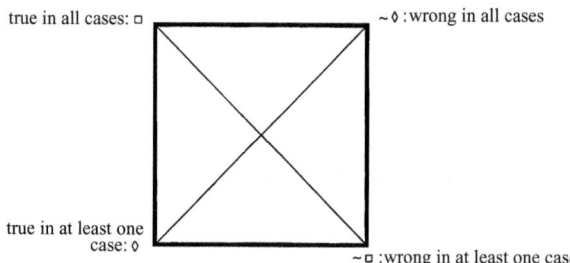

According to W. Lenzen, the possible worlds semantics which supports the interpretation of the square in *Elementa juris naturalis* lays over the following fundamental relations [23, §4].

NEC 1 $\Box p \leftrightarrow {\sim}\Box{\sim}p$
NEC 2 ${\sim}\Diamond p \leftrightarrow \Box{\sim}p$
NEC 3 $\Box p \rightarrow \Diamond p$
NEC 4 ${\sim}\Diamond p \rightarrow {\sim}\Box p$

The academic edition shows us that during the years 1687–1690, Leibniz resurrected his first juridical works, however passing from the mode of a square to the mode of a triangle,

---

[1]"Omnes ergo Modalium complicationes et transpositiones et oppositiones, ab Aristotle alisque in Logicis demonstratae ad haec nostra Juris Modalia non inutiliter transferri possint" [14, VI, 1: 465–485]. This text came to the fore thanks to Blanché, as it is remembered by Kalinowski, post-scriptum of ([11]: 207) (in the same book, see also chapters ii and iii, about the analogy between the alethic and deontic modalities). See further [10, 26].

which has as vertex what is obligatory (*officium*), allowed (*licitum*) and sin, that is, which is forbidden to commit (*peccatum*).[2]

So we obtain that what G. Kalinowski called the "triangle of contraries", with *licitum* in the position of bilateral allowed: both permission to do/not to do ([12]: 13).

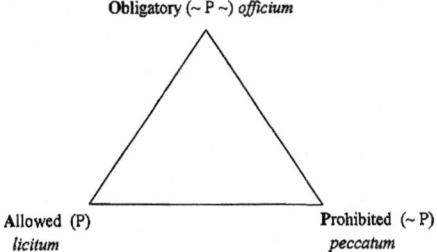

In this triangle, the lines express the contrariety relations, on the basis of which may obviously be written these interdefinibility formulas:

[1] $O = \sim(P \vee A)$
[2] $P \equiv \sim(O \vee A)$
[3] $I \equiv \sim(P \vee O)$

All we have to do is to split what is allowed with the result that the triangle returns to a square: **A** being that which is unilaterally allowed to do and **F** that which is unilaterally facultative, i.e. that which is allowed not to do.

Symmetrically, permission to do becomes a subaltern of the obligation, permission not to do representing a subaltern of prohibition.

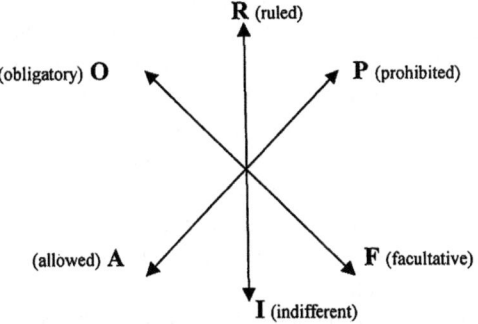

Adding **R**, the bilateral obligatory (ordered or prohibited) and renewing with the bilateral meaning of allowed under the mode of indifferent **I**, one constitutes a six rays star in which the arrows, in this case, express the *contradictory* relations, where we obviously recognize Blanché's hexagon if the vertices are linked.

A fourth interdefinissability formula completes the characterization of the deontic modalities:

[4] $F \equiv \sim(O \vee P)$

J.-L. Gardies, who intended to elaborate deontic algebra, however has advised to give up the term **R**, because of missing a "perceptive coherency" in respect of both the gathered

---

[2]*Aphorismi de felicitate, sapientia, caritate, justitia* [14, VI, 4-D: 2800].

contraries, further missing of usefulness other than marking the opposition with indifferent ([9]: 66).

Philosophers of law emphasize that it differently works at the other opposite of the hexagon, where it is, at the contrary, the unifying pole **I** which carries off considering the relevance with respect to the subcontraries [3, §12]. Indeed, it happens that a permission to do, as disconnected from a permission not to do, amounts to an obligation. And reciprocally, the permission not to do, being not associated to the permission to do, amounts to a prohibition, that is, in formulas:

$$\big((\mathbf{A}p \wedge \sim\!\mathbf{F}p) \to \mathbf{O}p\big) \to (\mathbf{A}p \leftrightarrow \mathbf{O}p) \quad \text{and} \quad \big((\mathbf{F}p \wedge \sim\!\mathbf{A}p) \to \mathbf{P}p\big) \to (\mathbf{F}p \leftrightarrow \mathbf{P}p)$$

Putting a bilateral allowed, **I** such as $\mathbf{I} \leftrightarrow (\mathbf{A}p \vee \mathbf{F}p)$ and on the bounce, the degeneration of the square in a deontic triangle, observable in the above Leibnizian extract, may be considered.

If we would translate the deontic triangle of contraries to the modal metaphysics, some confusion should follow between the modalities of possible and contingent.

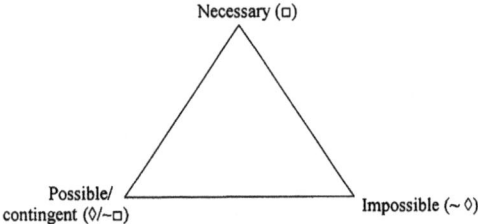

Does not yet it remain essential, under the scheme of modal metaphysics, to keep untouched the Apulean square? Yes in one sense, because the square displays the inferior vertex of the triangle; no in another sense, because the concept of existence introduces a dissymmetry between that which is possible and that which is contingent.

Whilst Leibniz called forth a combination of essences inside God's understanding during the best world calculus, he referred to the "country of possible".[3] Thus possible is characterized by a relation to existence without reciprocity, existence entailing possibility whereas possibility itself does not imply existence. Every possible strives for existence and this striving in any case amounts to an entailment, otherwise possible would not be in excess in respect with that which exists.

Such is the very definition of the necessitarianism attributed by Leibniz to Hobbes or Spinoza and even Descartes: if possible and existence coincide, all *entia* become necessary.[4] The existential import inherent to contingency cannot be represented onto the Apulean square. That is exactly that which is noticed by the equivalence under Leibniz's pen, between the existential truths and the contingent truths which bears on the order of realities coming to existence because of their consistency with the greater number of beings, i.e. their compossibility.[5]

---

[3]*De rerum originatione radicali* (1697) [16, 3: 302–318].

[4]*Origo veritatum contingentium* (1689) ([15]: 325–326).

[5]"Ajo igitur Existens esse quod cum plurimis compatibile est; seu Ens maxime possibile. Itaque omnia coexistentia aeque possibilia sunt" (*Generales inquisitiones de analysi notionum et veritate*, 1686, §73, [19]: 248).

According to the modal metaphysics, contingency inherits from possibility, given that each contingent being cannot lack of a previous possibility. This implication of contingency by possibility at its turn, cannot either be represented on the Apuleius' square, which at the contrary, discloses an implication of the subaltern:

Quia omne necessarium est possibile, omne impossibile est contingens seu potest non fieri.[6]

Theology finds among modalities the suitable operator likely to underline the Christian precariousness of the existence. So contingency and possibility tend to stretch one towards another: existing is said *"ens maxime possibile"*, namely possible being at the highest degree—because of the coherence with the greater number of beings.

That what is contingent shows beings deprived of all kind of necessity, and this feature may be well readable on the Apulean square with the contradictory relation. It is the modality of the precarious at the highest degree being (underlined with the theological doctrine of the continuous creation). The Apuleius' square cannot account for this underlying dynamism of a Leibnizian modality. Indeed Leibniz's possibility arises from this equality: possible = essence = quantity of perfection.

Thus we must admit all sorts of degrees of possible since there are degrees of perfection. Possibility means not only a modality; it must be understood as an intensive magnitude: it is a matter of "existentiability". Excluded from the Apulean square, existence nevertheless interjects on the axis of subcontraries as a limit in the sense of the infinite series theory: upper bound for the growth of degrees of possibilities, lower bound of the perfection of the created beings under contingency of the best world. But whatever their rank in the hierarchy of beings, whatever the degree of perfection, all the existents share the same precariousness in respect with the necessary being. Even natural laws are marked by the seal of contingency.

However, the preservation of contingency (similarly than permission of evil by a perfectly good being) deserves a particular explanation: how is it possible to save contingency of the world, given that contingent beings follow from a necessary being? How may we conceive an entailment where the Apulean square let us in front of a mere contradiction?

2 Is the K-Distributive Axiom Compatible with Leibniz's Modal Metaphysics?

To create the world consists in presenting freely into being that which existence is not necessary. How contingency of God's work (with the future contingents holding to human freedom) may be conceived as compatible with the absolute necessity of God's existence? How could the world not inherit of the modality of which it is the effect?

In spite of all the reservations that humanists and reformationists had kept against medieval logic, neither Calvin nor Luther could neglect, in order to save contingency, to appeal to the concept of conditional necessity devised by the theologians during the XIIIth century, from Boethius's reading of Aristotle's *Peri hermeneias*, Chap. 9, likely to receive three main interpretations ([30]: 161–162).

---

[6][14, VI, 4: 2759] (probably that is the rewriting *circa* 1679, of an extract soon given in [15], *Elementa* of 1671, quoted by Lenzen about the contraposition of the axiom NEC 3 to NEC 4).

1. In the first place, conditional necessity is merged with the necessity of the consequent: "if p is the case, thus it is necessary that q is the case" (or: $p \supset \Box q$).
2. A second interpretation refers to the necessity of the consequence: "it is necessary that, if p is the case, q is the case" (or: $\Box(p \supset q)$).

The first interpretation is tantamount to put a brute necessity, *de re*, even if this is formally hypothetical, since the second one, *de dicto*, offers an alternative to necessitarianism.

3. Further there is also a third interpretation, as necessity, *de re* again, here involves a temporal condition: "if p happens during the moment t, it is necessary that p happens during the moment t".

This reading was considered by Aristotle in his *De interpretatione* (19ª 23, [3]: 102) and then was reenacted by Boethius in his *Consolatio philosophiae* (V, vi, 27, [5]: 316–317) with the aim to preserve the contingency of human action in respect with the divine foreknowledge, in such a way that: "a conditioned necessity equals to the necessity of present".[7]

The Boethian version of the conditional necessity cannot be reduced to the position of a necessity of the consequence for the reason that, faithful to the Aristotelian logic of terms, Boethius could not conceive that the logical constants could bear on propositions themselves.

Dealing with the predetermination of contingent events, finally in a same sense, Calvin granted that one must not neglect, from an outlook similarly compatibilist:

> [...] the manners of speaking which would used in the schools, concerning the absolute necessity from the one which comes from another presupposed thing [*secundum quid*].

And he added in a same way:

> [...] similarly, about the consequent [*de consequentis*], and about the consequence [*de consequentiae*]" ([8]: 239).

Taking again advantage of the medieval logic of consequences, he went ahead, in his *Institution chrétienne*:

> [...] there is a mere or absolute necessity, and a necessity in some regard to. Similarly, there is a necessity of what follows from and of consequence" ([7]: I, XVI, 9).

Necessitarianism is avoided as long as modality bears on the *nexus*, the implicative link itself.

The entailment thus becomes comparable to C.I. Lewis' strict implication, given that one undertakes to express the relation of consequence in a way more acceptable for our intuition than it was allowed with the material implication of the Russellian propositional calculus. We know that Russell's material implication, noted "$p \supset q =_{\text{df.}} \sim(p \wedge \sim q)$", means that it is *actually* impossible to conjugate the truth of p with the falsity of q. Lewis' strict implication ("Diodorian" or "strict conditional") introduces a modal functor in order to signify for this time the *logical* impossibility that p may be true whilst a proposition as q is wrong. Thus, in accordance with the interdefinibility of the modal functors:

$$\rightarrow^3 pq =_{\text{df.}} \sim\Diamond(p \wedge \sim q) = \Box(p \wedge q)$$

---

[7] ([24]: 40) Boethius supported a similar version in his own commentary of Aristotle's *De interpretatione*.

The rebirth of modal logic in the XXth century thereby was associated to a new way of understanding the nature of implication.

Leibniz fighted himself too against the necessitarianist involvements of Christian theology, involvements which were soon contested by the reformationists, asserting the rebirth of the liberty of the Christians freed from the constraints of the Mosaic Law. He explained, at the occasion of one of his first essay of theodicy, *Confessio philosophi* (1673), the necessitarianism reasoning in this manner:

> God's existence is necessary; from this existence follow the sins nested in the [actual] series of things; that which follows from necessity is necessary; thus sins are necessary [14, VI, 3: 127].

Otherwise said, sins become necessary exactly in the same meaning according to which God is asserted to be necessary. Leibniz will improve towards a solution only but calling into question the univocity of modalities: there will be in one side a *per se* modality and in the other, a modality in respect with another thing.

Then in a first draft dated 1673, Leibniz appraised the modal situation by virtue of an analogy with propositional logic and syllogistic reasoning.

1. The first analogy disproves the solution of the compatibilist issue by referring to an hypothetical necessity: if it is true that from truth anything follows but that which is true, Leibniz explained, from *per se* necessary should thus follows only that wich is is *per se* necessary.

2. A second analogy pleads for this time in support of conditional necessity: just as in the syllogistic modes of Darapti (universal affirmative premises) and of Felapton (one universal negative premise and the other universal positive premise), a particular conclusion follows from merely universal premises. So why some contingent thing could not follow from some necessary thing?

Such analogies might bring to light a few uncertainties about the handling of logical modalities. However a more flattering hypothesis can be contrived: Leibniz could be aware that the distinction between necessity of the consequent and necessity of the consequence, that is, of $\Box(p \to q)$ and $(p \to \Box q)$, was not susceptible to overcome the necessitarianist issue.

Actually, we cannot deny that the combination of hypothetical necessity with necessity of the antecedent implies a necessity of the consequent:

[1] $\Box p$
[2] $\Box(p \supset q)$
[3] $\Box p \wedge \Box(p \supset q) \supset \Box q$

Exactly that is what we can find in propositional modal logic under the distributive axiom **K**, at the basis of the ordinary modal systems:

K $\Box(p \supset q) \supset (\Box p \supset \Box q)$

As yet are enough textual clues for admitting that Leibniz was actually aware of the modal axiom?

Some commentators as Adams, Sleigh and Griffin have agreed to identify an equivalent of the **K**-axiom in an extract dated of December 1675, *De mente, de universo, de Deo*, written between *Confessio philosophi* dated 1673 and some noticeable rectifications brought in 1677, following a series of discussions that Leibniz had in Hanover with the Danish geologist Stenon (who was meantime become theologian), rectifications whose result was, in the margins of *Confessio*, the explicitation "*per se*" linked to "necessary", read in the final writing, so that the bearing of a modality is divided in two.

The equivalent of **K**-axiom is enunciated thereby in Latin: "*Quicquid necessario incompatibile est, impossibile est*".[8] All which is uncompatible with that which is necessary is in a same way reckoned as impossible:

[1] $\Box p$
[2] $\sim(p \wedge q)$
[3] $\Box p \wedge \sim(p \wedge q) \supset \sim \Diamond q$

Concerning that which is given as equivalent in Leibniz's extract in [14, VI, 3: 464] to the **K**-axiom, its contraposition must be: all q compatible with p which is necessary, is necessary, and not "is possible", for the reason that otherwise, the **K**-axiom would no more represent any threat against contingency of the world.

[1] $\Box p$
[2] $(p \wedge q)$
[3] $\Box p \wedge (p \wedge q) \supset \Box q$

Indeed a mention of a quasi **K**-axiom concludes a whole paragraph in *De mente, de universo, de Deo* where Leibniz threw into relief, about impossible, than one had to distinguish two meanings:

> *Impossible* is a two-fold concept: that which does not have essence, and that which does not have existence, i.e., that which neither was, is, nor will be because it is incompatible with God, or, with the existence or reason which brings it about that things exist rather than do not exist. One must see if it can be proved that there are essences which lake existence, so that it cannot be said that nothing can be conceived which will not exist at some time in the whole eternity. All things which are, will be, and have been, constitute a whole. Whatever is incompatible with what is necessary is impossible.[9]

Thus impossibility has two roots:

1. the one which holds to the essence, in the case of something in itself contradictory;
2. the other which holds to existence, without however its essence is not in itself contradictory, but lacking of existence because of its incompatibleness with the contingent existence which flows from the necessary existence.

The consequence would be that this dividing in two of the meaning of "impossible" should bring to detail the principle put some lines lower, according to which all which is incompatible with what is necessary, is impossible!

Indeed, the impossibility in the sense (2) certainly does not imply the impossibility in the sense (1): there is possible independently of any perspective of actualization, non-Diodorian possibles of course compatible with the necessary existence, as Leibniz upheld in *Confessio philosophi* two years ago:

---

[8]*De mente, de universo, de Deo*, December 1675 [14, VI, 3: 464].

[9]"Impossibilis duplex notio, id quod essentiam non habet, et id quod Existentiam non habet seu quod nec fuit nec erit, quod incompatibile est deo, sive existentiae sive rationi quae facit ut res sint potius quam non sint. Videndum an demonstrari possit esse Essentias quae Existentia careant. Ne quis dicat nihil concipi posse quod non aliquando futurum sit in tota aeternitate. Omnia quae sunt erunt et fuerunt totum constituunt. Quicquid necessario incompatibile est, impossibile est" [14, VI, 3: 463–464].

So they are wrong, all the ones who declare absolutely impossible, that is to say by itself, all that which was not, is not, or will be.[10]

A thing remains possible functions of its own nature, because even if it would be not compatible with God's will, it is at least with her understanding.

But if possible is defined as that which does imply in itself any contradiction, it follows a reconsideration of the formula (2) and through it, of the **K**-axiom, as we can check reading a paramount passage, first partially disclosed by Grua, entitled *De libertate et necessitate*, composed, according the academic publishing, *circa* 1680–1684:

> I concede this, yet, such a thing remains possible in its nature, even if it is not possible with respect to the divine will, since we have defined as in its nature possible anything that, in itself, implies no contradiction, *even though its coexistence with God can in some way be said to imply a contradiction* [I underline]. But it will be necessary to use unequivocal meanings for words in order to avoid every kind of absurd locution.[11]

Thus Leibniz was led to reform the quasi **K**-axiom according to which all that which is incompatible with that which is necessary is impossible, since it is written in the quoted lines that it may be the case of a nature, possible in itself but contradictory in some way (*aliquo modo*) with God's existence as the root of the existence of the world.

Necessity must be understood morally or metaphysically, and this nuance cannot be expressed by the **K**-axiom. Hence, for something impossible, to be incompatible with that which is necessary may mean:

1. to be not compatible in respect with the understanding of the being by itself necessary (impossibility of essence);
2. to be not compatible in respect with the will of the being by itself necessary (impossibility of existence).

The consequence on Leibniz's modal thought was of an outstanding importance, since in a new version of the writing dated 1673, the philosopher could advance a dividing in two of the modalities, modalities *per se* in one hand, "unqualified modalities" (as named by Sleigh) on the other hand.

The threat represented by the distributive axiom now depends on the appreciation of the necessity *per se* (or *in se*, again *in sua natura*) which is not conveyed by the antecedent thanks to the implication. That which flows from God's necessary existence thus will be "*by itself contingent* and only hypothetically necessary".[12]

Nevertheless, the validity of the axiom $(\Box p \wedge \Box(p \supset q)) \supset \Box q$ would remain if each operator "$\Box$" means "metaphysically necessary": q is metaphysically necessary if and only if p is metaphysically necessary and if it is metaphysically necessary than p entails q.

---

[10]"Quare errant quicunque <absolutè id est per se> impossibile pronuntiant, quicquid nec fuit, nec est, nec erit" ([17]: 56).

[11]"Quia sua natura possibile definivimus, quod in se non implicat contradictionem, *etsi ejus coexistentia cum Deo aliquo modo dici possit implicare contradictionem*. Sed opus erit constantes adhibere vocabulorum significationes ut species omnis absurdae locutionis evitetur" [15, VI, 4: 1447]. This text appeared in ([14]: 289), but cut of the part of the sentence here underlined, which is of the highest importance for our purpose, happily restored in the academic edition (equally in its English translation in ([18]: 21)).

[12]"[…] indeed, here, we call *necessary* only but necessary *by itself*, namely that which possesses in itself the reason of its existence and of its truth, such are truths of geometry" ([17]: 57).

3 The Square of Modalities from the Outlook of a Proof Theory

If our analysis is right, the knowledge of the **K**-axiom does not more obstruct the solution of the compatibilist issue. Then the conceptual analysis at the ground of the *Characteristica universalis* demands an enlightenment concerning the nature of modalities: how can be the square of modalities expressed from a theory of demonstration?

Using the Apulean square as a metaphysical panoramic map, we see that Leibnizian modal metaphysics is built from the right low vertex, instead of the left upper vertex as Spinoza. This supremacy of possible even takes in Leibniz the shape of a sort of a modal optimism: everything must be held for possible until its impossibility is proved [16, 3: 444]. Leibniz's possibilism is understood from a redundancy of possible in respect with being. "Why is there something rather than nothing?" amounts to ask "Why is there possibility rather than nothing?" To be possible means to be, that is, postulates there are essences existing in some way into the divine understanding.[13]

The presumption of possibility corresponds to the mutual implication of possibility and the notion of to be not contradictory. In *Generales inquisitiones de analysi notionum et veritatum* (1686), this is the very criterion of non contradiction which provides a syntactical definition of the modalities, without the need of appealing to a theory of possible worlds (§§61, 67, 74, 130, 133). This is the criterion of non contradiction which implies the synonymy of to be true, to be *res* or *ens*. So Leibniz would perhaps be in a position to advance a new modal square.

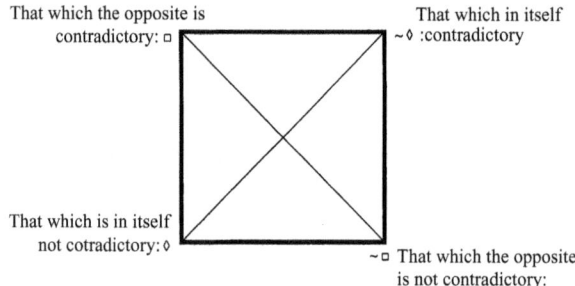

That which the opposite is contradictory: □

That which in itself ~◊ :contradictory

That which is in itself not cotradictory: ◊

~□ That which the opposite is not contradictory:

Leibniz intended to build a logical calculus under the form of an algebra of concepts where a term expresses either a concept or a proposition ([21]: 28–83). Then the square must be considered on two levels: level of the complex terms and level of the uncomplex terms, namely propositions.

Thereby, a term as A is possible if it is not contradictory, that is to say if, at the end of an analytic process, it does not contain <B non-B>. Therefore it results that an uncomplex term, absolutely simple, would not be otherwise than in itself possible. "To be impossible" means, for an uncomplex term, merely not to be (Gen. Inq. §32). In the order of complex terms, "impossible" signifies "to be wrong", Leibniz warned (Gen. Inq. §32[bis]).

The consequence is that a possible proposition, which implies any contradiction, must be true. So every true proposition must be reducible to the form of a universal affirmative proposition written $A = AB$. The sign of equality here refers to a relation of coincidence

---

[13]" […] pays des réalités possibles": to Arnauld, 1st July 1686 [16, 2: 55].

which covers, from an intensional point of view in logic, a relation of containing and contained (Gen. Inq. §83). Thus this logical equality, in order to be true, implies that the subject as A contains a predicate as B.

The universal negative proposition will be expressed by the inequality: $A \neq AB$. Quantification of subalterns further demands the symbol of a notion supposed well-known but indeterminate: $AY = B$ noting a particular affirmative proposition, and $AY \neq B$ noting a particular negative proposition (Gen. Inq. §§47–50).[14]

Everything being possible in virtue of a presumption, until its impossibility is not demonstrated, the ascription of modalities lays on a very demonstration. Thus, to demonstrate a proposition amounts to reduce some disparate terms, as well complex as uncomplex, until an identity is found.

Here takes place the famous Leibniz's law: a proposition keeps the same "*valor*" (his meaning) as we substitute one of its terms by a definition which analyzes this term, except in the case of contexts which we call today intensional, named "reflexive" by Leibniz.[15]

So that happens when occurs a reduplicative proposition (with expressions as: insofar inasmuch as, hold as...), which specifies the particular aspect under which the value of a subject must be taken.

> To demonstrate, indeed, is nothing else than to show a certain equation or coincidence of a predicate with a subject in a reciprocal proposition, resolving the terms of the proposition, and substituting, to the definite, the definition or a part of this-one.[16]

One succeeds in demonstrating the truth of the universal affirmative $A = AB$ if the analysis of A and of B discloses the same components. In this situation, $A = CDFG$ and $B = CDFG$. The substitution makes explicit a perfect coincidence, given the idempotent law $A = AA$:

$$AB = ACGDF = AA = A$$

However, the coincidence can also be only partial, i.e. if $B = CDF$. Nevertheless, the inclusion of the predicate into the subject will be demonstrated. In order to achieve the demonstration, it is sufficient to exhibit a whole or partial coincidence between a subject and its predicate, every proposition, even hypothetical, being reducible to a categorical form intensionally understood.

Likewise, a conditional proposition such as "if A is B entails C is D", can be rewritten under the form: "D-eity of C is enclosed into B-eity of A", to save the universality of the predicative schema (Gen. Inq. §§197–198, [19]: 301). In the depths, any contingent proposition may provide the matter of such a rewriting, where the antecedent of a condition would systematically express the creation of the best of all possible world by God.

Hence a big mistake would consist in holding the contingent propositions as undecidable statement, and that is exactly what it is categorically excluded by the principle de

---

[14]If A is "man" and B is "mortal", $A = AB$ expresses well the equality of the extensions "man" and "mortal man", hence the inclusion of "mortal man" in "man", thus, of "mortal" in "man" in virtue of the idempotent law. If then A is "man" and B is "wise", $AY = B$: every man associated to an undetermined quality is wise. If $Y = B$, this quality may be but wisdom, so all men are not wise.

[15]*General Notations* [14, VI 4-A: 752].

[16]*Note on freedom* (1689?) [14, VI 4-B: 1655–1656].

sufficient reason. Thus one must put the excluded middle in respect with demonstrability and refutability, without that this potentiality, for a human knowledge, is actualized in the case of contingent propositions.

It ensues that the modalities may systematically vary according to demonstrability, which one may show, with uncomplex terms (i.e. propositions), thanks to the following distribution.

1. A proposition is said *necessary* when the resolution to identity of the concepts of subject and of predicate happens in a *finite* number of steps in order to conclude with the coincidence of the values *salva veritate*.

Leibniz often wrote that are necessary the propositions of which the opposites may be reducible to some contradictory terms, or in other words, what it is necessary is that of which its opposite is impossible (Gen. Inq. §61, [19]: 237; see also §§67, 130, 133).

Thus we find again, following the way of a proof theory, the interdefinibility of the modal operators, which allows apagogical demonstrations of necessary propositions.

2. If one demonstrates that a proposition implies a contradiction (that is, if it is showed that A contains <B non-B>), this proposition is qualified as *impossible*.
3. A proposition is *possible* when it can be demonstrated that it implies in itself a contradiction.

"Possible" in its logical sense names as well necessary as contingent, since it partakes of non contradictory, even if possible must be distinguished from contingent as providing the matter of a demonstration.

As for sheer *possibilia* which are essences, or the first elements of what is thinkable, thus their name conceals mere necessity. They are just possibilities to be, not possibilities to be something else. So nothing seems amazing about the fact that statements concerning *possibilia* constitute only necessary propositions.

4. A proposition is *contingent* if one cannot demonstrate that neither it nor its negation implies a contradiction; it is never conceivable, indeed, to reduce it to some identical expressions (Gen. Inq. §56), no more than its opposite, given the infinite length of its analysis.

As for uncomplex terms, nothing seems amazing that they never are qualified of contingent. They can be only possible: contingency bears only on complex terms.

Thus it is impossible to take into account the negation of a contingent proposition with the aim to demonstrate the later in virtue of the apagogical way: obviously, the fact that Spinoza is not dead in Paris cannot prove that he actually died in The Hague! Both propositions remain of course contingent, one about the true is granted by an empirical knowledge (trusty because of agreeing testimonies), the other contingent and wrong.

Notwithstanding, there is now an issue to deal with: does modal metaphysics admit the usual logical convertibility of the contingent propositions from which it results that the *dictum* of one is exactly the negation of the *dictum* of the other[17]?

---

[17]By opposition to the *modum*, of which negation refers to what is necessary (Aristotle, *First Analytics*, I, chapter 13, 32ª-30, [2]: 61).

The possibility of such a logical operation is metaphysically supported by the very Aristotelian notion of accident, bounded by an only contingent link to its substance. Thus it is, in any way, an issue that a proposition as "it is contingent that Socrates is sitting down" is converted into "it is contingent that Socrates is not sitting down".

That occurring, how would it be possible to distinguish, as Leibniz presumably wished to do, the true contingent propositions, which are explicitly existential, from the false contingent propositions[18]?

A demonstration by resolution to identity of contingent propositions could take an infinite time, but the mere testimony of existence provides a valuable shortcut to establish their truthfulness—so the criterion of truth no more lays on the simple non contradiction.

Divorcing from modal logic, modal metaphysics binds us to reason in this manner: if it is contingent that p, thus p exists; thus it is not contingent than not-p, however it is possible that not-p—entailment which is not readable on the Apulean square.

The truth of a contingent proposition does not suspend the possibility of the truth of its negation, in the exact sense where its opposite is not in itself contradictory.[19] Otherwise if we keep convertibility, a proposition would be contingent whilst it would not be the case that the correspondent state of affairs would occur.

Leibniz summed up the whole thing in a very important passage at the §61 in *Generales inquisitiones*:

> Possible is that which one can show that the resolution will never meet a contradiction. Contingent and true are that which the resolution demands to be continued at the infinite. On the contrary contingent and false is that which one cannot demonstrate the falsity except by the fact that one cannot demonstrate its true. It seems doubtful that it suffices, to demonstrate a truth, to be sure that one never will meet any contradiction during its resolution. Because it results that all which is possible is true.[20]

Inside the frame of the algebra of concepts, modalities are inclined to be confused with the truth values ([13]: 259). However, in an anti-Spinozist modal metaphysics where God does not carry to existence all possibles, a proposition must be possible without to become likewise true ([17]: 57; Gen. Inq., note of §66, [19]: 243–245).

Possible is not reducible to the true as well as the true is not reducible to necessary: of course, if coincidence of the true with necessary does not make any trouble in mathematics and logic, that goes differently in the topic of natural sciences, history and politics.[21] In the order of contingent, a true proposition is so at the condition that it could be false too; otherwise this one would become necessary.

"Is contingent and true that which the resolution *demands* to be continued at the infinite" (Gen. Inq. §61): however some times, Leibniz upholded that a contingent true

---

[18]Gen. Inq. §§74, 134, 143, 144. "An existential proposition *tertii adjecti* is: *every man is* (namely *exists*) *exposed to the sin*; and this proposition certainly is existential namely contingent" (§144, [19]: 285).

[19]"However, in this case, this is a contingent proposition, namely it is possible that it would be true or that it would be false" (note of §66, [19]: 245).

[20][19]: 237, 239. The French translation is: "Est possible ce dont on peut montrer..." [ibid.]. However the corresponding Latin verb exactly is: "*demonstrare*", that appears to be more demanding (see also §§66, 134).

[21]This point is widely commentated by J. Bouveresse ([6]: 91).

proposition may be *proved*, whereas it appears that its resolution approaches indefinitely close to an identity without reaching it (Gen. Inq. §134, [19]: 277). But as Sleigh opportunely remembers, it is here convenient to enlarge the meaning of "proof" beyond the reducibility to an identity.

The theory of contingency lays on the idea that to have an *a priori* proof, for a proposition, is a wider property than to be likely of a demonstration. Sleigh sets forth an unpublished draft (L Br. 68-68v) extracted from a letter to Arnauld dated 14 July 1686, very explicit about this point:

> Contingent truths do not have demonstrations, properly speaking, however they must have *a priori* proofs ([21]: 88).

When Leibniz used referring to the notion of proof, he vacillated between three possible meanings:

1. the reducibility to identity, that is, a formal demonstration;
2. sequences of infinite identities where a coincidence is approached as a limit;
3. meta-proofs ([27]: 87).

According to the meaning (1), only necessary truths possesses a demonstration. Let a proposition p be an ordered pair with a first term which is the subject and a second term, predicate of p.

An identical proposition has for predicate a subset of the subject. Nowadays we would understand by a "meta-proof" of a proposition p:

(a) either the proof that exists a sequence S of ordered pairs with as first term p and as second term, a term which would be all over identical with p excepting a concept replaced by a set which constitutes its analysis step by step;

(b) or the proof that exists an infinite sequence S'. The perspective for obtaining a proof does not pertain to the proposition itself, but the process of demonstration itself becomes the topic of a proof.

However if a meta-proof remains of a finite length, to admit a meta-proof of contingent truths would amount to assert that there would exist a demonstration of the contingent truths, thanks to the middle excluded between (a) and (b).

In mathematics, since the posing of Cauchy's criterion in the XIXth century, the theory of series shows that there are indeed meta-proofs, as it can be demonstrated that some sequences of infinite length are convergent.

We know that it was not the current custom before: series were handled without the rule of checking their convergence. For instance, Leibniz succeeded in determining the area of the quarter of a circle with a radius equal to the unity by the mean of the following well-known series:

$$\frac{\pi}{4} = \sum_{k=1}^{\infty} \frac{(-1)^{k-1}}{2k-1} = 1 - \frac{1}{3} + \frac{1}{5} - \frac{1}{7} + \frac{1}{9} \cdots$$

In the absence of a general proof of convergence, nevertheless it was possible to determine step by step with a suited calculation that the error could be reduced to a ratio less than all given error.

This calculation step by step of the values of the series by the way here disclosed the slowness of the convergence. Leibniz was used to call upon the theory of irrational proportions and the geometrical example of asymptotes, in the aim of bringing an illustration to the process of an infinite proof in the demonstration of contingent propositions.

Thereof wouldn't Leibniz be a victim of the devil of analogy criticized by J. Bouveresse, in a different context [4]? Wouldn't Leibniz take in metaphysics much as liberty with the theory of series than the contemporary philosophers mocked by Bouveresse about Gödel's theorems?

So, there is all the more so as interesting to pay close attention to the §136 of *Generales inquisitiones* where Leibniz appraised what are the limits of his own analogy. The outcome indicated concerning the labyrinth of contingency meets the following objection formulated by Leibniz himself: is it not true that we have in mathematics some demonstrations about the behavior of an asymptote? However Leibniz was stuck by it: contingency holds as long as one cannot protract the asymptote or complete the calculation of a series, even if a certainty bearing on the mathematical knowledge that one cannot miss to have about this kinds of process, remains within human's reach.

Now the problem consists in knowing wether we have only but a certainty, insofar as an epistemic modality such as certainty does not in any way imply demonstrability (in the sense 1) of what it is certain. Indeed Leibniz assiduously explained that if it is certain that our world is the best, however it is in any way necessary that it is so, for the very reason that this proposition, "This actual world is the best", cannot be said demonstrable (*De contingentia*, [15]: 336).

We may think that Leibniz, having not at his disposal any conception of a meta-proof for the mathematical theory of series, could good-faith argue, by analogy, that there is any meta-proof for the contingency of a truth, and that would be rather some good news about the coherency of his doctrine of contingency.[22] However, if some meta-proofs of contingent truths could be given, a finite mind could come into possession of an *a priori* knowledge of the contingent truths.

Admittedly the Apulean square is not deducible from the square of epistemic modalities, but if there are demonstrations in mathematics dealing with the infinite (and Leibniz actually was aware of some of them, asserting against Descartes that the infinite is comprehensible) with all the features of the required necessity, Leibniz played with fire, as to say, eliciting as a model of contingency some mathematical processes, paradigms of necessary truths.

4 Conclusion: Towards a Possible Worlds Semantics

At the end of the first part of this inquity, we hope to better understand why possible worlds semantics then seemed to provide a more suitable expression of the modal notions in Leibniz's recent English-speaking exegesis, in order to ground more firmly the conception of contingency, not without generating a new metamorphosis of the Apulean square.

---

[22] Sleigh ([27]: 88) believes that Leibniz did not confess to Arnauld his mathematical analogy on the topic of the contingency considering the strong reluctances of his interlocutor for agreeing the containment theory; however nothing, in the *Discours de métaphysique* as in the correspondence with Arnauld, would be incoherent with this analogy.

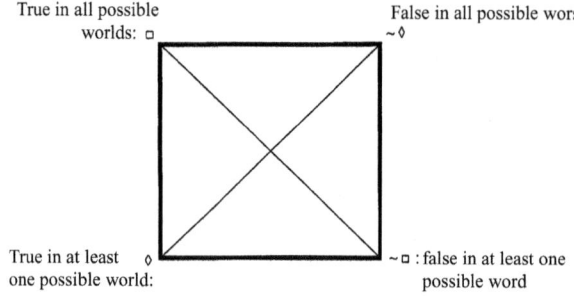

True in all possible worlds: □
False in all possible wors: ~◊
True in at least one possible world: ◊
~□ : false in at least one possible word

Instead of the modalities tending, as it was the case in the concept algebra, to express the truth values, quantification over possible worlds makes a restitution of the modal operators from the truth values.

Modal metaphysics had favored since Aristotle the axis necessary/contingency on the Apulean square, because this axis allowed reflecting the strength or at the contrary the weakness of the predication. Possible, *prima facie* without any ontological counterpart generated by the predicative link was doubly reappraised in Leibniz, by the theory of *prima possibilia* as well as by the theory of possible worlds.

References

1. Adams, R.M.: Leibniz. Determinist, Theist, Idealist. Oxford University Press, Oxford (1994)
2. Aristotle: First Analytics. Tricot, J. (French trans.). Organon III. Librairie philosophique Jules Vrin, Paris (1971)
3. Aristotle: De Interpretatione. Tricot, J. (French trans.). Organon II. Librairie philosophique Jules Vrin, Paris (1977)
4. Blanché, R.: Structures intellectuelles. Essai sur l'organisation systématique des concepts. Librairie philosophique Jules Vrin-reprises, Paris (1986)
5. Boethius: La consolation philosophique. Moreschini, C. (French edition), Vanpeteghem, É. (translated and annotated), Tilliette, J.-Y. (introduction). Le livre de Poche, Lettres gothiques, Paris (2005)
6. Bouveresse, J., Essais, V.: Descartes, Leibniz, Kant. Agone, Marseille (2006)
7. Calvin, J.: L'institution de la religion chrétienne. Kerygma/Farel, Marne-la-Vallée/Aix-en-Provence (1978)
8. Calvin, J.: De aeterna Dei praedestinatione. Neuser, W.H. (ed.), Fatio O. (French traduction). Scripta Ecclesiastica, vol. 1. Droz, Geneva (1998)
9. Gardies, J.-L.: Essai sur les fondements a priori de la rationalité morale et juridique. Bibliothèque de Philosophie du Droit, vol. 14. Librairie Générale de Droit et de Jurisprudence R. Pichon and R. Durand-Auzias, Paris (1972)
10. Gardies, J.-L., Kalinowski, G.: Un logicien déontique avant la lettre: Gottfried Wilhelm Leibniz. Arch. Rechts- Soz.philos. **60**, 79–112 (1974)
11. Kalinowski, G.: La logique des normes. Presses Universitaires de France, Paris (1972)
12. Kalinowski, G.: Un aperçu élémentaire des modalités déontiques. Langages **10**(43), 10–18 (1976)
13. Kalinowski, G.: La logique juridique de Leibniz. Stud. Leibnit. **9**, 168–189 (1977)
14. Leibniz, G.-W.: Sämtliche Schriften und Briefe, Hrsg. von der Berlin-Brandenburgischen und der Göttinger Akademie der Wissenschaften, vormals Preussische Akademie, danach Akademie der Wiss. der DDR, 8 Serien. Darmstadt, spätter Leipzig, Berlin (1923). Quoted: [14] followed by the number a series, volume and pagination
15. Leibniz, G.-W.: Textes inédits d'après les manuscrits de la bibliothèque provinciale de Hanovre. Grua, G. (published and annotated), 2 vols. Presses Universitaires de France, Paris (1948)
16. Leibniz, G.-W.: Philosophischen Schriften, 7 vol. G. Olms, Hildesheim (1968). Gerhardt, v.C.I. (herausg.). Quoted: [16] followed by the number of the book and pagination

17. Leibniz, G.-W.: La profession de foi du philosophe-Confessio philosophi. Belaval, Y. (established, presented, translated and annotated). Librairie philosophique Jules Vrin, Paris (1970)
18. Leibniz, G.-W.: Philosophical Essays. Ariew, R., Garber, D. (trans.). Hackett, Indianapolis (1989)
19. Leibniz, G.-W.: Recherches générales sur l'analyse des notions et des vérités. 24 thèses méta-physiques et autres textes logiques et métaphysiques. Rauzy, J.-B. (introductions and notes). Presses Universitaires de France, Paris (1998)
20. Lenzen, W.: Leibniz's calculus of strict implication. In: Srzednicki, J. (ed.) Initiatives in Logic (Reason and Argument I), pp. 1–35. Nijhoff, Dordrecht (1987)
21. Lenzen, W.: Das System der Leibniz'schen Logik. de Gruyter, Berlin (1990)
22. Lenzen, W.: Leibniz's logic. In: Gabbay, D.M., Woods, J. (eds.) Handbook of the History of Logic. The Rise of Modern Logic: From Leibniz to Frege, pp. 1–83 (2004)
23. Lenzen, W.: Leibniz on alethic and deontic modal logic. In: Berlioz, D., Nef, F. (eds.) Leibniz et les puissances du langage, pp. 341–362. Librairie philosophique Jules Vrin, Paris (2005)
24. Marenbon, J.: Le temps, l'éternité et la prescience de Boèce à Thomas d'Aquin. Librairie philosophique Jules Vrin, Paris (2005)
25. Russell, B.: A Critical Exposition of the Philosophy of Leibniz. Routledge, London (1996)
26. Schepers, H.: Zum Problem der Kontingenz bei Leibniz: Die beste der möglichen Welten. In: Collegium Philosophicum. Studien, Joachim Ritter zum 60, pp. 326–350. Geburstag, Basel (1965)
27. Sleigh, R.: Leibniz & Arnauld: A Commentary on Their Correspondence. Yale University Press, New Haven (1990)
28. Tarski, A.: Some observations on the concepts of ω-consistency and ω-completeness. In: Logic, Semantics, Metamathematics: Papers from 1923–1938, pp. 279–295 Hackett, Indianapolis (1983)
29. Von Wright, G.H.: On the logic of norms and action. In: Hilpinen, R. (ed.) New Studies in Deontic Logic, pp. 3–35. Reidel, Dordrecht (1981)
30. Vuillemin, J.: Nécessité ou contingence. L'aporie de Diodore et les systèmes philosophiques. Les éditions de Minuit, Paris (1984)

J.-P. Alcantara (✉)
Centre Georges Chevrier-UMR 5605, Université de Bourgogne, 4 boulevard Gabriel, 21 000 Dijon, France
e-mail: jeanpascal.alcantara@wanadoo.fr

Thinking Outside the Square of Opposition Box

Dale Jacquette

Abstract   The graphic meaning and formal implications of the canonical Aristotelian square of opposition are favorably compared with alternative ways of representing sub-syllogistic logical relations among the four categorical AEIO propositions of syllogistic logic. The canonical arrangement of AEIO propositions in the square is justified by an exhaustive survey of graphically nonequivalent alternatives. The conspicuous omission in a modified expanded Aristotelian square of inversions involving complements of subject terms is critically considered. Two of the four inversions are shown to be logically equivalent to two of the AEIO propositions, and are consequently discounted as offering no potential significant revision of the canonical square. The remaining two inversions are logically distinct from any of the AEIO propositions, and the advantages of adding them to an expanded square are explored. It is shown that the two AEIO-distinct inversions are inferentially, and as contraries, subcontraries, subalterns or contradictories, logically isolated from the AEIO propositions. From this it seems reasonable to conclude that no inversion makes a contribution to the subsyllogistic logical relations represented in an expanded version of the square. The investigation leaves the traditional Aristotelian square unchanged, but affords a deeper appreciation of the canonical square's optimal two-dimensional diagramming of subsyllogistic logical relations among the four fixed AEIO propositions, and of why inversions of the AEIO propositions are appropriately excluded from the square. The limits and disadvantages of subsyllogistic logic are made conspicuous in the use of contemporary classical logic to translate and verify in algebraic terms all potential square of opposition relations, as a formal basis for excluding inversions, where they are otherwise manifestly excluded without explanation from the canonical square.

Keywords   AEIO propositions · AEIO inversions · Canonical Aristotelian square of opposition · Graphic equivalences · Subsyllogistic logic · Surveyability

Mathematics Subject Classification   03A05 · 03B05 · 03B10 · 03B20 · 03B65

1 Canonical Square

The beauty and utility of the canonical square of opposition is largely attributable to its easy surveyability. In a glance we can take it all in and grasp the logical relations it represents among four types of quantified predications and their predicate term complements. The pragmatics of applied term logic in turn prepares propositions by paraphrase and

J.-Y. Béziau, D. Jacquette (eds.), *Around and Beyond the Square of Opposition,* 73–92
Studies in Universal Logic, DOI 10.1007/978-3-0348-0379-3_5, © Springer Basel 2012

syntactical manipulation into specific categories of quantified subject-predicate forms, so that all logical reasoning is theoretically representable in terms of logical relations among just those four types of propositions. We can see combinatorially how many possibilities there are, if all logical expressions and inferences can be formulated as truth-preserving relations between exactly these types of propositions, made up of a quantifier (all, some), a subject term, S, (optional) copula (are), and predicate term, P, or complement (internal negation, non-P) of a predicate term P. Constructing all four basic proposition types out of the different possible arrangements of four terms and predicate complementation begins to challenge surveyability, but remains systematically manageable with simple memory devices like a checklist of the combinatorial possibilities considered for their properties.[1]

The canonical square has this form, with AEIO propositions also expressed in contemporary first-order predicate-quantificational logic, respectively, for the two subject and predicate terms S and P: (A) All S's are P's; (E) All S's are non-P's; (I) Some S's are P's; (O) Some S's are non-P's:

Canonical Aristotelian Square of Opposition

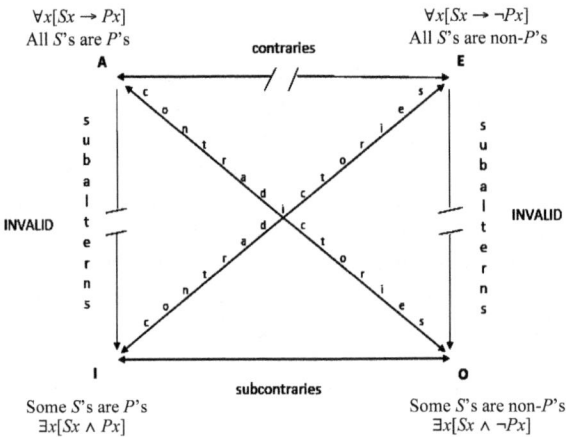

The four logical relations indicated in the square are still known and used today, although the original terminology has become somewhat antiquated. While it is traditionally considered that A logically implies I and E logically implies O, the inferences are not deductively valid if A is interpreted as $\forall x[Sx \rightarrow Px]$ and I is interpreted as $\exists x[Sx \wedge Px]$, and if E is interpreted as $\forall x[Sx \rightarrow \neg Px]$ and O is interpreted as $\exists x[Sx \wedge \neg Px]$. The inferences would hold only if A and E were interpreted as indicated, but for I interpreted as $\exists x[Sx \rightarrow Px]$ and O interpreted as $\exists x[Sx \rightarrow \neg Px]$. The latter symbolizations of I and O propositions evidently do not represent the propositions that Some S's are P's and Some S's are non-P's. Since the present purpose is to move as smoothly as possible between traditional and contemporary logical concepts depicted in the square of opposition, it is expedient in this version of the square to include the information both that A and I, E and O are traditionally interpreted as subalterns in the sense of A logically implying I and

---

[1] See Aristotle's *Prior Analytics* and *De Interpretatione* in [1], [24], [25], [27], [4], [5], [18, 17–65], [19].

E logically implying O, and that the inferences are deductively invalid as conventionally formalized.

Logical Relations in Canonical Square of Opposition

contraries   cannot both be true, but can both be false
contradictories cannot both be either true or false
subcontraries   cannot both be false, but can both be true
subalterns   one logically implies the other

Such easy surveyability generally comes at a price, and typically by oversimplifying the logical relations that actually obtain. No one is surprised today to hear that the canonical Aristotelian square of opposition is expressively limited in ways that in the past must have been accepted with a shrug in lieu of a superior kind of logic.[2]

The canonical square is an instrument for representing subsyllogistic logical reasoning between any combination of two among a total of four categorical propositions. Like any instrument, the square of opposition as a diagram of logical relations facilitates certain tasks but confers only limited usefulness. We need to understand the expressive and calculative power that the square of opposition represents within its limitations. One constraint is that the square itself does not teach us how to interpret and represent the AEIO structure of the basic propositions involved in a complex chain of syllogistic reasoning. If we do not know how to translate ordinary thinking and colloquial predicational and inferential expression into the most appropriate AEIO forms involving only subject and predicate terms and their converses in all and some quantifications, then knowing from the square's graphic depiction how the AEIO forms are logically interrelated as implications, contraries, subcontraries, and contradictions of one another will not support applications of the square to everyday thought and discourse. We must first know of a sentence in an argument whether it is an A, E, I, or O form of proposition, and this is not always obvious or easy to decide. For we do not always, in fact we probably seldom, think and express ourselves in terms of just these four categorical propositions. Perfectly reasonable arguments are sometimes encountered in raw, so to speak, *unsyllogismized*, thinking, speaking and writing, and they can turn on a logical connection between two propositions that on the face of things do not immediately belong to any one of the AEIO forms. An obvious example from a more contemporary algebraic rather than Aristotelian term logic standpoint is the inference: All S are either P or Q; No Q is P; therefore, some S are P only if no S are Q.

Here, then, is an expressive limitation of the square of opposition as a general instrumental aid in understanding logical inference. Clearly, there is a more basic propositional and predicate-quantificational logic underlying even Aristotelian syllogistic term logic, if we can validly reason but not logically justify a wide range of inferences involving the logic's categorical propositions built out of a relatively small number of terms. The square's limitation to AEIO propositions is already a conspicuous constraint on its usefulness in understanding logical relations in the most general sense, even among the AEIO propositions, let alone the unlimitedly many extra-AEIO propositions, that are encountered in scientific as well as most everyday reasoning. The canonical square of opposition, possessing at least these kinds of limitations, should make us wonder whether there are

---

[2][2, 3, 10, 17, 21].

others. How can we systematically study the scope and limits of the canonical square of opposition?

2 Logical Memory Device and Calculation Aid

The simplest answer as to why there is a *square* of opposition is that the idea of the diagram is to graphically display in two spatial dimensions the subsyllogistic logical interrelations between any two of the four categorical AEIO proposition types. Placing each one of the four AEIO propositions in a plane amounts to setting the corners for a four-sided polygon, of which a square is the simplest figure.

The diagram could in principle be drawn as a nonrectangular parallelogram or as a trapezoid, rhombus or other parallelogram or quadrilateral, but nothing is representationally gained thereby over the default square. It is how the four propositions stand logically to one another that further determines how the square is arranged and with which AEIO propositions in which corners displaying particular logical relations to one another within the square. The same logical inclusion and exclusion relations among the extensions of two or three (and more) predicates can be graphically represented in other ways also, for example, by Euler, Venn, or Lewis Carroll diagrams.[3] The AEIO proposition-limited syllogisms and logical basis for the square are already found in Aristotle. The square merely arranges them as further unanalyzable items, like the tokens on a game board, for which there are only so many possibilities to consider.[4]

The canonical square of opposition is evidently a *logical* device. But in what exact sense? Moreover, is it *only* or even *primarily* a logical device? We can use the square, once we have it available, as a shortcut way of checking on the inference relations between any two categorical propositions that could occur in any classical syllogism. Since a syllogism is a logical relation between three propositions, two assumptions and a conclusion, the canonical square of opposition as such constitutes at most a graphic representation of subsyllogistic logical relations. Despite its historical role in the teaching and application of a term logic that precedes syllogistic, there are no instructions for building the square step by step as itself a logical construction. With A propositions situated in the upper lefthand corner of the box, for example, *why not* put E propositions below A or diagonally across from A instead of at shoulder height? We are always presented with the square already dressed for dinner, so that we do not know from the square itself and cannot historically reconstruct exactly how the square came to be discovered in precisely the form it has. We cannot straightforwardly reverse-engineer the square to see how it was designed and assembled. We 'construct' the square on paper or by drawing it in the sand after making its acquaintance only in the sense of remembering how it is all supposed to go. We are accordingly led to ask to what extent if any the original Aristotelian

---

[3] [8, 9, 26].

[4] After Frege, Peano and Boole there is increasingly less interest in syllogistic logic, and so with the canonical Aristotelian square of opposition. See [16, especially 177–198, 404–427]. Boole [6, 7] in particular marks the turning point in logic, beginning with an expanded formulation of Aristotelian syllogistic logic, enhanced especially by William Hamilton to a four term logic, which Boole then converts to algebraic formulation at exactly the transitional midpoint between Aristotelian-Scholastic and contemporary symbolic logic.

square of opposition is a matter of logical inevitability or aesthetic graphic stylistic preference.

3 Square Topological Combinatorics

For each of 4 items to be arranged at the corners of a square, there will be 2 combinations for each planar paired (not diagonally opposed) items in the square, left or right of, or above or below, each other. For each of both of these combinations, there will always be 3 other items remaining in the square to enter into further combinations involving each distinct pair. It follows that there are $2 \times 3 = 6$ distinct combinations for each of 4 items, making a total of $6 \times 4 = 24$ possible arrangements of any 4 items (AEIO propositions of syllogistic logic) in a plane. Now we ask: Are all of these possibilities distinct, and, if so, in what sense? Here are the 24 combinations:

AEIO Graphic Topology

1.	A E I O	7.	E A I O	13.	I A E O	19.	O A E I
2.	A E O I	8.	E A O I	14.	I A O E	20.	O A I E
3.	A I O E	9.	E I A O	15.	I E A O	21.	O E A I
4.	A I E O	10.	E I O A	16.	I E O A	22.	O E I A
5.	A O E I	11.	E O A I	17.	I O A E	23.	O I A E
6.	A O I E	12.	E O I A	18.	I O E A	24.	O I E A

4 Graphic Equivalences and Basic Forms

Of these possibilities, we know that 1 is the Aristotelian choice for what is now known as the canonical square. We shall say that forms 4, 8, 11, 14, 17, 22 and 24, are *graphically equivalent* to 1, in the sense that they have the same relative configuration of propositions and relations as 1, if 1 were to be shown on its left side, or its right side, or upsidedown, plus the reversals or mirror-images of all of these repositionings, of what we can see to be essentially the same graphically equivalent square. Together with 1, these 7 forms suggest (what is easily confirmed), that of the 24 square forms, there are only $24/8 = 3$ graphically distinct nonequivalent forms, here, 1, 2 and 3, the first of which is the Aristotelian. The graphic equivalents to the Aristotelian canonical square of opposition 1 are indicated by sympathetic shading in the following diagram:

Graphic Equivalents to Canonical AEIO Square 1 (Original Aristotelian Square)

1. A E I O	7. E A I O	13. I A E O	19. O A E I
2. A E O I	8. E A O I	14. I A O E	20. O A I E
3. A I O E	9. E I A O	15. I E A O	21. O E A I
4. A I E O	10. E I O A	16. I E O A	22. O E I A
5. A O E I	11. E O A I	17. I O A E	23. O I A E
6. A O I E	12. E O I A	18. I O E A	24. O I E A

This leaves us with possibilities 2 and 3 to compare with the Aristotelian choice 1 (and their 7 equivalences each across the field) to see if his decision could be improved upon. Square arrangement 2 in turn is graphically equivalent to forms 5, 7, 9, 16, 18, 20, and 23, which we do not bother to illustrate; form 3 is graphically equivalent to forms 6, 10, 12, 13, 15, 19, and 20.

We thereby reduce the burden of evaluating alternative squares of opposition from 24 to 3. We take advantage of this more manageable set of choices to ask and hopefully answer whether the Aristotelian square is wisely chosen, whether there are any advantages to 1 over 2 and 3, or the reverse. In trying the experiment, we are nevertheless soon persuaded of the preferability of 1 over 2 or 3 for the most compact intelligible presentation of AEIO logical interrelations.

Noncanonical Alternative Square of Opposition 2

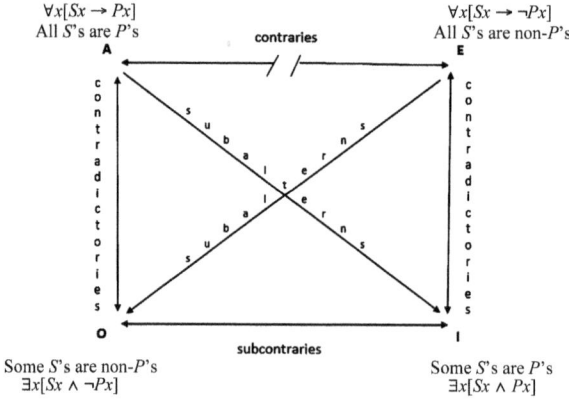

Combination 2 has the contradictories running up and down the vertical sides of the square, in place of the subalterns that now cross in the middle, and contraries and subcontraries remaining as in canonical square 1. There is nevertheless something more satisfying, perhaps more dramatic and memorable about placing the AEIO propositions as in square 1 with contradictories rather than subalterns crossing in the middle. This is particularly the case because we do not get double-headed arrows connecting subalterns as

we do for contradictories, making their opposition graphically expressed as more fundamental in square 1 than in square 2. Form 3 also features contradictories up and down the
square's vertical sides, but opposes contraries and subcontraries as double-headed arrows
crossing in the square's center. The visual disappointment here is that contraries and subcontraries are obviously different logical relations, so that their crossing in the middle of
the square does not place the propositions so related on the same footing, as in the consequently more stable appearing canonical square. We see this in the alternative version
presented here as square 3:

Noncanonical Alternative Square of Opposition 3

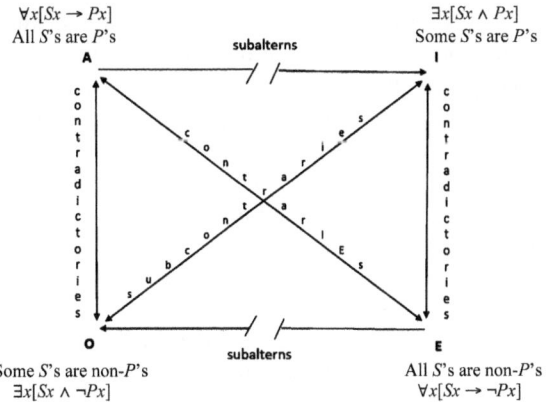

The AEIO symbolism by itself (in contrast with the entire square) is evidently at most
a memory device, rather than a logical instrument. The Aristotelian square 1 might also
be preferred over squares 2 and 3, by virtue of its alphabetical mnemonic. Square 1 has
the distinct advantage over 2 and 3 in this regard, for any language-users recognizing
the AEIO vowel group in exactly that order, and in whose culture the reading of text is
standardly scanned from left to right and top to bottom. Had noncanonical squares 2 or 3
been preferred for other reasons over canonical square 1, it obviously remains expedient
to simply rename the original propositions, making I E or O E, etc., since these labels are
only conventional anyway. The alphabetical mnemonic could remain while referring to at
least some different propositions than the canonical AEIO in the canonical square.

5 Utility of the Canonical Square

Although the AEIO propositions represent an interesting choice for a combinatorially
manageable construction and analysis of basic argument forms in a logic of four term
types, they do not exhibit the logical structures of all reasoning and argumentation.

- A: All S's are P's
- E: All S's are non-P's (No S's are P's)
- I: Some S's are P's
- O: Some S's are non-P's

The canonical square comprehends all four combinations of terms in two-term propositions with unnegated subject term and predicate or negated predicate term for the syllogistic quantifiers 'All' and 'Some'. The question remains how AEIO propositions are to be topographically understood in their interrelations. The canonical square offers one answer in the form of a two-dimensional display of the interesting subsyllogistic relations between AEIO proposition forms. We have already seen that the canonical square more appropriately represents exactly these relations than either of the two graphically nonequivalent alternatives.

The square, as previously observed, compactly *represents* the AEIO logical relations, but plays no part in the *logical development* of syllogistic logic. The square is in this sense entirely a logical afterthought, a two-dimensional graphic method of depicting syllogistic relations. It is the prosaic version of more abstractly graphic Euler and Venn or Lewis Carroll diagrams for the same choice of logical relations. The square, to conjure a Wittgensteinian analogy, can be used as something like a tides table. We consult it for convenience to remind ourselves of the subsyllogistic logical relations holding between any two AEIO propositions. Like the tides table, the square no more logically determines syllogistic relations than the tides table causally determines the tides. They are merely pictures of independently obtaining facts. Although the canonical square can be used as an external aid in syllogistic reasoning, reasoning generally is better conducted without external aids altogether, by internalizing a dispositional understanding of logical interrelations to the point where it becomes instinctive.

Internalization of logical principles that are not simply innate is largely a matter of learning through memory, practice and correction. The square provides a memory device for this purpose, and as such can be thought of largely as a crutch for students to lean on when they are first learning syllogistic logic and the subsyllogistic relations among the four categorical AEIO propositions represented by the square. The square can help students to achieve both of these conditions in reinforcing subsyllogistic principles. The square accordingly functions as a mnemonic device and fixed external check on our efforts to learn and apply subsyllogistic logical principles as a prelude to working with three-proposition syllogisms. With time, in the practical mastery of these subsyllogistic logical relations, we may hope that we can eventually construct and analyze logical inferences reliably correctly and without excessive deliberation, and in particular without having to stop and think in every instance about whether or not our thinking is correct.

The square itself, by virtue of its logical isomorphism with fundamental syllogistic logical relations, in addition to being a memory device and paper 'calculator' for checking and constructing two-proposition subsyllogistic AEIO inferential relations, is also the *graphic embodiment* of a commonly proposed formal Aristotelian subsyllogistic logic, at a comparatively shallow depth of formal analysis. Euler, Venn, and Lewis Carroll diagramming methods take us to a logically and semantically deeper two-dimensional graphic display of the *extensions of predicate terms* in classical syllogisms by virtue of which the AEIO subsyllogistics are logically interrelated as the square presents. They do so by portraying the objects comprehended by general subject and predicate terms in AEIO propositions, to which these later graphic techniques are logically limited, and by which the propositions are made true or made false. Properly applied, logic diagrams in Euler, Venn and Carroll, in comparison with the canonical square of opposition, provide an effective decision method for classical subsyllogistic, and even syllogistic inferences among the AEIO propositions.

6 Graphic Limitations of the Canonical Square

Although the square is a graphic device, there are still too many *words* in it for it to be as purely graphically informative as it might be, when compared with other two-dimensional diagrams of AEIO subsyllogistic relations. Specifically, the square contains logically unanalyzed terms for logical relations, and the AEIO propositional forms themselves are understood as any propositions of the form 'All *S*'s are *P*'s (non-*P*'s)' or 'Some *S*'s are *P*'s (non-*P*'s)'.

The square by itself only tells us, for example, *that* A and O are contradictories; that I is the subaltern of A, as is O of E, that A and E are contraries, and I and O subcontraries. Unfortunately, the square does not and cannot in and of itself tell us exactly *why* these relations hold. In contrast, contemporary algebraic logic does not need the kinds of memory devices associated with classical Aristotelian syllogistic logic. It has more powerful methods of dealing with the combinatorics of repeating significant terms occurring in logical expressions and reasoning chains of virtually unlimited inner complexity and length. Algebraic logic, moreover, is interpreted by means of a rigorously developed formal semantics, accessorized with equally powerful and practically useful decision methods for judging logical relations by calculation, rather than by consulting the information diagrammed in a memory chart of deductively valid or invalid inferences and other logical relations, such as the square provides.

Beyond memorizing truth tables for the five most useful (noble) truth-functions, working with contemporary algebraic logic is primarily a matter of syntactical skill in applying a convenient number of formal algorithms. In comparison to the freedom afforded by algebraic approaches to formal logic, the original Aristotelian square of opposition often requires reasoning to be forced into a logical straitjacket, accommodating inferences to a particularly non-ergonomic Procrustean Bed. The square has accordingly been relegated to the category of historical curiosity in the early development of symbolic logic. An important exception is the fact that familiarity with the square is useful for logicians today as a kind of *lingua franca*, when adapted as a shorthand to express logical relations in specialized applied logics with specialized domains. These may include the principles of a deontic, doxastic or temporal logic, or the like. Versions of the square, when the concepts in question cooperate, can make basic logical relations readily surveyable in a single glance, and thereby make it also easier to hold their components before the eye and mind for comparison in grasping their interrelations intuitively, conditioned by familiarization with the canonical square in its original subsyllogistic context. Thus, one occasionally sees square-like diagrams even in contemporary logical research, as well as the square itself being discussed as an historical topic. Contemporary uses of the square are often very cleverly done; they can be graphically highly informative as a map presenting a certain set of formal logical relations among concepts that can be squared off or otherwise polygonalized against one another in two- or even three-dimensional space in ways that neither Aristotle nor the medieval Scholastic tradition in logic anticipated.

Lest we conclude too quickly that the canonical square of opposition deserves its place on the dusty shelf of horse-and-buggy formal logical devices, and, at that, primarily as a memory device with minimal if any calculative potential, we should consider that the square can still teach us something interesting about classical logic that we might not otherwise have fully appreciated, even if only because in all likelihood we would not otherwise have thought about it in quite these terms.

7 Square Without Inversions

The first thing that commands attention when considering the square of opposition from a more contemporary algebraic perspective is that the diagram appears combinatorially incomplete in one crucial respect. What are conspicuously missing from the canonical square are the so-called *inverted* forms or *inversions* of AEIO:

- All/Some non-S is P
- All/Some non-S is non-P

The *inverses* (*versos*) of (*recto*) AEIO propositions are these:

- A(V): All non-S is P
- E(V): All non-S is non-P
- I(V): Some non-S is P
- O(V): Some non-S is non-P

Why are these propositions not included in or added to the square? Few logicians have missed them or wondered where they belong, with some important exceptions considered further below. Almost alone among contemporary historians of syllogistic logic, Robert N. McLaughlin, in his book, *On the Logic of Ordinary Conditionals*, offers the following puzzling comments about the AEIO Inversions in Aristotle's syllogistic:

> The question of inversion was not one that Aristotle had to consider directly. He doubted the equivalence of 'non-S' and 'not S' and his system did not admit of forms expressing negative terms. But there is some point in asking whether, had he accepted negative terms, he would have welcomed the inconsistency of 'All S is P' and 'All non-S is P'. He could have reasoned for the view that they are inconsistent by an argument similar to the one he used to show that if the statement A necessitates the statement B, then it cannot be true that the denial of A also necessitates B: "If 'A necessitates B' and 'Not A necessitates B' were both true, then not B would necessitate B". Aristotle considered the consequent absurd and therefore rejected the possibility of the antecedent. But if one allows negative terms, and if we accept the consistency of 'All S is P' and 'All non-S is P', a similar result follows. From 'All S is P' and 'All non-P is not S' and from this and 'All non-S is P' we obtain 'All non-P is P'. It seems likely that Aristotle would have regarded this consequent as nonsense and that he would have concluded that 'All S is P' is incompatible with 'All non-S is P' and that the former thus entails 'Some non-S is not P'.[5]

---

[5][20, 105–106]. On 105, McLaughlin rightly asks: 'Writers have also criticized the traditional process called 'inversion,' which licenses the inference from 'All S is P' to 'Some non-S is not P'—in our terms, from $s[p(N)]$ to $-(-s[p(N)])$ or $-s[-p(M)]$. Why must it be the case that if all S is P, some non-S is not P? After all, if it were the case that everything is P, then all things S and non-S alike would be P. But when 'All S is P' is taken to mean that a thing's being S warrants the conclusion that it is P, and when the problem is addressed from the point of view of OC [McLaughlin's theory of Ordinary Conditionals], the inference becomes entirely natural. Because $s[p(N)]$ means only that the presence of s ensures the presence of p.' McLaughlin's final remark suggests that N plays no role in the meaning of $s[p(N)]$, which would need to be explained. McLaughlin says that $s[p(N)]$ means that the 'presence' of s ensures the 'presence' of p, which is patently a metaphysical relation of ontic dependence or supervenience. It is a metaphysical relation, if 'presence' means as it appears 'existence', anyway, surely not just presence to us at a particular time, rather than the expected logical or semantic relation by which an inversion at least of an AEIO proposition might first be explicated. McLaughlin's interpretation supports an interesting and plausible interpretation by which, for example, the meaning of the A proposition, 'All whales are mammals', is directly bound to specific truth conditions by which the presence of whales ensures the presence of mammals, and, indeed, the presence of any whale ensures the presence of a mammal.

If not-B necessitates B because both A and not A necessitate B, it is presumably because B is independently logically necessary. One might argue, for example, that the proposition that $2 + 2 = 4$ is necessitated both by the proposition that $1 + 1 = 2$ and by the proposition that $1 + 1 \neq 2$, since the proposition that $2 + 2 = 4$ is necessarily true in either case. The fact that a proposition is conditionally true both on the grounds of a proposition and its negation might reasonably be taken as a formal semantic criterion for necessary truth in general, but not obviously as a justification for disallowing AEIO inversions containing negated or complement subject terms. *If* not-B necessitating B is absurd, which is not clearly true even in Aristotle's logic, and if not-B necessitating B follows logically from A necessitating B and not-A necessitating B, then presumably the argument's starting place, that A necessitates B and not-A necessitates B, must also be absurd. The logical connection from the imagined absurdity of not-B necessitating B, which seems in any case to be a propositional relation, to the impermissibility of non-B as a subject property term in the inversion of a categorical AEIO proposition, is as undemonstrated as it is intuitively implausible and unjustified in contemporary formal semantics. More pointedly, why should Aristotle have regarded the proposition in the consequent McLaughlin mentions as nonsensical? Does it not make sense, for example, to deduce from any permissible assumptions that all non-spatiotemporal entities are abstract?

Since there are exactly four categorical two-term AEIO propositions, there is combinatorially a total of 256 different syllogisms, of which 24 are deductively valid, and of which only a further 15 are deductively valid if subject terms in AEIO propositions are permitted to be empty in the sense of ostensibly referring to something nonexistent. Do the math if the number of terms to be considered in combination is increased by a factor of the negations of subject as well as predicate terms with AEIO propositions + their two distinct inversions. There must be 512 still surveyable combinations. Of the additional $512 - 128 = 384$, are there any logically interesting combinations? If in classical or scholastic times there was no practical way to identify two of the AEIO inversions as logically equivalent to two of the AEIO propositions, as is afforded algebraically, then it would have appeared, if logicians also had the arithmetical capacity without an algebra of exponential combinatorics to calculate the total, that there would be 2,048 syllogisms with all four AEIO propositions and all four AEIO inversions.

Few logicians have wondered or worried about the AEIO inversions. Have they been right not to bother with AEIO inversions? Or do the AEIO inversions have interesting logical properties that the canonical square simply does not include? We should not accept the answer that AEIO inversions are not incorporated in the traditional square because they are not AEIO in form. The picture is more complicated and hence logically more interesting, given, as we shall show, that two inverses of AEIO propositions are logically equivalent to two other AEIO propositions, while the other two inverses are not.[6]

---

[6]Lukasiewicz [12] does not even consider AEIO inversions in his second edition of *Aristotle's Syllogistic from the Standpoint of Modern Formal Logic*. Among the few available sources for the topic is [11]; a valuable study in its own right that is not concerned with the particular questions that I have been investigating. Prior [22, 23] discusses negative terms in Boethius's syllogistic logic, citing the latter's *Introductio ad Syllogismos Categoricos*, but does not mention inversions by that name. See especially [23, Part II, Chapter II—Categorical Forms with Negative, Complex, and Quantified Terms, 126–156; including §1 *Negative Terms in Boethius and de Morgan*, 126–134].

8 Case for AEIO Inversions

We argue first for the inclusion of AEIO inversions within an expanded syllogistic logic in this way. We note that we can construct a (hackneyed) syllogism in BARBARA:

1. All whales are mammals.
2. All mammals are warm-blooded animals.
   _____
3. All whales are warm-blooded animals.

In an expanded 'syllogistic' logic incorporating irreducible inverses, we should then presumably be able similarly to reason:

1. All non-abstract entities are spatiotemporal.
2. All spatiotemporal entities are extended.
   _____
3. All non-abstract entities are extended.

This first example works because it is already an instance of BARBARA, in which we have uniformly replaced the inverted subject term 'non-abstract' for '*S*', in effect allowing the inversion 'non-abstract' to play the same logical role in the syllogism as a positive subject term would do, but in which no negation, complementarity or contrariety is explicit. We can get away with such a deductively valid inference here only because there is no interplay between the predicates 'abstract' and 'non-abstract' in this particular syllogism. The 'non-' of predicate complementation in such instances does not matter syllogistically, just as it would not in standard first-order logic. We can simply replace the term 'non-abstract' uniformly throughout with a positive rather than negative term without jeopardizing the argument's logical validity.

Are there other inference forms to be considered in which AEIO and A(V), E(V), I(V) and O(V) inversions interact in more interestingly ways? Are the A(V), E(V), I(V), O(V) inversions of AEIO propositions logically independent of AEIO forms, or are at least some AEIO inversions logically reducible to AEIO forms? Reverting for convenience to the notation of contemporary first-order symbolic logic or the 'functional calculus', we find a quantificational form of contraposition in the following theorem, whereby an E(V) form reduces to a logically equivalent A form:

$$\forall x[\neg Sx \rightarrow \neg Px] \leftrightarrow \forall x[Px \rightarrow Sx]$$

Nevertheless, we also discover:

$$\forall x[\neg Sx \rightarrow Px] \leftrightarrow \forall x[\neg Px \rightarrow Sx]$$

in which the A(V) form is *irreducible* to any other AEIO form. Similarly:

$$\exists x[\neg Sx \wedge Px] \leftrightarrow \exists x[Px \wedge \neg Sx]$$

whereby I(V) *reduces* to a logically equivalent O form. Yet we also have as a theorem of standard first-order logic:

$$\exists x[\neg Sx \wedge \neg Px] \leftrightarrow \exists x[\neg Px \wedge \neg Sx]$$

in which O(V) by the same method is manifestly irreducible.

Inversion forms of AEIO, E(V) and I(V) are *reducible* (respectively, to A and O); whereas inversion forms A(V) and O(V) are *irreducible* to any of the AEIO categorical propositions. This result narrows the search space considerably for logical interrelations between A(V) and O(V), as the only AEIO inversions that need to be considered in their logical relations if any to the four AEIO forms.

9 Three Questions of Logical Interrelations

There are three questions to be addressed with respect to the logical interrelations between A(V) and O(V) with AEIO forms, inspired by the three categories of logical relations displayed and labeled in the canonical square of opposition. The questions are these:

(1) What deductively valid one-way logical *inferences* or *subalternations*, if any, obtain between A(V) and AEIO and between O(V) and AEIO?
(2) What *contradictories* or *logical inconsistencies*, if any, obtain between A(V) and AEIO and between O(V) and AEIO?
(3) What *contrarieties* or *subcontrarieties*, if any, obtain between A(V) and AEIO and between O(V) and AEIO?

We address these questions by applying standard predicate-quantificational logical methods to the relevant propositions to determine their logical relationships. Below for each category of question we present only the results, validated by algebraic logical decision methods and algorithms.

Question 1 Antitheorems
Answer to question (1): NONE.
There are exactly 16 cases to consider:

$\forall x[\neg Sx \rightarrow Px] \perp \forall x[Sx \rightarrow Px]$	—	A(V) does not imply A
$\forall x[\neg Sx \rightarrow Px] \perp \forall x[Sx \rightarrow \neg Px]$	—	A(V) does not imply E
$\forall x[\neg Sx \rightarrow Px] \perp \exists x[Sx \wedge Px]$	—	A(V) does not imply I
$\forall x[\neg Sx \rightarrow Px] \perp \exists x[Sx \wedge \neg Px]$	—	A(V) does not imply O
$\forall x[Sx \rightarrow Px] \perp \forall x[\neg Sx \rightarrow Px]$	—	A does not imply A(V)
$\forall x[Sx \rightarrow \neg Px] \perp \forall x[\neg Sx \rightarrow Px]$	—	E does not imply A(V)
$\exists x[Sx \wedge Px] \perp \forall x[\neg Sx \rightarrow Px]$	—	I does not imply A(V)
$\exists x[Sx \wedge \neg Px] \perp \forall x[\neg Sx \rightarrow Px]$	—	O does not imply A(V)
$\exists x[\neg Sx \wedge \neg Px] \perp \forall x[Sx \rightarrow Px]$	—	O(V) does not imply A
$\exists x[\neg Sx \wedge \neg Px] \perp \forall x[Sx \rightarrow \neg Px]$	—	O(V) does not imply E
$\exists x[\neg Sx \wedge \neg Px] \perp \exists x[Sx \wedge Px]$	—	O(V) does not imply I
$\exists x[\neg Sx \wedge \neg Px] \perp \exists x[Sx \wedge \neg Px]$	—	O(V) does not imply O
$\forall x[Sx \rightarrow Px] \perp \exists x[\neg Sx \wedge \neg Px]$	—	A does not imply O(V)
$\forall x[Sx \rightarrow \neg Px] \perp \exists x[\neg Sx \wedge \neg Px]$	—	E does not imply O(V)
$\exists x[Sx \wedge Px] \perp \exists x[\neg Sx \wedge \neg Px]$	—	I does not imply O(V)
$\exists x[Sx \wedge \neg Px] \perp \exists x[\neg Sx \wedge \neg Px]$	—	O does not imply O(V)

Question 2 Compossibilities
Answer to question (2): NONE.
There are exactly 8 cases to consider:

$\Diamond[\forall x[\neg Sx \rightarrow Px] \wedge \forall x[Sx \rightarrow Px]]$	—	A and A(V) are logically consistent
$\Diamond[\forall x[\neg Sx \rightarrow Px] \wedge \forall x[Sx \rightarrow \neg Px]]$	—	E and A(V) are logically consistent
$\Diamond[\forall x[\neg Sx \rightarrow Px] \wedge \exists x[Sx \wedge Px]]$	—	I and A(V) are logically consistent
$\Diamond[\forall x[\neg Sx \rightarrow Px] \wedge \exists x[Sx \wedge \neg Px]]$	—	O and A(V) are logically consistent
$\Diamond[\exists x[\neg Sx \wedge \neg Px] \wedge \forall x[Sx \rightarrow Px]]$	—	A and O(V) are logically consistent
$\Diamond[\exists x[\neg Sx \wedge \neg Px] \wedge \forall x[Sx \rightarrow \neg Px]]$	—	E and O(V) are logically consistent
$\Diamond[\exists x[\neg Sx \wedge \neg Px] \wedge \exists x[Sx \wedge Px]]$	—	I and O(V) are logically consistent
$\Diamond[\exists x[\neg Sx \wedge \neg Px] \wedge \exists x[Sx \wedge \neg Px]]$	—	O and O(V) are logically consistent

Question 3 Compossibilities
  Answer to question (3): NONE.
  There are exactly 8 cases to consider:

The cases in this final category are the same as those that arise for Question 2. Logical consistency establishes not only that the propositions are not contradictories, but also that they are not contraries or subcontraries of one another; they are not such that at most one of them can be true. The answer to Question 2 also shows that neither A(V) nor O(V) are contraries or subcontraries of any of the AEIO forms. The reason is that A(V) and O(V) are mutually contradictory, while A(V) and O(V) are each logically consistent with all four AEIO forms. It follows, then, that it is logically impossible for A(V) and any of the AEIO propositions to be conjointly true, and for O(V) and any of the AEIO propositions to be conjointly true. Thus, it is logically impossible for A(V) or O(V) to be subcontraries of any of the AEIO forms. The proof is an application of the standard modal theorem, $[\neg\Diamond[p \wedge q] \wedge \Diamond[q \wedge r]] \rightarrow \neg\Diamond[p \wedge r]$.

The result of inquiry is to show that inversions A(V) and O(V) are in every category of logical relationship represented in the canonical square of opposition as logically mutually independent. For the AEIO forms, it is logically just as though the A(V) and O(V) inversions simply did not exist. We represent the resulting square of opposition, revised and expanded to include the mutually contradictory inversions A(V) and O(V) in the following way, giving their relationship arrow a mere ghostly presence in grey, transecting the canonical square:

Revised and Expanded Square of Opposition with A(V) and O(V) Inversions

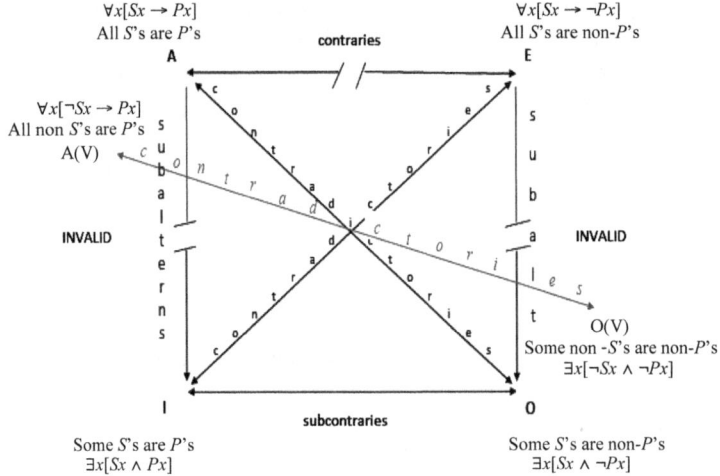

In terms of the logical relations of one-way implication or subalternation, contradiction, contrariety, and subcontrariety, in the canonical square, the logic of AEIO-irreducible categorical propositional inversions A(V) and O(V) are *logically altogether independent* of the AEIO syllogistic logic, *except* with respect to the *antitheorems* that can be demonstrated concerning the failure of implication or subaltern relations, contradictories and contraries, involving A(V) and O(V) *modulo* AEIO.

The logic of A(V), E(V), I(V), and O(V) does not have its own parallel square of opposition, in which all the square relations that obtain among AEIO obtain also among A(V),

E(V), I(V), and O(V). The obvious reason is that E(V) ↔ A and I(V) ↔ O. What is interesting is that there is *no positive logical relation* as represented in the canonical square between the AEIO and A(V), O(V) propositions. A(V), O(V) subsyllogistic logic exists in a logical space of its own, parallel to that of AEIO, with no logical interconnection linking any of the propositions in either canonical square or parallel inverse-expanded square, by any of the relations represented in the square, *except* that by which A(V) and O(V) are themselves contradictories, precisely as are A and O in the classical square, as we see clearly when we abstract the mutually contradictory AEIO inversions A(V) and O(V):

Diagram for Contradictory AEIO Inversions A(V) and O(V)

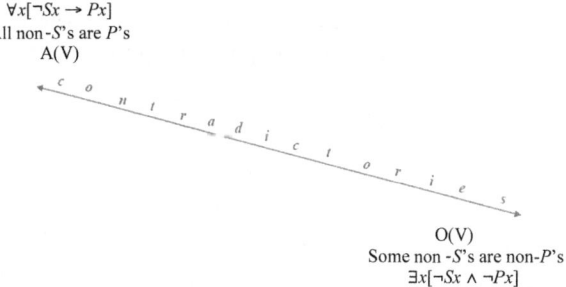

$$\forall x[\neg Sx \rightarrow Px]$$
All non-S's are P's
A(V)

O(V)
Some non-S's are non-P's
$$\exists x[\neg Sx \wedge \neg Px]$$

The two logics thus have a partially parallel but subsyllogistically independent internal logical structure, given that there are only two logically distinct inversions of AEIO propositions, A(V) and O(V). What is formally and philosophically curious is that the two logics do not touch one another by virtue of any recognized traditional syllogistic logical relation (subalternation, contradiction, contrariety, subcontrariety). The subsyllogistic logic of AEIO inversions is a shadow image of only one of the AEIO relations in the canonical square. The canonical square of opposition by implication establishes two parallel but logically isolated, non-interacting sub-subsyllogistic logics for four categorical propositions and two of their logically distinct irreducible inversions. The existence of such a partial parallelism between a term logic proposition and its inversion suggests interesting possible applications in the formal modeling of distinct computer software logics, mutually inaccessible memory cachés, and the like, in parallel dimensions, functioning logically altogether independently of one another within the same syllogistic logic.

10 Keynes-Johnson Octagon of Opposition

The possibility of AEIO inversions, while largely neglected in syllogistic logic, has not gone entirely unnoticed. Most prominently, J.N. Keynes in his (1906) *Studies and Exercises in Formal Logic: Including a Generalisation of Logical Processes in Their Application to Complex Inferences*, and W.E. Johnson in his three volume (1921–1924) *Logic*, remark on the occurrence of AEIO variants with negated subject terms.

Keynes, in the Preface to the Fourth Edition of his book, promises: 'the treatment of negative names has been revised'.[7] Keynes begins with AEIO propositions, which he

---

[7][15, v].

transforms by means of logical operations, generally contraposition, conversion and obversion, into formulas containing negated subjects not considered in the canonical square. He explains in §84 *Extension of the Doctrine of Opposition*: 'Two propositions may be *equivalent* or *equipollent*, each proposition being formally inferable from each other ... For example, as will presently be shown, *All S is P* and *All not-P is not-S* stand to each other in this relation.'[8] Elsewhere he writes: '*All S is P* and *Some not-S is not-P* are another pair of subalterns.'[9] 'For example, the contraries *All S is P* and *All not-S is P* are equivalent to *No not-P is S* and *All not-P is S*.'[10] Attributing the terminology to William Stanley Jevons's (1879) *Elementary Lessons in Logic*, Keynes defines 'Inversion' as '*a process of immediate inference in which from a given proposition another proposition is inferred having for its subject the contradictory of the original subject*'.[11]

Keynes presents the inversions designated as A′, E′, I′, O′ (in the above exposition, A(V), E(V), I(V), O(V)), in an 'octagon of opposition', possesing this form:

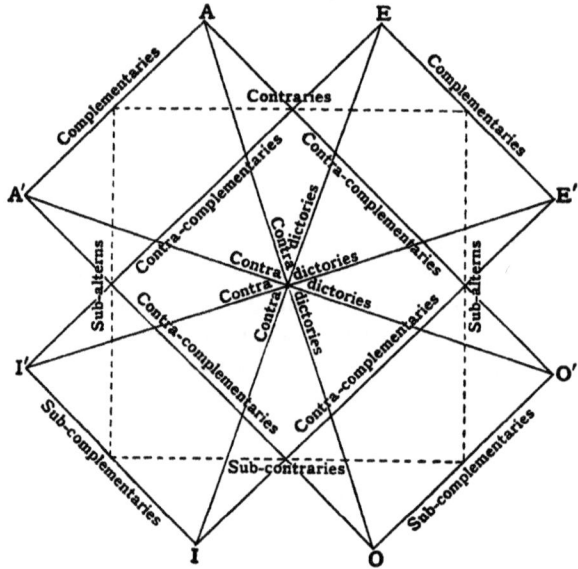

Keynes credits the octagon to Johnson, although presumably to another ancillary source than Johnson's *Logic*, where the octagon does indeed appear in Volume I, but which was only published in 1921, approximately fifteen years after Keynes' *Studies and Exercises in Formal Logic*.[12] What is interesting about the Keynes-Johnson octagon of opposition is that it includes inversions E′ and I′ (E(V) and I(V)) as contradictories, without collapsing these forms as logically equivalent, or, as Keynes also likes to say, equipollent, respectively, to A and O in the canonical square. These inversions are instead given their

---

[8][15, 147].

[9][15, 118].

[10][15, 119].

[11][15, 139]. Keynes refers to Jevons [13, 185–186].

[12][15, 144], [14, (vol. I) 142]. Keynes does not refer to a specific source prior to the publication of Johnson's 1921 from which he might have derived the octagon of opposition.

own places in the octagon, where they are further interrelated by the invented relations of 'Complementaries', 'Sub-complementaries', and 'Contra-complementaries', it would seem, to have something to call them. The octagon in its inclusion of E' and I' represents all four inversions at a relatively superficial level, in the sense of depicting logical relations between only a purely syntactical individuation of distinct propositions. Moreover, the Keynes-Johnson octagon does not graphically represent the important fact that E' is the contradictory of O, and that I' is the contradictory of A. The diagram preferred here is both simpler and more essential, if less geometrically engaging, in presenting the actual logical relations and their lack of classical square of opposition relations to the standard AEIO forms. Although Keynes recognizes the logical equivalence or equipollence between E' and A and between I' and O, the Keynes-Johnson octagon does not reductively illustrate the semantic equivalences of two of the inversions with the categorical propositions already included in the canonical square.[13]

11 Intriguesting (Intriguing + Interesting) Question

Can we produce a deductively valid genuine syllogism involving inversion forms A(V) or O(V) that satisfies the conditions for a canonical syllogism, and is not merely sub-syllogistic, with no logically unutilized inessential premise in deriving the syllogism's conclusion?

To answer this question, we must consider what an expanded classical Aristotelian syllogism allowing negated subject as well as predicate terms requires:

- Three propositions: two premises and one conclusion.
- Premises and conclusion consisting collectively of three terms, including (classically) negations of predicate terms and (non-classically) negations of subject terms.
- Premises constituted as propositions of two terms (major, minor or middle) each.
- Each premise having exactly one term in common with the conclusion, the major term of the major premise providing the conclusion's predicate term, and the minor term of the minor premise providing the conclusion's subject term, and where the middle term is the predicate term of the minor premise.

Experimenting briefly with the syntax for the expanded syllogistic logic with A(V) and O(V) inversions reveals a number of syllogisms that appear to offer opportunities for A and O inversions to participate in genuinely syllogistic inferences. Contrariwise, there appear to be no deductively valid genuine syllogisms containing inversions, reinforcing

---

[13] Keynes is explicit about the logical equivalence or equipollence of the E' (E(V)) and A and between I' (I(V)) and O inversions with their canonical categorical propositional forms. Johnson [14, (vol. I), 140] defines the inversions equivalent, using a short horizontal bar above subject and predicate terms to indicate their internal 'negations' or complements. Having presented the equivalences, however, he appears to deny that there are any equivalences between the inversions and the standard AEIO propositions, when he concludes immediately thereafter on the same page, that the enumeration provides 'a list of eight *distinct* [Johnson's emphasis] (i.e. non-equipollent) [Johnson's parenthetical clarification] general categoricals, as an extension of the usual four.' This certainly seems false, at least from a more contemporary logical perspective, as well as being at odds with Keynes' understanding.

the general conclusion that inversions lack syllogistic significance, and as such are appropriately excluded from the canonical square of opposition. Here are several representative and easily identified illustrations that appear to meet the requirements:

Canonical Syllogism with A(V) and A(O) Inversions

$$\frac{\begin{array}{ll} \forall x[\neg Sx \to Px] & \text{A(V)} \\ \exists x[Qx \wedge Sx] & \text{I} \end{array}}{\forall x[Qx \to Px] \quad \text{A}}$$

This argument is deductively invalid, but, by the above criteria, genuinely syllogistic. Similar inferences are available involving O(V) as a major premise. This example suffices to show that there are genuine syllogisms involving logically independent AEIO inversions A(V) and O(V).

Finally, as an example suggested by algebraic logic, building on the so-called paradoxes of strict implication, we may also consider:

Extrasyllogistic *Ex Falso Quodlibet* with A(V) and O(V)

$$\frac{\begin{array}{ll} \forall x[\neg Sx \to Px] & \text{A(V)} \\ \exists x[\neg Sx \wedge \neg Px] & \text{O(V)} \end{array}}{\forall \text{ny conclusion you fancy}}$$

This case also represents a deductively valid but technically extrasyllogistic as well as subsyllogistic inference including AEIO inversions A(V) and O(V).

12 Vindicating the Canonical Square of Opposition for the Relations It Proposes to Represent

The square of opposition is a logical device and memory aid. Its value, however, is only secondarily in the subsyllogistic logical relations it represents. For these are of interest primarily because they involve the propositions from which classical syllogisms are assembled. Adding inversions of the square's AEIO propositional forms to the square does not enhance its inferential range among the canonical + inversion syllogisms, except in a narrow technical sense, and then only if logically unutilized premises are permitted in a deductively valid *syllogism*, assuming for the moment that such constructions are allowed to count as genuine syllogisms.

Otherwise, the AEIO inversions are: (a) in two cases logically reducible to one of the AEIO forms (E(V) = A, I(V) = O); while (b) the other two cases, A(V) and O(V), are contradictories, and therefore contraries and subcontraries, a relation represented only implicitly, even in the canonical square, but are otherwise logically isolated from the AEIO forms in the square of opposition sense of not being subalterns of one another, nor of being contraries or subcontraries of one another. These relations are excluded in every case by virtue of their joint logical consistency, expressible as the logical possibility of their conjunction, which is generally a matter of purely formal decision-method decidable logical consistency.

There can be different systems of rules for logically valid syllogisms once we have identified a class of 'All' and 'Some' propositions of definite internal logical construction out of basic terms like those of the canonical syllogisms. The inverse axial crossing the canonical square, supported by independent logical inquiry, transcends the square as representing only a partial highly truncated mirror image of only one logical relation represented by the square, the contradiction between A and O, respectively, and A(V) and O(V). There is therefore no logical justification for including inversions graphically in the square of opposition. One-half of the AEIO inversions are already included under other logically equivalent designations, and the other half are mutually contradictory but otherwise logically isolated from each other and from all the AEIO forms.

With this conception, we are back where we started in a sense, but at least in the process we have learned *why* inversions are absent from the canonical square of opposition. It is interesting to speculate about why certain decisions were made in developing syllogistic logic, limiting syllogisms to propositions with two terms, and to inferences with three propositions including exactly one with a distributed middle term. The reasons undoubtedly have much to do with the combinatorial explosion that occurs when working with anything larger than three or maximally four subject and predicate terms. One nevertheless wonders, among other things, how Aristotle and the scholastic tradition in logic could have overlooked the Boolean operators and later truth-functions 'AND', 'OR' and (later) 'IF-THEN' as essential to the logical structure of propositions also found in scientific and everyday reasoning.

Thanks again to the surveyability of subsyllogistic relations represented in the canonical square of opposition, there is a relatively small number of cases to check individually. The task is easily accomplished in a relatively short amount of real time. When the necessarily individual checks are made, we find the results to be as we have reported them in summary here. What we know at the end of the inquiry is this:

- As far as the canonical square of opposition is concerned, the AEIO propositions are pre-given.
- There are only so many different ways in which four such tokens of distinct categories can be laid out on a plane.
- The most pleasing or satisfying arrangement that also lends itself to the most compact, convenient, and informationally complete presentation of subsyllogistic relations among the AEIO propositions is already found in the canonical Aristotelian square. (If we exhaustively consider the other arrangements, we see their disadvantages explicitly in comparison with the consequently preferred form of the canonical square.)
- The canonical square of opposition is correct not to include inversions of the AEIO propositional forms, which are either equivalent to one of the AEIO forms; or, while mutually logically contradictory, do not have any of the other logical relations to any of the AEIO propositions graphically depicted in the canonical square.

Acknowledgements I am grateful to an anonymous commentator for useful suggestions and criticisms in preparing the final version of this essay. Special thanks are due to my research assistant Sebastian Elliker at Universität Bern and Tina Marie Jacquette for digitalizing my hand-drawn diagrams.

References

1. Aristotle: The Complete Works of Aristotle. J. Barnes (ed.). Princeton University Press, Princeton (1984)
2. Bird, O.: Syllogistic and Its Extensions. Prentice-Hall, Englewood Cliffs (1964)
3. Boehner, P.: Medieval Logic. Manchester University Press, Manchester (1952)
4. Boethius, A.M.T.S.: Dialectica Aristotelis. Sebastian Gryphius, Lyons (1551)
5. Boethius, A.M.T.S.: Comentarii in librum Aristotelis (Peri Hermeneias [De Interpretatione]). Meiser, K.C. (ed.). B.G. Teubneri, Leipzig (1877)
6. Boole, G.: The Mathematical Analysis of Logic: Being an Essay Towards a Calculus of Deductive Reasoning. George Bell, London (1847)
7. Boole, G.: An Investigation of the Laws of Thought on Which Are Founded the Mathematical Theories of Logic and Probabilities. Walton and Maberley, London (1854)
8. Carroll, L.: Symbolic Logic [1896]. Bartley, W.W. III (ed.). Clarkson N. Potter, New York (1977)
9. Euler, L.: Elements of Algebra [1802]. Hewlett, J. (trans.). Springer, New York (1952)
10. Howard, D.T.: Analytical Syllogistics: A Pragmatic Interpretation of the Aristotelian Logic. Northwestern University Press, Evanston (1970)
11. Iwanus, B.: Remarks about syllogistic with negative terms. Stud. Log. **24**, 131–137 (1969)
12. Lukasiewicz, J.: Aristotle's Syllogistic from the Standpoint of Modern Formal Logic, 2nd edn., enlarged. Clarendon, Oxford (1955)
13. Jevons, W.S.: Elementary Lessons in Logic: Deductive and Inductive with Copious Questions and Examples and a Vocabulary of Logical Terms, 7th edn. Macmillan, London (1879)
14. Johnson, W.E.: Logic, 3 vols. Cambridge University Press, Cambridge (1921–1924)
15. Keynes, J.N.: Studies and Exercises in Formal Logic: Including a Generalisation of Logical Processes in Their Application to Complex Inferences. 4th edn., re-written and enlarged. Macmillan, London (1906)
16. Kneale, W., Kneale, M.: The Development of Logic. Oxford University Press, Oxford (1962)
17. Lejewski, C.: Aristotle's syllogistic and its extensions. Synthese **15**, 125–154 (1963)
18. Marenbon, J.: Boethius. Oxford University Press, New York (2003). For notes: 17–65
19. Martin, C.J.: The logical textbooks and their influence. In: Marenbon, J. (ed.) The Cambridge Companion to Boethius, pp. 56–84. Cambridge University Press, Cambridge (2009)
20. McLaughlin, R.N.: On the Logic of Ordinary Conditionals. SUNY, Albany (1990)
21. Peterson, P.L.: Syllogistic Logic and the Grammar of Some English Quantifiers. Indiana University Press, Bloomington (1988)
22. Prior, A.N.: The logic of negative terms in Boethius. Francisc. Stud. **13**, 1–6 (1953)
23. Prior, A.N.: Formal Logic, 2nd edn. Clarendon, Oxford (1962)
24. Smith, R.: Logic. In: Barnes, J. (ed.) The Cambridge Companion to Aristotle, pp. 27–65. Cambridge University Press, Cambridge (1995)
25. Thom, P.: The Logic of Essentialism: An Interpretation of Aristotle's Modal Syllogistic. Springer, Berlin (1996)
26. Venn, J.: Symbolic Logic [1894]. 2nd edn. Franklin Philosophy Monographs, New York (1971)
27. Woods, J., Irvine, A.: Aristotle's early logic. In: Gabbay, D., Woods, J. (eds.) Handbook of the History of Logic, vol. 1: Greek, Indian and Arabic Logic, pp. 27–100. Elsevier (North Holland), Amsterdam (2004)

D. Jacquette (✉)
University of Bern, Bern, Switzerland
e-mail: dale.jacquette@philo.unibe.ch

John Buridan's Theory of Consequence and His Octagons of Opposition

Stephen Read

Abstract One of the manuscripts of Buridan's *Summulae* contains three figures, each in the form of an octagon. At each node of each octagon there are nine propositions. Buridan uses the figures to illustrate his doctrine of the syllogism, revising Aristotle's theory of the modal syllogism and adding theories of syllogisms with propositions containing oblique terms (such as 'man's donkey') and with propositions of "non-normal construction" (where the predicate precedes the copula). O-propositions of non-normal construction (i.e., 'Some S (some) P is not') allow Buridan to extend and systematize the theory of the assertoric (i.e., non-modal) syllogism. Buridan points to a revealing analogy between the three octagons. To understand their importance we need to rehearse the medieval theories of signification, supposition, truth and consequence.

Keywords Octagons of opposition · Assertoric syllogism · Modal syllogism · Oblique syllogism · Signification · Buridan

Mathematics Subject Classification 01A35 · 03-03 · 03A05

John Buridan was born in the late 1290s near Béthune in Picardy. He studied in Paris and taught there as Master of Arts from the mid-1320s, and wrote his *Treatise on Consequences* sometime after 1335. He composed his *Summulae de Dialectica* (*Compendium of Dialectic*) from the late 1330s onwards, in eight (or nine) treatises with successive revisions into the 1350s. His *Sophismata*, the ninth treatise of the *Summulae* (sometimes presented as a separate work) has been much studied in the last fifty years. He wrote many other works, mostly commentaries on Aristotle and died around 1360.

Three intriguing figures occur in one of the manuscripts of Buridan's *Summulae*. They are Octagons. In addition, there is a particularly elaborate Square of Opposition. See Fig. 1.[1] In order to understand what doctrines Buridan is setting out in these figures, we need first to fill in some background in Aristotle's theory of the syllogism and in medieval theories of signification, supposition and consequence.

---

[1]The images come from Vatican ms. Pal.Lat. 994, ff. 6ra, $11[10]^v$, 7^r and 1^v, respectively. Reproduced by permission of the Biblioteca Apostolica Vaticana.

This work is supported by Research Grant AH/F018398/1 (Foundations of Logical Consequence) from the Arts and Humanities Research Council, UK.

Fig. 1 Buridan's square of opposition

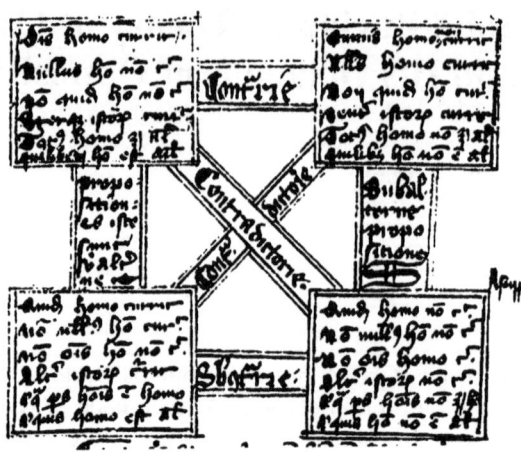

1 The Assertoric Syllogism

For Aristotle, according to Buridan, a syllogism consists of two premises, a major premise and a minor premise, made up of three terms. The major premise, containing the major term, is the first premise. The middle term appears in both premises. Aristotle's goal is to identify which pairs of premises, that is, figures, yield a conclusion:

> It seems to me that Aristotle takes a syllogism not to be composed of premises and conclusion, but composed only of premises from which a conclusion can be inferred. [9, Book III, ch. 4, p. 92]

Hence there are three figures:

- Figure 1 where the middle term is subject of one premise and predicate of the other;
- Figure 2 where the middle term is predicate of both premises; and
- Figure 3 where the middle term is subject in both premises.

Each proposition (or declarative sentence) is of one of four forms, *a-*, *e-*, *i-* and *o-*propositions, forming the traditional Square of Opposition. Validity of the Aristotelian syllogism is based on the *dictum de omni et nullo*:

> We say that one term is predicated of all of another when no examples of the subject can be found of which the other term cannot be asserted. In the same way, we say that one term is predicated of none of another. (*Prior Analytics* I 1, 24a28-30)

These establish the validity of the direct moods in the first figure: Barbara, Celarent, Darii and Ferio. The validity of all remaining syllogisms can be reduced to the direct moods in the first figure by conversion or reduction *per impossibile*:[2]

- Simple conversion: AiB implies BiA and AeB implies BeA

---

[2]The names of the moods in the medieval mnemonic record the reduction procedure (see, e.g., [8, p. 52]):

> Barbara Celarent Darii Ferio Baralipton
> Celantes Dabitis Fapesmo Frisesomorum;
> Cesare Camestres Festino Baroco; Darapti
> Felapton Disamis Datisi Bocardo Ferison.

- Conversion *per accidens*: AaB implies BiA[3]
- Reduction *per impossibile*: showing that the premises are incompatible with the contradictory of a putative conclusion.

2 Aristotle's Theory of the Modal Syllogism

Aristotle extended his theory of the syllogism to include the modal operators 'necessarily', 'contingently' and 'possibly'. He endorsed the K-principles:

> If, for example, one should indicate the premises by A and the conclusion by B, it not only follows that if A is necessary B is necessary, but also that if A is possible, B is possible. (*Prior Analytics* I 15, 34a22-24)

that is:

- from $\Box(A \to B)$ infer $\Box A \to \Box B$, and
- from $\Box(A \to B)$ infer $\Diamond A \to \Diamond B$

Aristotle also explicitly accepts the characteristic thesis of necessity: $\Box A \to A$ ("that which is of necessity is actual"—*De Interpretatione* 13, 23a21), and the thesis relating necessity and possibility: $\neg \Diamond A \leftrightarrow \Box \neg A$ ("when it is impossible that a thing should be, it is necessary ... that it should not be" and vice versa—22b5-6).

Mixed syllogisms allow premises with different modality. There are two Barbaras with one modal premise:

Necessarily all M are P	All M are P
All S are M	Necessarily all S are M
So necessarily all S are P	So necessarily all S are P

In *Prior Analytics* I 9, Aristotle says the first, Barbara LXL, is valid, but the second, Barbara XLL, is not. The natural interpretation (adopted by the medievals, and probably due to Peter Abelard) is that he reads 'Necessarily all M are P' *de re*, that is, as 'All M are necessarily P'. But that conflicts with I 3, where he says that 'Necessarily all M are P' converts accidentally to 'Necessarily some P are M', which seems only to be valid *de dicto*. So a major problem for the medievals was to try to produce a coherent account of the modal syllogism.

3 Signification

In *De Interpretatione* 1, Aristotle distinguished three kinds of terms, written, spoken and mental.

> Spoken [words] are signs (*notae*) of impressions (*passionum*) in the soul, and written of spoken. (16a3-4)

---

[3] We can also convert AeB *per accidens* to BoA, although Aristotle does not call this a conversion (since AeB converts to BeA). For BeA implies its subaltern BoA, for Aristotle writes at 26b15: "'P does not belong to some S' is ... true whether P applies to no S or does not belong to every S."

Boethius [4, p. 37] glossed this as: "thing, concept, sound and letter are four: the concept conceives the thing, spoken sounds are signs of the concept, and letters signify the sounds." But according to Augustine, words signify things, by means of the concepts. The medievals were left to ask: do words signify concepts or things? On the one hand, if words signified things, empty names would be impossible. But on the other, when I say Socrates runs, I mean that Socrates, not some concept of Socrates runs. The thirteenth century insight was that the concept is itself a sign. In the later words of Buridan:

> Categorematic words ... signify things by the mediation of their concepts, according to which concepts, or similar ones, they were imposed to signify. So we call the things conceived by those concepts 'ultimate *significata*' ... but the concepts we call 'immediate *significata*'. [6, IV 3, p. 253-4]

We can picture Buridan's theory of signification as follows:

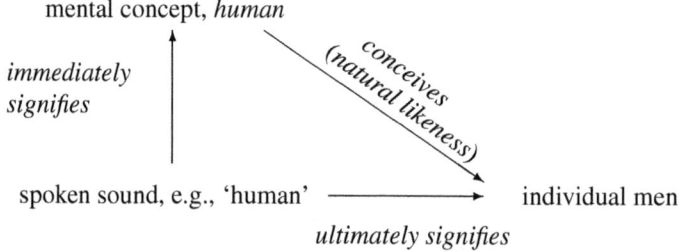

But what of the signification of propositions? Abelard had spoken of the *dictum* of a proposition, what it signified. We find this idea in the anonymous *Ars Burana*, dating from the 12th C.:

> We speak about the *dictum* of a proposition or of the significate of the proposition or of an *enuntiabile* ... For example: 'A human is an animal' is true because what it signifies is true, and that true thing is the *enuntiabile* ... Some enuntiabilia are of the present, some are of the past and some are of the future. [2, p. 208]

In the 14th C., Ockham tried to explain everything without recourse to such entities. What, he asked, is the object of knowledge? The question was the focus of the London debate in the 1320s between the Franciscans, William Ockham, Walter Chatton and Adam Wodeham. Ockham put forward his "Razor": "No plurality should be assumed unless it can be proved by reason, or experience, or by some infallible authority."[4] Accordingly, Ockham says that the object of knowledge is the mental proposition (*complexum*) that is, the *actus sciendi* is sufficient on its own, just as the *actus intellectus* is the only universal he recognizes. Chatton countered with his "Anti-Razor": if *n* things are not enough to explain something, posit $n + 1$. So Chatton claimed that the object of knowledge was the thing or things signified by the terms in the proposition.[5] Adam Wodeham's *via media* (1331) introduced the notion of the *complexe significabile*, things' being a certain way: the object of knowledge is the state of affairs (*modus rei*) signified by the proposition, the "complexly signifiable" (*complexe significabile*) that can only be signified in a complex way, that is, by a proposition.[6]

---

[4]See, e.g., [1, p. 1008].

[5]See, e.g., [12, p. 25].

[6]See [21].

The theory came to Paris through Gregory of Rimini (1344): we must recognize *enuntiabilia* or *complexe significabilia* as things: "every complex or incomplex signifiable, whether true or false, is said to be a thing (*res*) and to be something (*aliquid*)."[7] Buridan rejected all talk of *complexe significabilia*. It turns on taking truth to be a matter of signification, when it is really one of supposition:

> If we can explain everything by fewer, we should not ... posit many, because it is pointless to do with many what can be done with fewer. Now everything can be easily explained without positing such *complexe significabilia*, which are not substances, nor accidents, nor subsistent *per se*, nor inherent in anything else. Therefore, they should not be posited.[8]

4 Truth and Consequence

Buridan states his theory of truth in terms of supposition, not signification. First, we need to recall the main divisions of supposition:

- Material supposition: that of a term standing for itself or a similar expression, which it was not imposed to signify[9]
- Personal supposition: that of a term standing for the singulars of which it is truly predicated

In addition, there are various modes of personal supposition:

- Discrete supposition: that of a discrete, or singular term
- Common supposition, that of a common (or general) term, defined by the terminists using the notions of ascent and descent:
  - Determinate supposition: when one can descend to a disjunction of singulars, and ascend from any singular, with respect to that term; otherwise:
  - Confused and distributive supposition: when one can descend to an indefinite conjunction of singulars with respect to that term; otherwise:
  - Merely confused supposition (allowing descent to a disjunct term).

Thus, e.g., 'Some A is not B' is equivalent to 'This A is not B or that A is not B and so on for all the As' (i.e., '$\bigvee_i (A_i$ is not $B)$'), so 'A' has determinate supposition, and to 'Some A is not this B and some A is not that B and so on for all the Bs' (i.e., '$\bigwedge_j$ (Some A is not $B_j)$'), so 'B' has confused and distributive supposition—where $\bigvee \emptyset$ is false and $\bigwedge \emptyset$ is true.[10] Thus, e.g., 'Every A is B' is equivalent to 'This A is B and that A is B and so on for all the As' (i.e., '$\bigwedge_i (A_i$ is $B)$'), so 'A' has confused and distributive supposition, and to 'Every A is this B or that B and so on for all the Bs' (i.e., 'Every A is $\bigvee_j B_j$'), so 'B' has merely confused supposition).

Moreover, some words have the effect of widening or narrowing the supposition of other terms in a proposition. The past tense ampliates the subject to include past as well

---

[7] Cited in [21, p. 221].

[8] Buridan, *Commentary on Aristotle's Metaphysics*, cited in [21, p. 224].

[9] Simple supposition, of a term for a universal or concept, was subsumed by Buridan under material supposition.

[10] See [17].

as present supposita. For example, 'A white thing was black' means that something which is now white or was white in the past was black. The future tense ampliates the subject to include future as well as present supposita. Modal verbs ampliate the subject to possible supposita,[11] as do verbs such as 'understand', 'believe', and verbal nouns ending in '-ble': 'possible', 'audible', 'credible', 'capable of laughter'. Other expressions restrict the supposition of terms, e.g., qualifying 'horse' with the adjective 'white' restricts the supposition of 'horse' in 'A white horse is running' to white horses.

Buridan notes that some say a proposition is true if things are as it signifies they are (*qualitercumque ipsa significat ita est*).[12] He objects: if Colin's horse cantered well, 'Colin's horse cantered well' is true, but his horse is now dead, so nothing at all is signified. So perhaps a proposition is true if things are, were or will be as it signifies they are, were or will be. But, given his rejection of *complexe significabilia*, 'A human is a donkey' signifies humans and donkeys, which are indeed humans and donkeys, but is not true.[13] The moral is that to assign truth and falsity, we have to go beyond signification and consider supposition. A particular affirmative is true if subject and predicate supposit for the same; a particular negative is true if subject and predicate do not supposit for the same things; a universal affirmative is true if the predicate supposits for everything the subject supposits for; and so on.[14]

Turning to the theory of consequence, Buridan notes that some say a consequence is valid if the premise cannot be true without the conclusion's being true.[15] He objects: 'Every human runs, so some human runs' is valid, but it's possible for the premise to be true without the conclusion even existing, and so without it being true. We might therefore say that a consequence is valid if the premise cannot be true without the conclusion's being true, when both are formed together. But 'No proposition is negative, so no donkey is running' is not valid (for its contrapositive is not valid), but the premise cannot be true, and so cannot be true without the conclusion's being true. Finally, we could say that a consequence is valid if it is impossible that however the premise signifies to be it is not however the conclusion signifies, when they are formed together. Buridan accepts this, provided that 'however it signifies' is not taken literally, but in the sense previously given.

Buridan contrasts formal and material consequence, but gives them an unusually modern gloss:

> A consequence is called formal if it is valid in all terms retaining a similar form ... A material consequence, however, is one where not every proposition similar in form would be a good consequence, or, as it is commonly put, which does not hold in all terms retaining the same form; e.g., 'A human runs, so an animal runs', because it is not valid with these terms: 'A horse walks, so wood walks'. [9, I 4, p. 23]

Finally, he distinguishes absolute from *ut nunc* (or "as a matter of fact") consequence:

> Some material consequences are called simple consequences (*consequentiae simplices*) because they are simply good consequences, since it is not possible for the antecedent to be true the consequent being false. Others, which are not simply good, are called as-of-now consequences (*consequentiae ut nunc*) because it is possible for the antecedent to be true without the consequent, but

---

[11] As Aristotle noted: *Prior Analytics* I 13, 32b25-28.

[12] [9, I 1, p. 17].

[13] [6, *Sophismata* 2, sophism 5, pp. 847–848].

[14] [6, *Sophismata* 2, conclusion 14, p. 858].

[15] [9, I 3, p. 21].

are good as-of-now, because, things being as a matter of fact as they are, it is impossible for the antecedent to be true without the consequent. [9, I 4, p. 23]

So much for background. We can now consider what doctrines Buridan is illustrating in his figures.

5 Non-normal Form

Besides the further complication of the octagons, even the simple Square of Opposition in Pal.lat. 994 is more complex than usual, in containing several propositions at each node. (See Fig. 1.) The set of propositions at each node are equivalents, or at least of the same sort. For example, at the top left node we read:

- 'Every human is running' (*Omnis homo currit*)
- 'No human is not running' (*Nullus homo non currit*)
- 'Not any human is not running' (*Non quidam homo non currit*)
- 'Each of these is running' (*Uterque istorum currit*)
- 'The whole [of] human is an animal' (*Totus homo est animal*)
- 'Any human is an animal' (*Quilibet homo est animal*)

In the octagons, each of the eight nodes contains nine equivalent propositions (or again, of the same sort). For example, in the octagon of non-normal propositions (see Fig. 2), we read:

- 'Every *B* every *A* is' (*Omne B omne A est*)
- 'Every *B* no *A* is not' (*Omne B nullum A non est*)
- 'No *B* not every *A* is' (*Nullum B non omne A est*)
- ...

This "non-normal construction" (or "unusual way of speaking"—*de modo loquendi inconsueto*) consists in placing the predicate before the copula. Placing the predicate before the copula often has the effect of changing the sense: e.g., 'Some animal a human is not' means, not that some animal is not a human, but that some animal is not some human:

> I call it normal form where an external negation precedes the predicate, e.g., 'No *B* is *A*', 'Some *B* is not *A*'. But I call the form non-normal where the predicate precedes the negation and so is not distributed by it, e.g., 'Every *B* [some] *A* is not', 'Some *B* [some] *A* is not'. [9, I 8, p. 44]

This is an example of the increasing use in late medieval logic of regimented vernacular forms. What Buridan here calls the non-normal form is in fact the normal form used by, e.g., Apuleius and Boethius.[16] What Buridan does is to interpret the so-called non-normal form non-normally, so that, for example, *A* in 'Every *B* [some] *A* is not' and 'Some *B* [some] *A* is not' is not distributed by the negation, and so does not have confused and distributive supposition. One can see it as a further regimentation similar to that made by Aristotle in *Prior Analytics* I 46 when he distinguished 'me einai leukon' ('is not white') from 'einai me leukon' ('is not-white').

Using non-normal propositions, Buridan is able to extend accidental conversion (i.e., conversion *per accidens*) from universal propositions to particular propositions. Then

---

[16]See [16] and [19].

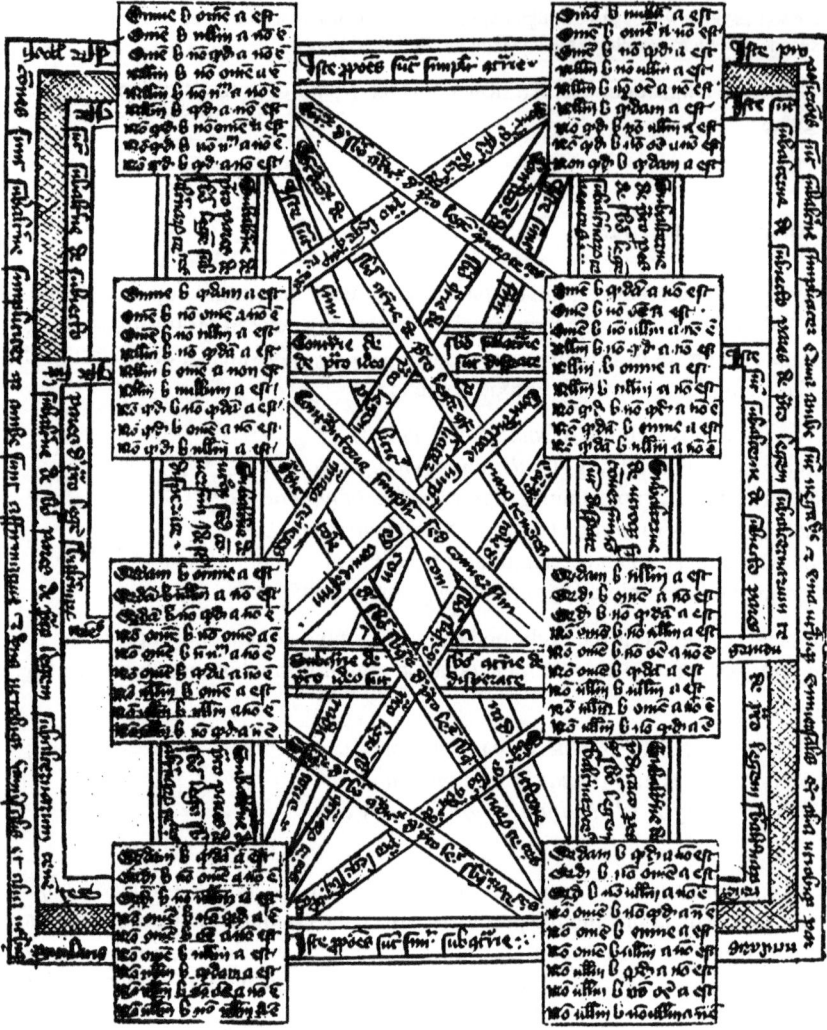

Fig. 2 Buridan's octagon of opposition for non-normal propositions

each categorical (better, "subject-predicate" or "predicative"—*propositio categorica*) assertoric proposition can be converted, either simply or accidentally:

- 'Every *A* is *B*' converts accidentally to 'Some *B* is *A*'
- 'No *A* is *B*' converts simply to 'No *B* is *A*' and accidentally to 'Some *B* is not *A*'
- 'Some *A* is *B*' converts simply to 'Some *B* is *A*'
- 'Some *A* is not *B*' converts accidentally to 'Every *B* [some] *A* is not' (e.g., 'Some animal is not a human, so every human [some] animal is not'—*quoddam animal non est homo, igitur omnis homo animal non est*)

  Accidental conversion involves changing the predicate into the subject and the subject into the predicate while preserving the quality but changing the quantity. [6, I 6, p. 51]

Buridan deploys these ideas in his theory of the syllogism. In the fourteenth century, the syllogism was brought under a general theory of consequence. Buridan treats the syllogism as a special case of formal consequence. He demonstrates all of these by *ecthesis* (or "setting-out"), a hypothetical method which Aristotle mentions for a couple of third-figure syllogisms.[17] But he introduces a novel approach to demonstrating the validity of assertoric syllogisms, noting that the following are severally necessary and jointly sufficient tests for syllogistic validity:

- No valid syllogism has two negative premises
- In every valid syllogism, the middle term is distributed (that is, has confused and distributive supposition) at least once
- A proposition with a distributed term entails the same proposition with the term undistributed, but never vice versa

These tests for validity were preserved in the traditional rules of the syllogism.[18] Then Buridan can go systematically through all combinations of premises and conclusion and identify the valid syllogisms.

These include syllogisms with "non-normal" propositions. Using such propositions, Buridan extends the assertoric syllogism to additional cases, so that there are eight valid moods in each of the first and second figures, and nine in the third.[19] Remember that here by 'mood', Buridan means 'combination of premises'. To the first figure (with the four perfect moods, Barbara, Celarent, Darii and Ferio), he adds:

- MiP, SoM: SPo (Some M is P, some S is not M, so some S some P is not)
- MaP, SoM: SPo (Every M is P, some S is not M, so some S some P is not)

and he counts Fapesmo (MaP, SeM: PoS) and Frisesomorum (MiP, SeM: PoS) as making up the eight (direct and indirect) valid moods in the first figure. Thus, e.g., Baralipton (in the mnemonic) and Barbari (not in the mnemonic) do not count for Buridan as additional to Barbara. So the order of the premises counts for Buridan, but whether the conclusion is direct or indirect does not: the crucial question is whether the premises yield a syllogistic conclusion (any conclusion) or not.

To the four traditional second-figure moods (Cesare, Camestres, Festino and Baroco), Buridan adds four more:

> If the premises of Festino and Baroco are transposed, there will be two more moods concluding only indirectly, which can be called 'Tifesno' and 'Robaco', which are proved by reduction to Festino and Baroco simply by transposing the premises. The other two moods can only be concluded according to the non-normal construction, namely, if both premises are particular, one affirmative and the other negative. [9, III i 4, p. 93]

To the six traditional third-figure moods (Darapti, Felapton, Disamis, Datisi, Bocardo and Ferison), he adds three more:

---

[17]Lagerlund [15, §7] and King [14, p. 71] mistakenly identify these principles as Aristotle's *dictum de omni et nullo*.

[18]See, e.g., [13, III 4, §114]. The necessity of the first was known even to Boethius [5, p. 316]. But Buridan seems to have been the first to recognize that the three conditions were jointly sufficient.

[19]Aristotle identified four direct and two indirect moods in the first figure, four in the second figure and six in the third figure.

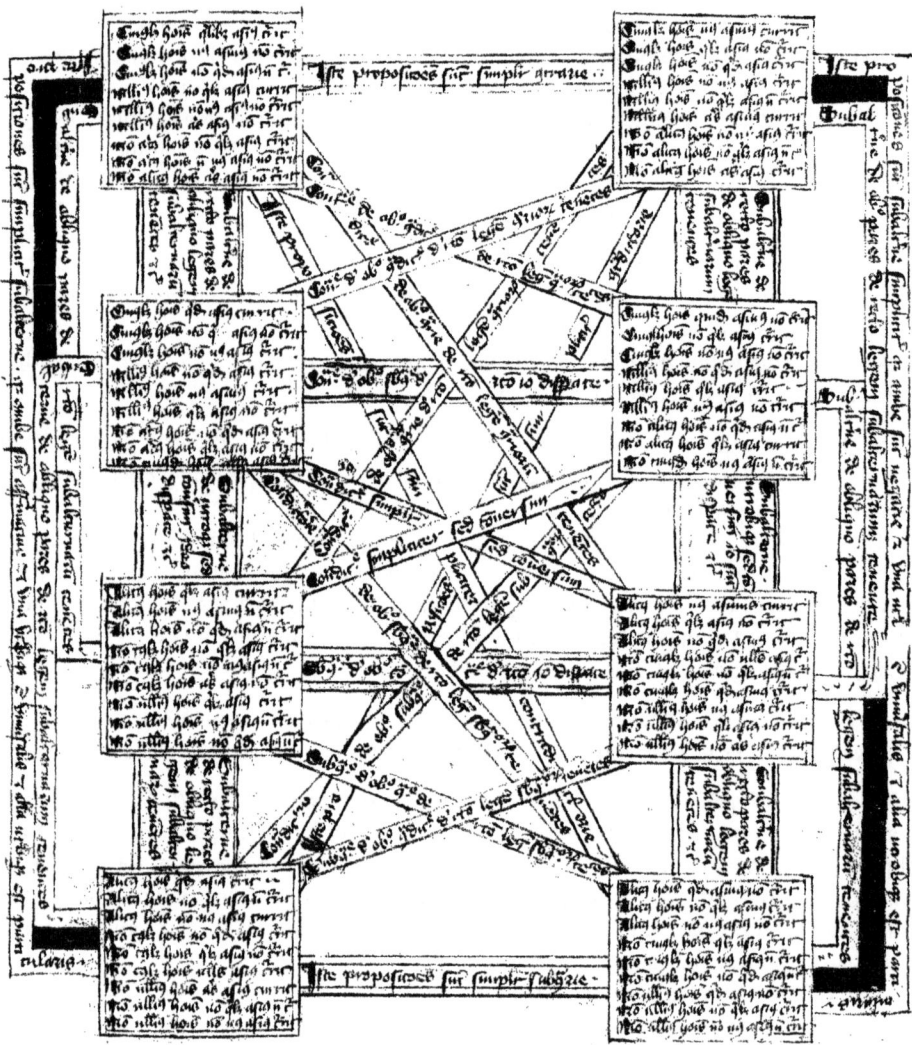

Fig. 3 Buridan's octagon of opposition for propositions containing oblique terms

Darapti, Disamis and Datisi are valid for concluding both directly and indirectly; but Felapton, Bocardo and Ferison are only valid for concluding directly. But conversion of these three, namely, Lapfeton, Carbodo and Rifeson, are valid for concluding only indirectly, and are reduced to direct ones by transposition of the premises. [9, III i 4, p. 93]

We can explain the validity of the consequences here by using Buridan's account of supposition and of distribution. For example, take the conversion of the O-form 'Some A is not B': 'A' has determinate supposition, and 'B' has confused and distributive supposition, since we can descend both to

'This A is not B or that A is not B and so on for all As'

(and ascend from any disjunct) and to

'Some A is not this B and some A is not that B and so on for all Bs'

Buridan claims the O-proposition converts accidentally to 'Every B [some] A is not': take the above disjunction, and descend on each disjunct 'This A is not B'. 'B' is distributed (by 'not'), so we end up with a disjunction of conjuncts of the form 'This A is not this B'. This disjunction entails the corresponding conjunction of negated disjuncts, that is,

$$\bigvee_{j} \bigwedge_{i} A_i \neq B_j \quad \text{(i.e., Some } A \text{ is not } B)$$

converts (partially, so it entails but is not entailed by) to

$$\bigwedge_{j} \bigvee_{i} B_j \neq A_i \quad \text{(i.e., Every } B \text{ [some] } A \text{ is not)}$$

The "non-normal" conclusions in Buridan's extra syllogisms are all of the form 'SPo', that is, 'Some S [some] P is not'. Here 'P' has determinate supposition (whereas in 'SoP', i.e., 'Some S is not P', 'P' has confused and distributive supposition on account of the preceding negation). Take this non-normal syllogism in the second figure:

PoM	Some P is not M
SiM	Some S is M
SPo	Some S [some] P is not

Only one premise is negative, and so is the conclusion (Buridan's condition 2). 'M' is distributed at least once—namely, in the first premise (Buridan's condition 6). Neither term is distributed in the conclusion—they both have determinate supposition (Buridan's condition 8). So by Buridan's conclusions 2, 6 and 8, the syllogism is valid.

6 Oblique Terms

Another example of syllogisms containing propositions in non-standard form takes propositions with oblique terms (see Fig. 3), e.g.:

> No human's donkey (i.e., no donkey of a human) is running
> A human is an animal
> So some animal's donkey is not running.[20]

An oblique term is any term in a proposition not in the nominative. The cases Buridan mostly considers are terms in the accusative and in the genitive, e.g.,

'Some human [it's] every horse [he or she] is seeing' (*Homo omnem equum est videns*—'*equum*' is in the accusative case)

'Every man's donkey is running' ('*Omnis asinus hominis currit*—'*hominis*' is in the genitive)

---

[20] Cf. *Prior Analytics* I 36.

Buridan notes that sometimes the whole subject is distributed, as in the second example here; but sometimes only the term in the nominative, or in the genitive is, as in 'Of every man a donkey is running' (*Cuiuslibet hominis asinus currit*)—'man' has confused and distributive supposition but 'donkey' has determinate supposition. The interesting syllogisms arise in this latter case, e.g., when the middle term is only part of the subject.

At each node of the octagon of propositions with oblique terms we find nine equivalent propositions (72 in all). E.g., at the top rightmost node we have:

- 'Of every human no donkey is running' (*Cuiuslibet hominis nullus asinus currit*)
- 'Of every human every donkey is not running' (*Cuiuslibet hominis quilibet asinus non currit*)
- 'Of every human not any donkey is running' (*Cuiuslibet hominis non quidam asinus currit*)
- 'Of no human [is it] not [that] no donkey is running' (*Nullius hominis non nullus asinus currit*)
- 'Of no human not every donkey is not running' (*Nullius hominis non quilibet asinus non currit*)
- 'Of no human is some donkey running' (*Nullius hominis aliquius asinus currit*)
- 'Not of any human is [it] not [that] no donkey is running' (*Non alicuius hominis non nullus asinus currit*)
- 'Not of any human is not every donkey not running' (*Non alicuius hominis non quilibet asinus non currit*)
- 'Not of any human is some donkey running' (*Non alicuius hominis aliquis asinus currit*)

There are 28 possible relations between the nodes $(= \frac{8!}{(8-2)!2!})$. Buridan's octagon with oblique terms draws the 24 logical relations which hold (there are also 4 disparate pairs):

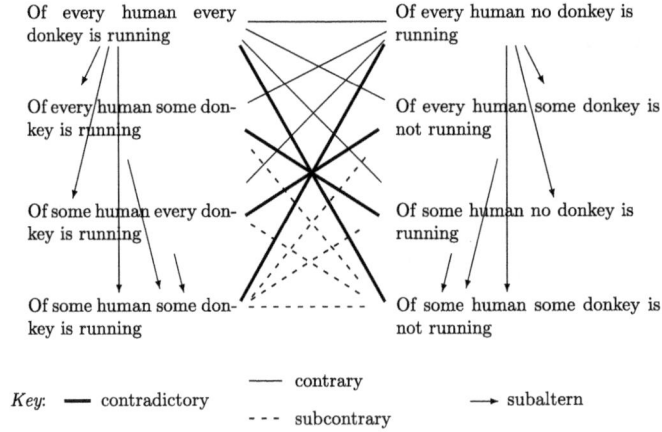

Buridan proceeds to discuss syllogisms with oblique terms:

Consider 'Every human's donkey is running' (i.e., 'Of every human some donkey is running'—*Cuiuslibet hominis asinus currit*): 'human' is distributed and 'donkey' is not distributed, nor [is] the whole 'human's donkey' (*hominis asinus*). It would be the same if I said 'A donkey of every human is running'. For here 'donkey' supposits determinately, because no cause of confused [supposition] precedes it. Nonetheless, 'donkey' does not supposit alone but restricted by the distributed oblique [term] following it; so the said proposition would not be true if no donkeys were running or no donkey running were a donkey of any human. [9, III ii 1, p. 99]

Just as the whole subject (e.g., 'human's donkey') may not be distributed, or have a single mode of supposition, so the subject may not be either extreme nor the middle term in a syllogism. In his *Treatise on Consequences*, Buridan's approach, like other medieval logicians, is to rephrase the proposition so that the middle term, or the extreme, appears as the subject in the nominative. E.g., 'Of some human a donkey is running' is equivalent to 'Some human is one of whom a donkey is running'.

But in the *Summulae*, Buridan seems happy to proceed without need for paraphrase. Consider:

> Of no human is a horse running (*Nullius hominis equus currit*)
> Of every human an animal is running (*Cuiuslibet hominis animal currit*)
> So of no human is [some] animal a human's horse (*Ergo nullius hominis animal est hominis equus*)

This is invalid, he claims, since 'animal' is distributed in the conclusion but was not distributed in the premises. As we saw in the octagon, 'Of no human is [some] animal ...' is equivalent to 'Of every human no animal ...',[21] that is,

$$\bigwedge_i \neg \bigvee_j \phi(i, j) \equiv \bigwedge_i \bigwedge_j \neg\phi(i, j)$$

What does follow, Buridan observes, is

> So of every human some animal is not a human's horse (*Ergo cuiuslibet hominis animal non est hominis equus*)

He writes:

> It is clear that in this syllogism, the middle term is 'running', the major extreme is 'human's horse', and the minor extreme is 'animal of a human', whatever you take to be the subject or the predicate in these propositions. [6, V 8, pp. 368–369]

Recall De Morgan famous criticism of traditional logic as incapable of explaining the validity of the following inference:[22]

> Every horse is an animal
> So every horse's head is an animal's head.

Buridan's theory can show that this is valid. It's an enthymeme, not formally valid, but valid *simpliciter*, that is, absolutely valid. The suppressed premise is the necessary truth: 'Every horse's head is a horse's head'. Then the argument has this form:

Every M is a P	Every horse is an animal
Every horse's head is an M's head	Every horse's head is a horse's head
So every horse's head is a P's head	So every horse's head is an animal's head.

'M' is distributed in the first premise, but 'P' is not distributed in the conclusion, nor is 'M' in the second premise.

---

[21] The octagon has 'Of no human is [some] donkey ...' and 'Of every human no donkey ...'

[22] [7, pp. 29 and 216]. Cf. [20]. Bochenski [3, p. 95] claims that Aristotle himself could have shown the inference valid.

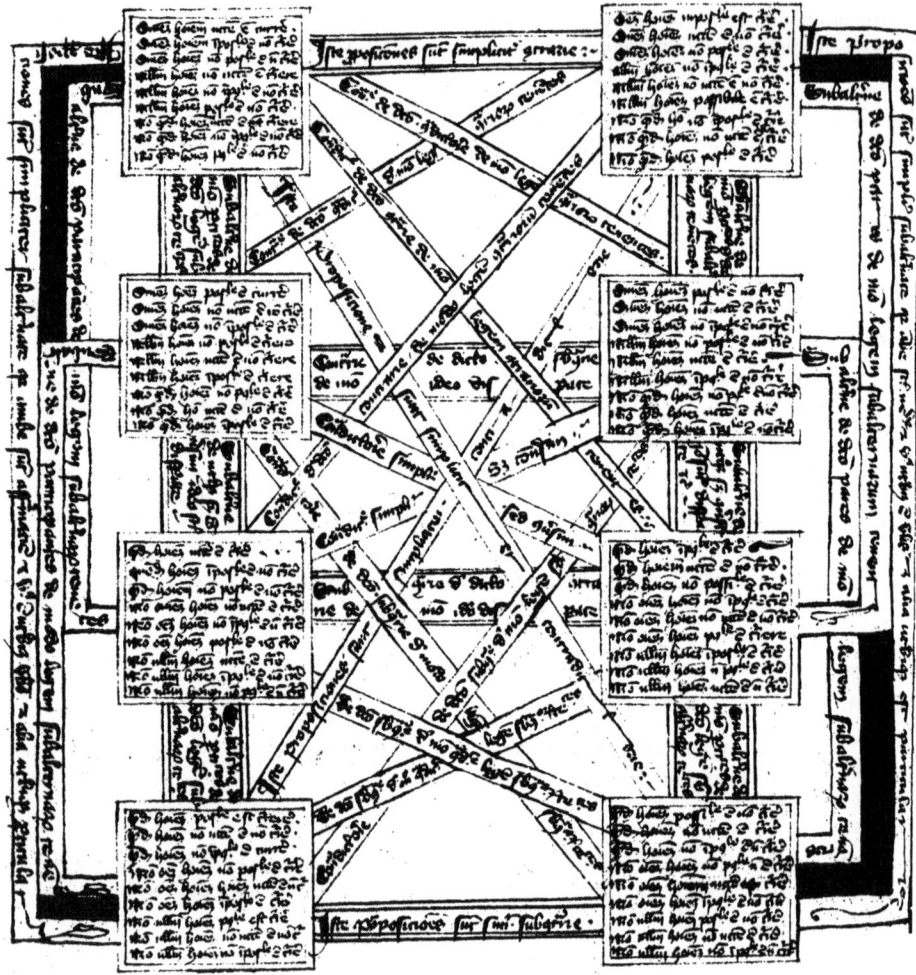

Fig. 4 Buridan's modal octagon

7 Modal Propositions

Buridan shows how the logical relations hold analogously between propositions with oblique terms, non-normal propositions, and modal propositions (see Fig. 4), e.g.,

Every human necessarily runs   is contrary to   Some human necessarily does not run

just as

Of every human every donkey runs   is contrary to   Of every human some donkey
                                                                                        does not run

He writes:

We match the oblique term before the nominative in the propositions with oblique terms, and the subject term of the non-normal propositions, with the subject term of modal propositions, and likewise, we match the nominative term of the propositions with oblique terms, and the predicate of the non-normal propositions, with the mode of modal propositions. In the three octagons, the careful observer will clearly see how the propositions with oblique terms distributable independently of their nominatives, or the non-normal propositions sharing both terms in the same order, are related to one another with respect to one or other of the laws of opposition. [6, I 5, p. 43]

Klima presents Buridan's octagons in a composite table:[23]

Modal	Oblique	Non-normal
1. Every human necessarily runs	Of every human every donkey runs	Every human every runner is
2. Every human possibly runs	Of every human some donkey runs	Every human some runner is
3. Every human necessarily does not run	Of every human every donkey does not run	Every human every runner is not
4. Every human possibly does not run	Of every human some donkey does not run	Every human some runner is not
5. Some human necessarily runs	Of some human every donkey runs	Some human every runner is
6. Some human possibly runs	Of some human some donkey runs	Some human some runner is
7. Some human necessarily does not run	Of some human every donkey does not run	Some human every runner is not
8. Some human possibly does not run	Of some human some donkey does not run	Some human some runner is not

According to Buridan's analysis of modal propositions, all modal propositions are either taken in the composite sense (that is, *de dicto*), or in the divided sense (that is, *de re*). Nothing follows from composite modal premises where the mode is 'possible' or 'contingent'. From composite modal premises of necessity, the conclusion follows in the same mode. Buridan claims that all divided modal propositions ampliate their subject. Thus, e.g., 'Every *A* is necessarily *B*' is true iff everything which is or might be *A* must be *B*. Similarly, 'Some *A* is possibly *B*' is true iff something which is or might be *A* might be *B*. We can override this ampliation by replacing the subject by a 'what is'-phrase, e.g., 'Everything which is *A* is necessarily *B*'.

Paul Thom [18, p. 17] uses the following key to represent the ampliation of terms and the forms of assertoric and modal categorical propositions. It makes the semantics explicit in the form: '$\rightarrow$' represents inclusion and '$|$' exclusion.

'Every *A* is *B*':	$\underline{a} \rightarrow b$	$\underline{a}$:	indicates existential import
'No *A* is *B*':	$a \mid b$	$a^\dagger$:	what is *A* or possibly *A*
'Some *A* is *B*':	$a \smile b$ (a overlaps b, i.e., $\exists d$,	a^*:	what is necessarily *A*
	$a \leftarrow d \rightarrow b$)		
'Not every *A* is *B*':	$a \nrightarrow b$		

---

[23] I have adapted this table from Klima's by interchanging nodes 2 and 5, and 4 and 7, in line with the mss. The 28 relations between the eight equivalence classes of modal propositions, analogous to those described in Sect. 6 between oblique propositions, are discussed clearly and at length in [11].

Using Thom's notation, we can set out Buridan's modal octagon as follows:[24]

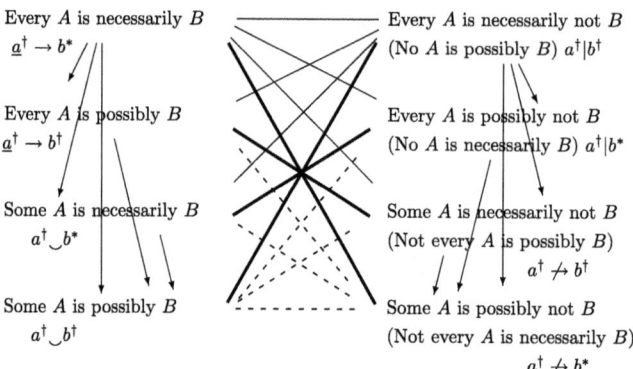

In a series of twenty-four theorems, Buridan takes us through all the various combinations of modal and assertoric propositions. E.g., Theorem 15 says:

> In the first figure, from an assertoric major and a minor of necessity there does not follow a conclusion of necessity, or even assertoric except in Celarent. [9, IV 2, p. 123]

In particular, Barbara XLL is invalid. Suppose God is, as it happens, creating:

> Every God is creating
> Any first cause is of necessity God
> But not every first cause is of necessity creating.

Celarent XLL is also invalid. But Celarent XLX is valid: $m|p$, $s^\dagger \to m^*$, so $s|p$. For $s \to s^\dagger \to m^* \to m|p$, so $s|p$.

In fact, Buridan also rejects Barbara LXL as invalid, by Theorem 16:

> In the first figure, from a major of necessity and an assertoric minor there is always a valid syllogism to a particular conclusion of necessity, but not to a universal conclusion of necessity. [9, IV 2, p. 124]

The reason is Buridan's interpretation of necessity propositions as ampliating their subject. From $m^\dagger \to p^*$ and $s \to m$, $s^\dagger \to p^*$ does not follow. For something could be s which could not be m. For example, suppose only donkeys are white. All donkeys necessarily bray (if it could be a donkey, it is a donkey). But although on this assumption, all actual white things are donkeys, not everything which could be white (say, a horse) could be a donkey, so the conclusion ('All white things necessarily bray') is false.

8 Conclusion

Medieval logicians extended Aristotle's theory of the syllogism to a general theory of consequence. Within consequence, they distinguished formal from material consequence,

---

[24]Buridan's modal octagon was described in detail in [10].

and absolute from *ut nunc* consequence. They developed a theory of properties of terms, in particular, of signification, supposition and ampliation, which played a central role in their theory of consequence. Much of their work on the modal syllogism was an attempt to correct and clarify Aristotle's theory. John Buridan's theory of consequence and of the syllogism was the most sophisticated and extensive in its theoretical organization. By the use of non-normal propositions, he was able to extend accidental conversion to allow O-propositions to convert, and to extend and systematize the assertoric syllogism. By giving a systematic account of propositions with oblique terms, he was able to show the validity of further consequences, such as De Morgan's notorious alleged counterexample. Finally, by making explicit Aristotle's brief remarks about the ampliation of terms in tensed and modal propositions, Buridan was able to give a more systematic account of the modal syllogism, and to exhibit a strong analogy between modal, oblique and non-normal propositions in his three octagons.

References

1. Adams, M.M.: William Ockham, vol. 2. University of Notre Dame Press, Notre Dame (1987)
2. Anonymous: Ars Burana. In: De Rijk, L.M. (ed.) Logica Modernorum, vol. II 2, pp. 175–213. Van Gorcum, Assen (1967)
3. Bochenski, I.M.: A History of Formal Logic. Thomas, I. (trans., ed.). Chelsea, New York (1962)
4. Boethius, A.M.S.: Commentarii in librum Aristotelis Peri Hermeneias, editio prima. Meiser, C. (ed.). Teubner, Leipzig (1877)
5. Boethius, A.M.S.: Commentarii in librum Aristotelis Peri Hermeneias, editio secunda. Meiser, C. (ed.). Teubner, Leipzig (1880)
6. Buridan, J.: Summulae de Dialectica. Klima, G. (Eng. trans.). Yale University Press, New Haven (2001)
7. De Morgan, A.: On the Syllogism and Other Logical Writings. Heath, P. (ed.). Routledge & Kegan Paul, London (1966)
8. De Rijk, L.M.: Peter of Spain: Tractatus. Van Gorcum, Assen (1972)
9. Hubien, H.: Iohanni Buridani: Tractatus de Consequentiis. Publications Universitaires, Louvain (1976)
10. Hughes, G.: The modal logic of John Buridan. In: Atti del Congresso Internazionale di Storia della Logica: La teorie delle modalità, pp. 93–111. CLUEB, Bologna (1989)
11. Karger, E.: John Buridan's theory of the logical relations between general modal formulae. In: Braakhuis, H.A.G., Kneepkens, C.H. (eds.) Aristotle's Peri Hermeneias in the Latin Middle Ages, pp. 429–444. Ingenium, Groningen (2003)
12. Keele, R.: Applied logic and mediaeval reasoning: iteration and infinite regress in Walter Chatton. Proc. Soc. Mediev. Log. Metaphys. **6**, 23–37 (2006)
13. Keynes, N.J.: Studies and Exercises in Formal Logic. Macmillan, London (1884)
14. King, P.: Jean Buridan's Logic: The Treatise on Supposition and the Treatise on Consequences. King, P. (trans., with a Philosophical Introduction). Reidel, Dordrecht (1985)
15. Lagerlund, H.: Medieval theories of the syllogism. In: Zalta, E.N. (ed.) The Stanford Encyclopedia of Philosophy (2010)
16. Londey, D., Johanson, C.: The Logic of Apuleius. Brill, Leiden (1987)
17. Priest, G., Read, S.: The formalization of Ockham's theory of supposition. Mind **86**, 109–113 (1977)
18. Thom, P.: Medieval Modal Systems. Ashgate, Farnham (2003)
19. Thomsen Thörnqvist, C.: Anicii Manlii Severini Boethii De Syllogismo Categorico. Acta Universitatis Gothoburgensis, Gothenburg (2008)
20. Wengert, R.G.: Schematizing De Morgan's argument. Notre Dame J. Form. Log. **15**, 165–166 (1974)
21. Zupko, J.: How it played in the rue Fouarre: the reception of Adam Wodeham's theory of the complexe significabile in the Arts Faculty at Paris in the mid-fourteenth century. Francisc. Stud. **54**, 211–225 (1997)

S. Read (✉)
Department of Philosophy, University of St Andrews, St Andrews KY16 9AL, Scotland, UK
e-mail: slr@st-and.ac.uk

Why the Fregean "Square of Opposition" Matters for Epistemology

Raffaela Giovagnoli

Abstract A relevant issue for epistemology is the distinction between expression of a "judgeable content" and "judgment". We would suggest that Frege's enterprise in the *Begriffsschrift* seems to be devoted to describe this difference from a formal point of view. For, he considers an expression as a simple or complex sign characterized by the function/argument structure. In our contribution, we try to elucidate Frege's effort by analyzing his variation of the Aristotelian "square of opposition" and by following his logical steps from an "epistemological" point of view which overcome the classical Aristotelian syllogism but plausibly respects the fundamental distinction between "judgeable content" and "judgment".

Keywords Square of opposition · Syllogism · Judgment · Judgeable content

Mathematics Subject Classification 03Axx

1 The Variation of the Aristotelian Square of Opposition

Frege formulated the modern predicate logic that incorporates the traditional theory of the syllogism but enriches it to give a useful account of sentences of multiple generality [1]. Traditionally, we say that every judgment is basically composed of a subject and a predicate such as 'Rossella is a dog'. This very structure was supposed to have fundamental implications for metaphysics and epistemology because it would be applicable also to general propositions such as 'Every man is mortal' and 'Some men are Greek' with 'every man' and 'some men' being considered to be subjects and 'mortal' and 'Greek' predicates of two sentences. As it is well known, Frege replaced the syllogism with a formal structure divided into 'function' and 'argument': "If in an expression (...) a simple or complex sign occurs in one or more places, and we consider it replaceable in one or more of these places by something, but everywhere by the same thing, then we call the part of the expression which hereby appears as unchangeable the function and the replaceable part its argument" [1, p. 16].

The distinction of function (the fixed part of an expression) and argument (the variable part of an expression) takes the important epistemological task to indicate when the argument is 'determinate' or 'indeterminate'. The fundamental 'indeterminateness' of the argument is evident in the case of a complex expression such as "being representable as a sum of four square" which is always s correct for any arbitrary positive whole number we put as argument.

J.-Y. Béziau, D. Jacquette (eds.), *Around and Beyond the Square of Opposition*, 111–116
Studies in Universal Logic, DOI 10.1007/978-3-0348-0379-3_7, © Springer Basel 2012

Let's show the Fregean notation and the new square of opposition he introduced [1, §12]

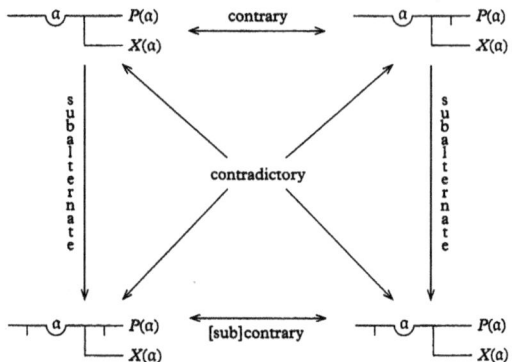

The concavity is the place for the gothic letter, which indicates the indeterminateness of the argument. The distinction of function and argument is relevant for general sentences as, for instance, to say that 'Everything is ϕ' is to say that 'the function is a fact whatever one might take as its argument'. Consequently, a general statement is a statement about a function. The change of terminology implies a change in the interpretation of the Aristotelian syllogism. The statement 'All men are mortal' is not a statement about all men but one about the function 'If a is a man, then a is mortal' and it specifies of it that it is a fact for anything we substitute for 'a'. In our previous example 'Everything is ϕ' could be 'Everything is mortal' and to say that the sentence 'a is mortal' turns into a true sentence whatever we substitute for a.

There are also important theoretical implications, which derive from Frege's notation about existential statements. Following the Kantian thought, Frege maintains that 'existence' is not a predicate or property of a thing. 'Being' is merely a 'positing' of a thing or of certain determinations as existing in themselves. Existence is for Frege a second-level concept namely a concept having instances or a concept that can be predicated truly of an object. It distinguishes itself from the concept of 'actuality' that identifies existence in a spatio-temporal field and enters into causal chains. In this sense, logic has not to do with the notion of actuality but with the ones of existence and objectivity.

2 The Priority of Judgment

From an epistemological point of view, we can say that Frege's original notation provides a view of the relation of the judgment with its parts completely different from traditional logic that rest on an 'aggregative' strategy. In this sense, judgment is formed by aggregation out of previously given constituent concepts. Frege inherits the Kantian conception according to which there could not be any combination of ideas unless there were already an original unity that made possible such combination. According to Sluga [2, p. 91], Kant anticipates the Fregean doctrine of concepts: "Concepts, as predicates of possible judgments, relate to some representation of a not yet determined object. Thus the concept of body means something, for instance, metal, which can be known by means of the concept. It is therefore a concept solely in virtue of its comprehending other representations, by means of which it can relate to objects. It is therefore the predicate of a possible judgment, for instance, 'every metal is a body' " [3, p. 69].

What remains unclear in syllogistic is the crucial problem of 'concept formation' as it starts from the assumption that concepts are formed by abstraction from individual things and that judgments express comparison of concepts like inferences. According to Frege, Boole's logic presents the same problem. In the letter to Marty [4] Frege maintains that for Boole only concepts exist, but the subordination of a concept under another concept is quite different from that of an individual's falling under a concept. Frege's theory introduces sophisticated logical relations among concepts that rests on the fundamental notion of 'unsaturedness'. Here, we can represent the falling of an individual under a concept by F(x), where x is the subject (argument) and F( ) is the predicate (function), and where the empty place in the parentheses after F indicates non-saturation.

This logical option aims at criticizing the Boolean distinction between 'primary propositions' and 'secondary propositions'. The former express the sorts of relations between concepts studied in Aristotelian logic (calculus of concepts) and the latter concern the sorts of relations between judgments studied in sentential logic (calculus of judgments). However, the two calculi are syntactically identical because they derive from the imposition of an interpretation onto what is originally an uninterpreted formalism, namely a pure syntactic abstract algebra [5]. For example, both contain expressions of the form 'A × B' or 'A + B'. In the calculus of concepts we must consider the letters as denoting classes or extensions of concepts and the operations as set-theoretically interpreted. In our case, multiplication is intersection and addition is union. The original idea of Boole is to reduce the calculus of judgments to that of concepts that is secondary propositions to primary propositions. In Boole words: "Let us take, as an instance for examination, the conditional proposition "If the proposition X is true, the proposition Y is true". An undoubted meaning of this proposition is, that the *time* in which the proposition X is true, is *time* in which the proposition Y is true [6, ch. XI, §5].

If we interpret this idea in an Aristotelian sense, we obtain a universal affirmative judgment namely 'All times at which X is true are times at which Y is true'.

The strategy of Boole was to treat sentential operators as expressing set-theoretic operations on sets of times. Frege rejects the Boolean reduction of the secondary propositions to the primary ones because there are 'organic' relationships between them that must be make logically explicit [5]. For instance, Boole's strategy cannot account for the relationship between universal affirmative propositions and hypothetical judgments. In Frege's notation these are represented as

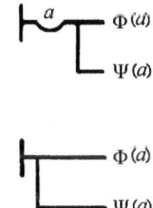

and

In Boolean logic we can represent both as, e.g., F × G = F even though the letters that occur in the two cases have nothing to do with one another.

Another problem in Boole's logic is the reduction of atomic sentences to universal affirmative propositions, namely to one sort of primary propositions: "To say, 'The sun shines', is to say, 'The sun is that which shines', and it expresses a relation between two classes of things, viz. 'the sun' and 'things which shine' " [6, ch. IV, §1].

Frege thinks that this account of atomic propositions is misconceived: "The relation of subordination of a concept under a concept is quite different from that of an individual's falling under a concept. It seems that logicians have clung too long to the linguistic schema of subject and predicate, which surely contains what are logically quite different relations. I regard it as essential for a concept that the question whether something falls under it has a sense" [7, pp. 100–101].

Frege's account of general propositions in the *Begriffsschrift* solves the Kantian problem of the unity of judgment in a logical form [2, ch. 3]. For instance, in the two sentences:

(a) The number 20 can be represented as the sum of four squares

and

(b) Every positive integer can be represented as the sum of four squares

we cannot consider 'the number 20' and 'every positive integer' as the subjects of their respective sentences. According to Frege, they are not concepts of the same rank. Thus we cannot assert 'in the same sense' of the number 20 what can be asserted of 'every positive integer'. Actually, the number 20 yields an independent idea whereas 'every positive integer' acquires its meaning from the 'context' of the sentence [1, p. 17].

In the *Begriffsschrift* we can find an anticipation of the notion of judgment presented in the later work *Function and Concept* [8]. The main point is that functions, concepts and relations are incomplete and require variables in their expression to indicate places of arguments. To establish a correspondence between function and concept Frege maintains that the linguistic form of an equation or identity is an assertoric sentence. It embeds a thought as its sense or, more precisely, we can say that it 'raises a claim' to have one. Generally speaking, the thought is true or false namely it possesses a truth-value that could be considered as the meaning of the sentence like the number 4 is the meaning of the expression '2 + 2' and London is the meaning of the expression 'the capital of England'.

The assertoric sentences can be decomposed into two parts: the 'saturated' and the 'unsaturated' one. For instance, in the sentence 'Caesar conquered Gaul' the second part is unsaturated and it must be filled up with a proper name (in our case Caesar) to give the expression a complete sense.

3 From Epistemology to Speech Acts

In this last section, I'd like to show that Frege's analysis of the pure thought opens an important space to a plausible view of the speaker's behavior in dialogical situation, which aims at reaching the truth. We have underscored the crucial difference between expression of a 'judgeable content' and 'judgment'. This distinction is further clarified by following some issues from his latest writings. When we perform an assertion we raise a claim to its validity namely we express a sentence that possesses a judgeable content. Consequently, our speech act will be either valid if our judgment will be recognized to have the truth-value 1 or not valid if it comes to have truth-value 0.

The notion of judgeable content plays a fundamental role not only in assertions but also in the speech acts of question and negation [9]. In the former, we ought to distinguish between our demand for a judgment that is simply connected with the use of words and the content or sense of the question that is subject to our judgment. This distinction is relevant for the possibility to acknowledge the truth of a thought or reject it as false. It also allows the impossibility to judge a thought such as 'two is bigger than three' for the grammatical subject is empty.

Another fundamental point is that because of the fact that the sense of a question is also inherent to an assertoric sentence that gives the answer to our question we must distinguish between the acts of grasping a sense and of judging also for the assertoric sentences. This is because the nature of a thought (namely a 'sense' or a 'judgeable content') does not consist in its being true namely, the content of a question is that as to which we must judge. If we perform the interrogative sentence 'Is the sun bigger than the Moon?' we cannot say that its sense is a truth otherwise we contradict the very nature of a question.

The distinction between judgeable content and judgment provides also an ontological thesis about the nature of the thoughts. Actually, science is the particular field that tends to discover truths about the world. But often the scientists simply perform the act of grasping a thought and this is not the same as performing a judgment. From this fact follows that it is possible to discover that one grasped a false thought, so we can admit the 'existence' of false thoughts. The fact that we can grasp false thoughts is exemplified by our possibility to distinguish between 'senseless' and 'falsity': "The teacher says 'Suppose a were not equal to b'. A beginner at once thinks 'What nonsense! I can see that a is equal to b'; he is confusing the senselessness of a sentence with the falsity of the thought expressed in it [9, p. 348].

A thought could be "complex' i.e. composed by other thoughts. For instance in conditionals (if . . . then . . .), we have a single acts of judgment and two thoughts that cannot be uttered 'assertively' in a separate way. A very clear example to explain the nature of false thought is the procedure of 'contraposition': "For thoughts consisting in an antecedent and a consequent there obtains the law that, without prejudice to the truth, the opposite of the antecedent may become the consequent, and the opposite of the consequent the antecedent" [9, p. 349].

The logical devices are not sufficient to explain the sense of expression such as 'Christ is immortal', 'Christ lives forever', 'Christ is mortal', 'Christ does not live for ever'. For this reason, we must suppose in a Kantian sense that the nature of thoughts is 'outside' logic. Consequently, negation namely the act of negating does not turn a true thought into a false one. The thought is self-sufficient but the negation needs to be completed by a thought. To express this thesis we write 'the negation of . . .' where the blank after 'of' indicates the place of the thought. The language of thought can therefore be linguistically expressed and we can put the definite article 'the' before negation so that 'the negation of' designates a single thing i.e. a thought (for instance, 'the negation of the thought that 3 is greater than 5').

4 Conclusion

Our analysis shows continuity between the *Begriffsschrift* and the latest writings of Frege. The distinction between judgeable content and judgment can be considered the most important result of Frege's epistemology. This idea emerged from our discussion of three fundamental points.

First, the distinction between function, namely the fixed part of an expression and argument, namely the variable part of it, plays the fundamental epistemological role to indicate when the argument is 'determinate' or 'indeterminate'. This very distinction is relevant for specifying a new notation of generality, which differs from the Aristotelian one and rests on a substitutional strategy.

Second, the change of the Aristotelian notation gives rise to a criticism of the aggregative strategy presented by Boole. Judgment is not formed by aggregation out of previously given constitutive concepts. Frege's notation admits differences of concept's level that overcome the classical distinction between subject and predicate. The logical form, composed of unsaturated and saturated parts could be considered by itself as predication; consequently we do not need to some additional factor to relate them predicationally.

Third, we pointed on the role of the difference between judgment and judgeable content in different kinds of speech acts. The functions of the use of linguistic devices together with the recognition of a content to be judged gives us the possibility to understand what is the subject of our judgment and thus the possibility to acknowledge the truth of a thought or reject it as false. Moreover, we notice that the same thought could be expressed in different way and underscore the role of assertions, questions and negations that represent the fundamental structure of dialogical situations and help the speakers to reach the truth.

References

1. Frege, G.: Begriffsschrift, eine arithmetischen nachbildete Formelsprache des reinen Denkens. Nebert, Halle (1879). In: Beaney, M. (ed.) The Frege Reader, pp. 47–78. Blackwell, Oxford (1997)
2. Sluga, H.: Gottlob Frege. Routledge & Kegan, London (1980)
3. Kant, I.: Critique of Pure Reason. Smith, N.K. (trans.). University of Edinburgh, Edinburgh (1929)
4. Frege, G.: Letter to Marty, 29.8.1882. In: Beaney, M. (ed.) The Frege Reader, pp. 47–78. Blackwell, Oxford (1997)
5. Heck, R.G., May, R.: The Function is Unsaturated. Brown University and University of California Davis (unpublished). http://Frege.brown.edu
6. Boole, G.: The Laws of Thought. Walton and Maberly, London (1854)
7. Frege, G.: Die Grundlagen der Arithmetik (1884). In: Beaney, M. (ed.) The Frege Reader, pp. 84–129. Blackwell, Oxford (1997)
8. Frege, G.: Funktion und Begriff. Vortrag gehalten in der Sitzung vom 9. Januar 1891 der Jenaischen Gesellschaft für Medizin und Naturwissenschaft. H. Pole, Jena. Trans.: Function and concept. In: Beaney, M. (ed.) The Frege Reader. Blachwell, Oxford (1997)
9. Frege, G.: Die Verneinung. Beitr. Dtsch. Ideal. **3** (1918). In: Beaney, M. (ed.) The Frege Reader, pp. 346–361. Blackwell, Oxford (1997)

R. Giovagnoli (✉)
Pontifical Lateran University, Piazza San Giovanni in Laterano, 00120 Vatican City, Italy
e-mail: raffa.giovagnoli@tiscali.it

Part II
Philosophical Discussion Around the Square of Opposition

Two Concepts of Opposition, Multiple Squares

John T. Kearns

Abstract This paper uses the resources provided by systems of illocutionary logic to characterize an opposition among assertive illocutionary acts that is parallel to incompatibility or inconsistency among statements. It is argued that opposition, or incoherence, among assertive illocutionary acts is more fundamental than inconsistency, and that our understanding of incoherence "underwrites" our conception of inconsistency. Diagrams are provided to illustrate relations of incoherence that are like such traditional diagrams as the square of opposition, which displays relations among statements.

Keywords Illocutionary logic · Speech acts · Logic of speech acts · Inconsistency · Incoherence

Mathematics Subject Classification 03B42 · 03B60 · 03B80

1 Language Acts and Rational Commitment

When we speak, write, or think with words, we are performing *speech acts*, or *language acts*. Some language acts are performed with whole sentences, and some whole-sentence acts are true or false. (What occurs as a whole and separate sentence on one occasion can occur as a clause in a larger sentence on another occasion, and the clause will still be used to perform a distinct sentential act when the larger sentence is used.) I will understand a *statement* to be a language act that employs a whole sentence or sentential clause, that is true or false; it is *true* if it "fits" the world, and *false* if it fails to fit. This is a stipulated meaning for the word 'statement,' because this word is often used as a near-synonym for 'assertion.'

Certain language acts performed with whole sentences are *illocutionary acts*, these are intentional acts, and are constituted by performing a sentential act with a certain, *illocutionary*, force. An illocutionary act is a complete intentional act (it is not simply an abstraction from a complete act), and it is the act a language user intends in performing a sentential act. Illocutionary acts include promises, threats, apologies, requests, assertions, and denials. Not every sentential act is used, by itself, to constitute an illocutionary act; some sentential acts are merely components of larger sentential acts which are used to constitute illocutionary acts.

As I have characterized statements, they are not illocutionary acts. But a statement can be performed with one or another illocutionary force, and will then constitute one or another kind of illocutionary act. Informally, we might regard the statement as a component of the illocutionary act, as the "content" of the act perhaps. More accurately, the statement is the illocutionary act abstractly conceived. However, not all statements are performed with illocutionary force. Anyway, the point of considering statements is to abstract away from considerations of illocutionary force.

Although discussions of illocutionary acts commonly regard them all as communicative acts, so that they are performed by agents and aimed at addressees, I am understanding illocutionary acts to constitute a larger class. For some illocutionary acts such as directives and promises, an addressee is essential. You can't give someone an order if there is no someone there. But I think what is essential to an assertion is that the speaker/writer/thinker produces (performs) a statement and accepts it as being or representing what is the case. A speaker can address her assertion to someone else, but I count it as an assertion if a speaker accepts a statement in the privacy of her office, without telling anyone.

Statements are true or false, and have truth conditions. Customary logical theories investigate statements, their logical forms, and semantic features that can be defined in terms of truth conditions and logical forms. A statement is logically true, for example, if it has a logical form such that the truth conditions associated with that form can't fail to be satisfied. A group of statements imply a further statement if it isn't possible to satisfy the truth conditions associated with the logical forms of the first statements without also satisfying the truth conditions associated with the logical form of the further statement.

Ordinarily, a person using language is performing illocutionary acts, whether she is addressing someone else, or just thinking things through on her own. If she is considering how things are, trying to determine what is the case, she might be asserting some statements and denying others. She can also *suppose* a statement to be the case, to determine what follows from doing this. In carrying out deductive arguments, it is common to make and discharge suppositions in ways that are familiar from systems of natural deduction. As well as devising logical systems for investigating statements and their truth conditions, we can develop systems of *illocutionary logic* for investigating illocutionary acts and certain of their features.

The subject, or field, of illocutionary logic was introduced by John Searle and Daniel Vanderveken in [2]; they have pursued what might be called a "top down" approach to this subject, developing a theory to characterize all illocutionary acts. They regard illocutionary logic as a kind of supplement, or appendix, to standard logic. I take a rather different approach, a "bottom up" approach that calls for developing many different systems of illocutionary logic. I regard illocutionary logic as a very broad field, one which encompasses standard logic as a proper part. A system of logic for *assertive* illocutionary acts (these include assertions, denials, and suppositions, among others) contains two levels, an *ontological*, or *ontic*, level that is concerned with statements and features of statements determined by their truth conditions, and an *epistemological*, or *epistemic*, level that is concerned with assertive illocutionary acts and their features. An epistemic-level illocutionary system deals with *speech-act* arguments which proceed from illocutionary acts as premises to illocutionary act conclusions.

In order to use language properly, and effectively, a person must understand the truth conditions of statements that she either encounters or produces. But it is her *recognition* of the truth conditions that matters, and not simply the truth conditions "by themselves." If statement A implies statement B, the person who asserts A may not know, she may not realize, that A implies B. The fact that she accepts A does not determine her to accept B, and need have no effect on her attitude toward B. However, once she comes to know that A implies B, then, so long as she continues to accept A, she is *committed, rationally committed*, to accept B. This commitment is conditional, for she is not required to think about B at all, or to adopt any position about B. But if the matter comes up, and she gives it some thought, then she is committed to accept B. It is *irrational* to accept A, realize that A implies B, and refuse to accept B.

Rational commitment can motivate a person to act, can rationally "require" that person to act. But it is important to distinguish *immediate* from *remote*, or mediate, commitment. If a person is immediately committed to doing X_1, and doing X_1 will immediately commit her to doing X_2, while doing X_2 will immediately commit her to doing X_3, and so on, down to, say, X_9, then doing X_1 may not immediately commit her to doing X_9. Doing X_1 may leave her indifferent with respect to X_9. It is immediate commitment that motivates a person to act, for immediate commitment is evident to the person for whom it is immediate—if she gives the matter her attention.

Asserting some statements, and denying others, will commit a person to assert still others, and to deny others. If someone begins by supposing one or more statements to be the case, and then reasons to a conclusion based on these suppositions, her conclusion also has the status of a supposition, and I call it a supposition. In constructing a deductive argument, a person traces the immediate commitments of her initial assertions, denials and suppositions to reach an assertion, denial, or supposition as conclusion. But commitment, and our "respecting" commitment, are important for all language use, and not simply for constructing deductive arguments. We can't appreciate the import of what people (including ourselves) say without taking account of rational commitment.

In a system of standard logic, we might determine that some statements imply a further statement. In a system of illocutionary logic, I use the expression 'logically require' for the relation between some illocutionary acts and a further act to which the first acts commit the speaker. *Logical requiring* is the illocutionary counterpart to the implication which links statements. In real life, where we say things and mean them, it is logical requiring which links the steps in our correct deductive arguments. The components of those arguments are illocutionary acts rather than simply statements.

Semantic inconsistency is a feature of statements which can't be true together. I use the word 'incoherence' for the illocutionary counterpart to semantic inconsistency. It is incoherent to both assert and deny a single statement, it is also incoherent to both suppose a statement to be the case and suppose that statement to be false. Illocutionary acts that commit a person to perform incoherent acts are themselves incoherent. It is a rational requirement for all of us to achieve coherence among our assertions and denials; incoherence is a sign of error, and not a license to infer no matter what conclusion. The requirement to achieve coherence also applies to suppositions, but it can be strategically useful to make incoherent suppositions, in order, say, to show that some statement supposed to be true must actually be false, or to show the reverse.

2 Complex Illocutionary Acts

When a statement is accepted, it is accepted as being, or representing, things as they are. For an act of asserting/accepting a statement to be successful, the statement must be true, and the person who accepts the statement must have adequate grounds for accepting it. But in making, and accepting, a statement A, a language user is not saying anything *about A*. She is not accepting the statement "It is true that *A*"; neither is she accepting the statement "I know that *A*" or "I have good grounds for *A*." (If she accepts *A*, then she is *committed* to accept "It is true that *A*," and, if she is epistemically responsible, she will also be committed to accept "I have good grounds for *A*.")

Although statements can either be asserted/accepted or denied, there is a sense in which statements are "designed" for being asserted. In a simple statement, where an expression is predicated of one or more objects, the predicate expression is associated with a criterion, and is truly predicated of an object or objects that satisfy this criterion. The whole "idea" of a predicate expression is *to be applied to objects which satisfy its criterion*. The statement constituted by predicating the expression of the object or objects is fully successful as a statement if it presents or represents things as they are. Asserting or accepting is what "comes naturally" to statements, because these assertions endorse the statements for their success.

But then, how shall we understand denials? Although we might say that in denying a statement we are rejecting that statement, what is really being rejected is the illocutionary act that the statement is designed for performing. To deny a statement is to *rule out* the assertion of that statement, to rule this out because the statement is deficient. There are different respects in which a statement can be deficient, but the simplest and most frequent deficiency is that the statement fails to fit the world. In many natural languages, it is common to "carry out" a denial by interfering with the formation of the statement that is denied. For example, someone might use the sentence 'Marseille is not in Corsica' to deny that Marseille is in Corsica. If they do, the 'not' is not functioning as the negation operator in familiar systems of propositional logic. Instead, the 'not' acts as an illocutionary-force indicating expression, which separates (*divides*) the act of referring to Marseille from the act of predicating *being in Corsica*.

In simple systems of illocutionary logic, I use the following *illocutionary operators*:

$\vdash$ the sign of assertion
$\dashv$ the sign of denial
$\llcorner$ the sign of positive supposition
$\neg$ the sign of negative supposition

and prefix them to sentences with no illocutionary operators so that the complete sentences represent illocutionary acts of the speaker/language user. Although I am the one using language as I talk and write about language acts and illocutionary logic, in designing an artificial language, I am providing expressions for the speakers of that language to use to indicate the force of their own illocutionary acts. I regard sentences in that language, when they are prefixed with illocutionary operators, as if they were produced by some other person than myself.

John Searle, and Frege before him, regards statement negation as prior to denial, so that the denial of a statement *A* has this significance:

$$\dashv A = (\text{def}) \vdash \sim A$$

However, I think that denial is prior to statement negation, and is in fact the source of significance for statement negation via these inference principles:

$$\begin{array}{cc}
\sim \textit{Introduction} & \sim \textit{Elimination} \\
\dashv A & \vdash \sim A \\
\hline
\vdash \sim A & \dashv A
\end{array}$$

(Similar principles apply to positive and negative supposition.) To negate a statement is to characterize the statement as being objectively unsuited for assertion (because of its failure to fit the facts).

Some illocutionary acts logically require others, and some illocutionary acts are logically incoherent. Logical requiring and incoherence are semantic features proper to illocutionary acts. These features affect the way we use language, for people frequently act on the basis of rational commitment, including the commitment to avoid and to eliminate incoherence. While statements which are semantically inconsistent might be regarded as opposing one another, I think that incoherent illocutionary acts are opposed in a more fundamental sense. Statements that are semantically opposed are statements that can't all be coherently accepted. It is this incoherence which we are rationally required to avoid or to eliminate.

A *simple* assertion or denial involves a single, simple statement. In [1] I have proposed (and argued) that we need to recognize *conjoined* or *conjunctive* assertions, where a single sentence is used to make two (or more) distinct and separate assertions, as well as *disjunctive* assertions in which a person disjunctively accepts two (or more) statements— "narrowing down" the way things are to the disjoined statements. This understanding brings assertive illocutionary acts "in line" with other forms of illocutionary acts, such as *directive* and *commissive* acts. For example, we might advise someone to do X *and* to do Y, we could also request her to do X *or* to do Y. I think that conjunctive and disjunctive assertions are logically prior to the statement-forming operators for conjunction and disjunction, and can be used to introduce and inferentially characterize the statement-forming operators in much the same way that the operator for denial was used earlier to characterize statement negation.

In making a conjunctive assertion, a person uses a sentence like this one:

> James was at the party, and Karen was also at the party.

to make two separate assertions rather than using the sentence to conjoin two statements, and then accept that conjunction. A conjunctive denial uses one sentence to make two separate denials:

> Ken wasn't at the party, and neither was Kathy.

With a disjunctive assertion or denial, a single sentence is used to perform an illocutionary act which "complexly disjoins" two statements, rather than performing a disjunctive statement with an illocutionary force.

If we use these operators for conjunctive assertion and denial, and for disjunctive assertion and denial:

Conjunctive		Disjunctive	
assertion	denial	assertion	denial
$\vdash_\wedge$	$\dashv_\wedge$	$\vdash_\vee$	$\dashv_\vee$

then we can use these expressions to represent the complex illocutionary acts:

$$\vdash_\wedge [A, B] \qquad \dashv_\wedge [A, B] \qquad \vdash_\vee [A, B] \qquad \dashv_\vee [A, B]$$

Conjunctive and disjunctive assertions could be used to introduce and inferentially characterize statement-forming operators as follows:

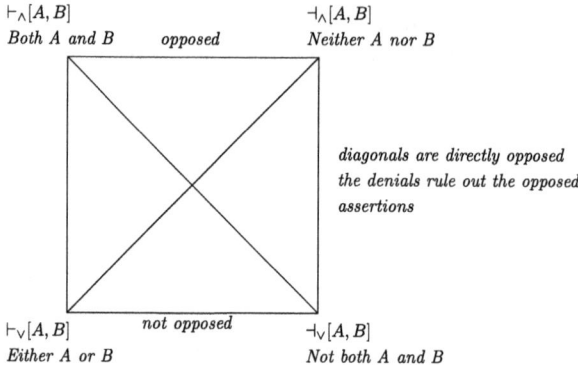

Fig. 1 Conjunctive-disjunctive square of opposition

$$\begin{array}{cc} \textit{\& Introduction} & \textit{\& Elimination} \\ \vdash_\wedge[A, B] & \vdash[A \, \& \, B] \\ \hline \vdash[A \, \& \, B] & \vdash_\wedge[A, B] \end{array}$$

$$\begin{array}{cc} \vee \textit{ Introduction} & \vee \textit{ Elimination} \\ \vdash_\vee[A, B] & \vdash[A \vee B] \\ \hline \vdash[A \vee B] & \vdash_\vee[A, B] \end{array}$$

The simple assertion of a statement: $\vdash A$, and its denial: $\dashv A$, are illocutionary acts that are exactly opposed to one another. An illocutionary act which consists in ruling out an assertion is exactly opposed to that assertion. Interestingly, a conjunctive denial: $\dashv_\wedge[A, B]$, is not exactly opposed to a conjunctive assertion: $\vdash_\wedge[A, B]$. The two acts are incoherent with one another, so they are opposed. But the conjunctive denial does not simply rule out the conjunctive assertion; it separately rules out the assertion of each conjunct. The conjunctive assertion of two statements is exactly opposed to the disjunctive denial: $\dashv_\vee[A, B]$, of those statements. And the conjunctive denial is exactly opposed to the disjunctive assertion: $\vdash_\vee[A, B]$. These four illocutionary acts give us the square of opposition shown in Fig. 1.

It isn't clear to me that English possess a simple and idiomatic expression for making disjunctive denials. Many of my undergraduate students, for example, recognize no difference between saying "not both" and "both not." ("Mary and Joan are not both philosophy majors" vs "Mary and Joan are both not philosophy majors.") In any case, our square provides a place for such acts. In this square each top corner logically requires the bottom corner beneath it.

Conjunctive and disjunctive assertions and denials, as well as conjunctive and disjunctive directives and commissives, are all examples of what I will call *complex* illocutionary acts. This label is somewhat misleading, because complex illocutionary acts do not contain other illocutionary acts as components. As we understand illocutionary acts, no illocutionary acts have illocutionary act components. (I have had no success in thinking of a more apt expression than 'complex illocutionary act.') The complex assertions can be used to introduce and characterize statement-forming operators for conjunction and disjunction; and those operators can be used to form complex *statements* which do con-

Fig. 2
Conjunctive-disjunctive
square of inconsistency

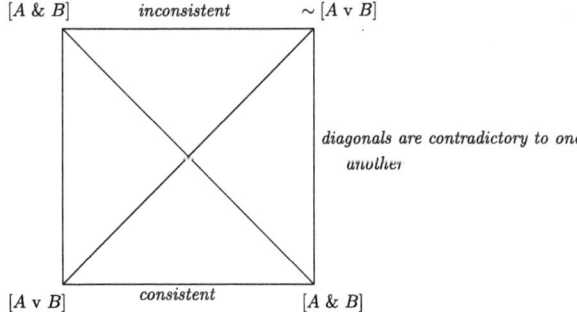

tain other statements as parts. We can use the statements to construct the different square shown in Fig. 2.

This is not a square of opposition, if we understand opposition to be a relation between illocutionary acts. It is a *square of inconsistency* or of *incompatibility*. But the assertions of these statements would be opposed. However, we should not think that all cases of incoherence involve statements which are inconsistent or incompatible. The assertions shown here (which figure in Moore's Paradox):

$$\vdash A \qquad \vdash \sim I \ believe \ A$$

are incoherent, even though the statements they "contain" are consistent (with one another):

I have also argued that we must distinguish conditional assertions from the assertion of conditional *statements*. I think that we don't so often make non-modal conditional statements in English, but we commonly do make conditional assertions and conditional denials. I represent the *assertion of B on the condition A* like this:

$$\vdash [A]/B$$

and the denial of B on the condition A like this: $\dashv [A]/B$.

A conditional assertion or denial establishes an inference principle (for the speaker) from asserting the antecedent to asserting, or to denying, the consequent. For the speaker, if the inference principle is in force for her, then, once she accepts the antecedent, she has an immediate commitment to assert, or to deny, the consequent. The assertion of B on the condition A is *conditionally opposed* to the denial of B on the condition A. English does not possess an idiomatic and convenient device for performing an illocutionary act which is exactly opposed to a conditional assertion, though perhaps we can regard a conjunctive assertion as exactly opposed to a conditional denial. It seems more-or-less idiomatic to deny someone's conjunctive assertion that both A and B are F by saying "If A is F, then B isn't."

This provides the incomplete square of opposition shown in Fig. 3.

Perhaps the incomplete square is better described as a triangle. But the missing corner marks a place which might be filled by an illocutionary act which is directly opposed to the conditional assertion. Although such acts are conceivable, English does not provide an idiomatic way to perform such an act. The conditional opposition between the top corners is not complete opposition, for both acts can legitimately be performed by a person who knows A to be false. The top left corner does not logically require the corner beneath it.

Fig. 3 Incomplete
conditional square of
opposition

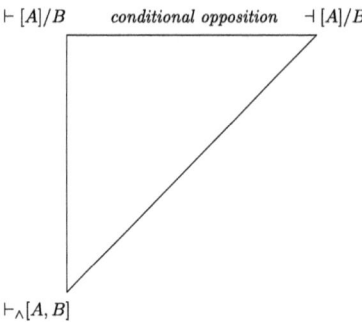

$\vdash [A]/B$ *conditional opposition* $\dashv [A]/B$

$\vdash_\wedge [A, B]$

A *universal assertion* is still another form of complex illocutionary act. Someone who asserts that, say, every F is a G, establishes an inference principle from accepting or supposing that some particular object is an F to accepting or supposing that that object is a G. Perhaps we could say that the speaker *asserts a thing to be a G on the condition of its being an F*. A universal assertion can be used to introduce and inferentially characterize an operator for forming universal statements, just as other illocutionary operators can provide a basis for introducing statement-forming operators. In English, a universal assertion (normally) involves an existential presumption. We don't assert a thing to be a G on the condition of its being an F unless we think there are Fs. There is no point in establishing an inference principle if we think it has no application.

We often deny a simple statement by dividing its predicate from its subject. In a universal denial, we make a "wholesale" division: we entirely divide the Fs from the Gs, for there is no F that can correctly be asserted to be a G. The universal denial establishes an inference principle from an assertion that some particular object is an F to a denial that that object is a G. A universal denial is opposed to a universal assertion, but it is not the exact opposite of a universal assertion. A universal denial is the exact opposite of an existential, or particular, assertion. An existential assertion that some F is a G establishes a kind of inference principle: the speaker is committed to the consequences of an arbitrary assertion that a particular F is a G—these are the consequences common to every assertion to the effect that some particular F is a G.

English does not seem to possess an idiomatic form for making a denial which is exactly opposed to a universal assertion. We can imagine circumstances in which sentences of the form

Not every F is a G.

were used to make such denials. One of these denials would have as a consequence the disjunctive assertion that either there are no Fs or some F is not a G. While they are possible, such denials don't figure in our actual practice.

English also lacks a form for making denials which are counterparts to existential assertions, such denials would rule out assertions of a kind we don't make in English: These would be assertions that every/any F is a G, which assertions are always acceptable if it is known or believed that there are no Fs. A sentence 'Some F is not a G' is not one that can appropriately be used to make such a denial, or any denial. A singular sentence 'George is not a plumber' can be used to make a denial, because when 'not' is used to block the predication of 'is a plumber' of George, this can serve to rule out the assertion that George is a plumber. In the sentence 'Some man is not a plumber,' the 'not' will

Fig. 4 Incomplete
universal-existential square of
opposition

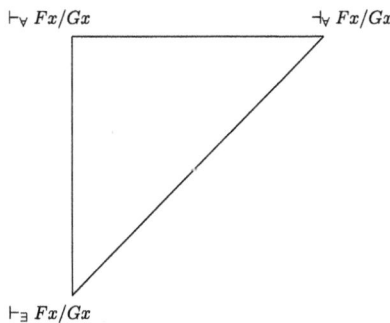

$\vdash_\forall Fx/Gx$ $\dashv_\forall Fx/Gx$

$\vdash_\exists Fx/Gx$

not block the predication of 'is a plumber' of some man—the sentence will not be used
to rule out the existential assertion "Some man is a plumber." The sentence 'Some man
is not a plumber' is suited for an existential assertion but not for any kind of denial. If,
say, 'Any F is a G' were used to make a universal assertion that is always in order if it
is known/believed that there are no Fs, then 'Not any F is a G' might be used for the
denial of this assertion, which denial would commit the speaker to assert that there is an
F which isn't a G. But this isn't how English works at present.

If we represent universal assertions and denials like this: $\vdash_\forall Fx/Gx$, $\dashv_\forall Fx/Gx$, and
existential assertions like this: $\vdash_\exists Fx/Gx$, we have another incomplete square (Fig. 4) for
the assertions and denials available in English.

I think the opposition that is most important to our use of language is the opposition between illocutionary acts which are incoherent with one another. The requirement to avoid
incoherence and to achieve coherence is a fundamental principle of rationality. There are
a variety of forms of complex assertions, and many of them give rise to oppositions which
can be organized as squares, complete or otherwise, though these squares don't have all
the features attributed to the traditional square of opposition.

References

1. Kearns, J.T.: Conditional assertion, denial, and supposition as illocutionary acts. Linguist. Philos. **29**,
   455–485 (2006)
2. Vanderveken, D., Searle, J.: Foundations of Illocutionary Logic. Cambridge University Press, Cambridge (1985)

J.T. Kearns (✉)
Department of Philosophy and Center for Cognitive Science, University at Buffalo, the State
University of New York, Buffalo, NY 14260, USA
e-mail: kearns@buffalo.edu

Does a Leaking O-Corner Save the Square?

Pieter A.M. Seuren

Abstract It has been known at least since Abelard (12th century) that the classic Square of Opposition suffers from so-called *undue existential import* (UEI) in that this system of predicate logic collapses when the class denoted by the restrictor predicate is empty. It is usually thought that this mistake was made by Aristotle himself, but it has now become clear that this is not so: Aristotle did not have the Conversions but only one-way entailments, which 'saves' the Square. The error of UEI was introduced by his later commentators, especially Apuleius and Boethius. Abelard restored Aristotle's original logic. After Abelard, some 14th- and 15th-century philosophers (mainly Buridan and Ockham) meant to save the Square by declaring the O-corner true when the restrictor class is empty. This 'leaking O-corner analysis', or LOCA, was taken up again around 1950 by some American philosopher-logicians, who now have a fairly large following. LOCA does indeed save the Square from logical disaster, but modern analysis shows that this makes it impossible to give a uniform semantic definition of the quantifiers, which thus become ambiguous— an intolerable state of affairs in logic. Klima (Ars Artium, Essays in Philosophical Semantics, Medieval and Modern, Institute of Philosophy, Hungarian Academy of Sciences, Budapest, 1988) and Parsons (in Zalta (ed.), The Stanford Encyclopedia of Philosophy, http://plato.standford.edu/entries/square/, 2006; Logica Univers. 2:3–11, 2008) have tried to circumvent this problem by introducing a 'zero' element into the ontology, standing for non-existing entities and yielding falsity when used for variable substitution. LOCA, both without and with the zero element, is critically discussed and rejected on internal logical and external ontological grounds.

Keywords Abelard · Aristotle · Boethius · Buridan · Existential import · Null element · O-Corner · Predicate negation · Restricted quantification · Square of Opposition

Mathematics Subject Classification Primary 03B65 · Secondary 03B50

1 Letting the O-Corner Leak

Some 14th-century nominalist philosophers, such as Buridan and Ockham and a bevy of 15th-century philosophers (Ashworth [1]), and then again some 20th-century American philosophers, notably Moody [4], Thompson [10], Klima [3] and Parsons [5, 6], were struck by the fact that the classic Square of Opposition (henceforth the *Boethian Square* or just the *Square*) is saved from the logical defect of *undue existential import* (UEI) if it proves possible to assign existential import only to the affirmative left-hand side of

J.-Y. Béziau, D. Jacquette (eds.), *Around and Beyond the Square of Opposition*, 129–138
Studies in Universal Logic, DOI 10.1007/978-3-0348-0379-3_9, © Springer Basel 2012

the Square but not to the negative right-hand side. This way, existential import depends on what is traditionally called the *quality* of the sentence type and not on its *quantity*: when the *restrictor* predicate F has a null extension ($[\![F]\!] = \emptyset$),[1] falsity is assigned to **A**- and **I**-type sentences but truth to **A***- and **I***-type sentences.[2] This proposal is often characterized as the proposal to let the **O**-corner 'leak', in that the **O**-corner (**I*** or **¬A**) drops its existential import. In fact, however, it is not just the **O**-corner that is made to 'leak' but also the **E**-corner (**A*** or **¬I**) of the Square. To avoid any possible negative connotations, the less suggestive acronym LOCA will be used instead of the full phrase 'leaking **O**-corner analysis'.

Many modern philosophers of this school, including Parsons, and to a lesser extent also Moody, claim that LOCA is the way the Square was interpreted by Aristotle and throughout the Middle Ages. This claim, however, is unsustainable. Aristotle proposed a logic, later elaborated by Abelard (Seuren, [8]: Ch. 5), without the defect of UEI. This Aristotelian-Abelardian system does not contain the classic Conversions (**A*** ≡ **¬I** and **I*** ≡ **¬A**) but only one-way entailments (**A** entails **¬I** and **I*** entails **¬A** but not vice versa; see Fig. 3). So there is no question of a 'leaking' **O**-corner there. As regards the proposals made by Buridan and Ockham, they were not only tentative and barely elaborated but also highly controversial in their day.[3] The extreme scarcity of passages in the 14th-century philosophical literature actually attesting LOCA shows the tenuousness of this claim as far as the 14th century is concerned (Seuren, [8]: 158–168).[4] Ashworth ([1]: 144) shows that throughout the 15th century, up till the early 16th century, there was a group of philosophers who saw a parallel between singular sentences with an empty subject term, such as (1a) and the corresponding existentially quantified sentence (1b), which were both considered false in the absence of any chimera.

(1) a. This chimera is a mammal.
    b. Some chimera is a mammal.
    c. This chimera is not a mammal.
    d. Some chimera is not a mammal.
    e. No chimera is a mammal.

---

[1] The standard double bracket notation is used for predicate extensions: $[\![F]\!]$ stands for all elements in the universe of entities **ENT** for which the predicate F delivers truth.

[2] The eight sentence types used are **A** (All F is G), **I** (Some F is G), their external negations **¬A** and **¬I**, their internal negations **A*** and **I*** (All/Some F is not G), and their external and internal negations **¬A*** and **¬I*** (Seuren, [8]: 31–37). The old **E**-corner is thus split up into **¬I** and **A*** and the old **O**-corner into **¬A** and **I***. **¬I** and **A*** and **¬A** and **I*** can thus be distinguished in systems where the members of each pair are not equivalent.

[3] "Ockham's *Summa Logicae* (*The Logic Handbook*), written ca. 1323, is a manifesto masquerading as a textbook" (King, [2]: 243).

[4] It is difficult to find a 14th-century text *actually stating* LOCA, instead of merely *playing* with the idea. One such text is Ockham, *Summa Logicae*, beginning of Chap. II.3 (http://individual.utoronto.ca/pking/resources/Ockham?Summa_logicae.txt):

> <Q>uandoque sufficit quod subiectum indefinitae vel particularis negativae pro nullo supponat. (<S>ometimes it suffices for the truth of a negative indefinite or particular proposition that the subject refers to nothing.)

For ample commentary and a full text of the passage in question, see Seuren ([8]: 163–166).

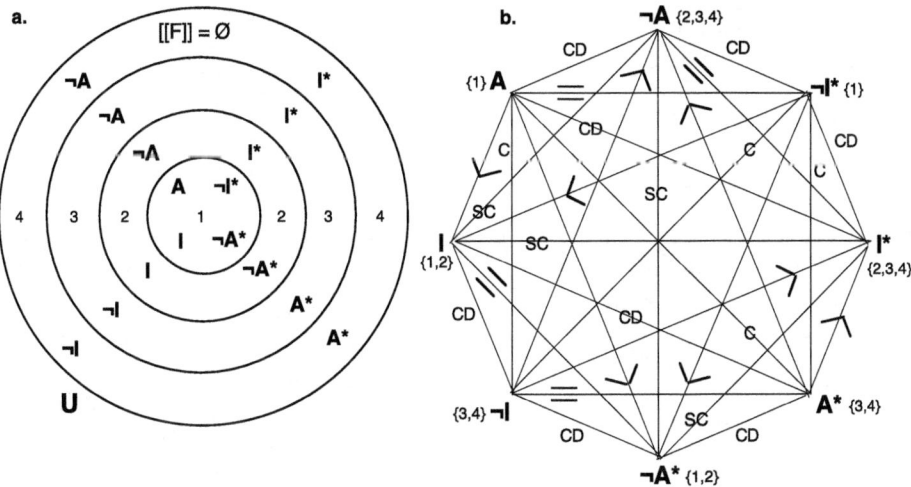

Fig. 1 The predicate logic in terms of LOCA

They reasoned that, just as (1c), the negation of (1a), is true, the apparent quantificational analog (1d) should, therefore, also be true. What these late medieval philosophers overlooked, however, is that whereas (1c) is the negation of (1a), (1d) is not the negation of (1b). The negation of (1b) is (1e), which destroys the analogy. After c. 1500 nothing is heard of this misguided analysis any more, until it was revived in America around 1950.

Apart from this historical claim, however, it is true that, on the face of it, LOCA saves the Square. We then get a logic that preserves all logical relations of the Square. That this is so is shown by Fig. 1, which shows the LOCA-based VS-model on the left and the corresponding octagon on the right,[5] which is identical with the octagon corresponding to the Square, shown in Fig. 2 (where space 4 in Fig. 2a has been left blank owing to UEI). This makes LOCA look interesting. The question is: is this tenable?

One immediate reason for not considering LOCA tenable is that it fails at least one test for logical soundness, the MODULO-*-PRINCIPLE. This principle says that the logic of quantifiers is independent of whether the *restrictor* or F-predicate and/or the *matrix* or G-predicate are or are not negated. That is, the semantics of the quantifiers must be the same whether or not the F- or G-predicate are negated. A violation of this principle means that the logic at hand is no longer definable in terms of the meanings of its operators. Thus, if, in cases where $\llbracket F \rrbracket = \emptyset$, All F is G and Some F is G are declared false but All F is not-G and Some F is not-G true, it follows that the semantic definitions of the universal and the existential quantifiers vary depending on whether a positive or a negative predicate

---

[5]For explicit exposés of Valuation Space (VS) modeling and corresponding polygonal representations of logical systems, see Seuren ([8]: 46–50; [9]: Ch. 6). Intuitively, the notions are clear enough. A VS model splits up the universe **U** of possible situations into classes of situations in which each sentence type of a logical system is true. The valuation space /S/ of a sentence S is the set of situations in which S is true. For example, in Fig. 2, /**A**/ = {1}, /**I**/ = {1, 2}, etc. A polygonal representation is a polygon with as many vertices as there are sentence types in the system (without vacuous repetitions of negations). Lines between vertices stand for logical relations (*equivalence* or =, *entailment* or >, *contradictoriness* or CD, *contrariety* or C, *subcontrariety* or SC).

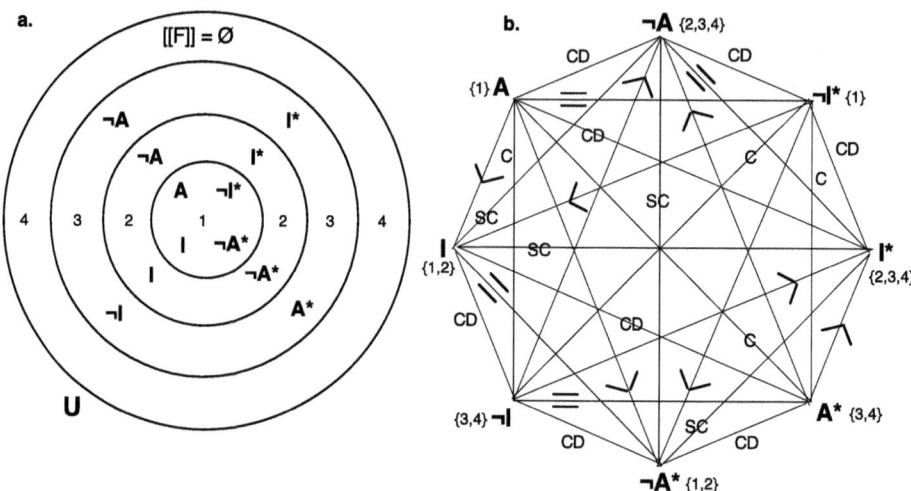

Fig. 2 The VS-model and the octagonal representation of the classic Square

is chosen. This makes it impossible to provide a uniform definition for the quantifiers concerned that is valid for all situations and for all values for the matrix G-predicate. It is shown below that this is borne out dramatically by Parsons' attempt at saving the Square.

No distinction can be made between the compound but single negative predicate not-P—with the so-called 'predicate negation'—and not P (without the hyphen), for any predicate P, as the only negation defined in the system is the logical negation ¬ and a 'predicate negation', as in not-P, is simply defined by saying that $[\![not\text{-}P]\!]$ is $[\![¬(P(x))]\!]$, the complement of $[\![P]\!]$, or $[\![\overline{P}]\!]$, in the universe of all entities **ENT**. The hyphen in not-P thus cannot make any logical difference, given the formal means at our disposal within the framework of a bivalent extensional logic. Yet in LOCA, Some F is not G is considered true but Some F is not-G (with the single predicate not-G) false, when $[\![F]\!] = \emptyset$, which means that the hyphen does make a logical difference there.[6]

The question is important because, in LOCA, the distinction between not-P and not P has stridently counterintuitive consequences, as it implies that *Some mermaids are unmarried* is false in this world but *Some mermaids are not married* is true. Moreover, a sentence like *Some mermaid is not a mermaid* would have to be considered true but *Some mermaid is a mermaid* false. This is, in fact, sufficient to reject out of hand this solution for the purpose of establishing a *natural* predicate logic. Yet it is worth our while to have a further look.

---

[6]When the logic turns presuppositional, the hyphen *will* make a logical difference. As Aristotle already observed, the only generalization that can be made for a predicate P and its negative counterpart not-P (or unP, or whatever morphological or lexical means the language has available for creating negative predicates) is that they make for *contrary* sentences, which cannot both be true, but can both be false. A relation of *contradiction* between sentences made with P and not-P is restricted to those predicates that are nongradable and have no presuppositional preconditions, such as *existent* and *nonexistent*. In presuppositional logic, this means that the predicate-internal negation can only be presupposition-preserving. But the not in not P, without the hyphen, can be the radical presupposition-canceling *not* that yields truth in cases of presupposition failure.

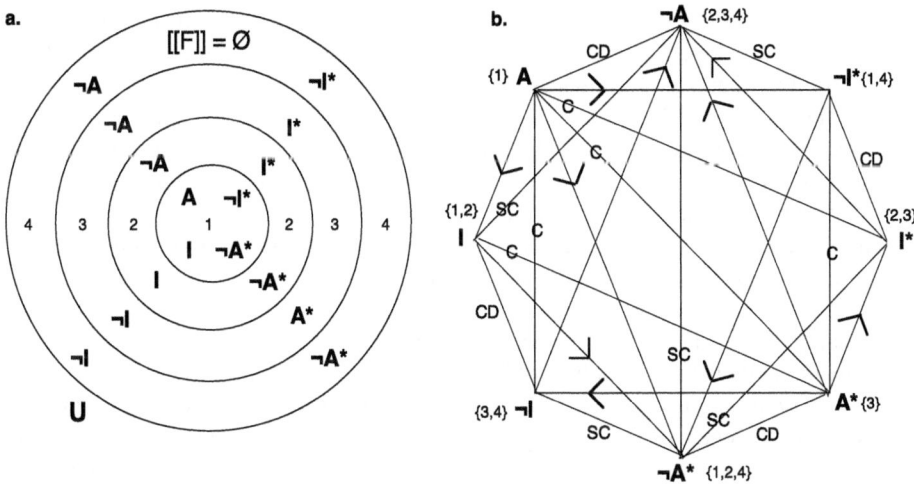

Fig. 3 The VS-model and the octagon for AAPL

Moody ([4]: 51–52) admits the counterintuitiveness of the proposal, adding that, therefore, the **O**-corner should be read as 'not all' rather than as 'some not', thereby implicitly, but, it seems, unknowingly, giving up the Boethian Square with its Conversions for the Abelardian Square with its one-way entailments (see Fig. 3). Thompson ([10]: 253) says, implicitly confessing to the misconception that logic is predestined for all eternity instead of being defined by the meanings of its operators:

> Even if we agree with the new defenders of Aristotle that the decision which leads to the modern analysis is repugnant to ordinary speech, we can still argue that this is more desirable than a decision repugnant to logical analysis itself.

Parsons declares ([6]: 5):

> What is important is that the logical notation be coherent and useful. If it does not perfectly match the usage of ordinary language, that is not on its own important for a system of logic. Indeed, if you are sure that ordinary language universal affirmatives should be false when their subject term is empty, then you may represent that fact by translating them into modern logical notation adding a conjunct. Instead of symbolizing 'Every A is B' by '$\forall x(Ax \rightarrow Bx)$', symbolize it as '$\exists x Ax \,\&\, \forall x(Ax \rightarrow Bx)$'.

Here again, both authors are, apparently, unaware that Aristotle did not have the Conversions but only one-way entailments, which, far from being 'repugnant to logical analysis itself', stays well within the range of natural intuitions and keeps the logic sound as well as free from undue existential import. Then, while Thompson uncritically takes it for granted that saving natural logical intuitions leads to logical disaster, Parsons overlooks the fact that adding a nonnull condition to the semantic definition of $\forall$ yields a different logic, Aristotelian-Abelardian predicate logic (AAPL), shown in Fig. 3.

But apart from the intuitively rebarbative character of the proposal, let us have a look at the more technical details. Klima ([3]: 18–2) and Parsons [6] are the only authors who actually present a logical *system* incorporating LOCA, rather than just *stating* that **I***-type sentences should be considered true when $[\![F]\!] = \emptyset$. Klima resorts to an old medieval device (revived, it seems, by the late 19th-century British philosopher Hugh MacColl, referred to in Russell, [7]: 491) of introducing a *null* or *zero element* ø standing for any nonexistent entity. Some late medieval philosophers (followed, apparently, by MacColl) used ø to solve what is now known as the problem of *reference failure*. They posited that the definite description *the unicorn* in a sentence like *The unicorn is running* refers to the null element ø when, as in the actual world, there are no unicorns. This made it possible to assign falsity to this sentence and thus to escape the predicament caused by reference failure, while maintaining strict bivalence.

Klima and Parsons, interpreting ø model-theoretically, posit that zero-element substitution in the range $[\![P]\!]$ of an extensional predicate P (no account is taken of intensional predicates) is admissible when $[\![P]\!] = \varnothing$, but leads not to truth but to falsity. This is at least remarkable because, by definition, substitution restricted to the extension $[\![P]\!]$ of a predicate P simply lists all members of $[\![P]\!]$ and thus always yields truth.

Around 1950, the machinery of variable substitution had become widely known and was seen as the center-piece of the semantics for the Russellian language of predicate logic, where it seemed to work well (provided **ENT** $\neq \varnothing$). At that time, the British philosopher Peter Geach was making headlines opposing the Russellian system of *unrestricted quantification*, so called because in this system every variable ranges over the whole of **ENT**. Geach proposed the theory of *restricted quantification*, where variables rotate not over the whole of **ENT** but only over $[\![F]\!]$, the extension of the restrictor predicate. This, however, landed Geach and his followers in the predicament that in cases where $[\![F]\!] = \varnothing$ no substitution is possible for the matrix predicate G(x), so that the truth value of propositions like ∃x:F(x) | [G(x)] or ∀x:F(x) | [G(x)] must remain undecided when $[\![F]\!] = \varnothing$.[7] Geach never solved this problem, but Klima saw that, with the help of medieval ø, these propositions turn out false—good enough, it would seem, for a respectable predicate logic.

Klima follows Geach's system of restricted quantification. This means that the variable in a formula like ∀x:F(x) | [G(x)], or 'for all x such that x is an F, G(x)', is allowed to rotate only over $[\![F]\!]$.[8] As has been said, substitution of ø is considered legitimate whenever $[\![F]\!] = \varnothing$, but it yields falsity.

Now consider the I*-type Some F is not G, formulated, in the language of restricted quantification, as (2):

(2) ∃x:F(x) | ¬[G(x)]

Suppose $[\![F]\!] = \varnothing$, which, in Klima's analysis, means that ø is the only admissible substitution. Substitution of ø for x makes G(x) false and thus ¬[G(x)] true. Therefore, one might feel justified to conclude that there is one substitution that produces truth for the

---

[7] A formula like ∀x:F(x) | [G(x)] is read as 'for all x such that x is an F, G(x)', and ∃x:F(x) | [G(x)] as 'for some x such that x is an F, G(x)'. In this restricted quantification notation, the part between the colon and the upright stroke is the *range* of the restricted quantifier (∀x or ∃x) and the part after the upright stroke is the *argument*. It should be noted that the use of a first-order logical *language* in no way implies the use of the a particular first-order *logical system*. When the Russellian language is used here, it is because it has been found satisfactory for the present purpose. The language of restricted quantification is used when it is found useful for that purpose. A given formal logical language can be used with different semantic interpretations according to the logical system under discussion. The important thing is that the set-theoretical basis underlying any logical system is made explicit in a convenient, consistent and unambiguous way.

[8] Klima's own notation is slightly different, but amounts to the same. I skip a discussion of Klima's elaborate formal and semantic description of his system, as it easily reduces to restricted quantification.

function $\neg[G(x)]$. And since the satisfaction of the existential quantifier $\exists$ requires that there be at least one truth-yielding substitution for x, the substitution of ø for x makes (2) true despite (or perhaps because of) the fact that $[\![F]\!] = \emptyset$.

The same result follows for **A***-type sentences formulated as:

(3) $\forall x{:}F(x) \mid \neg[G(x)]$

Again, $\neg[G(ø)]$ (with ø for x) is true, and since ø is the only admissible substitution, all admissible substitutions produce truth, which makes (3) true. Truth thus results vacuously for **I***-type and **A***-type sentences when $[\![F]\!] = \emptyset$.

Although this looks good at first sight, there are serious problems connected with the zero element ø. First, if ø makes for falsity under any (extensional) predicate, how about the predicate *be the zero element*? When I say "ø is the zero element," then this should be false despite the fact that ø is the zero element! But apart from this paradox, a set-theoretic problem arises when it is asked whether ø does or does not belong to the extension of a predicate P. If it does, why does it produce falsity? If it does not, why is it an admissible substitution within the range of $[\![P]\!]$? From the point of view of ontology, ø is likewise highly problematic. No ontology will come up with such an element on independent grounds. It is merely an artifice, introduced for the sake of getting logic out of trouble without sacrificing the Bivalence Principle.

Critical readers might well ask why the null *element* should be considered so objectionable while the null *set* is fully accepted and treated with respect. The answer is that the null *set* is solidly anchored in modern set theory and makes for straightforward and sound systems of predicate logic, such as modern standard predicate logic or AAPL, whereas the null *element* would upset set theory no end and prevents a uniform semantic definition of the quantifiers in all logical languages except the language of restricted quantification. Yet, the objectors may observe, ontologically speaking, the null set is every bit as questionable as the null element: no ontologist would come up with such a notion on independent grounds. And here we can only agree. In fact, when the logic is made intensional, allowing, as it should, for virtual or intensional entities, the null set is written out of the script altogether. The assumption into the ontology of virtual entities, along with the screening off of presupposition failure by means of a special radical negation, obviates the need for both the null set and the null element. But this aspect cannot be addressed in the present paper (for a full discussion see Seuren, [8]: Ch. 10).

Parsons [6] tries to achieve the same as Klima, but in a different and rather far-fetched formalism, which, however, again boils down to restricted quantification. For Parsons, all categorical sentences center around the identity formula $x = y$. Both variables must be bound by a quantifier (EVERY or SOME) for a proposition (sentence) to arise.[9] The negation (NOT) can be inserted in front of the whole sentence, between quantifiers, and in front of $x = y$. The substitution mechanism is again of the restricted kind: variable ranges are restricted to a given predicate extension. When either x or y in $x = y$ fails to find a value in their range, $x = y$ is reckoned to be false. Parsons uses the phrase "when nothing is assigned to x or to y," but this cannot be taken literally, because when an element **e** is substituted for y and nothing is substituted for x, all one has is a formula '$x = e$', which is a sentential function (or predicate) and thus cannot have a truth value. It is hard to see

---

[9]Parsons has a third quantifier NO, but, as far as I can make out, this NO is equivalent with NOT SOME, in his system.

how this obstacle can be removed without positing a zero element, with all the problems it brings along.[10] Since Parsons uses the hyphen ('−') as his symbol for null assignment, we may assume that he does indeed work with Klima's zero or null element, even if he does not say so. Given the typographical oddness of the hyphen notation, I will go on using ∅ for the null element. But let us follow Parsons and see where it takes us.

Given his formal system, (4) is a proposition, where F is the restrictor predicate and G the matrix predicate:

(4) (some F x) not (some G y) x = y
    (for some x that is F, it not so that for some y that is G, x = y, or: Some F is not G)

Now, according to Parsons, when $[\![F]\!] = \emptyset$, so that, in Parsons' phrase, 'nothing' is assigned to the variable x (though, in fact, ∅ is assigned to x), the part (some G y) x = y is false for any value $e \in [\![G]\!]$ for the variable y, simply because x has received the value ∅. More precisely, since the existential quantifier some is defined, for cases where $[\![F]\!] = \emptyset$, as requiring for truth that the scope of (some G y)—in this case x = y—is true with e for y and ∅ for x, and since x = y is always false when ∅ stands for either x or y, it follows that ∅ = y is false for any $e \in [\![G]\!]$ substituted for y, so that (some G y) ∅ = y is false. Since (some G y) ∅ = y is preceded by the negation, the part not (some G y) ∅ = y is true. The same procedure is now repeated for (some F x): since the scope of (some F x)—in this case not (some G y) ∅ = y—is true, the whole sentence (4) is true, even though $[\![F]\!] = \emptyset$.

Curiously, Parsons gives double definitions for the quantifiers, one for cases where the extension of the restrictor predicate is nonnull and one for cases where it is null. The universal quantifier every is defined as requiring for truth either, when $[\![F]\!] \neq \emptyset$, that every substitution for the variable bound by it delivers truth or, when $[\![F]\!] = \emptyset$, that its scope turns out true. Analogously, the existential quantifier some is defined as requiring for truth either, when $[\![F]\!] \neq \emptyset$, that at least one substitution for the variable bound by it delivers truth or, when $[\![F]\!] = \emptyset$, that its scope turns out true.

However, the distinction between cases where $[\![F]\!] \neq \emptyset$ and cases where $[\![F]\!] = \emptyset$ vanishes when it is admitted that null assignment amounts to the substitution of the null element ∅. Then EVERY can be defined uniformly as requiring that every admissible substitution for the variable bound by it delivers truth, and SOME as requiring that at least one admissible substitution delivers truth. This shows better than anything else that, as long as one stays within the bounds of a regular ontology without ∅, it is not possible, in Parsons' system, to provide the quantifiers with single uniform definitions respecting the MODULO-*-PRINCIPLE formulated above. But with the introduction of the ontologically vicious element ∅, uniform definitions become possible. LOCA is, therefore, spoiled by its ontology.

But does it save the Square? My answer is that it does not. According to Aristotle (*Prior Analytics* 25ª8–14), I and I! (Some G is F—the inversion of I) are equivalent. That this is so, is easily seen: I is true just in case $[\![F]\!] \cap [\![G]\!] \neq \emptyset$; I! is true just in case $[\![G]\!] \cap [\![F]\!] \neq \emptyset$;

---

[10]Parsons seems to make use of the ploy used by Odysseus in Homer's Odyssey, when he escaped from the Cyclops Polyphemus, whose single eye he had just pierced with a red-hot club, by telling the Cyclops that his name was 'No-one'. When the blinded Polyphemus called on his fellow Cyclopes for help and revenge, he told them that No-one had pierced his eye, which made the other Cyclopes think that Polyphemus had been punished by the gods. In an epilogue called 'epicycle', Parsons discusses the philosophical question of the reification of 'nothing', quoting various ancient, medieval and modern authors. But he does not come to any clear conclusion.

$\llbracket G \rrbracket \cap \llbracket F \rrbracket = \llbracket F \rrbracket \cap \llbracket G \rrbracket$; hence, I! is true just in case $\llbracket F \rrbracket \cap \llbracket G \rrbracket \neq \emptyset$. Likewise for the equivalence of I* (Some F is not-G) and I*! (Some not-G is F): I* is true just in case $\llbracket F \rrbracket \cap \llbracket \overline{G} \rrbracket \neq \emptyset$; I*! is true just in case $\llbracket \overline{G} \rrbracket \cap \llbracket F \rrbracket \neq \emptyset$; $\llbracket F \rrbracket \cap \llbracket \overline{G} \rrbracket = \llbracket \overline{G} \rrbracket \cap \llbracket F \rrbracket$; hence, I*! is true just in case $\llbracket F \rrbracket \cap \llbracket \overline{G} \rrbracket \neq \emptyset$. In Aristotelian-Boethian and Aristotelian-Abelardian logic, as well as in standard modern logic, all of which have a symmetrical existential quantifier, both I and I* are thus equivalent with their inversions I! and I*!, respectively. In these logics, the following principle holds:

- Under the existential quantifier, the restrictor term and the matrix term (range and argument) are freely interchangeable *salva veritate*.

In the ø-variant of LOCA, however, which is claimed by its authors to be the proper analysis of the Square, this does not hold—unless the function ¬G(x) is (arbitrarily) banned and only lexically negative predicates like non-G(x) are allowed. This is shown as follows. Take a situation Σ where $\llbracket F \rrbracket = \{a, b, c\}$ (a, b, c $\in$ **ENT**) and $\llbracket G \rrbracket = \emptyset$. Some F is not G (I*), analyzed as ∃x:F(x) | ¬G(x), is true in Σ, since a, b and c are the legitimate substitutions for x in G(x) and ¬G(a), ¬G(b) and ¬G(c) are all true, which satisfies the truth condition for I* as formulated above. However, Some not-G is F (I*!), analyzed as ∃x:¬G(x) | F(x), is false in Σ, since, given that $\llbracket G \rrbracket = \emptyset$, the only legitimate substitution for x in F(x) is the zero element ø yielding falsity for F(x). LOCA thus not only fails to save the Square, it also undermines the set-theoretic foundations of the entire system.

One could propose to modify the substitution conditions for ø in such a way that all elements **e** in **ENT** are admitted in G(x), since all **e** in **ENT** satisfy the range ¬G(x), which would make the range of the formula ∃x:¬G(x) | F(x) equal to **ENT**. This would make I*! true, just like I*. But now the problem reappears for a situation Σ', in which, again, $\llbracket F \rrbracket = \{a, b, c\}$ but $\llbracket G \rrbracket = $ **ENT**. Now I* is false but I*! is undecidable since now all elements in **ENT** yield falsity for the range ¬G(x) of ∃x:¬G(x) | F(x), which leaves this formula without any legitimate substitution for x in F(x) and hence with its truth value undecided.

Despite the contortions of Parsons' formal system, it is clear that, in the end, it reduces again to the simple system of restricted quantification enriched with the zero element as shown above in the examples (2) and (3). And this system is, in the end, untenable for the reasons given. We will, therefore, not regard LOCA as a viable avenue of research. It is a typical example of a logic that works in a purely technical sense but is vitiated by a paradoxical and vicious ontological element which makes the system run into logical aporia.

If the Square is to be saved—and this is indeed a worthy ambition—a totally different approach is to be followed, involving presupposition theory combined with an admission of virtual entities into the ontology and of a third truth value (radical falsity) into the logic.[11] This requires no zero element and no restricted quantification, and it steers clear of the crass violation of natural intuitions perpetrated in LOCA. The Square is fully saved as a default bivalent logical system for those situations where all presuppositions of the sentences processed in the logic are fulfilled. The situations not fulfilling presuppositional truth conditions are then covered by a second, radical, negation which requires special

---

[11] This is not the place to elaborate the presuppositional solution to UEI. For an up-to-date and full discussion, see Seuren ([8]: Ch. 10).

emphasis and a special intonation contour for the sentence as a whole, restoring truth for sentences suffering from presupposition failure.

The question is thus: what price is one prepared to pay for saving the Square? The choice is between, on the one hand, the introduction of an ontologically and set-theoretically untenable null element plus a strident violation of natural intuitions and, on the other, the admission of virtual entities into the ontology and of two kinds of falsity into the logic, in strict adherence to natural intuitions. The former may deserve a place in abstract logical space, but it is sterile and without any wider interest; the latter produces a logic that is organically embedded in the human ecology of language, mind and world and helps explaining crucial facts of language and cognition. It is clear where our preferences should lie.

References

1. Ashworth, E.J.: Existential assumptions in late medieval logic. Am. Philos. Q. **10**(2), 141–147 (1973)
2. King, P.: William of Ockham: Summa Logicae. In: Shand, J. (ed.) Central Works of Philosophy, vol. I, pp. 242–269. Acumen, Chesham (2005)
3. Klima, G.: Ars Artium. Essays in Philosophical Semantics, Medieval and Modern. Institute of Philosophy, Hungarian Academy of Sciences, Budapest (1988)
4. Moody, E.A.: Truth and Consequence in Mediæval Logic. North-Holland, Amsterdam (1953)
5. Parsons, T.: The traditional Square of Opposition. In: Zalta, E.N. (ed.) The Stanford Encyclopedia of Philosophy (October 1, 2006 revision). http://plato.standford.edu/entries/square/
6. Parsons, T.: Things that are right with the traditional Square of Opposition. Logica Univers. **2**, 3–11 (2008)
7. Russell, B.: On denoting. Mind **14**, 479–493 (1905)
8. Seuren, P.A.M.: The Logic of Language (= vol. II of Language from Within). Oxford University Press, Oxford (2010)
9. Seuren, P.A.M.: From Whorf to Montague. Oxford University Press, Oxford (to appear)
10. Thompson, M.: On Aristotle's Square of Opposition. Philos. Rev. **62**(2), 251–265 (1953)

P.A.M. Seuren (✉)
Max Planck Institute for Psycholinguistics, PO Box 310, 6500 AH Nijmegen, The Netherlands
e-mail: pieter.seuren@mpi.nl

The Right Square

Hartley Slater

Abstract It is shown that there is a way of interpreting the traditional Square of Opposition that overcomes the main historic problem that led to the present supremacy of modern predicate logic over Aristotelian Syllogistic: handing multiply quantificational, relational expressions like 'All boys love some girls'. The interpretation has other advantages, is plausibly Aristotle's own, original view, and was certainly known by several significant Aristotelian logicians in the late nineteenth century, and the early part of the twentieth century, when Syllogistic was falling from favour. That leads to a puzzle about why the proposed interpretation was not seen to overcome the problem of multiple generality at the time, and some points are made showing what might need to change before the interpretation is more widely accepted.

Keywords Square of Opposition · Conditional probability · Logically proper names · Epsilon terms

Mathematics Subject Classification Primary 03A05 · Secondary 03B48 · 03B10

1 A Historical Puzzle

The recent resurgence of interest in Aristotle's Square of Opposition is surprising in view of the disregard with which it has been held in modern logic for nearly a century. On the reading of the statement forms in the Square of Opposition given in most logic texts of today, the only traditional relations that hold are the contradictoriness of the A and O forms, and the contradictoriness of the E and I forms. The contrariety of A and E, the subcontrariety of I and O, and the implications of I by A, and of O by E, have gained no support in the tradition that has grown up since Frege's and Russell's day. But not only that makes the present resurgence of interest in the Square surprising, since, also, little if any defense of the traditional relations has been common, let alone agreed, in the present resurgence of interest. Yet, as we shall see in this paper, a much more viable one is easily obtained through a reading of Aristotle's initial views.

But that leaves us with a considerable historical puzzle. For, given the defense available, what is most surprising is that the appropriate analysis was lost sight of by the generality of logicians in the early twentieth century, when concentration on merely universal quantifiers and their negations took over public attention in modern logic. Syllogistic Logic reigned from Aristotle's day for many centuries, and it was only in the latter part of the nineteenth century that it started to lose its place. Extensions with negative and complex terms were developed around then, but it could, in the end, provide no competition to the complexities and rigour that modern logic has derived from the work of Frege and Peirce. By the middle of the twentieth century Syllogistic was no longer respected. One supposed difficulty with Aristotelian Syllogistic, for instance, was its seeming restriction to monadic predicate logic. In this area the polyadic logics developed by Peirce and Frege were taken to win out. How could Aristotelian Syllogistic handle relational expressions like 'All boys love some girls', for example? That was, and continues to be, thought a great stumbling block. But not only can the appropriate interpretation easily overcome this problem, it also solves a substantial problem that contemporary predicate logic has made no attempt to address.

2 Aristotle's Understanding

Let us look at some central aspects of the problem of Existential Import, as it was understood in the early twentieth century.

An extended analysis of the possibilities of defending the traditional relations in the Square of Opposition by means of a rendering in the language of modern logic is to be found in P.F. Strawson *Introduction to Logical Theory* [6]. This book appeared just about the time when Aristotelian Logic was losing its traditional pre-eminent public place. On page 167 Strawson gives the standard modern view, but then he considers two further possibilities, on pages 169 and 173, the latter of which does verify all the traditional relations—though at a severe cost. The standard rendering has the A form as a universal quantification '$(x)(Sx \supset Px)$', the E form as a universal quantification '$(x)(Sx \supset \neg Px)$', the I form as an existential quantification '$(\exists x)(Sx \& Px)$', and the O form as an existential quantification '$(\exists x)(Sx \& \neg Px)$'. This reproduces just the contradictories, as before. By adding existential import just to the universal forms, i.e. by conjoining each of them with '$(\exists x)Sx$', Strawson then gets his second interpretation:

$$A \ (x)(Sx \supset Px) \ \& \ (\exists x)Sx$$
$$E \ (x)(Sx \supset \neg Px) \ \& \ (\exists x)Sx$$
$$I \ (\exists x)(Sx \& Px)$$
$$O \ (\exists x)(Sx \& \neg Px).$$

By adding further expressions of existential import to the universal forms, and then the negatives of all such additions to the particular forms, he gets his third interpretation:

$$A \ (x)(Sx \supset Px) \ \& \ (\exists x)Sx \ \& \ (\exists x)\neg Px$$
$$E \ (x)(Sx \supset \neg Px) \ \& \ (\exists x)Sx \ \& \ (\exists x)Px$$
$$I \ (\exists x)(Sx \& Px) \ v \ (x)\neg Sx \ v \ (x)\neg Px$$
$$O \ (\exists x)(Sx \& \neg Px) \ v \ (x)\neg Sx \ v \ (x)Px.$$

The second interpretation saves the contrariety of A and E, and the implication relations between A and I, and between E and O, but loses the contradictoriness of A and O, and E and I, and also the subcontrariety of I and O. The third interpretation fares better formally, because, as before, it does render all the traditional relations within the Square valid. But it loses out completely on natural language plausibility, since, for instance, Strawson has this to say about it: 'It is quite implausible to suggest that if someone says "Some students of English will get Firsts this year", it is a sufficient condition of his having made a true statement, that no one at all should get a First. But this would be a consequence of accepting the above interpretation for I' [6, 173].

Strawson took all this to be a part of his motivation for introducing his pre-suppositional approach, for which he is famous. But that move, whatever its other merits, is not justified on the basis of the other analyses Strawson presented, since there is further analysis available that delivers all the traditional relations within the Square, and which does not suffer any natural language implausibility. What might have persuaded Strawson to neglect this further analysis is the fact that, in addition to considering representations of the Square of Opposition he also was concerned to assess the validity of other Laws of the Syllogism on the various interpretations. So he was concerned to assess laws such as Conversion, Conversion *per Accidens*, Obversion, and Contraposition, as well as the 24 valid Syllogistic moods themselves. Only the last of his proffered interpretations justifies the totality of all of these laws, and that is quite implausible as we have seen. Even the alternative interpretation we shall now consider will not render valid all of the traditional 'Laws of the Syllogism,' so if the search was to find a way to do that, then this fourth formal analysis would fare no better in total than the two interpretations Strawson first considered. Nevertheless there is a very good reason why this further interpretation is to be preferred.

For, in direct response to Strawson's story, Manley Thompson, in the *Philosophical Review* for 1953 [7], put forward a further account of Aristotle's four forms. This was Aristotle's view of the situation according to Thompson, although others have had some doubts (see [2, 169], for instance). Certainly the account was well backed up by the passages from Aristotle himself that Thompson presented, and indeed this further interpretation was recommended by several notable Aristotelian logicians in the immediately preceding period such as Keynes, Johnson, and Popper [2, 169 n1].

One Aristotelian logician who kept to the further interpretation even wrote a whole text book using a version of it. That was Lewis Carroll, author of the 'Alice' books. In professional life Carroll was Charles Dodgson, a logician at the University of Oxford, and he produced several significant articles on logic, as well as a substantial book. On Carroll's analysis it is the positive forms A and I that carry existential import while their contradictories do not. Thus he says [1, 74]:

> A Proposition of Relation, beginning with "All" asserts (as we already know) that "*All* Members of the Subject are members of the Predicate." This evidently contains, as a *part* of what it tells us, the smaller Proposition "*Some* Members of the Subject are Members of the Predicate." [Thus, the Proposition "*All* bankers are rich men" evidently contains the smaller Proposition "*Some* bankers are rich men."] The question now arises "What is the *rest* of the information which this Proposition gives us?"

After some investigation he concludes:

> Hence a Proposition of Relation, beginning with "All" is a *Double* Proposition, and is *equivalent* to (i.e. gives the same information as) the *two* propositions *Some* Members of the Subject are Members of the Predicate; *No* Members of the Subject are members of the Class whose Differentia is *contradictory* to that of the Predicate. [Thus, the Proposition "*All* bankers are rich men" is a *Double* Proposition, and is equivalent to the *two* Propositions (1) "Some bankers are rich men"; (2) "No bankers are poor men."]

That means that all the relations in the Square of Opposition are validated, along with the 24 Syllogistic moods, although, in tune with many of Aristotle's remarks, not all the laws of Obversion, or the laws of Contraposition, for instance, hold, on the following account:

$$A (x)(Sx \supset Px) \& (\exists x)(Sx \& Px)$$
$$E (x)(Sx \supset \neg Px)$$
$$I (\exists x)(Sx \& Px)$$
$$O (\exists x)(Sx \& \neg Px) \vee (x)(Sx \supset \neg Px).$$

Thus SEP does not imply SA¬P, and SOP does not imply SI¬P; and neither does SAP imply ¬PA¬S, or SOP imply ¬PO¬S (see [7, 262–265]). The lack of implication in the latter case, for instance, is because the O form is now read with an external negation: 'Not all Ss are Ps', in place of 'Some Ss are not Ps', which has an internal or predicate negation. 'Not all Ss are Ps', as can be seen, is the disjunction of 'Some Ss are not Ps' and 'No Ss are Ps' (or 'There are no Ps'). What has come down to us from antiquity is a mix of this account together with additions probably provided by Boethius amongst others (see [2, 126f]), and it is the totality of this whole tradition that Strawson was trying to justify, without success. Indeed it is not justifiable at all, once one takes a more rigorous look at it than the medievalists evidently took.

What additionally substantiates the above as Aristotle's reading, according to Thompson, is that he applied the principle of it to other cases. For Aristotle applied the distinction between external and internal negations to singular statements, saying that both 'Socrates is well' and 'Socrates is ill' would be false if 'Socrates does not exist' was true [7, 254-5]. Taking 'Socrates' to be a 'disguised description', in the manner of Russell (cf. [3]), this shows that Russell's analysis of definite descriptions followed both Aristotle's and Carroll's line of analysis. In fact it is well known that Russell thought highly of Lewis Carroll's work, so it could well have been one inspiration for Russell's Theory of Descriptions. Thus, for Russell, 'The king of France is bald' entails 'Some king of France is bald'. But 'It is not the case that the king of France is bald' does not entail 'The king of France is not bald'. The former contains an external negation, and in it 'the king of France' has a 'secondary occurrence' as a result. The latter contains an internal, or predicate negation, and in it 'the king of France' has a 'primary occurrence' as a result.

3 Multiple Generality

So how does this account overcome the standard objection that Aristotelian Syllogistic cannot be extended to relational expressions? It is because it parallels a *probabilistic* account. For a probabilistic analysis of the Aristotelian forms also supports the variant reading. Thus $pr(Yx/Xx) = 1$ (i.e. 'All Xs are Ys') entails $Pr(Yx/Xx) > 0$ (i.e. 'Some Xs are Ys'). But $pr(Yx/Xx) \neq 1$ (i.e. 'Not all Xs are Ys') does not entail $Pr(\neg Yx/Xx) > 0$ (i.e. 'Some Xs are not Ys'), since it is possible that $pr(Xx) = 0$ (i.e. 'there are no Xs') in which case the probability is not defined. That means that the $pr(Yx/Xx)$ is certainly not 1, although it is not greater than 0, since it cannot be given a value at all. The probabilistic analysis is applicable also, of course, to many other quantifiers. Thus

Most Xs are Ys

can be represented as

$$pr(Yx/Xx) > 1/2,$$

and

Few Xs are not Ys

(where by 'Few Xs' I do not mean 'Few, if any, Xs') can be represented as

$$pr(\neg Yx/Xx) < 1/2.$$

This latter facility is a feat that standard predicate logic has no means of tackling, and yet it is supported by everyday facts about many other quantifiers. Indeed there is very strong supporting evidence for the probabilistic interpretation of the A and O forms, when one considers other quantifiers. For 'Almost all Xs are Ys', 'Most Xs are Ys', and 'A lot of Xs are Ys' surely all entail 'Some Xs are Ys'. Also a negative form like 'Not a lot of Xs are Ys', for instance, unlike the related positive form 'A few Xs are not Ys', allows it to be possible that no Xs are Ys (or that there are no Xs at all). So, unlike with the form where there is an internal negation ('A few Xs are not Ys'), there is no entailment from the form with the external negation ('Not a lot of Xs are Ys'), to 'Some Xs are not Ys'. All of this parallels the behaviour of 'all' and 'some not' on the probabilistic interpretation.

But not only can probabilistic analyses handle cases involving 'all' and 'some not' appropriately, they also can be easily generalised to *relations* involving these as well as many other quantifiers, which is a feat quite out of the question using just the devices in modern logic. Thus not only is

All boys love some girls,

re-expressible as

$$pr([pr(Lxy/Gy) > 0]/Bx) = 1,$$

but

Most boys love few girls

is

$$pr([pr(Lxy/Gy) < 1/2]/Bx) > 1/2.$$

(If probabilities are given in terms of proportions of existent cases, there is no difficulty with these nested probabilities.) So, whatever it was that made Carroll, Keynes, Johnson, Popper, and many others, miss these generalisations, clearly on the right interpretation of the Square there is a distinct possibility of Aristotelian Syllogistic competing much more strongly with the current paradigm, and maybe even regaining the crown that it lost.

4 Proper Names

But it can only do that if the current paradigm gives ground on the logic of individual statements. For, despite Russell, proper names like 'Socrates' are not treated as disguised descriptions in standard texts on modern logic, which leads to the internal relation between 'Socrates is well' and 'Socrates is ill' being unrepresentable. Only each of these expressions separately, together with their contradictories is part of the modern system, becoming, for instance, 'Ws', 'Is', '¬Ws', '¬Is'. If 'Socrates' is taken to be a disguised description, on the other hand, we can exhibit the four Aristotelian interrelated forms thus:

Socrates is well: $(\exists x)(Sx \& Wx)$, Socrates is ill: $(\exists x)(Sx \& \neg Wx)$,
Socrates is not well: $(x)(Sx \supset \neg Wx)$, Socrates is not ill: $(x)(Sx \supset Wx)$,

all under the condition that there is at most one S, i.e. that $(y)(z)((Sy \& Sz) \supset y = z)$.

In fact, from the standpoint of Russell, the use of ordinary proper names in place of variables is logically impossible, since the only name that could replace a variable would be a 'logically proper name' in Russell's terminology, i.e. a name for something whose existence cannot be questioned. But the text-book tradition seems to have forgotten this more strict understanding of the matter. And Russell cannot escape blame for the mix-up. For in *Principia Mathematica* [4], not only were logically proper names not provided, but also descriptive terms were allowed into what otherwise would be the place for logically proper names, in such expressions as '$(\exists y)(y = \iota x Kx)$'. So what else were Russell's followers to use as substitutes for variables but ordinary proper names and the like?

The reprehensibility of the text-book tradition, and especially its leaders, is heightened once it is realised that there is in fact no difficulty in finding the required logically proper names in classical logic. I have pointed this out in a large number of publications starting in 1986 (for instance, most recently [5]). For one classical theorem is:

$$(\exists x)\big((\exists y)Ky \supset Kx\big),$$

so the existence of certain objects is guaranteed in classical logic, and in the epsilon calculus they are given names: epsilon terms. Thus

$$(\exists y)Ky \supset K\epsilon x Kx,$$

is a theorem in the epsilon calculus, and

$$(\exists y)(y = \epsilon x Kx),$$

containing what Russell called a 'complete' term for an individual [3], is also a theorem there, while

$$(\exists y)(y = \iota x Kx),$$

containing Russell's 'incomplete' iota term still isn't. Certainly it is contingent whether anyone is *a sole king of France* (i.e. $(\exists y)(y = \iota x Kx)$), but what is not contingent is that there is something that is *a sole king of France if anything is* (i.e. $(\exists y)(y = \epsilon x Kx)$).

It is because of the ready availability of 'logically proper names' in this straightforward way that it is particularly unfortunate that Russell allowed his iota expressions to be thought of as individual terms in his system. For from this it too easily seems that contingently descriptive terms—and even ordinary names, and indeed any other singular term or phrase that is only contingently applicable—can take the place of variables. It is this, therefore, that primarily needs to be corrected in current predicate logic if there is to be a recovery of the Aristotelian system. For it is the common use of ordinary proper names as substitutes for individual variables in modern logic that has led to the abandonment of the Aristotelian four fold logic for contingent individuals.

References

1. Bartley, W.W. III (ed.): Lewis Carroll's Symbolic Logic. C.N. Potter, New York (1977)
2. Prior, A.N.: Formal Logic. Oxford University Press, Oxford (1962)
3. Russell, B.: On denoting. Mind **14**, 479–493 (1905)
4. Russell, B., Whitehead, A.N.: Principia Mathematica. Cambridge University Press, Cambridge (1910)
5. Slater, B.H.: Hilbert's epsilon calculus and its successors. In: Gabbay, D., Woods, J. (eds.) Handbook of the History of Logic, vol. 5. Elsevier, Burlington (2009)
6. Strawson, P.F.: Introduction to Logical Theory. Methuen, London (1952)
7. Thompson, M.: On Aristotle's square of opposition. Philos. Rev. **62**, 251–265 (1953)

H. Slater (✉)
University of Western Australia, Crawley, WA 6009, Australia
e-mail: hartley.slater@uwa.edu.au

Oppositions and Opposites

Fabien Schang

Abstract A formal theory of oppositions and opposites is proposed on the basis of a non-Fregean semantics, where opposites are negation-forming operators that shed some new light on the connection between opposition and negation. The paper proceeds as follows.

After recalling the historical background, oppositions and opposites are compared from a mathematical perspective: the first occurs as a relation, the second as a function. Then the main point of the paper appears with a calculus of oppositions, by means of a non-Fregean semantics that redefines the logical values of various sorts of sentences. A number of topics are then addressed in the light of this algebraic semantics, namely: how to construct value-functional operators for any logical opposition, beyond the classical case of contradiction; Blanché's "closure problem", i.e. how to find a complete structure connecting the sixteen binary sentences with one another.

All of this is meant to devise an abstract theory of opposition: it encompasses the relation of consequence as subalternation, while relying upon the use of a primary "proto-negation" that turns any relatum into an opposite. This results in sentential negations that proceed as intensional operators, while negation is broadly viewed as a difference-forming operator without special constraints on it.

Keywords INRC Group · Negation · Opposite-forming operators · Opposition · Proto-negation

Mathematics Subject Classification Primary 03B35 · Secondary 03B05 · 03B65

1 The Historical Background of Logical Oppositions

To begin with, everyone knows the "famous square of opposition", but what it consists in is less clear. Moreover, almost everyone wrongly equates this polygon with the entire theory of opposition. Let us see this into detail.

1.1 Aristotle's Square?

The theory of opposition is to be traced back from Aristotle's logical corpus [1–3], where the Stagirite wanted to examine the whole list of necessary conclusions that could be derived from any pair of sentences varying with respect to their quantity (universal or

J.-Y. Béziau, D. Jacquette (eds.), *Around and Beyond the Square of Opposition*, 147–173
Studies in Universal Logic, DOI 10.1007/978-3-0348-0379-3_11, © Springer Basel 2012

Fig. 1 The logical squares
and triangles of contraries for
quantified and modal
sentences

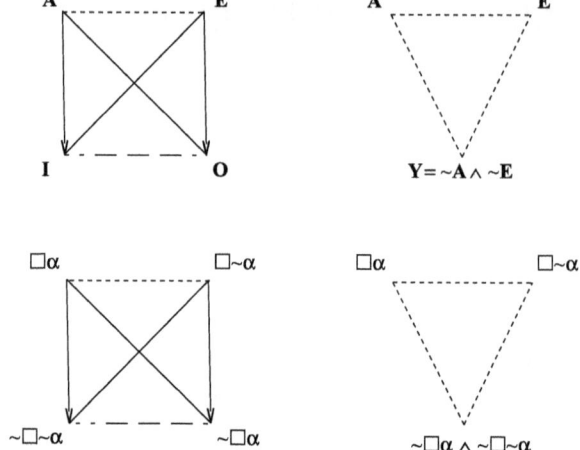

particular) or quality (affirmative or negative). But contrary to the current expression of "Aristotelian square", Aristotle never used and wouldn't have accepted the logical square as a proper diagram for logical oppositions.[1]

One main reason why this theory hardly went beyond traditional logic is that Aristotle sticked to two sorts of opposition, viz. contradiction and contrariety. The former follows the laws of contradiction and excluded middle: two contradictory sentences cannot be true together and cannot be false together, while the latter follows the second law only (two contrary propositions can be false together). Furthermore, why bother with a theory of opposition if the only proper task of logic is to state what follows from what in a consequence relation? Be that as it may, the history of logic can testify to the relative poverty of developments about logical opposition taken in isolation. Despite a special attention to opposition in the medieval works (Abelardus, Buridan, Burley, Ockam, Paul of Venice, Peter of Spain, William of Sherwood), all seems to show that the theory of opposition is nothing but a minor section within the area of logic.

1.2 Beyond and Beneath the Square

Nevertheless, a certain revival of the subject-matter occurred during the fifties under the impetus of two French philosophers: Augustin Sesmat [18] and Robert Blanché [5], who independently gave rise to an extension of the well-known Aristotelian square of opposition **AEIO**. In a nutshell, Blanché claimed that the ambiguous meaning of possibility in the Aristotelian modal syllogistics justifies a further distinction between pure or one-sided and two-sided possibility (contingency): the former means that a possible proposition is not impossible but can be necessary, whereas the latter means that a possible proposition is

---

[1] Actually, the geometrical presentation of logical oppositions has been inspired by Apuleius ($\cong$123–170 C.E.): in his *Peri Hermeneias*, he added the symmetric relation of subcontrariety as paralleling the line of contrariety; this didn't still result in a square, but a crossed polygon. Then Boethius (480–525 C.E.) turned this quadrilateral into a genuine square by introducing the further relation of subalternation.

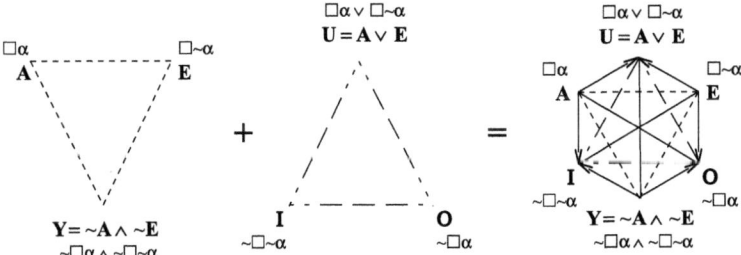

Fig. 2 The two logical triangles and in Blanché's logical hexagon

neither impossible nor necessary. Blanché added that possibility and existential quantification share a common feature in the so-called "reflexive" or natural (informal) reading: their sentential expression is expected to exclude necessity and universality, respectively, so that Blanché and Sesmat favored the use of triangles rather than the Aristotelian square (including the unnatural one-sided possibility **I**). (See Fig. 1.)

Opposition essentially proceeds by means of the classical (or standard) negation $\sim$, given the dual relation between universal and existential quantifiers ($\sim\forall\sim\,\leftrightarrow\,\exists$) or between necessity and possibility ($\sim\Box\sim\,\leftrightarrow\,\Diamond$). It entails that the above square of quantification assumes a relation of logical consequence from $\forall$ to $\exists$. That this relation fails whenever the subject term is empty has been largely discussed in the literature under the name of "existential import", contributing to the decline of traditional logic; but it could be replied that such a failure is not so much due to the logical square as to its lexicalization. Indeed, the logical square can be seen to be fully valid when only negative sentences are used to express the vertices **O** and **I**.[2]

As opposition is closely related to negation, a care for symmetry led Blanché to supplement his triangle of contraries with an opposite triangle of subcontraries whose sixth uppermost edge **U** is an unnatural expression of determinacy; **U** corresponds to the contradictory of two-sided possibility $\mathbf{Y} = \sim\mathbf{U}$ and eventually extends the debated square into a logical hexagon of oppositions **AUEIYO**. (See Fig. 2.)

However, a strict combination of the two above triangles should not lead to the right-sided hexagon as presented with its arrowed lines; these correspond to the fourth opposition of subalternation. At the same time, Aristotle never mentioned the latter in his logical works about opposition, and Jean-Yves Béziau supported this point in [4] in order to leave subalternation aside and reduce the hexagon to a more restricted star of logical oppositions. (See Fig. 3.)

1.3 Is Subalternation an Opposition?

Why not to include subalternation among the logical oppositions? It is usually assumed that logical opposition refers to incompatibility and means that two opposite sentences cannot be true together. Now given that such is the case in a subalternation, where the

---

[2]See e.g. [16] in this respect, where an illocutionary reading of statements entails that the denial of the universal affirmative **A** is not equivalent with the assertion of its particular negative **O**. See also [8] for a classical defence of the square.

Fig. 3 Béziau's logical star
of oppositions

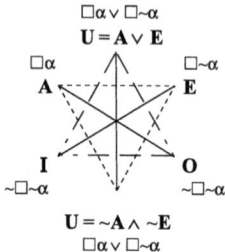

subaltern is necessarily true whenever its superaltern is, it naturally follows that Aristotle could not include it among his theory of opposition because the latter is closely related to negation.

But if so, why not to exclude subcontrariety by the same token? Two subcontrary propositions cannot be false together but can be true together; and if a necessary condition for being an opposite is to differ in truth-value, then subalternation also fulfills this requirement since a subaltern can be true whereas its superaltern is false. [4] follows Aristotle's stringent view of opposition as incompatibility, however, which states that two opposite sentences can never be true together. While noting that subcontraries are merely "verbal" oppositions according to [3, 4] still includes them into the class of oppositions whether for indirect or constructive reasons: indirect, in the sense that subcontraries amount to *contradictories of contraries* and are thus defined in terms of direct or genuine opposites; constructive, in the sense that refusing subcontrariety leads to incomplete polygons that cannot yield a full geometrical presentation of logical oppositions.

In the light of these preliminaries, the paper wants to give a broader analysis of logical oppositions. Rather than restricting opposition to a relation of incompatibility between two propositions, a more comprehensive presentation should include some other sorts of syntactic object (beyond quantified sentences) and relation (beyond incompatibility). For this purpose, let us abstract from the Aristotelian context and construct a purely formal theory of logical opposition.

2 Opposition and Opposites

To begin with, it is worthwhile noticing that no opposition can occur without opposites: any two concepts, terms, propositions or whatever else express an *opposition* if and only if either of these is the *opposite* of the other one. However, no genuine theory of opposites has ever been proposed in support of the current theory of opposition. In order to do justice to this logical primacy of opposites over oppositions, let us consider the difference between the two related concepts from a mathematical point of view.

2.1 Two Mathematical Notions

Take the opposition between black and white as an example. It is well known that these stand into a relation of contrariety, which entails that black is the contrary of white just as white is the contrary of black. Furthermore, such an opposition between concepts can be transferred to the sentential level by saying that the related sentences "This table is black" and "This table is white" cannot be true together but can be false together (whenever the table is neither white nor black). The mathematical difference between an opposition and an opposite is that the former is a relation, the latter a function. In the sentence "Tom and Jim are mutual brothers", brotherhood is the expression of a relation between the two individuals Tom and Jim. In logical symbols:[3]

$$(\exists x)(\exists y)B\big(T(y), J(x)\big)$$

If this sentence is true, then it is the case that both Jim is Tom's brother and Tom is Jim's brother; this means that the two-place relation of brotherhood can be translated into a function of being brother of. It results in two synonymous sentences as follows:

$$(\exists x)(\exists y)\big(J(x) = B(T(y))\big)$$
$$(\exists x)(\exists y)\big(T(y) = B(J(x))\big)$$

By analogy, the same logical reformulation can be applied to the higher-order sentence "The sentences α and ψ are mutual contradictories" once quantification over sentential variables is allowed:

$$(\exists \alpha)(\exists \psi)C(\alpha, \psi)$$

Again, this entails that ψ is the contradictory of α and, conversely, that α is the contradictory of ψ:

$$(\exists \alpha)(\exists \psi)\big(\psi = C(\alpha)\big)$$
$$(\exists \alpha)(\exists \psi)\big(\alpha = C(\psi)\big)$$

Oppositions and opposites can be further described in a syntactic and semantic way.

Syntactically speaking, opposition is a sentence-forming relation attached to predicate variables if its arguments are concepts, or sentence variables if its arguments are sentences; it is usually presented as a two-place predicate, but more than two arguments can be related by it.[4] As to the opposite, it is a one-place function that yields another concept if attached to a concept, or another sentence if attached to a sentence. Only concepts and sentences are the syntactic objects concerned with oppositions and opposites: it does not make sense to say that a given individual is logically opposed to another one, so that we restrict the theory of opposition to predicate or sentence variables as its arguments.

Semantically speaking, a relation of opposition proceeds as a mapping from $V \times V$ to V (where V is a set of truth-values). In the sentence "Black and white are contrary to each other", for instance, contrariety is a relation that is attached to a pair of propositions like {this table is black, this table is white} and makes the whole sentence true in V. By contrast, a function of opposite is a mapping from V to V that turns the truth-value associated to a sentence into another one in V. In the sentence "ψ is the contrary of α",

---

[3] Tom and Jim are normally treated as individual constants, but they are taken here to be predicates following the Quinean paraphrase. This helps to compare these with the next opposite-forming operators, given that both grammatical categories proceed as one-place predicates or *functors*.

[4] Returning to Blanché's triangle of contraries, this can be seen as either a set of three 2-place relations ({$\Box\alpha, \Box\sim\alpha$}, {$\Box\alpha, \Diamond\alpha \wedge \Diamond\sim\alpha$} and {$\Box\sim\alpha, \Diamond\alpha \wedge \Diamond\sim\alpha$}), or a single 3-place relation ({$\Box\alpha, \Box\sim\alpha, \Diamond\alpha \wedge \Diamond\sim\alpha$}).

contrariety is attached to the truth-value of α and results into the truth-value of ψ. The entire sentence is true only if either α is false whenever ψ is true, or α and ψ are both false.

2.2 Opposition and Truth-Values

Nevertheless, such a purely formal presentation could be blamed for not taking the sentential content into account: formal logic is topic-neutral, but the traditional precondition for logical opposition is that opposite sentences to differ with respect to quality or quantity while being about the *same* subject and predicate. Thus "Every raven is black" and "Some raven is not black" are properly opposed to each other by differing only with respect to quality and quantity, the former affirming blackness of every raven and the latter negating it of some of these. Consequently, a purely formal characterization fails by reducing the relation of opposition to a relation between contentless truth-values. We challenge the content-dependent approach of opposition in the following, and the motivation for a purely formal theory of oppositions and opposites will be detailed in the next section.

More generally, a logical opposition is meant to be a relation between a sentence α and its opposite ψ. Letting OP stand for the relation of opposition and O for the opposite-forming function (or operator), it thus follows from it this general logical form:

$$OP(\alpha, \psi) = OP\big(\alpha, O(\alpha)\big)$$

2.3 Towards a Calculus of Logical Oppositions

That an opposition is taken to be logical should also mean that it deals with some given calculus upon values.

Truth-values are the expected values for a logic of opposition, and the appropriate calculus will be discussed later on. Let us recall that Aristotle also mentioned several additional oppositions between concepts outside his logical works, especially in [1]: "double" and "half" are said to be opposed by correlation, but nothing logical occurs in this opposition and the same holds for the oppositions by contrariety ("black" and "white") and privation ("blind" and "sighted"). It could strike one as surprising to say now that opposition by contrariety is not a logical opposition, given that it is a crucial component of the logical square. But again, contrariety is a logical relation only when applied to sentences rather than concepts: only the former can be said to be true or false.

What of its calculus? In standard sentential logic, some oppositions are expressed by truth-*functions*, i.e. binary connectives that assign truth-values to complex sentences in V. For instance, conjunction is a two-place sentential function that assigns truth to a whole sentence whenever its two arguments are also true, and falsity otherwise. Negation is the only one-place function (or operator) at hand in standard logic, since it turns any true single sentence into a false one and conversely.

Can it be said that each logical connective corresponds to a relation of opposition, and negation to an opposite-forming function? It seems so, while noting at the same time that standard logic cannot give a complete calculus for oppositions and opposites.

Fig. 4 The four logical
oppositions and their
corresponding four binary
connectives

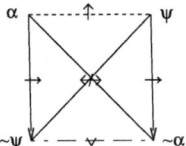

On the one hand, to each relation of opposition corresponds one of the sixteen standard connectives.[5] (See Fig. 4.)

This can be easily depicted by means of the standard truth-tables, where the various truth-value assignments for any combination of sentential arguments characterize the resulting oppositional relation between these. Thus α and ψ are opposed to each other through the truth-conditions of the standard connectives. (See Fig. 5.)

On the other hand, nothing is said here about the remaining twelve connectives of standard logic. Which opposition is established by an arbitrary standard connective $\oplus$ upon its arguments α and ψ in the whole sentence $\alpha \oplus \psi$ cannot be answered unless a mechanical procedure is found for this purpose. A least information is given by the above logical square, however: subcontraries have been previously said to be the contradictories of contraries, and this is shown by the opposition between the contrary pairs $\{\alpha, \psi\}$ and the subcontrary pairs $\{\sim\alpha, \sim\psi\}$. This entails two things: $\sim\alpha$ and $\sim\psi$ are the contradictories of α and ψ, respectively; classical negation $\sim$ is a contradictory-forming operator, since the relation $OP(\alpha, \sim\alpha)$ is a contradiction.

Let us symbolize the four traditional oppositions with capital letters, such that $\{CT, CD, SCT, SB\}$ is the set of logical oppositions OP including the four elements of contrariety (CT), contradiction (CD), subcontrariety (SCT), and subalternation (SB). Two questions remain unanswered thus far.

Firstly, classical negation $\sim$ is a contradictory-forming operator in the relation $CD(\alpha, \sim\alpha)$; but what is O for the other three relations? O is a unary operator, but standard logic includes only one such non-trivial function that is classical negation; this means that another theory has to be constructed to identify the other opposite-forming operators.

Secondly, every relation of opposition is symmetrical except for subalternation, given that $SB(\alpha, \psi) \neq SB(\psi, \alpha)$. How to account for this peculiarity?

contrariety

α	ψ	$\alpha \uparrow \psi$
T	T	F
T	F	T
F	T	T
F	F	T

contradiction

α	ψ	$\alpha \leftrightarrow \psi$
T	T	F
T	F	T
F	T	T
F	F	F

subcontrariety

α	ψ	$\alpha \vee \psi$
T	T	T
T	F	T
F	T	T
F	F	F

subalternation

α	ψ	$\alpha \rightarrow \psi$
T	F	F
T	T	T
F	T	T
F	F	T

Fig. 5 The four binary connectives and their truth-tables

---

[5]The number of the logical connectives in any logical system is $m^{(m^n)}$, where m is the number of truth-values and n is the number of arguments combined by n-ary connectives. In the standard logic of binary connectives, $m = n = 2$, hence a total number of $2^{(2^2)} = 2^4 = 16$ connectives. Now some alternative cases may be entertained, by changing either the number of truth-values or the number of combined sentences. Piaget focused upon the second option in [13] by considering the $2^{(2^3)} = 2^8 = 256$ ternary connectives within the standard two-valued logic.

A non-standard semantics will be presented in the next section, in order to answer these questions and even more.

3 Question-Answer Semantics

A first step towards the formalization of oppositions has been suggested by Alessio Moretti's n-opposition theory (N.O.T.) in his so-called "Aristotelian P^Q-semantics" [11], where P refers to the number of answers and Q to the number of questions in a question-answer game.

3.1 The Legacy of N.O.T.

This game was meant to characterize the Aristotelian theory of opposition by means of a combinatorial set of ordered questions and answers.

Thus every given opposition results from a set of two ordered answers to questions about the compossibility of two arbitrary sentences. Let Q_1: "Can α and ψ be both false?", Q_2: "Can α and ψ both true?", and let P $= 1$ and P $= 0$ the two sorts of answers to be given accordingly: "yes" or "no", respectively. It results in a combinatorial set of four ordered pairs of answers, namely: [1|0], [0|0], [0|1], and [1|1]. Moretti argues that each of these pairs characterizes a logical opposition: contrariety, contradiction, subcontrariety, and subalternation.

While giving rise to a new relevant research field, this semantics suffers from two main defaults: on the one hand, a calculus for the P^Q-semantics is still to be found since no operator has been suggested to turn any paired value into another one; on the other hand, the characterization of subalternation as [1|1] is too broad since it includes the biconditional $\alpha \leftrightarrow \psi$ and cannot explain why ψ cannot be false once α is true in the relation SB(α, ψ).

Another question-answer game is purported to fulfill these two requirements: Question-Answer semantics (thereafter: **QAS**), which proceeds as an algebraic semantic frame.

3.2 Disjunctive Normal Forms and Logical Values

First and foremost, **QAS** is a non-Fregean semantics: the *sense* of any sentence α is an ordered finite set of n questions $\mathbf{Q}(\alpha) = \langle \mathbf{q}_1(\alpha), \ldots, \mathbf{q}_n(\alpha) \rangle$ (where $n \geq 1$) about this sentence; its *reference* is not a variable truth-value but another sort of constant logical value, viz. an ordered finite set of answers $\mathbf{A}(\alpha) = \langle \mathbf{a}_1(\alpha), \ldots, \mathbf{a}_n(\alpha) \rangle$ to the corresponding questions.[6]

Just as in Moretti's semantics, there are m^n different logical values for a specific set of sentences (where m is the number of the different sorts of answers available for each

---

[6]For a more detailed presentation and application of this semantics, see e.g. [15–17].

of the n questions) and, accordingly, $2^4 = 16$ logical values for any two-place sentential function $\alpha \oplus \psi$. The $n = 4$ ordered questions relate to the Disjunctive Normal Form of any such complex sentence, so that the following expresses the sense of a binary sentence by the truth-conditions of its arguments α and ψ: $\mathbf{Q}_1$: "Can "α and ψ be both true?", $\mathbf{Q}_2$: "Can α be true and ψ be false?", $\mathbf{Q}_3$: "Can α be false and ψ be true?", and $\mathbf{Q}_4$: "Can α and ψ be both false?". Symbolizing again a yes-answer by 1 and a no-answer by 0, we thus obtain a set of 16 logical values that refer to the meaning of a binary sentence $\Phi = \alpha \oplus \psi$. Its logical value $\mathbf{A}(\Phi) = \langle \mathbf{a}_1(\Phi), \dots, \mathbf{a}_n(\Phi) \rangle$ results in the following combinatorial set (omitting the brackets for sake of convenience):

(1) $\mathbf{A}(\bot) = 0000$ (antilogy);
(2) $\mathbf{A}(\alpha \wedge \psi) = 1000$ (conjunction);
(3) $\mathbf{A}(\alpha \nrightarrow \psi) = 0100$ (negated conditional);
(4) $\mathbf{A}(\alpha \nleftarrow \psi) = 0010$ (negated reverted conditional);
(5) $\mathbf{A}(\alpha \downarrow \psi) = 0001$ (rejection, or negated disjunction);
(6) $\mathbf{A}(\alpha) - 1100$ (truth of α);
(7) $\mathbf{A}(\alpha \nleftrightarrow \psi) = 0110$ (negated biconditional);
(8) $\mathbf{A}(\sim\alpha) = 0011$ (negated truth of α);
(9) $\mathbf{A}(\alpha \leftrightarrow \psi) = 1001$ (biconditional);
(10) $\mathbf{A}(\psi) = 1010$ (truth of ψ);
(11) $\mathbf{A}(\sim\psi) = 0101$ (negated truth of ψ);
(12) $\mathbf{A}(\alpha \vee \psi) = 1110$ (disjunction);
(13) $\mathbf{A}(\alpha \leftarrow \psi) = 1101$ (reverted conditional);
(14) $\mathbf{A}(\alpha \rightarrow \psi) = 1011$ (conditional);
(15) $\mathbf{A}(\alpha \uparrow \psi) = 0111$ (incompatibility, or negated conjunction);
(16) $\mathbf{A}(\top) = 1111$ (tautology).

It is well known that the half of these values stand for the classical negation of the remaining half: (1) $= \sim(16)$, (2) $= \sim(15)$, (3) $= \sim(14)$, (4) $= \sim(13)$, (5) $= \sim(12)$, (6) $= \sim(8)$, (7) $= \sim(9)$, and (10) $= \sim(11)$. This means that one half of the above list already results from the application of one contradictory-forming operator to one complex sentence from the other half.

3.3 Piaget's INRC Group and Opposites

Jean Piaget showed in [14] that some further opposite-forming operators also help to turn a binary sentence into another one in addition to the function of classical negation. According to his so-called "theory of reversibility", here is the INRC Group that proceeds as follows in the line of Klein's *Vierergruppe*. For every binary sentence Φ of standard logic and its logical value $\mathbf{A}(\Phi) = abcd$, there are four sorts of higher-order operators to be applied as follows:

Identity $\mathbf{A}(\mathrm{I}(\Phi)) = abcd$
Inversion $\mathbf{A}(\mathrm{N}(\Phi)) = a'b'c'd'$
Reciprocity $\mathbf{A}(\mathrm{R}(\Phi)) = dcba$
Correlation $\mathbf{A}(\mathrm{C}(\Phi)) = d'c'b'a'$

Fig. 6 Piaget's matrix for the
INRC Group

X \ Y	I	N	R	C
I	I	N	R	C
N	N	I	C	R
R	R	C	I	N
C	C	R	N	I

Just as the operator of inversion N corresponds to the truth-function of standard negation, the operator of identity I corresponds to the trivial truth-function of affirmation, turning its argument into itself. And just as the truth-function of affirmation is to be strictly distinguished from a yes-answer in **QAS**, the truth-function of classical negation applied to sentences differs from the metalinguistic operation of *denial*: denying a given answer $\mathbf{a}_i(\Phi) = x$ (where $x = 1$ or 0) results in x' and means that the initial yes- or no-answer is reverted into a no- or yes-answer. Actually, denial is logically prior to negation in the sense that it helps to define the truth-functional operation of standard negation in **QAS**, where the latter consists in denying each component $\mathbf{a}_i(\Phi)$ of a logical value $\mathbf{A}(\Phi)$ through the operator N.

3.4 Non-truth-functional, but Value-Functional Operators

In addition to the truth-functional affirmation I and negation N, the remaining two operators of reciprocity R and correlation C don't appear within the standard theory of truth-functions.

A conjecture is to the effect that these are *non-truth-functional* versions of negation, or non-classical negations whose properties can be made value-functional within our non-Fregean semantics, i.e. a Boolean algebra of non-Fregean logical values.

For one thing, the set {I, N, R, C} includes a group of four commutative and involutive opposite-forming operators whose relative product results in another operator. That is, for any such operations X, Y, Z we have $XY = YX = Z$ and $XX = YY = ZZ = I$. This accounts for the standard law of double negation, when $X = N$: $NN(\Phi) = \Phi$, and the same holds for the remaining three operators. (See Fig. 6.)

Furthermore, the properties of INRC help to answer to our preceding three questions about OP.

Firstly, it turns out that R proceeds alternatively as a contrary-, contradictory- or subcontrary-forming operator, depending upon which value its argument Φ is given: R is a contrary-forming operator when the logical value of Φ contains only one yes-answer, a contradictory-forming operator when applying to two yes-answers, a subcontrary-forming operator upon three yes-answers, and even collapsing with I when its argument includes only yes- or no-answers (i.e. when α is either a tautology or an antilogy).[7]

---

[7]E.g. $R(1000) = 0001$, and $OP(1000, 0001) = OP(\alpha \wedge \psi, \alpha \downarrow \psi) = CT(\Phi, \Psi)$; $R(1100) = 0011$, and $OP(1100, 0011) = OP(\alpha, \sim\alpha) = CD(\Phi, \Psi)$; $R(1110) = 0111$, and $OP(1110, 0111) = OP(\alpha \vee \psi, \alpha \uparrow \psi) = SCT(\Phi, \Psi)$; $R(1111) = 1111$, and $OP(1111, 1111) = OP(\top, \top) = SB(\Phi, \Psi)$; $R(0000) = 0000$, and $OP(0000, 0000) = OP(\bot, \bot) = SB(\Phi, \Psi)$. See note 13 about the last two cases.

Secondly, SCT can be seen as a sort of mixed double negation applied to both the whole sentence and its arguments: the definition of subcontraries as "contradictories of contraries" amounts to the equation $SCT(\alpha, \psi) = CT(N(\alpha), N(\psi))$, where the relation of contrariety CT includes two complex arguments with contradictory-forming operators N.[8] It is in this sense that subcontrariety occurs as an "indirect" opposition: its arguments α and ψ can be both true, but its syntactic construction calls for negation-forming operators in O.

Thirdly, the asymmetry of SB entails that no operator O can be said to be a subalternation-forming operator *per se*: although $OP(\alpha, C(\alpha)) = SB(\alpha, \psi)$ for some α, a necessary condition for subalternation is this relation of opposition to be *ordered* since $OP(C(\alpha), \alpha) \neq SB(\psi, \alpha)$. Given that ordering cannot be accounted for by the operation itself, this means that something more must be added to the characterization of subalternation.

Actually, a Boolean algebra of logical oppositions shows that subalternation cannot be properly defined in usual terms of composable truth-values. Let $\cap$ and $\cup$ be the Boolean operations of meet and join such that $1 \cap x = 0 \cup x = x$, $0 \cap x = 0$, $1 \cup x = 1$, and $A(\alpha) \circ A(\psi) = \langle a_1(\alpha) \circ a_1(\psi), \ldots, a_n(\alpha) \circ a_n(\psi) \rangle$ for any $\circ \in \{\cap, \cup\}$. Then we have the four possible cases of logical opposition $OP(\alpha, \psi)$ in **QAS**: α and ψ can be both true only if there is at least one common yes-answer about their truth-conditions, and they can be both false if they share at least one common no-answer. This can be expressed algebraically as follows, where join and meet proceed pointwisely upon the elements of each logical value:

$OP(\alpha, \psi) = CT(\alpha, \psi)$	iff	$A(\alpha) \cap A(\psi) = A(\bot)$ and $A(\alpha) \cup A(\psi) \neq A(\top)$
$OP(\alpha, \psi) = CD(\alpha, \psi)$	iff	$A(\alpha) \cap A(\psi) = A(\bot)$ and $A(\alpha) \cup A(\psi) = A(\top)$
$OP(\alpha, \psi) = SCT(\alpha, \psi)$	iff	$A(\alpha) \cap A(\psi) \neq A(\bot)$ and $A(\alpha) \cup A(\psi) = A(\top)$
$OP(\alpha, \psi) = NCD(\alpha, \psi)$	iff	$A(\alpha) \cap A(\psi) \neq A(\bot)$ and $A(\alpha) \cup A(\psi) \neq A(\top)$

3.5 Consequence in Opposition

The fourth relation NCD is a weaker relation of mere *non-contradiction*, corresponding to the general case in which the two opposite sentences can be both true or both false. Subalternation generally appears as a subcase of non-contradiction, but it requires another condition to be satisfied.

Blanché states it in his own vocabulary of truth-cases and indeterminate sentences (those including three truth-cases, i.e. yes-answers):

> The indeterminates are implied by the determinates whose truth-case they contain: since each has three truth-cases, the latter is thus implied by three determinates $(\ldots)$[9]

---

[8]Note the logical difference between "contraries of contradictories" and "the contrary of the contradictory": the first plural expression includes one relation SCT and two functions N in the form $CT(N(\alpha), N(\psi))$, whereas the second singular expression includes no relation and two functions R and N in the form $R(N(\alpha))$.

[9][7, p. 137]. Let $A(\alpha \wedge \psi) = 1000$ be a case of determinate with only one truth-case, i.e. the yes-answer $a_1(\alpha) = 1$; accordingly, this sentence entails the three indeterminates $A(\alpha \vee \psi) = 1110$, $A(\alpha \leftarrow \psi) =$

Fig. 7 Czeżowski's logical
hexagon for binary sentences

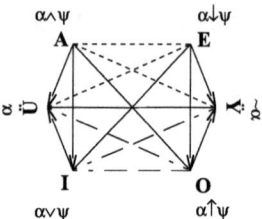

In other words, $SB(\alpha, \psi)$ is true if and only if every truth-case for α is also a truth-case for ψ; but the converse needn't hold, and this does account for the asymmetry of the relation of subalternation unlike that of non-contradiction. This characterization of subalternation as a *containment* relation between the truth-cases of α and ψ can be rephrased as follows:

$$OP(\alpha, \psi) = SB(\alpha, \psi) \qquad \text{iff} \qquad \mathbf{A}(\alpha) \cap \mathbf{A}(\psi) = \mathbf{A}(\alpha) \text{ and } \mathbf{A}(\alpha) \cup \mathbf{A}(\psi) = \mathbf{A}(\psi)$$

It clearly appears now why SB is an asymmetric relation: the truth of $SB(\alpha, \psi)$ entails that each truth-case of α is also a truth-case of ψ, whereas the truth of $SB(\psi, \alpha)$ would require the converse to hold and this cannot be the case unless α and ψ express one and the same sentence. Unlike the other three logical oppositions, subalternation should not be formed by a standard connective: while being frequently confused with the conditional $\rightarrow$, the criterion of containment that characterizes SB makes it similar to the metalinguistic relation of entailment or *consequence*. In other words, a parallel can be made between the general properties of subalternation and Tarski's logical consequence as truth-preservation. That is, for any α, ψ:

$$\alpha \models \psi \qquad \text{iff} \qquad \nu(\psi) = T \text{ whenever } \nu(\alpha) = T$$
$$SB(\alpha, \psi) \qquad \text{iff} \qquad \mathbf{a}_i(\psi) = 1 \text{ whenever } \mathbf{a}_i(\alpha) = 1$$

There are at least two reasons why NCD has never been mentioned thus far as a proper logical opposition. On the one hand, most of the literature on logical opposition merely focused upon the logical square and its limited stock of sentences where any two sentences that can be both true and both false happen to stand into a relation of subalternation. However, not every pair of such non-contradictories collapses into SB with some more complex logical polygons like Tadeusz Czeżowski's hexagon [9], for instance.[10] (See Fig. 7.)

Taking again the binary sentences of standard logic, the above relation between α and ψ is a case in point: these can be both true and both false, but not every truth-case of α is also a truth-case of ψ. On the other hand, the content-dependent view of logical opposition seems to reject that α and ψ could be properly opposed to each other; but again, our purely formal and embracing presentation of logical opposition wants to go beyond these

---

1101, and $\mathbf{A}(\alpha \rightarrow \psi) = 1011$. Note that this containment relation can be extended to what Blanché called as "semi-determinates", i.e. the binary sentences with two truth-cases or yes-answers: the determinate $\mathbf{A}(\alpha \nrightarrow \psi) = 0100$ entails the two semi-determinates $\mathbf{A}(\alpha) = 1100$ and $\mathbf{A}(\alpha \nleftrightarrow \psi) = 0110$, for instance.

[10]Initially, Czeżowski's hexagon was about quantified sentences and consisted in supplementing universal $\{\mathbf{A}, \mathbf{U}, \mathbf{E}\}$ and particular $\{\mathbf{I}, \mathbf{Y}, \mathbf{O}\}$ with singular sentences $\{\ddot{\mathbf{U}}, \ddot{\mathbf{Y}}\}$ ($Sa \wedge Pa$, $Sa \wedge \sim Pa$). But again, our purely formal approach of oppositions abstracts from the content of the sentences and uniquely takes the logical values into account.

preconditions and includes the cases in which any two sentences are compatible while differing by their content.

Let us explore further the algebraic properties of oppositions and opposites, by returning upon the binary connectives of standard logic and developing some new opposite-forming operators in the following.

4 A Theory of Opposites

It is recognized in [7] that [14] developed a famous problem of standard logic, namely: the interconnection between the standard connectives, and its visual presentation within an extended polygon of logical oppositions. Initially, a number of interpreted squares depicted the oppositions between binary sentences instead of quantifiers or modalities. It has been noted earlier that some of the related connectives embed the four standard relations of opposition, but they occur as arguments in these interpreted squares.

4.1 Functional Completeness

It is well known that *functional completeness* obtains in standard sentential logic: each logical connective can be defined in terms of basic ones by a number of intermediary steps. Thus incompatibility can be defined as a negated conjunction, for instance. But that does not mean that any such connective can be defined by *any* other one: although the single connective of incompatibility suffices to define any other connective, conjunction and negation are usually taken to be the two basic connectives for a definition of the remaining ones: rejection can be translated as a conjunction of negated sentences, disjunction as a negation of negated conjuncts, and so on.

At the same time, one hardly sees how any connective could be turned directly into any other one: how to turn rejection into tautology, for instance? This is not an impossible task, however: standard negation contributes to functional completeness, but its limited properties prevent one from having a direct definition between any two standard connectives. Such is the goal of the present section, in the light of the theory of opposites.

4.2 Blanché's "Closure Problem"

Let us call by "closure problem" the following one: how to relate any elements of a exhaustive set to each other into a closed logical structure? To begin with, a logical hexagon cannot suffice to structure a complete set of relations between binary sentences: for every set of n sentences, the number of oppositions between these corresponds to the decreasing sequence $(n - 1) + (n - 1) - 1 + \cdots + 1$ or, equivalently, $(n \times (n - 1))/2$. Since there are $n = 16$ elements in the set of binary sentences, a closed structure of their logical oppositions should include $(16 - 1) + (15 - 1) + \cdots + 1 = 120$ (or, equivalently, $(16 \times 15)/2 = 120$) instances of relation $OP(\alpha, \psi)$.

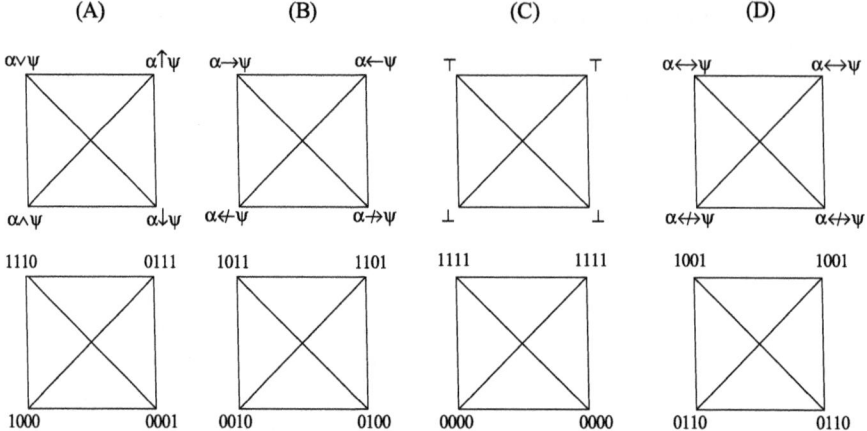

Fig. 8 Piaget's four quaterns and their logical values in **QAS**

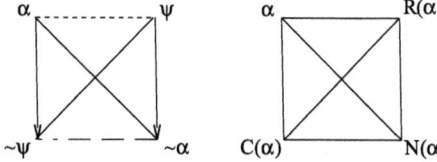

Fig. 9 Logical square of
oppositions, logical quatern
of opposites

Blanché noted in [6, 7] that [14] failed to find a closed structure in this respect:

> This problem has hardly been addressed thus far and never, to our knowledge, in a fully satisfying
> way. A quadratic shape of the four junctions can be found in Bocheński, but the idea has not been
> pushed farther. More developed are Piaget and Gottschalk's investigations. Unfortunately, they
> remain incomplete in that, while they organize the connectives into some well structured groups,
> these very groups remain isolated from each other without being into a whole structure.[11]

Indeed, [7] split the various logical interconnections into four independent quadratic
structures: the logical *quaterns*, supplemented with their logical values in **QAS**. (See
Fig. 8.)

[6, 7] blamed these squares from being plainly wrong: not only are the contraries and
subcontraries inverted in (A) and (B), but several sentences are opposed to themselves
in (C) and (D). Actually, Blanché was wrong himself by taking these figures for logical
squares of opposition: these are *quaterns* of opposites, where the relation obtaining between the edges does not count as much as the operators by means of which they are
constructed. Here is a way to depict the difference of perspective between a square of
opposition and a quatern of opposites. (See Fig. 9.)

Despite this, [7] relevantly proposed a more structured polygon of oppositions that
helps to link some of Piaget's quaterns. This polygon is called by him a "double hexadic
structure", where two logical hexagons are connected in an indefinitely repeated sequence
(see Fig. 10). As the whole structure includes a restricted set of 10 binary sentences, it
cannot contain more than 45 relations of opposition instead of the expected 120 ones.

---

[11][7, p. 122] (the author's translation).

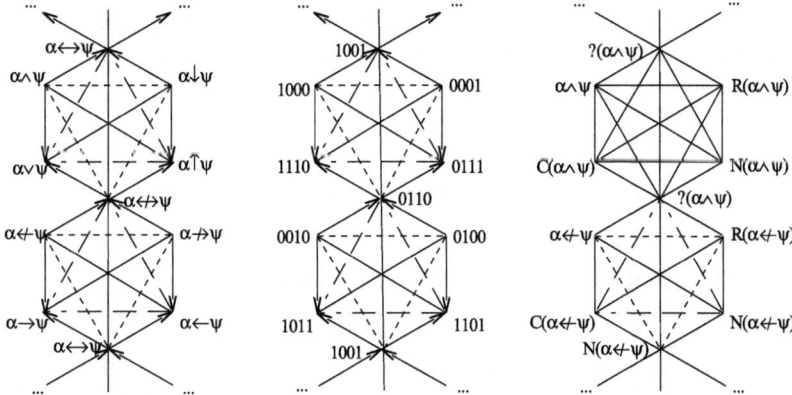

Fig. 10 Blanché's double helix of logical oppositions, their corresponding logical values in **QAS**, and their incomplete opposite-forming operators among INRC

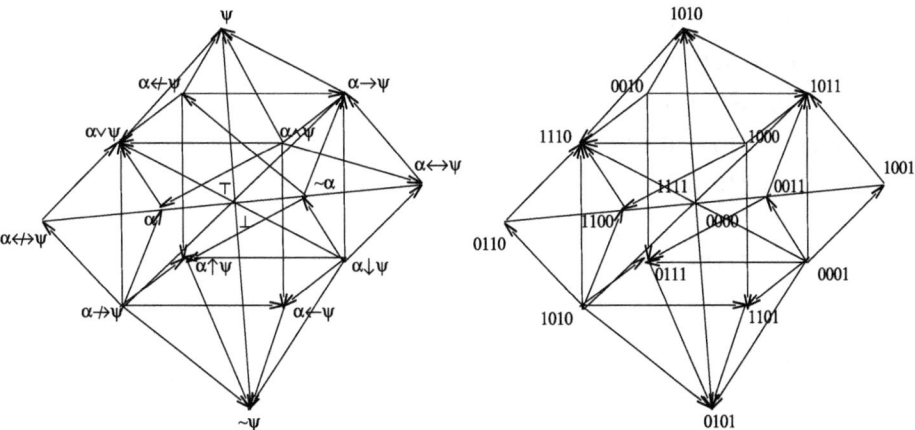

Fig. 11 Pellissier's tetraicosahedron of logical oppositions for binary sentences, and their corresponding logical values in QAS

Moreover, it can be seen that the "functional" counterpart of this polygon contains some blanks: not every relation OP can be constructed by means of an operator O in the right-sided structure, as witnessed by its unknown functions? $(\alpha \oplus \psi)$.

Actually, a solution to the closure problem has been found after [7]: each of the 16 binary sentences can be related to each other within a more complex polygon than Blanché's double hexadic structure. This solution has been discovered by Régis Pellissier in [12], illustrated by Dominique Luzeaux, Jean Sallantin and Christopher Dartnell in [10], and improved by Hans Smessaert in [20] with a closed structure that is a 3D tetraicosahedron (Fig. 11).[12]

---

[12]The choice of such a 3D polygon is motivated by a basic construction rule in [11], namely: a geometrical structure of logical oppositions is to be constructed so that its contradictory lines must intersect each other in the center of the polygon without overlapping each other.

Here is a complete list of the 120 relations $OP(\alpha, \psi)$ between any binary sentences α and ψ, according to the logical values of the binary sentences and the algebraic definitions of the oppositions in **QAS**.

- $CT(\alpha, \psi)$ (18 instances):

  $\{1000, 0100\}$; $\{1000, 0010\}$; $\{1000, 0001\}$; $\{1000, 0110\}$; $\{1000, 0011\}$; $\{1000, 0101\}$;
  $\{0100, 0010\}$
  $\{0100, 0001\}$; $\{0100, 0011\}$; $\{0100, 1001\}$; $\{0100, 1010\}$;
  $\{0010, 0001\}$; $\{0010, 1100\}$; $\{0010, 1001\}$; $\{0010, 0101\}$;
  $\{0001, 1100\}$; $\{0001, 0110\}$; $\{0001, 1010\}$

- $CD(\alpha, \psi)$ (8 instances):
  For every α: $\{\alpha, N(\alpha)\}$
- $SCT(\alpha, \psi)$ (18 instances):

  $\{1100, 1011\}$; $\{1100, 0111\}$;
  $\{0110, 1101\}$; $\{0110, 1011\}$;
  $\{0011, 1110\}$; $\{0011, 1101\}$;
  $\{1001, 1110\}$; $\{1001, 0111\}$;
  $\{1010, 1101\}$; $\{1010, 0111\}$; $\{1010, 1110\}$; $\{1010, 1011\}$;
  $\{1110, 1101\}$; $\{1110, 1011\}$; $\{1110, 0111\}$;
  $\{1101, 1011\}$; $\{1101, 0111\}$;
  $\{1011, 0111\}$

- $SB(\alpha, \psi)$ (65 instances)[13]
  For every α: $\langle \bot, \alpha \rangle$; $\langle \alpha, \top \rangle$

  $\langle 1000, 1100 \rangle$; $\langle 1000, 1001 \rangle$; $\langle 1000, 1010 \rangle$; $\langle 1000, 1110 \rangle$; $\langle 1000, 1101 \rangle$; $\langle 1000, 1011 \rangle$
  $\langle 0100, 1100 \rangle$; $\langle 0100, 0110 \rangle$; $\langle 0100, 0101 \rangle$; $\langle 0100, 1110 \rangle$; $\langle 0100, 1101 \rangle$; $\langle 0100, 0111 \rangle$
  $\langle 0010, 0110 \rangle$; $\langle 0010, 0011 \rangle$; $\langle 0010, 1010 \rangle$; $\langle 0010, 1110 \rangle$; $\langle 0010, 1011 \rangle$; $\langle 0010, 0111 \rangle$
  $\langle 0111, 0011 \rangle$; $\langle 0001, 1001 \rangle$; $\langle 0001, 0101 \rangle$; $\langle 0001, 1101 \rangle$; $\langle 0001, 1011 \rangle$; $\langle 0001, 0111 \rangle$
  $\langle 1100, 1110 \rangle$; $\langle 1100, 1101 \rangle$
  $\langle 0110, 1110 \rangle$; $\langle 0110, 0111 \rangle$
  $\langle 0011, 1011 \rangle$; $\langle 0011, 0111 \rangle$
  $\langle 1001, 1011 \rangle$; $\langle 1001, 1011 \rangle$
  $\langle 1010, 1110 \rangle$; $\langle 1010, 1011 \rangle$
  $\langle 0101, 1101 \rangle$; $\langle 0101, 0111 \rangle$

---

[13]$OP(\alpha, \alpha)$ is not counted among the range of SB, because it is assumed there that the two relata are different sentences. But strictly speaking, the relation of self-identity can be seen as a kind of subalternation since it satisfies its definition: $\mathbf{A}(\alpha) \cap \mathbf{A}(\alpha) = \mathbf{A}(\alpha)$ and $\mathbf{A}(\alpha) \cup \mathbf{A}(\alpha) = \mathbf{A}(\alpha)$. See note 6.

- NCD(α, ψ) (12 instances)

  $\{1100, 0110\}$; $\{1100, 1001\}$; $\{1100, 1010\}$; $\{1100, 0101\}$

  $\{0110, 0011\}$; $\{0110, 1010\}$; $\{0110, 0101\}$

  $\{0011, 1001\}$; $\{0011, 1010\}$; $\{0011, 0101\}$

  $\{1001, 1010\}$; $\{1001, 0101\}$

4.3 Global and Local Opposites

This is not the whole story, however, since the closure problem for oppositions can be extended to the case of opposites as follows: how to extend the exhaustive range of opposite-forming operators such that, for *any* two binary sentences α, ψ of standard logic, there is a O such that $\psi = O(\alpha)$? Piaget could not have thought of the double hexadic structure suggested by Blanché, because the latter does not stand for a polygon of *opposites* (based on the operators O) but, rather, a polygon of *oppositions* (based on the relations OP). Actually, no operator from the INRC Group can be used to construct the relata of Blanché's hexagons such that, for every side X of the square **AEIO**, $O(X) = U$ and $O(X) = Y$.

A solution to this problem requires an extension of the set O to new opposite-forming functions beyond Piaget's subset of the three non-trivial elements $\{N, R, C\}$. For this purpose, let us make a distinction between global and local operations.

A function O is said to be *global* if and only if it applies the same sort of transformation to every element $\mathbf{a}_i(\alpha)$ of a logical value $\mathbf{A}(\alpha)$. Thus N is a global operation that inverts every element of $\mathbf{a}_i(\alpha)$ (from 1 to 0, and conversely); R is a global operation that proceeds by permutation for every such element; and C is a composed operation of inversion and permutation, following the equation $C = NR$. N and R are primitive with respect to C, since permutation and inversion are basic operations that help to define the operation of C (while the converse is not the case).

A function is said to be *local* if and only if it is not global, i.e. it applies the same sort of transformation to some element(s) $\mathbf{a}_i(\alpha)$ of $\mathbf{A}(\alpha)$ but not all. For every $\mathbf{A}(\alpha) = abcd$, we have an inversion function $N^k{}_j$ (where k is the number of inverted elements $\mathbf{a}_i(\alpha)$ in $\mathbf{A}(\alpha)$, and j the jth case of the kth subset) such that, for any pair of binary sentences α, ψ, $\psi = N^k{}_j(\alpha)$. This results in a difference between inversion and contradiction, since only the Piagetian *global* inversion function N is still a contradictory-forming operator as depicted in the following:

- Global inversion: $N = N^4{}_1$

$$\mathbf{A}(N(\alpha)) = \mathbf{A}(N^4{}_1(\alpha)) = a'b'c'd'$$

- Local inversions:

$\mathbf{A}(N^1{}_1(\alpha)) = a'bcd$; $\mathbf{A}(N^1{}_2(\alpha)) = ab'cd$; $\mathbf{A}(N^1{}_3(\alpha)) = abc'd$; $\mathbf{A}(N^1{}_4(\alpha)) = abcd'$

$\mathbf{A}(N^2{}_1(\alpha)) = a'b'cd$; $\mathbf{A}(N^2{}_2(\alpha)) = ab'c'd$; $\mathbf{A}(N^2{}_3(\alpha)) = abc'd'$;

$\mathbf{A}(N^2{}_4(\alpha)) = a'bcd'$; $\mathbf{A}(N^2{}_5(\alpha)) = a'bc'd$;

$\mathbf{A}(N^2{}_6(\alpha)) = ab'cd'$;

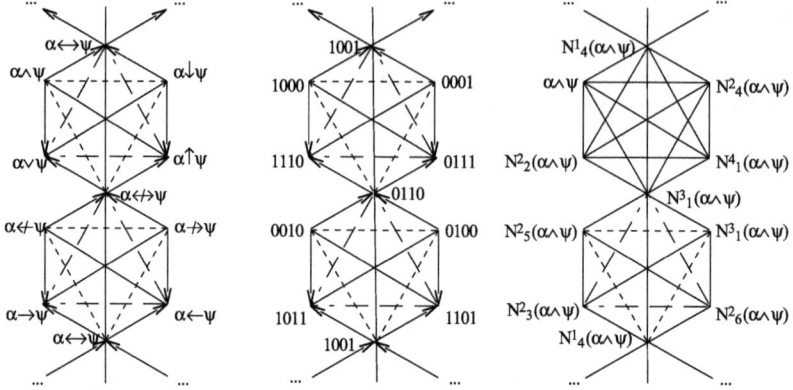

Fig. 12 A complete set of local opposite-forming operators $N^k{}_j$ for Blanché's double helix of logical oppositions

$$\mathbf{A}\left(N^3{}_1(\alpha)\right) = a'b'c'd;\ \mathbf{A}\left(N^3{}_2(\alpha)\right) = ab'c'd';\ \mathbf{A}\left(N^3{}_3(\alpha)\right) = a'b'cd';$$
$$\mathbf{A}\left(N^3{}_4(\alpha)\right) = a'bc'd'$$

Accordingly, it can be seen that the generic operation of inversion can solve the closure problem for opposite-forming operators by means of local applications. Returning to Blanché's double helix, the unknown functions are then to be replaced by new local functions in the right-sided polygon of opposites. (See Fig. 12.)

It must be noted that the use of local functions entails a complete relativization of the correspondences between opposites and oppositions: every relation of opposition OP can be constructed by means of any local opposition $O^k{}_j$. While this holds only for some sentences α and some operators of INRC, in the sense that $OP(\Phi, R(\Phi)) = OP(\Phi, N(\Phi))$ or $OP(\Phi, C(\Phi)) = OP(\Phi, N(\Phi))$ only when $\Phi \in \{\alpha, \psi, \sim\alpha, \sim\psi, \bot, \top\}$, this is the case for every binary sentence Φ whenever O is a local function. The preceding case has been restricted to the localization of inversion N, but the same process can be safely applied to the other two local functions of reciprocity R and correlation C. Taking conjunction as an arbitrary starting point, here is a tetraicosahedron of opposites that exhaustively describes the way in which any relation of opposition $OP(\alpha \oplus \psi, O(\alpha \oplus \psi))$ can be constructed by means of any local operator $N^k{}_j$. (See Fig. 13.)

Following Piaget's matrix for the group of global functions INRC, each of these functions equates with the relative product of two other ones $O^k{}_j \times O^k{}_j$. Here is a more fine-grained matrix for this group of commutative local functions, such that $XY = YX$. (See Fig. 14.)

The preceding patently shows how opposition proceeds from negation. But it is not clear which sort of negation one talks about in this theory of logical opposites. Actually, the enlarged group $O^k{}_j$ contains a set of negation-forming operators without any special constraint: it can be considered as a set of "proto-negations", i.e. unary operators that give rise to proper negations without being logical negations themselves (see Sect. 5.5). But what do they amount to exactly, if so?

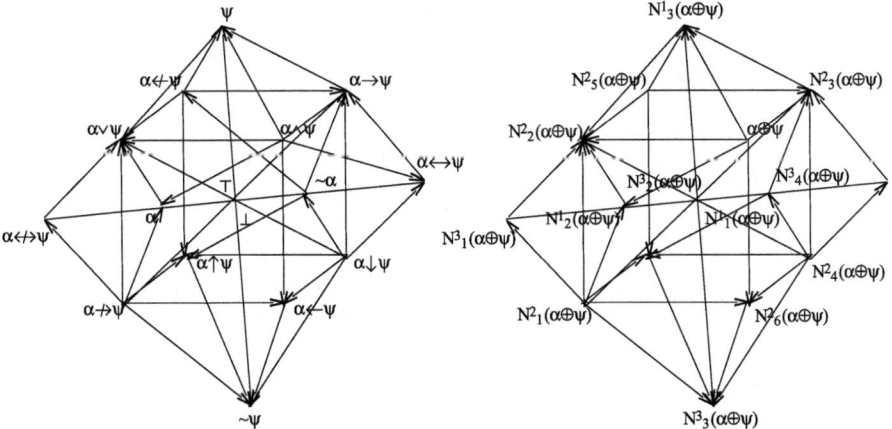

Fig. 13 A complete set of local functions $N^k{}_j$ for Pellissier's tetraicosahedron of binary connectives, where the basic argument is $\alpha \wedge \psi$

	$N^1{}_1$	$N^1{}_2$	$N^1{}_3$	$N^1{}_4$	$N^2{}_1$	$N^2{}_2$	$N^2{}_3$	$N^2{}_4$	$N^2{}_5$	$N^2{}_6$	$N^3{}_1$	$N^3{}_2$	$N^3{}_3$	$N^3{}_4$	$N^4{}_1$
$N^1{}_1$	I	$N^2{}_1$	$N^2{}_5$	$N^2{}_4$	$N^1{}_2$	$N^3{}_1$	$N^3{}_4$	$N^1{}_4$	$N^1{}_3$	$N^3{}_3$	$N^2{}_2$	$N^4{}_1$	$N^2{}_6$	$N^2{}_3$	$N^3{}_2$
$N^1{}_2$	$N^2{}_1$	I	$N^2{}_2$	$N^2{}_6$	$N^1{}_1$	$N^1{}_3$	$N^3{}_2$	$N^3{}_3$	$N^3{}_1$	$N^1{}_4$	$N^2{}_5$	$N^2{}_3$	$N^2{}_4$	$N^4{}_1$	$N^3{}_4$
$N^1{}_3$	$N^2{}_5$	$N^2{}_2$	I	$N^2{}_3$	$N^3{}_1$	$N^1{}_2$	$N^1{}_4$	$N^3{}_4$	$N^1{}_1$	$N^3{}_2$	$N^2{}_1$	$N^2{}_6$	$N^4{}_1$	$N^2{}_4$	$N^3{}_3$
$N^1{}_4$	$N^2{}_4$	$N^2{}_6$	$N^2{}_3$	I	$N^3{}_3$	$N^3{}_2$	$N^1{}_3$	$N^1{}_1$	$N^3{}_4$	$N^1{}_2$	$N^4{}_1$	$N^2{}_2$	$N^2{}_1$	$N^2{}_5$	$N^3{}_1$
$N^2{}_1$	$N^1{}_2$	$N^1{}_1$	$N^3{}_1$	$N^3{}_3$	I	$N^2{}_5$	$N^4{}_1$	$N^2{}_6$	$N^2{}_2$	$N^3{}_3$	$N^1{}_3$	$N^3{}_4$	$N^2{}_6$	$N^3{}_2$	$N^2{}_3$
$N^2{}_2$	$N^3{}_1$	$N^1{}_3$	$N^1{}_2$	$N^3{}_2$	$N^2{}_5$	I	$N^2{}_6$	$N^4{}_1$	$N^2{}_1$	$N^3{}_4$	$N^1{}_1$	$N^1{}_4$	$N^3{}_4$	$N^3{}_3$	$N^2{}_4$
$N^2{}_3$	$N^3{}_4$	$N^3{}_2$	$N^1{}_4$	$N^1{}_3$	$N^4{}_1$	$N^2{}_6$	I	$N^2{}_5$	$N^2{}_4$	$N^2{}_2$	$N^3{}_3$	$N^1{}_2$	$N^3{}_1$	$N^1{}_1$	$N^2{}_1$
$N^2{}_4$	$N^1{}_4$	$N^3{}_3$	$N^3{}_4$	$N^1{}_1$	$N^2{}_6$	$N^4{}_1$	$N^2{}_5$	I	$N^2{}_3$	$N^2{}_1$	$N^3{}_2$	$N^3{}_1$	$N^1{}_2$	$N^1{}_3$	$N^2{}_2$
$N^2{}_5$	$N^1{}_3$	$N^3{}_1$	$N^1{}_1$	$N^3{}_4$	$N^2{}_2$	$N^2{}_1$	$N^2{}_4$	$N^2{}_3$	I	$N^4{}_1$	$N^1{}_2$	$N^3{}_3$	$N^3{}_2$	$N^1{}_4$	$N^2{}_6$
$N^2{}_6$	$N^3{}_3$	$N^1{}_4$	$N^3{}_2$	$N^1{}_2$	$N^3{}_3$	$N^3{}_4$	$N^2{}_2$	$N^2{}_1$	$N^4{}_1$	I	$N^3{}_4$	$N^1{}_3$	$N^1{}_1$	$N^3{}_1$	$N^2{}_5$
$N^3{}_1$	$N^2{}_2$	$N^2{}_5$	$N^2{}_1$	$N^4{}_1$	$N^1{}_3$	$N^1{}_1$	$N^3{}_3$	$N^3{}_2$	$N^1{}_2$	$N^3{}_4$	I	$N^2{}_4$	$N^2{}_3$	$N^2{}_6$	$N^1{}_4$
$N^3{}_2$	$N^4{}_1$	$N^2{}_3$	$N^2{}_6$	$N^2{}_2$	$N^3{}_4$	$N^1{}_4$	$N^1{}_2$	$N^3{}_1$	$N^3{}_3$	$N^1{}_3$	$N^2{}_4$	I	$N^2{}_2$	$N^2{}_1$	$N^1{}_1$
$N^3{}_3$	$N^2{}_6$	$N^2{}_4$	$N^4{}_1$	$N^2{}_1$	$N^2{}_6$	$N^3{}_4$	$N^3{}_1$	$N^1{}_2$	$N^3{}_2$	$N^1{}_1$	$N^2{}_3$	$N^2{}_2$	I	$N^2{}_2$	$N^1{}_3$
$N^3{}_4$	$N^2{}_3$	$N^4{}_1$	$N^2{}_4$	$N^2{}_5$	$N^3{}_2$	$N^3{}_3$	$N^1{}_1$	$N^1{}_3$	$N^1{}_4$	$N^3{}_1$	$N^2{}_6$	$N^2{}_1$	$N^2{}_2$	I	$N^1{}_2$
$N^4{}_1$	$N^3{}_2$	$N^3{}_4$	$N^3{}_3$	$N^3{}_1$	$N^2{}_3$	$N^2{}_4$	$N^2{}_1$	$N^2{}_2$	$N^2{}_6$	$N^2{}_5$	$N^1{}_4$	$N^1{}_1$	$N^1{}_3$	$N^1{}_2$	I

Fig. 14 A matrix for the commutative group of local functions $N^k{}_j$

5 Opposition and Negation

It is taken to be granted that opposition is closely related to negation, since the relation $OP(\alpha, N(\alpha))$ relies upon the opposite-forming operator N. Moreover, it can be shown that N corresponds to the classical negation: $OP(\alpha, N(\alpha)) = CD(\alpha, \psi)$. But what about the other opposite-forming operators beyond N?

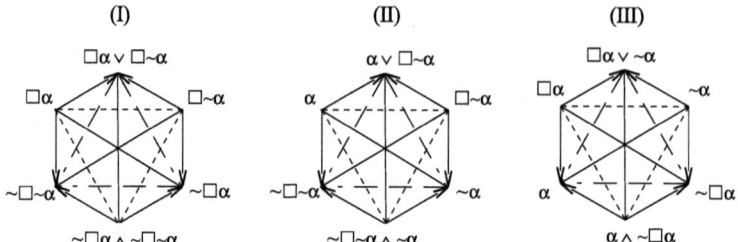

Fig. 15 Three sorts of logical negation within three hexagons of opposition for alethic modalities: classical negation (**I**), paracomplete negation (**II**), and paraconsistent negation (**III**)

5.1 A Theory of Negations?

Hartley Slater famously claimed in [19] that there cannot be true contradictions, insofar as every such negation should result in a relation of incompatible truth-values. It has been variously replied that paraconsistent negation needn't rely upon *dialetheia*, and [4] developed this objection by arguing that a modal hexagon implicitly includes three distinct negations in the classical (I), paracomplete (II), and paraconsistent (III) hexagons. (See Fig. 15.)

Following [20], **QAS** helps to algebraize the language of modal atomic sentences $\clubsuit\alpha$ by means of an alternative question-answer game: instead of questions about the truth-conditions of $\alpha \oplus \psi$, Smessaert's Modal Quantified Algebra MQA2 considers modalities $\clubsuit$ as generalized quantifiers and asks questions about how often an atomic sentence α is true or false. It results in an ordered set of four alternative questions $\mathbf{Q_1}$: "Is α always false?", $\mathbf{Q_2}$: "Is α sometimes (but not always) false?", $\mathbf{Q_3}$: "Is α sometimes (but not always) true?", and $\mathbf{Q_4}$: "Is α always true?", together with their $2^4 = 16$ ordered answers:

(1) $\mathbf{A}(\bot) = 0000$;
(2) $\mathbf{A}(\Box{\sim}\alpha) = 1000$;
(3) $\mathbf{A}({\sim}\alpha \wedge {\sim}\Box{\sim}\alpha) = 0100$;
(4) $\mathbf{A}(\alpha \wedge {\sim}\Box\alpha) = 0010$;
(5) $\mathbf{A}(\Box\alpha) = 0001$;
(6) $\mathbf{A}({\sim}\alpha) = 1100$;
(7) $\mathbf{A}({\sim}\Box\alpha \wedge {\sim}\Box\alpha) = 0110$;
(8) $\mathbf{A}(\alpha) = 0011$;
(9) $\mathbf{A}(\Box{\sim}\alpha \vee \Box\alpha) = 1001$;
(10) $\mathbf{A}(\Box{\sim}\alpha \vee (\alpha \wedge {\sim}\Box\alpha)) = 1010$;
(11) $\mathbf{A}(({\sim}\alpha \wedge {\sim}\Box{\sim}\alpha) \vee \Box\alpha) = 0101$;
(12) $\mathbf{A}({\sim}\Box\alpha) = 1110$;
(13) $\mathbf{A}({\sim}\alpha \vee \Box\alpha) = 1101$;
(14) $\mathbf{A}(\Box{\sim}\alpha \vee \alpha) = 1011$;
(15) $\mathbf{A}({\sim}\Box{\sim}\alpha) = 0111$;
(16) $\mathbf{A}(\top) = 1111$

These logical values of the modal sentences shed some light upon the connection between logical negations and oppositions. Here are the algebraic counterparts of the above hexagons, where each modal sentence is replaced by its logical value in **QAS**. (See Fig. 16.)

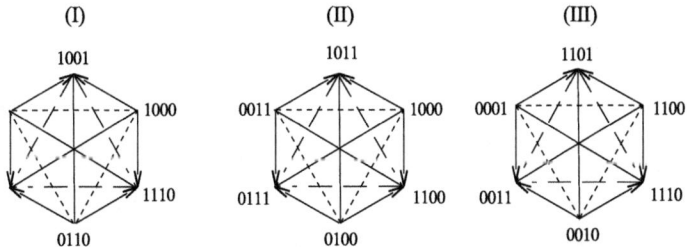

Fig. 16 The three logical hexagons and their corresponding logical values in **QAS**

Now according to [4], (I) is a classical logical hexagon because it essentially exemplifies the classical negation through the relation of contradiction $\mathrm{CD}(\Box\alpha, \sim\Box\alpha) = \mathrm{CD}(\Box\alpha, \mathrm{N}(\Box\alpha))$; (II) expresses the paracomplete negation through the contrary relation $\mathrm{CT}(\alpha, \Box\sim\alpha)$, provided that $\Box$ stands for the necessity from S4; and (III) exemplifies the paraconsistent negation through the relation of subcontrariety $\mathrm{SCT}(\alpha, \sim\Box\alpha)$, assuming that $\Box$ is the necessity from S5. How to account for such a correlation between oppositions and negations?

5.2 Non-standard Negations Are Local Opposites

To begin with, it is argued in [4] that a negation is *paracomplete* whenever a sentence and its paracomplete negation can be both false, *paraconsistent* if these can be both true. Such criteria match with the main theorems of intuitionistic and paraconsistent logics as well as the definitions of contrariety and subcontrariety.

Moreover, a difference made between the three preceding hexagons is that only the first negation in (I) can be constructed by means of a Piaget's opposite-forming operator. Two remarks can be addressed to this comparison between oppositions and negations, however. For one thing, each of the aforementioned negations can be constructed by means of a local opposite function instead of the global ones: thus

$$\mathrm{CD}(\Box\alpha, \sim\Box\alpha) = \mathrm{OP}\big(\Box\alpha, \mathrm{N}^4{}_1(\Box\alpha)\big)$$
$$\mathrm{CT}(\alpha, \Box\sim\alpha) = \mathrm{OP}\big(\alpha, \mathrm{N}^3{}_4(\alpha)\big)$$
$$\mathrm{SCT}(\alpha, \sim\Box\alpha) = \mathrm{OP}\big(\alpha, \mathrm{N}^3{}_3(\alpha)\big)$$

At the same time, negation needn't be equated with a modal unary operator: although [4] identified paracomplete and paraconsistent negations with the contrary and subcontrary relations of the hexagons (II) and (III), it can be replied that a proper paracomplete negation of any sentence α needn't be rendered as $\Box\sim\alpha$, just as its proper paraconsistent negation needn't be $\sim\Box\alpha$. Actually, a paracomplete negation is any opposite-forming operator O such that α and $\mathrm{O}(\alpha)$ cannot be both true, and a paraconsistent negation is such that these can be both true. In the semantics of **QAS**, this means that the negations of α are variously defined in terms of their compatible answers $\mathbf{a}_i(\alpha)$:

- Classical negation: $\sim$

$$\mathrm{OP}(\alpha, \psi) = \mathrm{OP}(\alpha, \sim\alpha) = \mathrm{CD}(\alpha, \psi) \qquad \text{iff} \qquad \mathbf{a}_i(\alpha) = 1 \Leftrightarrow \mathbf{a}_i(\sim\alpha) = 0$$

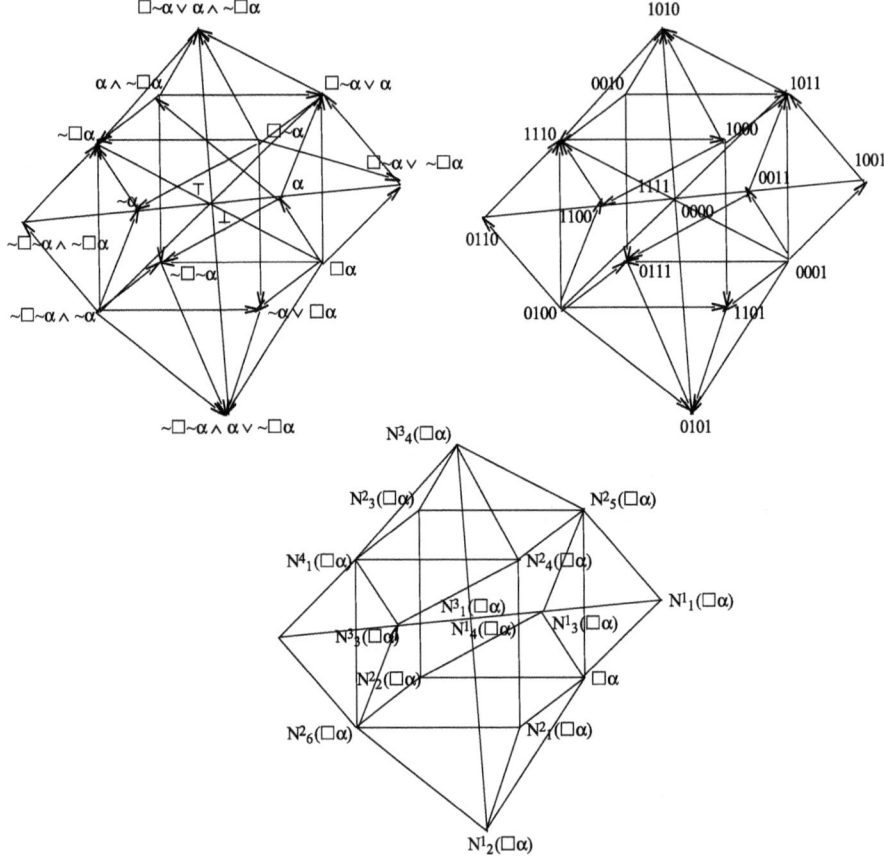

Fig. 17 A complete set of local functions $N^k{}_j$ for Pellissier's tetraicosahedron of alethic modalities, where the basic argument is $\Box\alpha$

- Paracomplete negation: $\neg$

    $$OP(\alpha, \psi) = OP(\alpha, \neg\alpha) = CT(\alpha, \psi) \qquad \text{iff} \qquad \mathbf{a}_i(\alpha) = 1 \Rightarrow \mathbf{a}_i(\neg\alpha) = 0$$

- Paraconsistent negation: $-$

    $$OP(\alpha, \psi) = OP(\alpha, -\alpha) = SCT(\alpha, \psi) \qquad \text{iff} \qquad \mathbf{a}_i(\alpha) = 0 \Rightarrow \mathbf{a}_i(-\alpha) = 1$$

Therefore, each of the three preceding hexagons actually contains the three sorts of negation among their oppositional relations since each of the contrary or subcontrary relations relies upon the use of an opposite-forming operator O, such that the opposites α and $O(\alpha)$ cannot be both true or both false. This entails that there can be more than three exemplifications of negation in the modal hexagons, in the sense that there can be more than three ways not to be both true or false. Hence the seven modal hexagons that can be added to the preceding three ones, thus yielding a combined tetraicosahedron of modalities.[14] (See Fig. 17.)

---

[14] See [11], where the logical tetracosahedron is presented as a gathering of one logical cube and ten logical hexagons. Six of these hexagons are Blanché's, Béziau's and Smessaerts's *strong* hexagons, i.e.

Again, the connection made between negation and modalities is merely convenient but not crucial: the syntactic translation of negations holds beyond the sole fragment of modal sentences and equally holds for the preceding language of binary non-modal sentences.

5.3 Non-standard Negations Are Intensional Functions

The main difference between contradiction and the three other relations of opposition is that only the former is a *truth-function*: to each truth-value α corresponds only one truth-value for the bijection $N(\alpha)$, whereas there can be several values associated to the injection $R(\alpha)$.

This accounts for the *non-truth-functional* behavior of the non-standard negations constructed by means of R, insofar as these proceed as *intensional* rules by associating a *set* of values to an initial value. This helps to explain why paracomplete and paraconsistent negations are formed by means of R without being identifiable with it; the validity of $RR = I$ shows evidence of this difference, since the iteration of a paracomplete negation does not result in the corresponding affirmation: $\neg\neg\alpha \nvDash \alpha$, i.e. not $SB(\neg\neg\alpha, \alpha)$.

More generally, the intensional application of $\sim$ (or $-$) to α maps α to the set of formulas that cannot be true (or false) together with α. This equally holds for $\sim$, but with the crucial difference that the classical negation $\sim$ maps to a *single* value and accounts for its involutivity.

But just as the intensional modalities in a possible-world semantics, R can be extensionalized by changing the set of truth-values into a set of alternative logical values. This has been done with **QAS**, so that this non-Fregean semantics turns O into a set of intensional functions similar with the modal operators of a Fregean (truth-functional) semantics.

Using opposites as negations, contrary-forming operators proceed as paracomplete negations of either modal or non-modal sentences, e.g. in the relation $OP(\alpha \nleftrightarrow \psi,$ $\alpha \downarrow \psi) = OP(\alpha \nleftrightarrow \psi, N^2{}_3(\alpha \nleftrightarrow \psi)) = OP(\alpha \nleftrightarrow \psi, \neg(\alpha \nleftrightarrow \psi))$ or in $OP(\Box\alpha, \sim\alpha \wedge \sim\Box\alpha) = OP(\Box\alpha, N^2{}_6(\Box\alpha)) = OP(\alpha \nleftrightarrow \psi, \neg(\alpha \nleftrightarrow \psi))$; subcontrary-forming operators proceed as paraconsistent negations, e.g. in the relation $OP(\alpha \to \psi, \alpha \nleftrightarrow \psi) = OP(\alpha \to \psi, N^3{}_3(\alpha \to \psi)) = OP(\alpha \to \psi, -(\alpha \to \psi))$ or in $OP(\sim\Box\sim\alpha, \sim\alpha \vee \Box\alpha) = OP(\sim\Box\sim\alpha, N^2{}_5(\sim\Box\sim\alpha)) = OP(\sim\Box\sim\alpha, -(\sim\Box\sim\alpha))$.

It is worthwhile recalling that, except for the contradictory-forming operator $N^4{}_1$, no single operator O (whether global or, *a fortiori*, local) results in only one relation of opposition: the nature of $OP(\alpha, O(\alpha))$ crucially depends upon the value of the argument α it applies upon. To compare with the preceding examples, $N^2{}_3$ is not a contrary- but subcontrary-forming operator when applied to $\alpha \vee \psi$ or $\sim\alpha \wedge \sim\Box\alpha$. And conversely, $N^2{}_5$ is not a subcontrary- but contrary-forming operator when applied to $\alpha \nleftrightarrow \psi$ or $\Box\alpha$.

Finally, some sentences are such that their logical values prevent them from standing into a contrary or subcontrary relation; this relies upon the number of *inversions* applied by the local function $N^k{}_j$. Letting i be this number such that $i(x) = x'$ (where $x \in \{0, 1\}$), here are the resulting oppositions $OP(\alpha, O(\alpha))$ for the 16 sentences $A(\alpha) = \langle a_1(\alpha), a_2(\alpha), a_3(\alpha), a_4(\alpha)\rangle$:

---

such that $A(U) = A(A) \cup A(E)$ and $A(Y) = A(\sim I) \cap A(\sim E)$; and the remaining four ones are Pellissier's *weak* hexagons such that $A(U) \neq A(A) \cup A(E)$ and $A(Y) \neq A(\sim I) \cap A(\sim E)$.

- $i = 1$: $OP(\alpha, O(\alpha)) = SB(\alpha, \psi)$ for every α
- $i = 2$: $OP(\alpha, O(\alpha)) = CT(\alpha, \psi) \times 2$, $SCT(\alpha, \psi) \times 2$, $NC(\alpha, \psi) \times 4$, $SB(\alpha, \psi) \times 8$
- $i = 3$: $OP(\alpha, O(\alpha)) = CT(\alpha, \psi) \times 6$, $SCT(\alpha, \psi) \times 6$, $SB(\alpha, \psi) \times 4$
- $i = 4$: $OP(\alpha, O(\alpha)) = SB(\alpha, \psi) \times 2$, $CD(\alpha, \psi)$ for every α

This characterization of SB accounts for the peculiar fact that CD and SB don't differ when $\alpha = \bot$ and $\psi = \top$. For SB does not state that its two opposite terms can be true and false together, which would be incompatible with CD. Rather, SB states that every truth-case (or yes-answer) of the superaltern α is also a truth-case (or yes-answer) of the subaltern ψ. But given that this definition of SB doesn't require for the superaltern to include at least one truth-case (or yes-answer), the same situation appears here as with the empty class: its set of truth-cases (or yes-answers) is empty and, thus, is included into every other set of truth-cases (or yes-answers); hence $\mathbf{A}(\bot)$ is included into $\mathbf{A}(\psi)$ for every ψ. Consequently, $\bot$ and $\top$ satisfy both CD and SB without entailing any paradoxical situation but a peculiar case in which two opposites satisfy two different relations of opposition.

5.4 Subalternation and Double Negations

Subalternation has not been mentioned yet in this section about negation, although it always occurs in the above oppositions and irrespective of i. Now assuming that O proceeds as an operator of negation, why not to conclude from it that subalternation equally results from the relation of a sentence and one of its negations?

Actually, the functional approach to oppositions through opposite-forming operators justifies the general reluctance to consider subalternation as a proper opposition: the latter concept is closely related to negation, whereas subalternation is used to preserve truth and proceeds as a sort of mixed double negation or weak *affirmation*. This clearly appears in the light of Piaget's INRC Group of global operators O, where correlation C helps to construct subalternation and is defined as the product of inversion N and reciprocity R: $C = NR = RN$; C is a subalternation-forming operator, N a contradictory-forming operator that behaves like a classical negation, and R an operator whose behavior crucially depends upon the value of its argument. Let $y(\alpha)$ be the number of yes-answers to $\mathbf{Q}(\alpha)$; it results in a number of distinctive unary operators formed by R according to $y(\alpha)$:

- Affirmation
  $R(\alpha) = \alpha$ with $y(\alpha) = 0$,   or $y(\alpha) = 2$ and $\mathbf{a}_1(\alpha) = \mathbf{a}_4(\alpha)$,   or $y(\alpha) = 4$
- Classical negation
  $R(\alpha) = \sim\alpha$ with $y(\alpha) = 2$ and $\mathbf{a}_1(\alpha) \neq \mathbf{a}_4(\alpha)$
- Paracomplete negation
  $R(\alpha) = \neg\alpha$ with $y(\alpha) = 1$
- Paraconsistent negation
  $R(\alpha) = -\alpha$ with $y(\alpha) = 3$

According to the nature of the unary operator formed by R, it is easily verified that the relation of subalternation SB needn't result from the operator $C = NR$:

$OP(\alpha, C(\alpha)) = OP(\alpha, \sim\alpha)$ when $y(\alpha) = 0$, or $y(\alpha) = 2$ and $\mathbf{a}_1(\alpha) = \mathbf{a}_4(\alpha)$

$OP(\alpha, C(\alpha)) = OP(\alpha, \sim\neg\alpha)$ when $y(\alpha) = 1$

$OP(\alpha, C(\alpha)) = OP(\alpha, \sim\sim\alpha) = OP(\alpha, \alpha)$ when $y(\alpha) = 2$ and $\mathbf{a}_1(\alpha) \neq \mathbf{a}_4(\alpha)$

$OP(C(\alpha), \alpha) = OP(\sim - \alpha, \alpha)$ when $y(\alpha) = 3$

$OP(\alpha, C(\alpha)) = OP(\sim\alpha, \alpha)$ when $y(\alpha) = 4$

These constructions betray two exceptions in Piaget's matrix INRC (see Sect. 3).

Firstly, the contradictory-forming operator N can also yield a relation of subalternation when applied to $\perp$ (i.e. with $y(\alpha) = 0$) or $\top$ (i.e. with $y(\alpha) = 4$). It means that two opposite sentences can stand into a relation of subalternation without being non-contradictory to each other, in the sense that they are contradictories when $y(\alpha) = 0$ or $y(\alpha) = 4$. This special feature has not been taken seriously by Blanché, but it should not be neglected in a complete formal theory of oppositions and opposites.

Secondly, the cases where $y(\alpha) = 2$ and $\mathbf{a}_1(\alpha) = \mathbf{a}_4(\alpha)$ are such that $NR(\alpha) = NI(\alpha) = N(\alpha)$, so that $C(\alpha) = N(\alpha)$ with these two values.

Apart from these peculiarities, subalternation typically results from the standard negation of a non-standard negation (or the converse).

Finally, the intensional aspect of the non-standard negations $\bullet = \{\neg, -\}$ entails that these cannot be safely commuted; they can be so only with the extensional, standard negation $\sim$. That is:

$$\text{For every } \bullet \in \{\neg, -\} : \sim\bullet\alpha = \bullet\sim\alpha, \text{ but } \neg-\alpha \neq -\neg\alpha$$

5.5 A Theory of "Proto-negations"!

All of this could relevantly shed some new light on the connection between opposition and negation.

On the one hand, the view of subalternation as a mixed double negation could relevantly justify the usual rejection of subalternation by whoever considers opposition to be essentially related to negation while excluding the opposite operator of affirmation. But such a mixed double negation needn't lead to a plain affirmation, depending upon the value of its arguments.

On the other hand, a reference to Piaget's INRC Group has shown that any opposition somehow resorts to negation through the application of opposite-forming operators: although subcontrariety and subalternation admit their opposites α and ψ to be true together, ψ results from the application of a kind of "proto-negation" O to α such that $O(\alpha) = \psi$. Whether global or local, the family of O turns the Disjunctive Normal Form of any sentence into another one and this seems to be how opposition is always related to negation as *change*: opposition essentially works as a change- or difference- rather than an incompatibility- or falsity-forming operator in the light of **QAS**, turning yes-answers into no-answers (or conversely) in every case (with $N^4{}_1 = N$) or in some case(s) (with R or C, or any $N^k{}_j$, where $k \neq 4$).

6 Conclusion and Prospects

Let us recapitulate the main theses of this paper:

(i) Opposition is a relation OP(α, ψ) between two opposite terms α and ψ, the second of which results from the first one by applying an opposite-forming operator O such that O(α) = ψ; the properties of OP and O can be investigated within such a non-Fregean semantics as **QAS**, where the logical value of a sentence is an ordered set of answers to corresponding questions about it.

(ii) With respect to the theory of binary sentences $\alpha \oplus \psi$, the logical value of a sentence characterizes its Disjunctive Normal Form; Piaget's INRC Group roughly expresses the various properties of O, where the operation N is a contradictory-forming operator corresponding to the standard negation.

(iii) The case of correlation C has led to the result that subalternation is a special subcase of opposition: the compossibility of truth and falsity for its opposite terms is a necessary, but not sufficient condition for it. The definition of subalternation in terms of yes-answers-containment shows that opposition includes the Tarskian relation of entailment among its components.

(iv) It is possible to turn any given sentence into any other one by means of O, provided that O includes a more fine-grained set of local functions. For this purpose, Piaget's three global operations N, R, C have been collapsed into a unique set of 16 local operations $N^k{}_j$ which correspond to proto-negations: they are the basic operations by means of which logical negations can be constructed.

(v) Following Béziau's translation of logical negations within modal logic, it has been shown that the connection between negations and oppositions is sufficient for an algebraic definition of standard and non-standard negations. Subcontrariety and subalternation deserve to be included among the proper range of oppositions, in the sense that they essentially proceed from the application of opposite-forming operators to their argument(s).

A number of problems remains to be investigated within this new theory of opposites, among which the following ones: Is there an informal meaning for the opposite-forming operators O? Can the various features of O be generalized for an arbitrary logical value? What about the counterparts of local reciprocity $R^k{}_j$ or local correlation $C^k{}_j$, in addition to the functions of local inversion $N^k{}_j$? How to characterize the logical value of any structured sentence, in addition to the binary or modal unary sentences; what is the logical value of a modal binary sentence like $\Box(\alpha \land \psi)$, for instance? What of many-valued oppositions? Non-standard negations have been defined within the standard frame of a Boolean algebra, but what about the converse; that is: how to define classical negation within a non-standard frame; does this require the use of a non-Boolean algebra?

A lot is still to be accomplished before giving a more substantial role to this introductory theory of opposites within the area of logic.

References

1. Aristotle: Metaphysics
2. Aristotle: On Interpretation

3. Aristotle: Prior Analytics
4. Béziau, J.-Y.: New light on the square of opposition and its nameless corner. Log. Investig. **10**, 218–233 (2003)
5. Blanché, R.: Sur l'opposition des concepts. Theoria **19**, 89–130 (1953)
6. Blanché, R.: Sur la structuration du tableau des connectifs interpropositionnels binaires. J. Symb. Log. **22**, 178 (1957)
7. Blanché, R.: Structures intellectuelles (Essai sur l'organisation systématique des concepts). Vrin, Paris (1966)
8. Chatti, S., Schang, F.: Import, or not import? How to handle negation inside the square. Submitted paper
9. Czeżowski, T.: On certain peculiarities of singular propositions. Mind **64**, 392–395 (1955)
10. Luzeaux, D., Sallantin, J., Dartnell, C.: Logical extensions of Aristotle's square. Logica Univers. **2**, 167–187 (2008)
11. Moretti, A.: The Geometry of Logical Opposition. PhD thesis, University of Neuchâtel (2009)
12. Pellissier, R.: Setting *n*-opposition. Logica Univers. **2**, 235–262 (2008)
13. Piaget, J.: Essai sur les transformations des opérations logiques (Les 256 opérations ternaires de la logique bivalente des propositions). Presses Universitaires de France, Paris (1952)
14. Piaget, J.: Traité de logique (Essai de logistique opératoire). Armand Colin (1972) (1st edition, 1949)
15. Schang, F.: Relative charity. Rev. Bras. Filos. **233**, 159–172 (2009)
16. Schang, F.: Questions and answers about oppositions. In: Béziau, J.-Y., Payette, G. (eds.) New Perspectives on the Square of Opposition. Peter Lang, Bern (2011)
17. Schang, F.: Two Indian dialectical logics: saptabhaṅgī and catuṣkoṭi. J. Indian Counc. Philos. Res. **27**, 45–75 (2011)
18. Sesmat, A.: Logique—II. Les raisonnements, la logistique. Hermann, Paris (1951)
19. Slater, H.: Paraconsistent logics? J. Philos. Log. **24**, 451–454 (1995)
20. Smessaert, H.: On the 3D visualization of the logical relations. Logica Univers. **3**, 212–231 (2009)

F. Schang (✉)
LHSP Henri Poincaré (UMR7117), Université de Lorraine, Nancy, France
e-mail: schang.fabien@voila.fr
e-mail: fabien.schang@univ-nancy2.fr

Pluralism in Logic: The Square of Opposition, Leibniz' Principle of Sufficient Reason and Markov's Principle

Antonino Drago

Abstract According to the present pluralism in mathematical logic, I translate from classical logic to non-classical logic the predicates of the classical square of opposition. A similar unique structure is obtained. In order to support this new logical structure, I investigate on the rich legacy of the non-classical arguments presented by ingenuity by several authors of scientific theories. A comparative analysis of their ways of arguing shows that each of these theories is severed in two parts; the former one proves a universal predicate by an *ad absurdum* proof. This conclusion of every theory results to be formalised by the A thesis of the new logical structure. Afterwards, this conclusion is changed in the corresponding affirmative predicate, which in the latter part plays the role of a new hypothesis for a deductive development. This kind of change is the same suggested by Leibniz' principle of sufficient reason. Instead, Markov's principle results to be a weaker logical change, from the intuitionist thesis I in the affirmative thesis I. The relevance of all the four theses of the new logical structure is obtained by studying all the conversion implications of intuitionist predicates. In the Appendix, I analyse as an example of the above theories, Markov's presentation of his theory of real numbers.

Keywords Square of opposition · Non-classical logic · Doubly negated statements · Logical principles

Mathematics Subject Classification Primary 01A20 · Secondary 03A05 · 03B20

1 The Present Pluralism in Logic

Even few decades ago the kinds of logic which differ from the classical one were called 'deviant logic'.[1] Instead, at present time a plethora of kinds of logic appear to be relevant at least in some particular situations.[2]

In philosophy we can trace back this logical pluralism to the ancient times. A recent study on Aristotle's logic concludes: "Thus the first logic was at least in the spirit of an intuitionist logic" [43], which is the most relevant non-classical logic. In last century the mathematical logicians introduced a formal pluralism in mathematical logic. The works by Glyvenko, Kolmogorov and Goedel started to put on the same par of classical logic the

---

[1] The last instance of this qualification is the otherwise excellent text by [23].

[2] I supported this view by the paper [11]. About the pluralism in science, see my paper [8].

J.-Y. Béziau, D. Jacquette (eds.), *Around and Beyond the Square of Opposition*, 175–189
Studies in Universal Logic, DOI 10.1007/978-3-0348-0379-3_12, © Springer Basel 2012

intuitionist one [21]. Some more studies established a borderline between classical logic and almost all kinds of non-classical logic; this borderline is constituted by the failure of, better than the law of the excluded middle, the law of double negation [18, 22, 37, 38, 42]; hence, a doubly negated sentence which is not equivalent to the corresponding affirmative sentence (= DNS) belongs to non-classical logic,[3] whose a first instance is intuitionist logic.

Let us remark that a doubly negated sentence may be a DNS for several reasons; within scientific theories the most important reason is the lack of operative-experimental evidence supporting the corresponding affirmative sentence. This test is easily applied to sentences belonging to scientific theories (which usually include, beyond affirmative sentences about experimental data, also DNSs for theoretical reasons). Hence, this test when applied to a scientific text reveals the author's choice on the kind of mathematical logic governing his theory. Notice that even a sole, essential DNS plays a discriminating role in an argument or in a text, its presence entails that this argument or this text is governed by non-classical logic. This crucial test, by relying on the evidence of a sentence, links the variety of the kinds of logic to reality in an unprecedented way. No surprise if in the following some important consequences will result.

In the following Sect. 2 the three translations from classical logic to non-classical logic are exploited for defining a SO-like. In order to support such a logical structure, in Sect. 3 I will investigate on the rich legacy of the arguments manifested by the original texts of past scientific theories, through their use of DNSs belonging to non-classical logic. In fact, each of these theories is severed in two parts; in the former part several DNSs play essential roles. In this part a final *ad absurdum* proof states a doubly negated predicate of universal validity on all the cases at issue. In Sect. 4 the conclusions of the more relevant theories are quoted and then formalised in logical terms; all them result to be equivalent to a non-classical predicate which is the same of the predicate formalising Leibniz' principle of sufficient reason. In Sect. 5 I illustrate how the author changes this conclusion in the corresponding affirmative predicate, that he then considers as a new hypothesis-axiom from which in the latter part of the theory he deductively develops all consequences according to classical logic; this change is formally the same as the change from the principle of sufficient reason to the corresponding affirmative statement. In Sect. 6 I will take an advantage from Markov's paper on the theory of constructive mathematics, for investigating on what justifies an author of the above theories to change the kind of logic. Markov performed a similar change of a doubly negated existential predicate, which is both a decidable one and proved by an *ad absurdum* proof, in the corresponding affirmative predicate (Markov's principle). In Sect. 7 this change is shown to be a change of the I thesis of the SO-like in the corresponding I of SO, whereas the principle of sufficient reason, together with all previous changes, does the same for the stronger A thesis. In Sect. 8 the list of all valid conversion implications between the couple of intuitionist predicates shows the relevance of the four theses of the SO-like among all the intuitionist predicates.

---

[3]Let us recall that the relationships of an affirmative proposition of propositional classical logic with the corresponding DNS of intuitionist logic is assured by a well-known theorem on their relationship—i.e. once a formula A is true in classical logic, then evenly ¬¬A is true in the latter logic. Instead, to translate a DNSs in an affirmative sentences, owing to the failure of the law of the double negation, is a problematic move [41, 56ff]. Notice also that in intuitionist logic an *ad absurdum* proof ends by a DNS; to change it in an affirmative sentence is allowed by classical logic only [18, p. 27].

Table 1 The versions of the four theses of the SO in both classical and non-classical logic

	Classical SO	Goedel SO^g	Kolmogorov SO^k	Kuroda SO^q
A	$\forall x\,A(x)$	$\forall x\,\neg\neg A(x)$	$\neg\neg\forall x\,\neg\neg A(x)$	$\neg\neg\forall x\,\neg\neg A(x)$
E	$\neg\exists x\,A(x)$	$\neg\neg\forall x\,\neg A(x)$	$\neg\exists x\,\neg\neg A(x)$	$\neg\exists x\,A(x)$
I	$\exists x\,A(x)$	$\neg\forall x\,\neg A(x)$	$\neg\neg\exists x\,\neg\neg A(x)$	$\neg\neg\exists x\,A(x)$
O$_1$	$\exists x\,\neg A(x)$	$\neg\forall x\,\neg\neg A(x)$	$\neg\neg\exists x\,\neg A(x)$	$\neg\neg\exists x\,\neg A(x)$
O$_2$	$\neg\forall x\,A(x)$	$\neg\forall x\,\neg\neg A(x)$	$\neg\forall x\,\neg\neg A(x)$	$\neg\forall x\,\neg\neg A(x)$

In the Appendix Markov's paper will be analysed as an example of the above theories through its DNSs.

2 Translations of the Square of Opposition in Non-classical Logic

Let us recall that by introducing in various ways double negations Glyvenko-Goedel, Kolmogorov and Kuroda suggested easy rules for translating predicates of classical logic in the corresponding ones in both intuitionist logic and minimal logic. The first translation adds two negations to each prime and substitutes (in shortened notation) $\neg\forall\neg$ for the existential quantifier $\exists$ (plus the negation of the de Morgan version of the LEM). The second translation is obtained "by simultaneously inserting $\neg\neg$ in front to all subformulas of X (including X itself)"; the third one is obtained by inserting "$\neg\neg$ after each occurrence of $\forall$ and in front to the entire formula".[4]

Let us apply the above three translations to the four predicates of the classical SO (recall that a triple negation is equivalent to a single negation; the classical thesis O is considered in both versions: "Some S is not P" and "Not every S is P"). I present all them by means of Table 1, where A(x) summarises "S is P".[5]

I call respectively SO^g, SO^k, SO^q the three SO-like so obtained.

Which relationships among the corresponding predicates of the three translations of SO in non-classical logic? The answer is not easy because whereas the classical predicate logic is a calculus, the intuitionist predicate logic not.

However, there exist some equivalences theorems between the different translations of classical logic to intuitionist predicate logic.[6] In this logic P^g $\leftrightarrow$ P^k (where P is any predicate). Moreover, the Corollary 3.6 of the previous reference states that negative predicates

---

[4][41, pp. 57–59]. In a previous paper [12] I introduced a square of opposition including at the same time two classical predicates (the theses A and E) and two non-classical predicates (the theses I and O) which differ from the traditional ones because the copula 'is' is changed in 'it is not true that it is not', or 'is equivalent'; this addition of double negations is not enough to represent one of the above-mentioned translations; however the results of the Sects. 7 and 8 of this paper agree with the above-mentioned translations.

[5]According to modern logicians Aristotle' square of opposition did not take in account the existential import of the theses. A more accurate investigation by authoritative scholars showed that instead Aristotle did it [33].

[6]The same translations hold true for minimal predicate logic, i.e. when the logical law *ex falso quodlibet* $(P \rightarrow (\neg P \rightarrow Q)$, where P and Q are whatsoever predicates) is weakened in the law $P \rightarrow (\neg P \rightarrow \neg Q)$,

Table 2 The equivalence relationships among the translations in intuitionist logic of the theses of the classical SO

$$A^g \leftrightarrow (3.8, 3.6) \ A^k = A^q$$
$$E^g \leftrightarrow (3.8) \ E^k \leftrightarrow (3.8) \ E^q$$
$$I^g \leftrightarrow (11, 12) \ I^k \leftrightarrow (3.8) \ I^q$$
$$O_1{}^g \leftrightarrow (D) \ O_1{}^k = O_1{}^q$$
$$O_2{}^g = O_2{}^k = O_2{}^q$$

(i.e. the predicates without $\exists$ and v and whose prime is negative) enjoys the property $P \leftrightarrow \neg\neg P$ [41, p. 59]. In the following these result will be recalled by merely writing respectively 3.8 and 3.6. In addition, we refer to the list of implications offered by two classical presentations of intuitionism.[7] As a result, all the three translations of a thesis of SO are mutually equivalent.

Table 2 summarises the mutual relationships among the three translations of each thesis of the classical square of opposition in intuitionist logic.

Result 1 The three doubly negated translations of the four theses of SO in intuitionist logic give a unique SO-like which is baptised SO^{gkq}.

3 The Use of Non-classical Logic in the Original Texts of Some Past Scientific Theories

In order to discover the possible use of this SO-like logical structure I will follow Troelstra's suggestion for a similar question: "to gain further insight into the acquisition of mathematical [and scientific] experience by historical studies" [40, p. 223]. I will investigate on the scientific theories because they represent thinking structures at the highest level as possible of the logical rationality. I will explore the rich legacy of scientific theories in the aim to find out the commonly accepted patterns of arguing suggested by their founders. In other terms, I will deal with an experimental logic whose experimental data are the arguing patterns of some past scientific theories.

Along forty years I investigated on the original texts illustrating scientific theories. I discovered *several theories instantiating a specific use of non-classical logic in their respective presentations since they rely on some essential DNSs.* These theories are the following ones: L. Carnot's theories of calculus and geometry, Lagrange's mechanics, Lavoisier's chemistry, Avogadro's atomic theory, S. Carnot's thermodynamics, Lobachevsky's non-Euclidean geometry, Galois' theory of groups, Klein's Erlangen

---

hence the latter logic is strictly weaker than intuitionist logic [41, p. 57]. One may guess that an experimental scientist is not allowed to deduce *ex falso* everything, included the true; but the false only; hence, his logic is the minimal one. In such a case the above translations in minimal predicate logic give distinct results from those in intuitionist logic when one claims that the copula "is" of a thesis of SO is better translated in logical terms by an equivalence, i.e. a double implication. In such a case the three translations of the theses of SO in minimal predicate logic are not equivalent in both E and I theses. However, according to the next footnote 18, the following results seem hold true for the minimal SO-like too even in the latter case.

[7]Reference [25] lists 14 valid implications among couples of predicates. In the following an implication between two predicates will be denoted by putting in round bracket its number in the list of this section [18, p. 29].

program, Poincaré's theory of integer numbers, Einstein's theory of special relativity, Planck's theory of quanta, Kolmogorov's foundation of minimal logic, Church's thesis, Markov's theory of constructive functions.

In each text presenting one of these theories the author argues in a little noticed way. By ingenuity he makes use of DNSs so that their mere sequence gives the logical thread of the author's entire illustration.[8] Moreover, a comparative analysis on the kind of development shared by the above scientific theories using DNSs, shows that each theory does not begin by stating some axioms from which to draw deductions; that is, it is not organised according to that ideal model of an apodictic (= deductive) theory, which was first suggested by Aristotle, then applied by Euclid in geometry, subsequently confirmed by Newton in mechanics and eventually improved by Hilbert to the model of an axiomatic theory.[9] Instead, such a theory presents as first a universal problem which at that time was unsolvable by current scientific techniques; to look for a solution requires to find out a new scientific method; which the development of the theory goes to discover.

In the texts written by the above-mentioned scientists a former part concludes, by means of an *ad absurdum* proof, a doubly negated predicate of an universal validity on all the cases considered by the theory.[10] Of course, this way of organising a theory is manifestly a non-deductive one. I call such a specific organisation of a theory a problem-based organisation (PO).

By summarising, in his construction of a PO theory a scientist follows a specific logical strategy; he argues by means of DNSs belonging to non-classical logic and he wants to prove by an *ad absurdum* proof an universal predicate.

Let us recall Leibniz' logico-philosophical principle of sufficient reason:

> Two are the principles of the human mind: the principle of non-contradiction and the principle of the sufficient reason ..., [that is] <u>nothing</u> is <u>without</u> reason, or everything has its reason, although we are not always capable of discovering this reason ...[11]

The above sentence ("Nothing is <u>without</u> reason") is a DNS, as the same Leibniz explains why by means of the next sentence: Leibniz wants to conclude "everything has a reason", but he underlines that not always we have sufficient evidence for affirming with certainty this reason; hence, only the doubly negated sentence holds true. Hence, the

---

[8] In the past, this way of arguing joining together the DNSs was ignored. Rather some philosophers argued by considering at the same time the three values of a sole proposition A, i.e. A, ¬A and ⌐⌐A. Cusanus claimed to argue through an opposite's coincidence of A and ⌐A; Hegel through an almost mechanical addiction of a negation to a ⌐A sentence. See my paper [15].

[9] It is by appealing to the past experiences of scientific theories that I improved the unsuccessful Beth's research for a non-deductive way of organising a mathematical theory [2]. D'Alembert first stressed that there exists two kinds of theory organisation; he suggested that, beyond the "rational" kind, an "empirical" one exists [6]. Subsequently, for illustrating these two different models L. Carnot devoted two pages of each his two books on mechanics [3, 4]; he claimed to develop his theory in an "empirical" way. In past century both H. Poincaré [34, 35] and independently A. Einstein again suggested two similar kinds of theory organisation [27, 32].

[10] Some of the above-mentioned authors wrote texts which are less structured in logical terms; e.g. L. Carnot's calculus, Lavoisier's chemistry, Galois' theory of groups, Klein's Erlangen program, Einstein's theory of special relativity all lack of *ad absurdum* theorems.

[11] [29]. Here and in the following, emphasis is added for manifesting to the reader the two negations within a doubly negated sentence.

above logical strategy may be considered as summarised by Leibniz' principle of suffi-cient reason.

4 Qualifying in Formal Terms the Predicate Concluding a Problem-Based Theory

Let us list the universal conclusions of some PO theories (in order to facilitate the reader, I will add to two conclusions their translations in more plain words).

Lobachevsky: "... without leading to any contradiction in the results"; i.e. No contradiction results.[12]

S. Carnot: "... no change of temperature inside the bodies employed for obtaining the motrice power of heat occurs without a change in the volume".[13]

Poincaré: "If the absence of contradiction of a syllogism whose number is entire implies the absence of contradiction of the following one, one has to not fear any contradiction for each syllogism whose number is entire".[14]

Kolmogorov: "None of the conclusions of ordinary mathematics that are based on the use outside the domain of the finitary .... can be regarded as firmly established" (and also: "No contradiction from the use of the principle of excluded middle").[15]

We remark that each above conclusion is formalised by a same logical formula:

$$\neg\exists S \text{ is } \neg P.$$

Some authors achieved conclusions of a different kind. Avogadro: "[All] The propor-tions among the quantities in the combinations of the substances do not seem depend other than both the relative number of molecules which combine themselves and the number of the composed molecules which result from them",[16] i.e., All proportions

---

[12][30, Proposition 19]. This book was analysed through its DNSs by my paper [14]. Lobachevsky's main text puts the problem of how much parallel lines to a straight line exist. In order to obtain evidence for his guess—i.e. two parallel lines—he proves through DNSs five theorems, most of which are *ad absurdum* theorems. At the end of the prop. 22, shown by an *ad absurdum* theorem, he concludes that his supposition with respect to Euclid's hypothesis, receives an equivalent evidence in both *all* points and in *all* figures in the space.

[13][5, p. 23]. S. Carnot's thermodynamics puts the problem of the maximum efficiency in the heat/work conversions; in order to solve it, he looks for a new method by arguing through DNSs about his celebrated cycle of four transformations. The list of DNSs ends by means of his well-known *ad absurdum* theorem about the maximum efficiency in *all* heat/work conversions. Carnot's book was analysed through its DNSs by the paper [17].

[14][36, p. 187]. Poincaré criticism to Formalists' attempt to prove by finitist means the principle of math-ematical induction concludes by essentially the following DNS: "... does not exist contradiction" for all entire numbers. The current version of this principle changes it in the corresponding affirmative predicate. See my paper [7].

[15][28]. Kolmogorov's foundation of the minimal logic argues by means DNSs and an *ad absurdum* theo-rem stating the above conclusion. Afterwards, Kolmogorov thinks that nothing opposes to *always* deduc-tively argue by means of pseudotruths from the axioms of the type A of Hilbert's formalisation of logic. See also [10].

[16][1, p. 58]. The subsequent sentence ("Hence, it is necessary thus to admit that [there it is not true that do not exist] there exist simple relationships also among the volumes of [all] the gaseous substances and the number of the simple or composed molecules which compose them) constitutes the celebrated

among the quantities in the combinations of the substances do not <u>depend</u> <u>other</u> <u>than</u> ...".

Kleene's statement on Church's thesis: "Every general recursive function <u>cannot</u> <u>conflict</u> with the intuitive notion which is supposed to complete ..." [26, pp. 318–319].

Each of the above statements is formalised by the following formula:

$$\forall S \text{ are } \neg\neg P.$$

It is a remarkable fact that the latter formula is equivalent to the former formula (6 and 7); both represent the A^{gkq} thesis.

More in general, this fact gives a very important conclusion: a PO theory, searching a new method capable to solve a basic problem, argues according to a sequence of DNSs aimed to eventually obtain an instance of the predicate A^{gkq}.[17]

Result 2 A PO theory is a goal-oriented logical theory to state an universal predicate concluding the last *ad absurdum* proof of its former part. This predicate is formalised by the thesis A^{gkq} of the SO^{gk}.[18]

Let us add that a Leibniz' principle of sufficient reason ("<u>Nothing</u> is <u>without</u> reason") may be formalised by calling S an "event" and P "connected <u>to some events</u>"; the following formula is obtained:

$$\neg\exists S \text{ is } \neg P$$

Result 3 Even the principle of sufficient reason is represented by the same predicate of all previous conclusions of the PO theories, i.e. by a thesis A^{gkq}.

Hence, Leibniz' logico-philosophical principle appears to be the specific principle addressing an author of a PO theory in his non-classical arguing for achieving the final conclusion of.

5 The Change of Kind of Logic and Leibniz' Principle

The universal nature of the last DNS suggests the author to have ended his inductive arguing and to accept the corresponding affirmative predicate as a new hypothesis for the subsequent part of the theory, to be deductively developed in classical logic. In fact,

---

"Avogadro's law" on the molecular constitution of whatsoever kind of matter. An analysis of the paper through its DNSs is given by [16].

[17] Reference [20] studies a similar problem in various kinds of logic, but at the propositional level only

[18] Remarkably, A^{gkq} is the same in both minimal logic and intuitionist logic. Notice that one may guess that when an author of a PO theory obtains, by arguing in minimal logic, the universal predicate A^{gkq}, he governs by this final DNS the entire universe of his logical arguing with respect to the basic problem. At this stage, he, when meeting a false sentence, can consider it as belonging to a purely theoretical universe; hence, he can apply the intuitionist law on the false, according to which everything of this theoretical universe follows from it. In other words, when he achieves this predicate, at the same time he implicitly changes the kind of logic from the minimal one to the intuitionist one.

the author changes this predicate in the corresponding affirmative predicate; which in the latter part is considered as an hypothesis, from which a lot of theorems are deductively drawn in classical logic.

It is remarkable that both the first instance of a PO theory in the history of mathematics—Lobachevsky's non-Euclidean geometry—, and the first instance of a PO theory in the history of modern theoretical physics—Einstein's special relativity—originated two scientific revolutions. Each of these celebrated authors declared a change of the organisation of his theory. Lobachevsky wrote in his most relevant work: "[My supposition of two parallel lines] can likewise be *admitted* [as a principle-axiom for the following, deductive part of my theory] <u>without</u> leading to <u>any contradiction</u> in the results and [deductively] *founds* a new geometry ...." (emphasis added) [30, Proposition 19]. Einstein wrote in his celebrated paper: "We will raise this conjecture (the substance of which will be hereafter called the "[axiom-]principle of relativity")...".[19]

In the following, I will investigate on the question, how qualify in formal terms this change of kind of logic, occurring in the texts of the PO theories.

Each of the above theories makes use of two kinds of logic; in the former part, the non-classical logic and in the latter part, the classical logic. In fact, the author changes the non-classical predicate A^{gkq}, which is representative of the new method discovered by the former part of the PO theory, in a classical predicate A, which is representative of just the beginnings of a deductive theory.

Result 4 The authors of PO theories changed their universal conclusions in the corresponding affirmative predicates of classical logic: $A^{gkq} \Rightarrow A$.

Let us now recall that in order to obtain classical logic from intuitionist logic there exist four ways, each constituted by the addition of one of the following logical features to the intuitionist logic: (i) the law of the excluded middle, (ii) the law of the double negation, (iii) the dilemma and (iv) the change of a conclusion of an *ad absurdum* argument $\neg\neg T$ in an affirmative T [24, 39].

We recognise that all scientists of the quoted PO theories, practiced the last way, by adding a qualification; the predicate $\neg\neg T$ is universal in nature with respect to all the problems involved by the basic problem.

Let us now recall Leibniz's statement for the principle of sufficient reason. He too stated as a first sentence a universal DNS which subsequently changed in the corresponding affirmative sentence "... everything has its reason" (although afterwards he remarked that this sentence may be unsupported by sufficient evidence). Let us call this change PSR°. Remarkably, this change also is represented by the same formula $A^{gkq} \rightarrow A$ formalising the final move of a PO theory. Notice in addition that even Leibniz' principle of sufficient reason PSR° is implicitly justified by an *ad absurdum* argument: "It is <u>absurd</u> to <u>reject</u> it"; which, in its turn, implicitly relies on the principle "It is <u>impossible</u> that the reality is <u>not</u> rational".

Result 5 The above logical strategy is summarised by Leibniz' first two sentences of his version of the principle of sufficient reason. Hence, Leibniz' PSR° may be considered as

---

[19][19]. An analysis of the text through its DNSs is given by my paper [13].

the logico-philosophical scheme inspiring the change occurring in the final part of a PO theory.[20]

6 Markov's Principle in His PO Theory on Constructive Mathematics

Fortunately, a further qualification of this change in logic was suggested by a recent Markov's paper (1962) founding a theory of constructive mathematics [31]. Let us analyse this paper.

A former part (pp. 1–5) includes several DNSs (40; see the Appendix). By scrutinising them, one sees that the first three DNSs present the main problem of the paper, i.e. how the rational numbers may be extended to the real numbers according to the constructive method; that is, without the "use of the abstractions of actual infinity" and "the so-called pure existence theorems".

In the middle of the paper (p. 6) Markov illustrates the constructive way to build mathematics. Actually, he refers to Church's thesis on all algorithms; we saw that this thesis—as Kleene writes it—is a DNS. In the following part of Markov's paper few DNSs (8 out the 48 DNSs in the entire paper) occur and moreover in an occasional way; that means that the logical sequence of the DNSs connected one to another is terminated by the previous universal statement and he is substantially following classical logic. In conclusion, also Markov followed the same theoretical organisation, PO, as the previously mentioned authors did.

This paper deserves further attention because he adds a novelty to the constructive theory of real numbers. At the end of p. 4 he introduces in loose terms a specific problem, when the application of an algorithm A to a word P of an alphabet has an end in a finite number of steps. He introduces his solution by the following words (DNSs nos. 27, 28, 35):

> I consider it [is] possible [= is not true that it is false] to apply here an argument "by contradiction", i.e. to assert that the [read: every] algorithm A is applicable to the word P if the assumption that the process of applying A to P continues indefinitely leads to a contradiction. ... [In other words,] If we assert on the basis of the proved impossibility of the indefinite continuation of a given procedure that this procedure ends, then this yields a perfectly well-defined method of construction ...

i.e. the algorithm is applicable. In other terms, he claims that this predicate is equivalent to its affirmative version (Markov principle, in short MP). In the following of the paper he considers the latter one as a new axiom-principle for the development of the theory in a deductive way (in particular in the middle of p. 7).

---

[20]The two requirements on the predicate on which MP is applied suggests that PRS° would have to be applied according to the same requirements, i.e. on a predicate which is decidable one and it is obtained as the result of an *ad absurdum* theorem. Such requirements surely would avoid all criticisms to the application of PSS°, first of all the criticism to be a metaphysical principle. In the past, a great debate aimed to clarify the use of the PSR°. From the above we conclude that PSR is not a heuristic principle for validating a mere guess on an isolated event, but an architectural principle, to be applied to an entire PO theory (see my paper [9]). Its theory-dependence explains why in the ancient times, when an analysis on PO theories was premature, it was ignored; and why in modern times, when scholars devoted little attention to both non-deductive theories and non-classical logic, it was misinterpreted.

Let us analyse the above quotation. In fact, Markov changes a doubly negated predicate of intuitionist logic—"the assumption that the process applying A to P continues indefinitely leads to a <u>contradiction</u>...", i.e. it is contradictory that the process has <u>no</u> end—in the corresponding affirmative predicate of classical logic—"the algorithm is applicable to the word P". Markov claims that this change is a valid logical step, although apparently it is supported neither by the constructive method he declared in the beginnings of his paper, nor in logical terms by intuitionist logic, where the double negation law fails.[21]

7 Qualifying in Formal Terms Markov's Principle Changing Predicates from Intuitionist Logic to Classical Logic

Let us scrutinise Markov's justifications for the change (DNS no. 30). Apart his "intuition [which] finds it sufficiently clear" and apart the advantage that "arguments of this type make it possible to construct a constructive mathematics that is well able to serve contemporary natural science", let us consider the third justification: "I see <u>no</u> reasonable basis for <u>rejecting</u> it [= the resulting affirmative predicate]". It is easy to recognise that, as suggested by Dummett, [18, p. 19] Markov's claim constitutes an intuitive application of the principle of sufficient reason, concluded by its affirmative version—I accept it. In fact, the logical structure of the above justification is the same of that of PSR: $\neg\exists S$ is $\neg P$.

The usual interpretation of Markov's paper concludes that his claim concerns a decidable, doubly negated existential predicate which ends an *ad absurdum* proof—or, in Markov's terms, it is proved *by contradiction*.

Without the former requirement on the predicate—to be a decidable one—, Anselm's proof of God's existence would be a decisively valid one. Without the latter requirement— to be the conclusion of an *ad absurdum* proof—, the application of this principle to two co-planar straight lines, conceived as ever more prolonged segments, would erroneously state that they always have a meeting point.[22]

However, for supporting his change Markov does not exhibit a general *ad absurdum* proof on his predicate; he appeals to its possibility only; nor textbooks suggest the logical origin of Markov's both requirements. Instead, the two requirements receive support by the past experience of arguing in scientific PO theories, as illustrated in the above. In particular, within a PO theory this requirement is obvious; indeed, a PO theory, in order to achieve a final result which is not assured in experimental terms, has to rely its arguments on concrete, decidable objects.

---

[21] [41, p. 27], [42, p. 274]: "... a patently non-intuitionist principle". In fact, he applies a classical law to a specific non-classical predicate.

[22] It is just after a chain of *ad absurdum* proofs (propositions nos. 17–22) that Lobachevsky stated the existence of two parallel lines, conceived as ever more prolonged segments, as an alternative hypothesis to the Euclidean one. Markov's two requirements enlightens the implicit requirements of the common move performed by all the above authors of a PO theory, which of course concerns a method which has to rely on decidable predicates; also the common move occurring at the end of the former parts of OP theories is implicitly justified by <u>not</u> having reason for <u>excluding</u> the affirmative conclusion. Moreover, notice that the same implicit *ad absurdum* argument that justifies PRS° holds true for Markov's move: "... <u>otherwise</u> the reality is <u>absurd</u>".

Result 6 Markov principle, being a tentative application of Leibniz' principle of sufficient reason, is a similar change of that occurring inside a PO theory; but it is not the result of a specific theory about the algorithms.

MP is usually formalised as follows:

$$\neg\neg\exists x\, A(x) \rightarrow \exists x\, A(x). \tag{MP}$$

We recognise that in intuitionist logic the former member $\neg\neg\exists x\, A(x)$ is the non-classical I^{gkq}, whereas the latter member is the thesis I of SO.

Markov's principle states no more than the existence of one value of a predicate, whereas PRS° states that all values of a predicate hold true. Owing to the relationship between theses A and I, the change from I^{gkq} to I is weaker than the change of PRS°.

In particular, MP is not enough to change the logic of a PO theory, because the addition of an existential affirmative predicate does not change the entire set of the predicates in the classical ones.[23]

Result 7 Markov's change of an existential predicate from non-classical logic to the corresponding affirmative predicate of classical logic concerns the thesis I^{gkq}. This change is weaker than PRS°; it alone is not enough for changing the kind of logic.

8 The Relevance of the Four Theses of the Square of Opposition in Intuitionist Predicate Logic

Both minimal predicate logic and predicate intuitionist logic have no calculus and a completeness theorem (stating that if P is valid, P is derivable) either. Being the logical framework of a PO theory so dubious, its sequence of DNS constructed by its author forms a highly creative arguing, possibly facilitated by the specific subject of his theory. However I will prove that to refer to SO^{gkq} enlightens author's arguing in a PO theory.

First of all, it is a remarkable fact that two above two kinds of change in logical predicates concern the left part ("AffIrmo") of the SO^{gkq}.

Moreover, the four theses of SO^{gkq} enjoy a specific logical property. Let us consider in intuitionist logic all conversion relationships between a total predicate and an existential predicate. The only equivalence relationships are the following ones:

$$\forall\neg \leftrightarrow (3 \text{ and } 4)\neg\exists; \quad \forall\neg\neg \leftrightarrow (6 \text{ and } 7)\neg\exists\neg;$$

$$\neg\forall\neg \leftrightarrow (11 \text{ and } 12)\neg\neg\exists; \quad \neg\forall\neg\neg \leftrightarrow \neg\neg\exists\neg;$$

the last one relationship being obtained by external negation of the second one relationship. Remarkably, these four equivalence relationships concern just the four theses of SO^{gkq}: A^{gkq}, I^{gkq}, E^{gkq}, O^{gkq}.

Result 8 Only the four theses of the non-classical square of opposition SO^{gkq} enjoy the property of convertibility. Hence, the square of opposition SO^{gkq} enjoys a formal relevance in non-classical predicate logic too.

---

[23] Also Markov states that by means of the DNSs nos. 32, 33, 34.

9 Conclusions

According to Brouwer, mathematics precedes logic. In this vein the present paper discovers some logical features by pondering on the logical experiences of some mathematised theories of the past. It proves that there exists a historical tradition of a way of arguing in non-classical predicate logic; this tradition is formalised by a logical framework which is similar to the classical SO, includes Leibniz' principle, addressing the arguing of the non-deductive theory to a conclusive non-classical thesis A^{gkq}, which then is changed in the affirmative thesis A. The logical framework includes also Markov's principle, which does the same but weaker change in the thesis I^{gkq}. However, all four non-classical theses corresponding to those of the classical SO and moreover are shown to be the most relevant predicates of this logical framework.

These results support the first quotation [43] about Aristotle's logic with respect to the non-classical predicate logic. As a consequence, logical pluralism is traced back to ancient times and the Aristotle's suggestion of a square of opposition has to be intended as a seminal hint for capturing the basic predicates of all kinds of logic.

Acknowledgements   I acknowledge two anonymous referees for an improvement of my previous logical framework, and David Braithwhaite for the correction on my English text.

Appendix: The analysis through the double negations of [31]

Notice that modal words are equivalent, *via* S4 translation, to DNSs; these words are wave underlined.

(1) Recently the constructive trend in mathematics has been significantly developed. Its goal is to base all investigations on constructive objects and to carry them out within the bounds of the abstraction of potential realizability and without use of the abstraction of actual (= not potential) infinity; ...

(2) ... it rejects the so-called pure existence [= not constructive] theorems, ...

(3) ... since the existence of an object with given properties is considered proved only when a potentially realizable method for the construction of an object with those properties bas been indicated ....

(4) ... We do not define the concept of a constructive object, but rather only clarify it.

(5) In constructive mathematical theories we limit ourselves to the consideration of constructive objects of some standard type, which frees us from the necessity of formulating a general definition of a constructive object.

(6) The *abstraction of identity* is used here in a natural way; we identify the words (1) and (2); we abstract away from their having any differences; we say that they are the same word ....

(7) ... When considering words in a given alphabet we are forced into an abstraction of another kind—into the abstraction of potential existence ....

(8) ... It consists in abstracting away from the practical limits of our possibilities in space, time and material when it comes to the existence of words ....

(9) ... We cannot write on a given blackboard of a given dimension words of arbitrary [= unlimited] length ....

(10) ... We abstract away from this practical impossibility and begin to argue as if this were possible.

(11) This does not at all mean that we begin to consider the "sequence of naturals" as an infinite "object." ...

(12) ... Such a consideration would involve an abstraction of an actual infinity, ...

(13) ... taking us beyond the limits of constructive mathematics and into something characteristic of the so-called "classical" mathematics.

(14) In "classical" mathematics there have been many "pure existence theorems," which consist in assertions about the "existence" of objects with certain properties even despites a complete ignorance of means to construct such an object.

(15) Constructive mathematics rejects such propositions [on pure existence] ...

(16) ... In constructive mathematics the existence of an object with certain properties is only considered proved when a potentially realizable method has been given for the construction of an object with the given properties.

(17) This understanding of disjunction does not permit one to take as true the law of the excluded middle: "P or not P.". ...

(18) ...2. The formulation and development of the constructive trend took place on the basis of work that appeared in the 1930s which made precise the concept of an algorithm, freeing this concept from vagueness and subjectivity....

(19) As we know, this vague concept of a was made precise in the 30s in the work of several men, who took different approaches: Church, Kleene, Turing, Post. The theories constructed by these men—Kleene's theory of recursive functions, Church's calculus of λ-conversion, the theory of Turing machines and Post's theory of finite combinatory processes—turned out to be equivalent to one another and to lead to essentially the same formulation of the concept of an algorithm ...

(20) ... New formulations of this concept, also equivalent to the previous ones, were constructed one after another by other authors, ...

(21) ... and even in the present time new theories of algorithms are continually being published that are equivalent to the previous theories ....

(22) ... It is not necessary for us at this time to look into these theories to try to find the best one.

(23) The algorithm also determines the end of the procedure which may or may not occur.

(24) The theory of normal algorithms is constructed which in the framework of abstract potential existence ....

(25) ... The words in the alphabet A under consideration and the schemes of the normal algorithms in A are potentially realizable constructive objects ....

(26) ... The procedure itself of applying a normal algorithm to a given word is considered by us to be a potentially realizable procedure.

(27) I consider it [is] possible to apply here an argument "by contradiction",

(28) ... i.e. to assert that the algorithm A is applicable to the word P if the assumption that the process of applying A to P continues indefinitely leads to a contradiction.

(29) If I defend this means of argument here, it is not because I find it without error according to my intuition, ...

(30) ... but rather, firstly, because I see no reasonable reason for rejecting it, ...

(31) ... and secondly, because arguments of this type make it possible to construct a constructive mathematics that is well able to serve contemporary natural Science ....

(32) ... I insist that this does <u>not</u> go <u>beyond</u> <u>the</u> <u>bounds</u> of the constructive direction: ...

(33) ... the abstraction of <u>actual</u> infinity is <u>not</u> made, ...

(34) ... existence continues to coincide with a <u>potentially</u> realizable construction ...

(35) ... If we assert on the basis of the proved <u>impossibility</u> of the <u>indefinite</u> continuation of a given procedure that this procedure ends, then this yields a perfectly well-defined method of construction: continue the process until its completion ....

(36) ... The circumstance that the number of steps <u>cannot</u> be <u>bounded</u> "in advance", here changes nothing of importance ....

(37) ... It is even <u>doubtful</u> that the requirement that this number be bounded in advance will ever be formulated precisely and objectively ....

(38) ... It is <u>not</u> <u>difficult</u> to see that this method for proving the applicability of an algorithm ...

(39) ... <u>allows</u> one to justify the following method of argument ....

(40) ... *Let P a property, ad let there be an algorithm that decide for every natural number n whether or not n has the property P. If the proposition that <u>no</u> number has the property P leads to a <u>contradiction</u> there is a natural number with the property P.*

The analysis of this list of DNS is easy. The list presents one only *ad absurdum* proof; it is exposed wordily by the DNSs 29 and 30. Hence, there is one unit of arguing only. The following DNSs are aimed to illustrate the result obtained. After the DNS 40 the development of the theory assumes the last DNS as a new hypothesis.

References

1. Avogadro, A.: Essay d'une manière de determiner les masses relatives des molecules élémentaires des corps... J. Phys. Chim. Hist. Nat. **73**, 58–76 (1811)
2. Beth, E.W.: Foundations of Mathematics, ch. I, 2. Harper, New York (1959)
3. Carnot, L.: Essai sur le machines en général, pp. 101–103. Defay, Dijon (1783)
4. Carnot, L.: Principes de l'équilibre et du mouvement, pp. xii–xv. Deterville, Paris (1803)
5. Carnot, S.: Réflexions sur la puissance motrice du feu, p. 23. Blanchard, Paris (1824)
6. Le Ronde d'Alembert, J.: Elémens. In: Le Ronde d'Alembert, J., Diderot, D. (eds.) Éncyclopédie Française, p. 17 (1754)
7. Drago, A.: Poincaré vs. Peano and Hilbert about the mathematical principle of induction. In: Greffe, J.-L., Heinzmann, G., Lorenz, K. (eds.) Henri Poincaré. Science et Philosophie, pp. 513–527. Blanchard/Springer, Paris/Berlin (1996)
8. Drago, A.: Mathematics and alternative theoretical physics: the method for linking them together. Epistemologia **19**, 33–50 (1996)
9. Drago, A.: The birth of an alternative mechanics: Leibniz' principle of sufficient reason. In: Poser, H., et al. (eds.) Leibniz-Kongress. Nihil Sine Ratione, vol. I, pp. 322–330. Berlin (2001)
10. Drago, A.: A.N. Kolmogoroff and the relevance of the double negation law in science. In: Sica, G. (ed.) Essays on the Foundations of Mathematics and Logic, pp. 57–81. Polimetrica, Milano (2005)
11. Drago, A.: Four ways of logical reasoning in theoretical physics and their relationships with both the kinds of mathematics and computer science. In: Cooper, S.B., et al. (eds.) Computation and Logic in Real World, pp. 132–142. Univ. Siena Dept. of Math. Sci., Siena (2007)
12. Drago, A.: The square of opposition and the four fundamental choices. Logica Univers. **2**, 127–141 (2008)
13. Drago, A.: Lo scritto di Einstein sulla relatività. Esame secondo il modello di organizzazione basata su un problema. Atti Fond. G. Ronchi **64**, 113–131 (2009)
14. Drago, A.: A logical analysis of Lobachevsky's geometrical theory. Atti Fond. G. Ronchi **64**(4), 453–481 (2010). Italian version in: Period. Mat., Ser. VIII **4**, 41–52 (2004)
15. Drago, A.: The coincidentia oppositorum in Cusanus (1401–1464), Lanza del Vasto (1901–1981) and beyond. Epistemologia **33**, 305–328 (2010)

16. Drago, A., Oliva, R.: Atomism and the reasoning by non-classical logic. Hyle **5**, 43–55 (1999)
17. Drago, A., Pisano, R.: Interpretation and reconstruction of Sadi Carnot's Réflexions through non-classical logic. Atti Fond. G. Ronchi **59**, 615–644 (2004). Italian version in: G. Fis. **41**, 195–215 (2000)
18. Dummett, M.: Principles of Intuitionism. Clarendon, Oxford (1977)
19. Einstein, A.: Zur Elektrodynamik bewegter Koerper. Ann. Phys. **17**, 891–921 (1905)
20. Gabbey, D., Olivetti, N.: Goal-oriented deductions. In: Gabbey, D., Guenthner, F. (eds.) Handbook of Philosophical Logic, vol. 9, pp. 134–199. Reidel, Boston (2002)
21. Goedel, K.: Collected Works, papers 1932 and 1933f. Oxford University Press, Oxford (1986)
22. Grize, J.B.: Logique. In: Piaget, J. (ed.) Logique et connaissance scientifique. Éncyclopédie de la Pléiade, pp. 135–288 (206–210). Gallimard, Paris (1970)
23. Haak, S.: Deviant Logic. Oxford University Press, Oxford (1978)
24. Hand, M.: Antirealism and Falsity. In: Gabbey, D.M., Wansing, H. (eds.) What is Negation?, pp. 185–198 (188). Kluwer, Dordrecht (1999)
25. Heyting, A.: Intuitionism. An Introduction, Sect. 2. North-Holland, Amsterdam (1971)
26. Kleene, C.: Introduction to Metamathematics, pp. 318–319. Van Nostrand, Princeton (1952)
27. Klein, M.J.: Thermodynamics in Einstein's thought. Science **57**, 505–516 (1967)
28. Kolmogorov, A.N.: On the principle 'tertium non datur'. Mat. Sb. **32**, 646–667 (1924/1925). Engl. translation in: van Heijenoorth, J. (ed.) From Frege to Goedel, pp. 416–437. Harvard University Press, Cambridge (1967)
29. Leibniz, G.W.: Letter to Arnauld, 14-7-1686, Gerh. II, Q., Opusc. 402, 513
30. Lobachevsky, N.I.: Geometrische Untersuchungen zur der Theorien der Parallellineen. Finkl, Berlin (1840). Engl. translation as an Appendix to: Bonola, R.: Non-Euclidean Geometry. Dover, New York (1955)
31. Markov, A.A.: On constructive mathematics. Tr. Mat. Inst. Steklova **67**, 8–14 (1962). Also in: Am. Math. Soc. Transl. **98**(2), 1–9 (1971)
32. Miller, A.I.: Albert Einstein's Special Theory of Relativity, pp. 123–142. Addison-Wesley, Reading (1981)
33. Parsons, T.: Things that are right wit the traditional square of opposition. Logica Univers. **2**, 3–11 (2008)
34. Poincaré, H.: La Science et l'Hypothèse, Ch. "Optique et Electricité". Hermann, Paris (1903)
35. Poincaré, H.: La Valeur de la Science, Ch. VII. Flammarion, Paris (1905)
36. Poincaré, H.: Science et Méthode, p. 187. Flammarion, Paris (1912)
37. Prawitz, D.: Meaning and proof. The conflict between classical and intuitionist logic. Theoria **43**, 6–39 (1976)
38. Prawitz, D., Melmnaas, P.-E.: A survey of some connections between classical intuitionistic and minimal logic. In: Schmidt, A., Schuette, H. (eds.) Contributions to Mathematical Logic, pp. 215–229. North-Holland, Amsterdam (1968)
39. Tennant, N.: Natural Logic, p. 57. Edinburgh University Press, Edinburgh
40. Troesltra, A.: Remarks on intuitionism and the philosophy of mathematics. In: Costantini, D., Galavotti, M.C. (eds.) Nuovi Problemi della Logica e della Filosofia della Scienza, pp. 211–228 (223). CLUEB, Bologna (1991)
41. Troelstra, A., van Dalen, D.: Constructivism in Mathematics, vol. 1. North-Holland, Amsterdam (1988)
42. van Dalen, D.: Intuitionist logic. In: Gabbey, D.M., Guenthner, F. (eds.) Handbook of Mathematical Logic, vol. II, pp. 225–449 (226 and 230). Reidel, Dordrecht (2002)
43. Woods, J., Irvine, A.: Aristotle's early logic. In: Gabbey, D.M., Woods, J. (eds) Handbook of History of Logic, vol. 1, pp. 27–99 (66). Elsevier, New York (2004)

A. Drago (✉)
University of Pisa, Pisa, Italy
e-mail: drago@unina.it

Part III
The Square of Opposition and Non-classical Logics

The Square of Opposition in Orthomodular Logic

H. Freytes, C. de Ronde, and G. Domenech

Abstract In Aristotelian logic, categorical propositions are divided in Universal Affirmative, Universal Negative, Particular Affirmative and Particular Negative. Possible relations between two of the mentioned type of propositions are encoded in the square of opposition. The square expresses the essential properties of monadic first order quantification which, in an algebraic approach, may be represented taking into account monadic Boolean algebras. More precisely, quantifiers are considered as modal operators acting on a Boolean algebra and the square of opposition is represented by relations between certain terms of the language in which the algebraic structure is formulated. This representation is sometimes called the modal square of opposition. Several generalizations of the monadic first order logic can be obtained by changing the underlying Boolean structure by another one giving rise to new possible interpretations of the square.

Keywords Square of opposition · Modal orthomodular logic · Classical consequences

Mathematics Subject Classification 03G12 · 06C15 · 03B45

1 Introduction

In Aristotelian logic, categorical propositions are divided into four basic types: *Universal Affirmative, Universal Negative, Particular Affirmative* and *Particular Negative*. The possible relations between each two of the mentioned propositions are encoded in the famous *Square of Opposition*. The square expresses the essential properties of the monadic first order quantifiers ∀, ∃. In an algebraic approach, these properties can be represented within the frame of monadic Boolean algebras [8]. More precisely, quantifiers are considered as modal operators acting on a Boolean algebra while the Square of Opposition is represented by relations between certain terms of the language in which the algebraic structure is formulated. This representation is sometimes called *Modal Square of Opposition* and is pictured as follows:

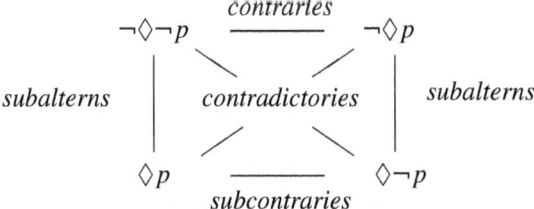

J.-Y. Béziau, D. Jacquette (eds.), *Around and Beyond the Square of Opposition,* 193–200
Studies in Universal Logic, DOI 10.1007/978-3-0348-0379-3_13, © Springer Basel 2012

The interpretations given to $\Diamond$ from different modal logics determine the correspond-ing versions of the modal Square of Opposition. By changing the underlying Boolean structure we obtain several generalizations of the monadic first order logic (see for exam-ple [9]). In turn, these generalizations give rise to new interpretations of the Square.

The aim of this paper is to study the Square of Opposition in an orthomodular structure enriched with a monadic quantifier known as *Boolean saturated orthomodular lattice* [5]. The paper is structured as follows. Section 2 contains generalities on orthomodular lat-tices. In Sect. 3, the physical motivation for the modal enrichment of the orthomodular structure is presented. In Sect. 4 we formalize the concept of classical consequence with respect to a property of a quantum system. Finally, in Sect. 5, logical relationships be-tween the propositions embodied in a square diagram are studied in terms of classical consequences and contextual valuations.

2 Basic Notions

We recall from [1, 12, 13] some notions of universal algebra and lattice theory that will play an important role in what follows. Let $\mathcal{L} = \langle \mathcal{L}, \vee, \wedge, 0, 1 \rangle$ be a bounded lattice. An element $c \in \mathcal{L}$ is said to be a *complement* of a iff $a \wedge c = 0$ and $a \vee c = 1$. Given a, b, c in $\mathcal{L}$, we write: $(a, b, c)D$ iff $(a \vee b) \wedge c = (a \wedge c) \vee (b \wedge c)$; $(a, b, c)D^*$ iff $(a \wedge b) \vee c = (a \vee c) \wedge (b \vee c)$ and $(a, b, c)T$ iff $(a, b, c)D$, $(a, b, c)D^*$ hold for all permutations of a, b, c. An element z of a lattice $\mathcal{L}$ is called *central* iff for all elements $a, b \in \mathcal{L}$ we have $(a, b, z)T$ and z is complemented. We denote by $Z(\mathcal{L})$ the set of all central elements of $\mathcal{L}$ and it is called the *center* of $\mathcal{L}$.

A *lattice with involution* [11] is an algebra $\langle \mathcal{L}, \vee, \wedge, \neg \rangle$ such that $\langle \mathcal{L}, \vee, \wedge \rangle$ is a lattice and $\neg$ is a unary operation on $\mathcal{L}$ that fulfills the following conditions: $\neg \neg x = x$ and $\neg(x \vee y) = \neg x \wedge \neg y$. An *orthomodular lattice* is an algebra $\langle \mathcal{L}, \wedge, \vee, \neg, 0, 1 \rangle$ of type $\langle 2, 2, 1, 0, 0 \rangle$ that satisfies the following conditions

1. $\langle \mathcal{L}, \wedge, \vee, \neg, 0, 1 \rangle$ is a bounded lattice with involution,
2. $x \wedge \neg x = 0$,
3. $x \vee (\neg x \wedge (x \vee y)) = x \vee y$.

We denote by $\mathcal{OML}$ the variety of orthomodular lattices. Let L be an orthomodular lattice and $a, b \in L$. Then a commutes with b if and only if $a = (a \wedge b) \vee (a \wedge \neg b)$. A non-empty subset A is called a *Greechie set* iff for any three different elements of A, at least one of them commutes with the other two. If A is a Greechie set in L then $\langle A \rangle_L$, i.e. the sublattice generated by A, is distributive [7]. *Boolean algebras* are orthomodular lattices satisfying the *distributive law* $x \wedge (y \vee z) = (x \wedge y) \vee (x \wedge z)$. We denote by **2** the Boolean algebra of two elements. Let A be a Boolean algebra. Then, as a consequence of the application of the *maximal filter theorem* for Boolean algebras, there always exists a Boolean homomorphism $f : A \rightarrow \mathbf{2}$. If $\mathcal{L}$ is a bounded lattice then $Z(\mathcal{L})$ is a Boolean sublattice of $\mathcal{L}$ [13, Theorem 4.15].

3 Modal Propositions About Quantum Systems

In the usual terms of quantum logic [2, 10], a property of a system is related to a subspace of the Hilbert space $\mathcal{H}$ of its (pure) states or, analogously, to the projector operator onto that subspace. A physical magnitude $\mathcal{M}$ is represented by an operator $\mathbf{M}$ acting over the state space. For bounded self-adjoint operators, conditions for the existence of the spectral decomposition $\mathbf{M} = \sum_i a_i \mathbf{P}_i = \sum_i a_i |a_i\rangle\langle a_i|$ are satisfied. The real numbers a_i are related to the outcomes of measurements of the magnitude $\mathcal{M}$ and projectors $|a_i\rangle\langle a_i|$ to the mentioned properties. Thus, the physical properties of the system are organized in the lattice of closed subspaces $\mathcal{L}(\mathcal{H})$. Moreover, each self-adjoint operator $\mathbf{M}$ has associated a Boolean sublattice $W_{\mathbf{M}}$ of $L(\mathcal{H})$ which we will refer to as the spectral algebra of the operator $\mathbf{M}$. More precisely, the family $\{\mathbf{P}_i\}$ of projector operators is identified as elements of $W_{\mathbf{M}}$. Assigning values to a physical quantity $\mathcal{M}$ is equivalent to establishing a Boolean homomorphism $v : W_{\mathbf{M}} \to \mathbf{2}$ which we call *contextual valuation*. Thus, we can say that it makes sense to use the "classical discourse"—this is, the classical logical laws are valid—within the context given by $\mathbf{M}$.

Modal interpretations of quantum mechanics [3, 4, 15] face the problem of finding an objective reading of the accepted mathematical formalism of the theory, a reading "in terms of properties possessed by physical systems, independently of consciousness and measurements (in the sense of human interventions)" [4]. These interpretations intend to consistently include the possible properties of the system in the discourse establishing a new link between the state of the system and the probabilistic character of its properties, namely, sustaining that the interpretation of the quantum state must contain a modal aspect. The name modal interpretation was used for the first time by B. van Fraassen [14] following modal logic, precisely the logic that deals with *possibility* and *necessity*. The fundamental idea is to interpret "the formalism as providing information about properties of physical systems". A physical property of a system is "a definite value of a physical quantity belonging to this system; i.e., a feature of physical reality" [3] and not a mere measurement outcome. As usual, definite values of physical magnitudes correspond to yes/no propositions represented by orthogonal projection operators acting on vectors belonging to the Hilbert space of the (pure) states of the system [10].

The proposed modal system [5, 6] over which we now develop an interpretation of the Square of Opposition, is based on the study of the "*classical consequences*" that result from assigning values to a physical quantity. In precise terms, we enriched the orthomodular structure with a modal operator taking into account the following considerations:

(1) Propositions about the properties of the physical system are interpreted in the orthomodular lattice of closed subspaces of $\mathcal{H}$. Thus, we retain this structure in our extension.

(2) Given a proposition about the system, it is possible to define a context from which one can predicate with certainty about it together with a set of propositions that are compatible with it and, at the same time, predicate probabilities about the other ones (Born rule). In other words, one may predicate truth or falsity of all possibilities at the same time, i.e. possibilities allow an interpretation in a Boolean algebra. In rigorous terms, for each proposition p, if we refer with $\Diamond p$ to the possibility of p, then $\Diamond p$ will be a central element of a orthomodular structure.

(3) If p is a proposition about the system and p occurs, then it is trivially possible that p occurs. This is expressed as $p \leq \Diamond p$.

(4) Let p be a property appertaining to a context $\mathcal{M}$. Assuming that p is an actual property (for example the result of a filtering measurement) we may derive from it a set of propositions (perhaps not all of them encoded in the original Hilbert lattice of the system) which we call *classical consequences*. For example, let q be another property of

the system and assign to q the probability $prob(q) = r$ via the Born rule. Then equality $prob(q) = r$ will be considered as a classical consequence of p. In fact, the main characteristic of this type of classical consequences is that it is possible to simultaneously predicate the truth of all of them (and the falsity of their negations) whenever p is true. The formal representation of the concept of classical consequence is the following: A proposition t is a classical consequence of p iff t is in the center of an orthomodular lattice containing p and satisfies the property $p \leq t$. These classical consequences are the same ones as those which would be obtained by considering the original actual property p as a possible one $\Diamond p$. Consequently $\Diamond p$ must precede all classical consequences of p. This is interpreted in the following way: $\Diamond p$ is the smallest central element greater than p.

From consideration 1, it follows that the original orthomodular structure is maintained. The other considerations are satisfied if we consider a modal operator $\Diamond$ over an orthomodular lattice $\mathcal{L}$ defined as

$$\Diamond a = Min\{z \in Z(\mathcal{L}) : a \leq z\}$$

with $Z(\mathcal{L})$ the center of $\mathcal{L}$. When this minimum exists for each $a \in \mathcal{L}$ we say that $\mathcal{L}$ is a *Boolean saturated orthomodular lattice*. We have shown that this enriched orthomodular structure can be axiomatized by equations conforming a variety denoted by $\mathcal{OML}^\Diamond$ [5]. More precisely, each element of $\mathcal{OML}^\Diamond$ is an algebra $\langle \mathcal{L}, \wedge, \vee, \neg, \Diamond, 0, 1 \rangle$ of type $\langle 2, 2, 1, 1, 0, 0 \rangle$ such that $\langle \mathcal{L}, \wedge, \vee, \neg, 0, 1 \rangle$ is an orthomodular lattice and $\Box$ satisfies the following equations:

S1 $x \leq \Diamond x$
S2 $\Diamond 0 = 0$
S3 $\Diamond\Diamond x = \Diamond x$
S4 $\Diamond(x \vee y) = \Diamond x \vee \Diamond y$
S5 $y = (y \wedge \Diamond x) \vee (y \wedge \neg \Diamond x)$
S6 $\Diamond(x \wedge \Diamond y) = \Diamond x \wedge \Diamond y$
S7 $\neg\Diamond x \wedge \Diamond y \leq \Diamond(\neg x \wedge (y \vee x))$

Orthomodular complete lattices are examples of Boolean saturated orthomodular lattices. We can embed each orthomodular lattice $\mathcal{L}$ in an element $\mathcal{L}^\Diamond \in \mathcal{OML}^\Diamond$ see [5, Theorem 10]. In general, $\mathcal{L}^\Diamond$ is referred as a *modal extension of* $\mathcal{L}$. In this case we may see the lattice $\mathcal{L}$ as a subset of $\mathcal{L}^\Diamond$.

4 Modal Extensions and Classical Consequences

We begin our study of the Square of Opposition analyzing the classical consequences that can be derived from a proposition about the system. This idea was suggested in the condition 4 of the motivation of the structure $\mathcal{OML}^\Diamond$. In what follows we express the notion of classical consequence as a formal concept in $\mathcal{OML}^\Diamond$. We first need the following technical results:

Proposition 4.1 *Let L be an orthomodular lattice. If A is a Greechie set in L such that for each $a \in A$, $\neg a \in A$ then, $\langle A \rangle_L$ is Boolean sublattice.*

Proof It is well known from [7] that $\langle A \rangle_L$ is a distributive sublattice of L. Since distributive orthomodular lattices are Boolean algebras, we only need to see that $\langle A \rangle_L$ is closed by $\neg$. To do that we use induction on the complexity of terms of the subuniverse generated by A. For $comp(a) = 0$, it follows from the fact that A is closed by negation. Assume validity for terms of the complexity less than n. Let τ be a term such that $comp(\tau) = n$. If $\tau = \neg\tau_1$ then $\neg\tau \in \langle A \rangle_L$ since $\neg\tau = \neg\neg\tau_1 = \tau_1$ and $\tau_1 \in \langle A \rangle_L$. If $\tau = \tau_1 \wedge \tau_2$, $\neg\tau = \neg\tau_1 \vee \neg\tau_2$. Since $comp(\tau_i) < n$, $\neg\tau_i \in \langle A \rangle_L$ for $i = 1, 2$ resulting $\neg\tau \in \langle A \rangle_L$. We use the same argument in the case $\tau = \tau_1 \vee \tau_2$. Finally $\langle A \rangle_L$ is a Boolean sublattice. $\qquad\square$

Since the center $Z(\mathcal{L}^\Diamond)$ is a Boolean algebra, it represents a fragment of discourse added to $\mathcal{L}$ in which the laws of the classical logic are valid. Thus the modal extension $\mathcal{L} \hookrightarrow \mathcal{L}^\Diamond$ is a structure that rules the mentioned fragment of classical discourse and the properties about a quantum system encoded in $\mathcal{L}$. Let W be a Boolean sub-algebra of $\mathcal{L}$ (i.e. a context). Note that $W \cup Z(\mathcal{L}^\Diamond)$ is a Greechie set closed by $\neg$. Then by Proposition 4.1, $\langle W \cup Z(\mathcal{L}^\Diamond) \rangle_{\mathcal{L}^\Diamond}$ is a Boolean sub-algebra of $Z(\mathcal{L}^\Diamond)$. This represents the possibility to fix a context and compatibly add a fragment of classical discourse. We will refer to the Boolean algebra $W^\Diamond = \langle W \cup Z(\mathcal{L}^\Diamond) \rangle_{\mathcal{L}^\Diamond}$ as a *classically expanded context*. Taking into account that assigning values to a physical quantity p is equivalent to fix a context W in which $p \in W$ and establish a Boolean homomorphism $v : W \to \mathbf{2}$ such that $v(p) = 1$, we give the following definition of classical consequence.

Definition 4.2 Let $\mathcal{L}$ be an orthomodular lattice, $p \in \mathcal{L}$ and $\mathcal{L}^\Diamond \in \mathcal{OML}^\Diamond$ a modal extension of $\mathcal{L}$. Then $z \in Z(\mathcal{L}^\Diamond)$ is said to be a *classical consequence* of p iff for each Boolean sublattice W in $\mathcal{L}$ (with $p \in W$) and each Boolean valuation $v : W^\Diamond \to \mathbf{2}$, $v(z) = 1$ whenever $v(p) = 1$.

$v(p) = 1$ implies $v(z) = 1$ in Definition 4.2 is a relation in a classically expanded context that represents, in an algebraic way, the usual logical consequence of z from p. We denote by $Cons_{\mathcal{L}^\Diamond}(p)$ the set of classical consequences of p in the modal extension $\mathcal{L}^\Diamond$.

Proposition 4.3 *Let $\mathcal{L}$ be an orthomodular lattice, $p \in \mathcal{L}$ and $\mathcal{L}^\Diamond \in \mathcal{OML}^\Diamond$ a modal extension of $\mathcal{L}$. Then we have that*

$$Cons_{\mathcal{L}^\Diamond}(p) = \{z \in Z(\mathcal{L}^\Diamond) : p \le z\} = \{x \in Z(\mathcal{L}^\Diamond) : \Diamond p \le z\}$$

Proof $\{z \in Z(\mathcal{L}^\Diamond) : p \le z\} = \{z \in Z(\mathcal{L}^\Diamond) : \Diamond p \le z\}$ follows from definition of $\Diamond$. The inclusion $\{z \in Z(\mathcal{L}^\Diamond) : \Diamond p \le z\} \subseteq Cons_{\mathcal{L}^\Diamond}(p)$ is trivial. Let $z \in Cons_{\mathcal{L}^\Diamond}(p)$ and suppose that $p \not\le z$. Consider the Boolean sub-algebra of $\mathcal{L}$ given by $W = \{p, \neg p, 0, 1\}$. By the maximal filter theorem, there exists a maximal filter F in $W^\Diamond$ such that $p \in F$ and $z \notin F$. If we consider the quotient Boolean algebra $W^\Diamond / F = \mathbf{2}$, the natural Boolean homomorphism $f : W^\Diamond \to \mathbf{2}$ satisfies that $f(p) = 1$ and $f(z) = 0$, which is a contradiction. Hence $p \not\le z$ and $z \in Cons_{\mathcal{L}^\Diamond}(p)$. $\qquad\square$

The equality $Cons_{\mathcal{L}^\Diamond}(p) = \{x \in Z(\mathcal{L}^\Diamond) : \Diamond p \le z\}$ given in Proposition 4.3 states that the notion of classical consequence of p results independent of the choice of the context W in which $p \in W$. In fact, each possible classical consequence $z \in Z(\mathcal{L}^\Diamond)$ of p is only determined by the relation $\Diamond p \le z$. Thus $Z(\mathcal{L}^\Diamond)$ is a fragment of the classical discourse added to $\mathcal{L}$ which allows to "predicate" classical consequences about the properties of the system encoded in $\mathcal{L}$ independently of the context. It is important to remark that the contextual character of the quantum discourse is only avoided when we refer to "classical consequences of properties about the system" and not when referring to the properties in themselves, i.e. independently of the choice of the context. In fact, in our modal extension, the discourse about properties is genuinely enlarged, but the contextual character remains a main feature of quantum systems even when modalities are taken into account.

5 Square of Opposition: Classical Consequences and Contextual Valuations

In this section we analyze the relations between propositions encoded in the scheme of the Square of Opposition in terms of the classical consequences of a chosen property about the quantum system. To do this, we use the modal extension. Let $\mathcal{L}$ be an orthomodular lattice, $p \in \mathcal{L}$ and $\mathcal{L}^{\Diamond} \in \mathcal{OML}^{\Diamond}$ a modal extension of $\mathcal{L}$. We first study the proposition $\neg\Diamond\neg p$ denoted by $\Box p$. Note that $\Box p = \neg\Diamond\neg p = \neg Min\{z \in Z(\mathcal{L}^{\Diamond}) : \neg p \le z\} = Max\{\neg z \in Z(\mathcal{L}^{\Diamond}) : \neg p \le z\} = Max\{\neg z \in Z(\mathcal{L}^{\Diamond}) : \neg z \le p\}$. Considering $t = \neg z$ we have that

$$\Box p = \neg\Diamond\neg p = Max\{t \in Z(\mathcal{L}^{\Diamond}) : t \le p\}$$

When W is a Boolean sublattice of $\mathcal{L}$ such that $p \in W$ (i.e. we are fixing a context containing p), $\Box p$ is the greatest classical proposition that implies p in the classically expanded context $W^{\Diamond}$. More precisely, if p is a consequence of $z \in \mathcal{L}^{\Diamond}$ then $\Box p$ is consequence of z.

Remark 5.1 It is important to notice that Proposition 4.3 allows us to refer to classical consequences of a property of the system independently of the chosen context. But in order to refer to a property which is implied by a classical property, we need to fix a context and consider the classically expanded context.

Lemma 5.2 *Let $\mathcal{L}$ be an orthomodular lattice, $p \in \mathcal{L}$ and $\mathcal{L}^{\Diamond}$ be a modal extension of $\mathcal{L}$. If $\Diamond p \wedge \Diamond\neg p = 0$ then $p \in Z(\mathcal{L})$.*

Proof $\Diamond p$ and $\Diamond\neg p$ are central elements in $\mathcal{L}^{\Diamond}$. Taking into account that $1 = p \vee \neg p \le \Diamond p \vee \Diamond\neg p$, if $\Diamond p \wedge \Diamond\neg p = 0$ then $\neg\Diamond p = \Diamond\neg p$ since the complement is unique in $Z(\mathcal{L}^{\Diamond})$. Hence $\Box p = \neg\Diamond\neg p = \Diamond p$, $p \in Z(\mathcal{L}^{\Diamond})$ and $p \in Z(\mathcal{L})$. □

Now we can interpret the relation between propositions in the Square of Opposition. In what follows we assume that $\mathcal{L}$ is an orthomodular lattice and $p \in \mathcal{L}$ such that $p \notin Z(\mathcal{L})$, i.e. p is not a classical proposition in a quantum system represented by $\mathcal{L}$. Let $\mathcal{L}^{\Diamond}$ be a modal extension of $\mathcal{L}$, W be a Boolean subalgebra of $\mathcal{L}$, i.e. a context, such that $p \in W$ and consider a classically expanded context $W^{\Diamond}$.

- $\neg\Diamond\neg p$ *contraries* $\neg\Diamond p$

$\neg \Diamond p = \neg \Diamond \neg \neg p = \Box \neg p$. Thus, the *contrary proposition* is the greatest classical proposition that implies p, i.e. $\Box p$, with respect to the greatest classical proposition that implies $\neg p$, i.e. $\Box \neg p$, in each possible classically expanded context containing $p, \neg p$.

In the usual explanation, two propositions are contrary iff they cannot both be true but can both be false. In our framework we can obtain a similar concept of contrary propositions. Note that $\Box p \wedge \Box \neg p \leq p \wedge \neg p = 0$. Thus, there is not a maximal Boolean filter containing $\Box p$ and $\Box \neg p$. Hence there is not a Boolean valuation $v : W^{\Diamond} \to \mathbf{2}$ such that $v(\Box p) = v(\Box \neg p) = 1$, i.e. $\Box p$ and $\Box \neg p$ "cannot both be true" in each possible classically expanded context.

Since $p \notin Z(\mathcal{L})$, by Lemma 5.2, $\Diamond p \wedge \Diamond \neg p \neq 0$. Then there exists a maximal Boolean filter F in $W^{\Diamond}$ containing $\Diamond p$ and $\Diamond \neg p$. $\Box p \notin F$ otherwise $\Box p \wedge \Diamond \neg p \in F$ and $\Box p \wedge \Diamond \neg p = \Diamond(\Box p \wedge \neg p) \leq \Diamond(p \wedge \neg p) = 0$ which is a contradiction. With the same argument we can prove that $\Box \neg p \notin F$. If we consider the natural homomorphism $v : W^{\Diamond} \to W^{\Diamond}/F \approx \mathbf{2}$ then $v(\Box p) = v(\Box \neg p) = 0$, i.e. $\Box p$ and $\Box \neg p$ can both be false.

- $\Diamond p$ *subcontraries* $\Diamond \neg p$

The *sub-contrary proposition* is the smallest classical consequence of p with respect to the smallest classical consequence of $\neg p$. Note that sub-contrary propositions do not depend on the context.

In the usual explanation, two propositions are sub-contrary iff they cannot both be false but can both be true. Suppose that there exists a Boolean homomorphism $v : W^{\Diamond} \to \mathbf{2}$ such that $v(\Diamond p) = v(\Diamond \neg p) = 0$. Consider the maximal Boolean filter given by $Ker(v)$. Since $Ker(v)$ is a maximal filter in $W^{\Diamond}$, $p \in F_v$ or $\neg p \in F_v$. If $p \in F_v$ then $v(\Diamond p) = 1$ which is a contradiction, if $\neg p \in F_v$ then $v(\Diamond \neg p) = 1$ which is a contradiction too. Hence $v(\Diamond p) \neq 0$ or $v(\Diamond \neg p) \neq 0$, i.e. they cannot both be false. Since $p \notin Z(\mathcal{L})$, by Lemma 5.2, $\Diamond p \wedge \Diamond \neg p \neq 0$. Then there exists a maximal Boolean filter F in $W^{\Diamond}$ containing $\Diamond p$ and $\Diamond \neg p$. Hence the Boolean homomorphism $v : W^{\Diamond} \to W^{\Diamond}/F \approx \mathbf{2}$ satisfies that $v(\Diamond p) = v(\Diamond \neg p) = 1$, i.e. $\Diamond p$ and $\Diamond \neg p$ can both be true.

- $\neg \Diamond \neg p$ *subalterns* $\Diamond p$ and $\neg \Diamond p$ *subalterns* $\Diamond \neg p$

The notion of sub-contrary propositions is reduced to the relation between $\Box p$ and $\Diamond p$. The *subaltern proposition* is the greatest classical proposition that implies p with respect to the smallest classical consequence of p.

In the usual explanation, a proposition is subaltern of another one called *superaltern*, iff it must be true when its superaltern is true, and the superaltern must be false when the subaltern is false. In our case $\neg \Diamond \neg p = \Box p$ is superaltern of $\Diamond p$ and $\neg \Diamond p = \Box \neg p$ is superaltern of $\Diamond \neg p$. Since $\Box p \leq p \leq \Diamond p$, for each valuation $v : W^{\Diamond} \to \mathbf{2}$, if $v(\Box p) = 1$ then $v(\Diamond p) = 1$ and if $v(\Diamond p) = 0$ then $v(\Box p) = 0$.

- $\neg \Diamond \neg p$ *contradictories* $\Diamond \neg p$ and $\Diamond p$ *contradictories* $\neg \Diamond p$

The notion of *contradictory proposition* can be reduced to the relation between $\Diamond p$ and $\Box \neg p$. The contradictory proposition is the greatest classical proposition that implies $\neg p$ with respect to the smallest classical consequence of p. In the usual explanation, two propositions are contradictory iff they cannot both be true and they cannot both be false. Due to the fact that $ker(v)$ is a maximal filter in $W^{\Diamond}$, each maximal filter F in $W^{\Diamond}$ contains exactly one of $\{\Diamond p, \neg \Diamond p\}$ for each Boolean homomorphism $v : W^{\Diamond} \to \mathbf{2}$, $v(\Diamond p) = 1$ and $v(\neg \Diamond p) = 0$ or $v(\Diamond p) = 0$ and $v(\neg \Diamond p) = 1$. Hence, $\Diamond p$ and $\Diamond p$ cannot both be true and they cannot both be false.

References

1. Burris, S., Sankappanavar, H.P.: A Course in Universal Algebra. Graduate Text in Mathematics. Springer, New York (1981)
2. Birkhoff, G., von Neumann, J.: The logic of quantum mechanics. Ann. Math. **37**, 823–843 (1936)
3. Dieks, D.: The formalism of quantum theory: an objective description of reality. Ann. Phys. **7**, 174–190 (1988)
4. Dieks, D.: Quantum mechanics: an intelligible description of reality? Found. Phys. **35**, 399–415 (2005)
5. Domenech, G., Freytes, H., de Ronde, C.: Scopes and limits of modality in quantum mechanics. Ann. Phys. **15**, 853–860 (2006)
6. Domenech, G., Freytes, H., de Ronde, C.: Modal type orthomodular logic. Math. Log. Q. **55**, 287–299 (2009)
7. Greechie, R.: On generating distributive sublattices of orthomodular lattices. Proc. Am. Math. Soc. **67**, 17–22 (1977)
8. Halmos, P.: Algebraic logic I, monadic Boolean algebras. Compos. Math. **12**, 217–249 (1955)
9. Hájek, P.: Metamathematics of Fuzzy Logic. Kluwer, Dordrecht (1998)
10. Jauch, J.M.: Foundations of Quantum Mechanics. Addison-Wesley, Reading (1968)
11. Kalman, J.A.: Lattices with involution. Trans. Am. Math. Soc. **87**, 485–491 (1958)
12. Kalmbach, G.: Orthomodular Lattices. Academic Press, London (1983)
13. Maeda, F., Maeda, S.: Theory of Symmetric Lattices. Springer, Berlin (1970)
14. van Fraassen, B.C.: A modal interpretation of quantum mechanics. In: Beltrametti, E.G., van Fraassen, B.C. (eds.) Current Issues in Quantum Logic, pp. 229–258. Plenum, New York (1981)
15. van Fraassen, B.C.: Quantum Mechanics: An Empiricist View. Clarendon, Oxford (1991)

H. Freytes (✉)
Universita degli Studi di Cagliari, Via Is Mirrionis 1, 09123, Cagliari, Italy
e-mail: hfreytes@gmail.com

H. Freytes
Instituto Argentino de Matemática, Saavedra 15, Buenos Aires, Argentina

C. de Ronde
Departamento de Filosofía "Dr. A Korn", Universidad de Buenos Aires-CONICET, Buenos Aires, Argentina

C. de Ronde
Center Leo Apostel and Foundations of the Exact Sciences, Vrije Universiteit Brussel, Krijgskundestraat 33, 1160 Brussels, Belgium
e-mail: cderonde@vub.ac.be

G. Domenech
Instituto de Astronomía y Física del Espacio, CC 67, Suc 28, 1428 Buenos Aires, Argentina
e-mail: domenech@iafe.uba.ar

No Group of Opposition for Constructive Logics: The Intuitionistic and Linear Cases

Baptiste Mélès

Abstract The aim of this work is to show that no group of opposition can be built for constructive logics like intuitionistic or linear logics. Moreover, the attempt to apply the square of opposition to linear logic shows that the lack of subcontrariety and the asymmetry of contradiction are not equivalent properties. The former property, not the latter, is the general reason why there can be no group of opposition for constructive logics.

Keywords Square of opposition · Linear logic · Intuitionistic logic

Mathematics Subject Classification Primary 03F52 · Secondary 03B47 · 03B20

1 Introduction

1.1 Argument

The purpose of this paper is to show that:

1. there can be no group of oppositions for constructive logics, for example intuitionistic and linear logics, even though there can be one for classical logic;
2. the lack of subcontrariety provides a finer definition of constructivity than the asymmetry of contradiction, since both properties can be distinguished in some constructive logical systems—linear logic in this case—even though they can not be in classical and intuitionistic logics.

In order to prove it, we will use the sequent calculus systems for classical logic (LK), intuitionistic logic (LJ) and classical linear logic (CLL).

1.2 Definitions

The logical relations of the square of opposition will be defined as follows:

The author would like to thank Élisabeth Schwartz, Emmanuel Cattin, Camille and Baptiste Debrabant, Marion Duquerroy, Sébastien Gandon, Jean-Baptiste Joinet, Nicolas Mascot and Fabien Schang.

- *contrariety*: B is A's contrary if every atomic assertion of A is negated in B. Following this definition, $\neg P \wedge \neg Q$ is the contrary of $P \wedge Q$, and $\forall x \neg P$ is the contrary of $\forall x\, P$. This relation is not always symmetrical; for instance, in intuitionistic logic, $\neg P$ is P's contrary, but P is not $\neg P$'s contrary, since $\nvdash_{LJ} \neg\neg P \to P$;
- *subalternation*: B is A's subaltern if $A \to B$ is provable; this relation is generally not symmetrical, since implication is not commutative;
- *contradiction*: B is A's contradictory if $\neg A \to B$ is provable. Contradiction is not necessarily symmetrical, since $\nvdash_{LJ} (\neg A \to B) \to (\neg B \to A)$;
- *subcontrariety*: B is A's subcontrary if $A \vee B$ is provable; if disjunction is commutative, which is generally the case, this relation is symmetrical.

Every relation in the square will be represented by a simple arrow ($\to$), and by a double arrow ($\leftrightarrow$) when it is symmetrical. Then, a complete square would be as follows:

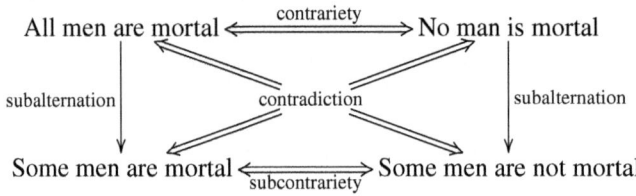

All men are mortal $\xleftrightarrow{\text{contrariety}}$ No man is mortal

subalternation | contradiction | subalternation

Some men are mortal $\xleftrightarrow[\text{subcontrariety}]{}$ Some men are not mortal

2 Classical Logic (LK)

2.1 The Complete Square

Following these definitions, the square of classical logic is obviously complete, what can be demonstrated in propositional calculus as well as in predicate calculus:

$$A \wedge B \Longleftrightarrow \neg A \wedge \neg B \qquad \forall x\, A \Longleftrightarrow \forall x \neg A$$

$$A \vee B \Longleftrightarrow \neg A \vee \neg B \qquad \exists x\, A \Longleftrightarrow \exists x \neg A$$

This means that all following statements are provable (in predicate calculus, let us assume, as usual, that the set of x's is not empty):

$\vdash_{LK} \neg\neg A \to A$	$\leftarrow$-contrariety	$\vdash_{LK} \forall x \neg\neg A \to \forall x\, A$
$\vdash_{LK} A \wedge B \to A \vee B$	left-subalternation	$\vdash_{LK} \forall x\, A \to \exists x\, A$
$\vdash_{LK} \neg A \wedge \neg B \to \neg A \vee \neg B$	right-subalternation	$\vdash_{LK} \forall x \neg A \to \exists x \neg A$
$\vdash_{LK} \neg(A \wedge B) \to \neg A \vee \neg B$	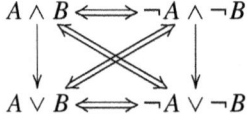-contradiction	$\vdash_{LK} \neg\forall x\, A \to \exists x \neg A$
$\vdash_{LK} \neg(A \vee B) \to \neg A \wedge \neg B$	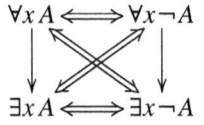-contradiction	$\vdash_{LK} \neg\exists x\, A \to \forall x \neg A$
$\vdash_{LK} \neg(\neg A \wedge \neg B) \to A \vee B$	$\diagup$-contradiction	$\vdash_{LK} \neg\forall x \neg A \to \exists x\, A$
$\vdash_{LK} \neg(\neg A \vee \neg B) \to A \wedge B$	$\diagdown$-contradiction	$\vdash_{LK} \neg\exists x \neg A \to \forall x\, A$
$\vdash_{LK} (A \vee B) \vee (\neg A \vee \neg B)$	subcontrariety	$\vdash_{LK} \exists x\, A \vee \exists x \neg A$

The $\rightarrow$-contrariety

$$\vdash_{LK} \neg(A) \rightarrow \neg A$$

$$\vdash_{LK} \forall x \neg(A) \rightarrow \forall x \neg A$$

is true by definition; from now on, it will not even be mentioned any more. The $\leftarrow$-contrariety is nothing but the involutivity of classical negation, $A \leftrightarrow \neg\neg A$, which is proved as follows (one of the rules, $\neg_l$, will later appear as problematic):

$$
\cfrac{
  \cfrac{
    \cfrac{
      \cfrac{\overline{A \vdash A}\ \text{ax}}{A, \neg A \vdash}\ \neg_l
    }{A \vdash \neg\neg A}\ \neg_r
  }{\vdash A \rightarrow \neg\neg A}\ \rightarrow_r
  \qquad
  \cfrac{
    \cfrac{
      \cfrac{
        \cfrac{\overline{A \vdash A}\ \text{ax}}{\vdash A, \neg A}\ \neg_r
      }{\neg\neg A \vdash A}\ \boxed{\neg_l}
    }{\vdash \neg\neg A \rightarrow A}\ \rightarrow_r
  }{}
}{\vdash A \leftrightarrow \neg\neg A}\ \wedge_r
$$

The four types of contradiction ($\searrow$, $\nearrow$, $\swarrow$, $\nwarrow$) reveal the perfect duality of conjunction and disjunction in classical logic. Some rules play a very important role in the proofs, as we shall see later:

$\searrow$-contradiction

$$
\cfrac{
  \cfrac{
    \cfrac{
      \cfrac{
        \cfrac{
          \cfrac{\overline{A \vdash A}\ \text{ax}}{A, B \vdash A}\ \text{weak}_l
          \qquad
          \cfrac{\overline{B \vdash B}\ \text{ax}}{A, B \vdash B}\ \text{weak}_l
        }{A, B \vdash A \wedge B}\ \wedge_r
      }{\neg(A \wedge B), A, B \vdash}\ \neg_l
    }{\neg(A \wedge B) \vdash \neg A, \neg B}\ \boxed{\neg_r} \times 2
  }{\neg(A \wedge B) \vdash \neg A \vee \neg B}\ \boxed{\vee_r}
}{\vdash \neg(A \wedge B) \rightarrow \neg A \vee \neg B}\ \rightarrow_r
$$

$\swarrow$-contradiction

$$
\cfrac{
  \cfrac{
    \cfrac{
      \cfrac{
        \cfrac{\overline{A \vdash A}\ \text{ax}}{A \vdash A, B}\ \text{weak}_r
      }{\vdash A, B, \neg A}\ \neg_r
      \qquad
      \cfrac{
        \cfrac{\overline{B \vdash B}\ \text{ax}}{B \vdash A, B}\ \text{weak}_r
      }{\vdash A, B, \neg B}\ \neg_r
    }{\vdash A, B, \neg A \wedge \neg B}\ \wedge_r
  }{
    \cfrac{\neg(\neg A \wedge \neg B) \vdash A, B}{\neg(\neg A \wedge \neg B) \vdash A \vee B}\ \boxed{\vee_r}
  }\ \boxed{\neg_l}
}{\vdash \neg(\neg A \wedge \neg B) \rightarrow A \vee B}\ \rightarrow_r
$$

$\nearrow$-contradiction

$$
\cfrac{
  \cfrac{
    \cfrac{
      \cfrac{
        \cfrac{\overline{A \vdash A}\ \text{ax}}{A \vdash A, B}\ \text{weak}_r
      }{A \vdash A \vee B}\ \vee_r
    }{
      \cfrac{\neg(A \vee B), A \vdash}{\neg(A \vee B) \vdash \neg A}\ \boxed{\neg_r}
    }\ \neg_l
    \qquad
    \cfrac{
      \cfrac{
        \cfrac{\overline{B \vdash B}\ \text{ax}}{B \vdash A, B}\ \text{weak}_r
      }{B \vdash A \vee B}\ \vee_r
    }{
      \cfrac{\neg(A \vee B), B \vdash}{\neg(A \vee B) \vdash \neg B}\ \boxed{\neg_r}
    }\ \neg_l
  }{\neg(A \vee B) \vdash \neg A \wedge \neg B}\ \boxed{\wedge_r}
}{\vdash \neg(A \vee B) \rightarrow \neg A \wedge \neg B}\ \rightarrow_r
$$

↖-contradiction

$$\dfrac{\dfrac{\dfrac{\dfrac{\dfrac{\dfrac{\overline{A \vdash A}\ \text{ax}}{A, B \vdash A}\ \text{weak}_l}{\vdash \neg A, \neg B, A}\ \neg_r \times 2}{\vdash \neg A \vee \neg B, A}\ \vee_r}{\neg(\neg A \vee \neg B) \vdash A}\ \boxed{\neg_l} \qquad \dfrac{\dfrac{\dfrac{\dfrac{\overline{B \vdash B}\ \text{ax}}{A, B \vdash B}\ \text{weak}_l}{\vdash \neg A, \neg B, B}\ \neg_r \times 2}{\vdash \neg A \vee \neg B, B}\ \vee_r}{\neg(\neg A \vee \neg B) \vdash B}\ \boxed{\neg_l}}{\neg(\neg A \vee \neg B) \vdash A \wedge B}\ \boxed{\wedge_r}}{\vdash \neg(\neg A \vee \neg B) \rightarrow A \wedge B}\ \rightarrow_r$$

One can easily notice that:

- every left-right contradiction (↘ or ↗) requires $\neg_r$;
- every right-left contradiction (↙ or ↖) requires $\neg_l$;
- every top-bottom contradiction (↘ or ↙) requires $\vee_r$;
- every bottom-top contradiction (↗ or ↖) requires $\wedge_r$.

Subalternation can be proved, mainly due to weakening:

$$\dfrac{\dfrac{\dfrac{\dfrac{\overline{A \vdash A}\ \text{ax}}{A, B \vdash A, B}\ \boxed{\text{weak}_l, \text{weak}_r}}{A \wedge B \vdash A, B}\ \wedge_l}{A \wedge B \vdash A \vee B}\ \vee_r}{\vdash A \wedge B \rightarrow A \vee B}\ \rightarrow_r \qquad \dfrac{\dfrac{\dfrac{\dfrac{\dfrac{\dfrac{\overline{A \vdash A}\ \text{ax}}{\neg A, A \vdash}\ \neg_l}{\neg A \vdash \neg A}\ \neg_r}{\neg A, \neg B \vdash \neg A, \neg B}\ \boxed{\text{weak}_l, \text{weak}_r}}{\neg A \wedge \neg B \vdash \neg A, \neg B}\ \wedge_l}{\neg A \wedge \neg B \vdash \neg A \vee \neg B}\ \vee_r}{\vdash \neg A \wedge \neg B \rightarrow \neg A \vee \neg B}\ \rightarrow_r$$

Finally, subcontrariety is true, using weakening and $\vee_r$:

$$\dfrac{\dfrac{\dfrac{\dfrac{\dfrac{\overline{A \vdash A}\ \text{ax}}{A, B \vdash A, B}\ \boxed{\text{weak}_r, \text{weak}_l}}{\vdash A, B, \neg A, \neg B}\ \neg_r \times 2}{\vdash A \vee B, \neg A \vee \neg B}\ \boxed{\vee_r} \times 2}{\vdash (A \vee B) \vee (\neg A \vee \neg B)}\ \boxed{\vee_r}$$

One can therefore summarize the square of opposition in more modern terms:

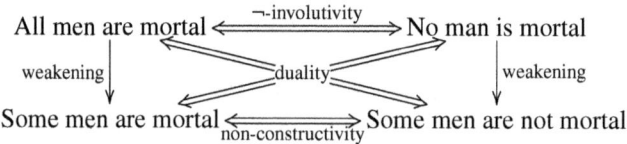

2.2 The Group of Opposition

Let us see what are the requirements for the square of opposition to be considered as a group.

As such, it seems not to be possible, since one of the relations, namely subalternation, is not symmetrical; if some mappings have no inverse element, it can not be a group. But with a light redefinition of subalternation, this problem can be get round. Say for instance that subalternation is the mapping exchanging conjunction and disjunction, and it suddenly becomes symmetrical, involutive. The classical square of opposition may then be seen as a complete graph, i.e. a graph where every vertex is linked to every other vertex, and obviously to itself:

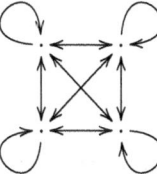

This graph is a group, since all group axioms are respected:

1. *closure*: the target of every arrow belongs to the set;
2. *identity element*: due to the principle of identity $A \rightarrow A$, for all statement A, there always is an arrow $\iota : A \rightarrow A$:

3. *inverse element*: every arrow has an inverse element, which is itself, for all relations are symmetrical (involutive); i.e. for each opposition f, $f \circ f = \iota$:

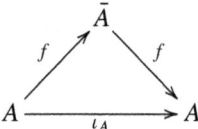

4. *associativity*: given two consecutive arrows f and g, there always is a shortcut $h = g \circ f$, i.e. producing the same result:

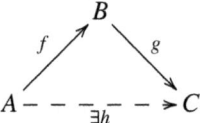

As a consequence, there is—retrospectively—nothing surprising if Piaget recognized here the INRC group, which is isomorphic to the Klein group $(\mathbb{Z}_2 \times \mathbb{Z}_2, +)$:

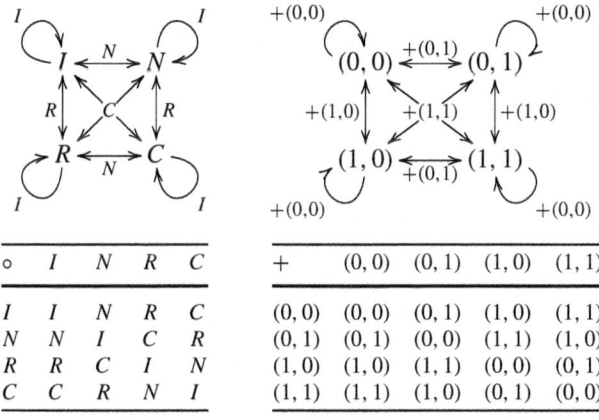

$\circ$	I	N	R	C
I	I	N	R	C
N	N	I	C	R
R	R	C	I	N
C	C	R	N	I

$+$	$(0,0)$	$(0,1)$	$(1,0)$	$(1,1)$
$(0,0)$	$(0,0)$	$(0,1)$	$(1,0)$	$(1,1)$
$(0,1)$	$(0,1)$	$(0,0)$	$(1,1)$	$(1,0)$
$(1,0)$	$(1,0)$	$(1,1)$	$(0,0)$	$(0,1)$
$(1,1)$	$(1,1)$	$(1,0)$	$(0,1)$	$(0,0)$

Starting from this fact, one can legitimately wonder whether the set of oppositions of every logical system can be expressed as a group. Let us show that it is not the case, particularly when the system is constructive, which is the case with intuitionistic and linear logics.

3 Intuitionistic Logic (LJ)

3.1 The Intuitionistic Incomplete Square

In intuitionistic logic, the square is clearly incomplete:

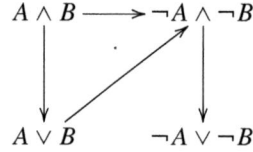 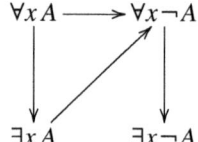

Actually, a lot of formulae are not provable:

$$\nvdash_{LJ} \neg\neg A \to A \qquad\qquad \leftarrow\text{-contrariety} \qquad\qquad \nvdash_{LJ} \forall x \neg\neg A \to \forall x A$$
$$\vdash_{LJ} A \wedge B \to A \vee B \qquad\qquad \text{left-subalternation} \qquad\qquad \vdash_{LJ} \forall x A \to \exists x A$$
$$\vdash_{LJ} \neg A \wedge \neg B \to \neg A \vee \neg B \qquad\qquad \text{right-subalternation} \qquad\qquad \vdash_{LJ} \forall x \neg A \to \exists x \neg A$$
$$\vdash_{LJ} \neg(A \vee B) \to \neg A \wedge \neg B \qquad\qquad \nearrow\text{-contradiction} \qquad\qquad \vdash_{LJ} \neg\exists x A \to \forall x \neg A$$
$$\nvdash_{LJ} \neg(A \wedge B) \to \neg A \vee \neg B \qquad\qquad \searrow\text{-contradiction} \qquad\qquad \nvdash_{LJ} \neg\forall x A \to \exists x \neg A$$
$$\nvdash_{LJ} \neg(\neg A \wedge \neg B) \to A \vee B \qquad\qquad \nearrow\text{-contradiction} \qquad\qquad \nvdash_{LJ} \neg\forall x \neg A \to \exists x A$$
$$\nvdash_{LJ} \neg(\neg A \vee \neg B) \to A \wedge B \qquad\qquad \nwarrow\text{-contradiction} \qquad\qquad \nvdash_{LJ} \neg\exists x \neg A \to \forall x A$$
$$\nvdash_{LJ} (A \vee B) \vee (\neg A \vee \neg B) \qquad\qquad \text{subcontrariety} \qquad\qquad \nvdash_{LJ} \exists x A \vee \exists x \neg A$$

First, intuitionistic negation is not involutive. Only $A \to \neg\neg A$ can be proved, but not the converse. Let us write "!!!" whenever a sequent is not provable:

$$\cfrac{\cfrac{\cfrac{\cfrac{\overline{A \vdash A}\ ^{\text{ax}}}{A, \neg A \vdash}\ ^{\neg_l}}{A \vdash \neg\neg A}\ ^{\neg_r}}{\vdash A \to \neg\neg A}\ ^{\to_r} \qquad \cfrac{\overline{\neg\neg A \vdash A}\ ^{!!!}}{\vdash \neg\neg A \to A}\ ^{\to_r}}{\vdash A \leftrightarrow \neg\neg A}\ ^{\wedge_r}$$

Then, only one of the four types of contradiction is provable:

$\nearrow$-contradiction

$$\cfrac{\cfrac{\cfrac{\cfrac{\cfrac{\overline{A \vdash A}\ ^{\text{ax}}}{A \vdash A \vee B}\ ^{\vee_r}}{\neg(A \vee B), A \vdash}\ ^{\neg_l}}{\neg(A \vee B) \vdash \neg A}\ ^{\neg_r} \qquad \cfrac{\cfrac{\cfrac{\overline{B \vdash B}\ ^{\text{ax}}}{B \vdash A \vee B}\ ^{\vee_r}}{\neg(A \vee B), B \vdash}\ ^{\neg_l}}{\neg(A \vee B) \vdash \neg B}\ ^{\neg_r}}{\neg(A \vee B) \vdash \neg A \wedge \neg B}\ ^{\wedge_r}}{\vdash \neg(A \vee B) \to \neg A \wedge \neg B}\ ^{\to_r}$$

The three other proofs actually fail quite quickly:

⬊-contradiction

$$\frac{\overline{\neg(A \wedge B) \vdash \neg A \vee \neg B}^{\ !!!}}{\vdash \neg(A \wedge B) \rightarrow \neg A \vee \neg B} \rightarrow_r$$

⬋-contradiction

$$\frac{\overline{\neg(\neg A \wedge \neg B) \vdash A \vee B}^{\ !!!}}{\vdash \neg(\neg A \wedge \neg B) \rightarrow A \vee B} \rightarrow_r$$

⬉-contradiction

$$\frac{\overline{\neg(\neg A \vee \neg B) \vdash A \wedge B}^{\ !!!}}{\vdash \neg(\neg A \vee \neg B) \rightarrow A \wedge B} \rightarrow_r$$

Why is there in intuitionistic logic only one valid type of contradiction? The classical $\vee_r$ and $\neg_l$ rules can not be taken as such in intuitionistic logic, for they would prevent, during the bottom-up search of a proof, the right part of the sequent from having at most one formula, which is the fundamental characteristic of intuitionistic sequent calculus LJ as it was defined by Gentzen [1, 2].[1] Since the ⬋-contradiction and the ⬉-contradiction use $\neg_l$, their classical-like proof fails in intuitionistic sequent calculus. For the same reason, since the ⬊-contradiction and (once again) the ⬋-contradiction use $\vee_r$, their classical-like proofs also fail. Only the ⬈-contradiction remains; its proof is essentially the same as in classical logic. This is why there is no left-right symmetry in the intuitionistic square. Let us call this the intuitionistic *gap in contradiction*.

Subalternation is still provable, for LJ maintains weakening:

$$\frac{\frac{\frac{\frac{\overline{A \vdash A}^{\ ax}}{A, B \vdash A}^{\boxed{weak_l}}}{A \wedge B \vdash A}^{\wedge_l}}{A \wedge B \vdash A \vee B}^{\vee_r}}{\vdash A \wedge B \rightarrow A \vee B} \rightarrow_r$$

$$\frac{\frac{\frac{\frac{\frac{\overline{A \vdash A}^{\ ax}}{\neg A, A \vdash}^{\neg_l}}{\neg A \vdash \neg A}^{\neg_r}}{\neg A, \neg B \vdash \neg A}^{\boxed{weak_l}}}{\neg A \wedge \neg B \vdash \neg A}^{\wedge_l}}{\frac{\neg A \wedge \neg B \vdash \neg A \vee \neg B}{\vdash \neg A \wedge \neg B \rightarrow \neg A \vee \neg B}^{\vee_r}} \rightarrow_r$$

Finally, subcontrariety is obviously not provable, since there is no *tertium non datur*:

$$\frac{\overline{\vdash (A \vee B) \vee (\neg A \vee \neg B)}^{\ !!!}}{}$$

Let us call this the intuitionistic *lack of subalternation*.

---

[1] The classical "multiplicative" rule $\frac{\Gamma \vdash A, B}{\Gamma \vdash A \vee B} \vee_r$ can not be used, and has to be replaced by the "additive" rule $\frac{\Gamma \vdash A}{\Gamma \vdash A \vee B} \vee_r$. But the issue is not symmetrical: the classical multiplicative rule $\frac{\Gamma, A, B \vdash C}{\Gamma, A \wedge B \vdash C} \wedge_l$ can be maintained in intuitionistic logic.

3.2 No Group for Intuitionistic Logic

Is it possible to build a group of opposition for the incomplete square of intuitionistic logic? Following the same method as for the classical square, it can not be done. Since the graph is

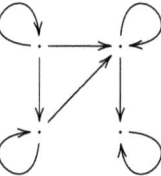

the following group axioms are not respected:

- *inverse element*: no relation is symmetrical—except, of course, identity; therefore, the set of intuitionistic oppositions is not a group;
- *associativity*: the composition of oppositions is generally not associative. Even though the $\rightarrow$-contrariety is a shortcut for the $\nearrow$-contradiction of a subalternation

$$A \wedge B - \overset{\exists}{\longrightarrow} \neg A \wedge \neg B$$
$$\downarrow \qquad \nearrow$$
$$A \vee B$$

there is no shortcut for the subalternation of a $\nearrow$-contradiction:

$$\neg A \wedge \neg B$$
$$\nearrow \qquad \downarrow$$
$$A \vee B \underset{\neg \exists}{\rightsquigarrow} \neg A \vee \neg B$$

As a consequence, the structure of this set of opposition is not even an associative monoid. At most, it is a non-associative monoid—probably the least interesting algebraic structure that can be imagined;

- *closure*: but it is not even a monoid, since there is no closure; for example, the contrary of $\neg A \wedge \neg B$ is not even defined inside the square.

3.3 The Trivial Intuitionistic Square

Is it at least possible to modify some definitions, in order to get a complete square, as we did with respect to subalternation in the classical square? Of course, one could draw a perfect square, *modulo* some translations, namely changing each formula A into $\neg\neg A$:

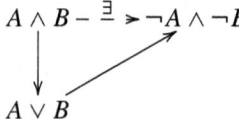

But this solution is trivial, since its only ground is the Gödel-translation of classical logic into intuitionistic logic [3]. There is also nothing more than the classical square.

Therefore, even though there can be a group for intuitionistic logic, it has to cancel its very specificity, namely constructivity. Concerning intuitionistic logic, the philosophical and algebraic interests can not be conciliated: one has to choose between them. Either you get a constructive system, or you get an interesting algebraic structure.

But what is precisely causing the intuitionistic square to be incomplete? It first seems to be constructivity, and *only* constructivity; this would imply a strong correlation between the lack of subcontrariety and the gap in contradiction. But as we shall see, even though constructivity is responsible for the lack of subcontrariety, it is not for the gap in contradiction. Both can be separated: linear logic will be used to prove it.

4 Classical Linear Logic (CLL)

4.1 Principles of Linear Logic

The main feature of linear logic is to control severely weakening and contraction; for this reason, it belongs to the so-called *substructural logics* (see Appendix A). Because of this control, there are two conjunctions, "times" ($\otimes$) and "with" (&). The first one behaves like an accumulation of resources, the second one like a choice between resources.

$$\frac{\Gamma, A, B \vdash \Delta}{\Gamma, A \otimes B \vdash \Delta} \otimes_l \qquad\qquad \frac{\Gamma \vdash A, \Delta \qquad \Gamma' \vdash B, \Delta'}{\Gamma, \Gamma' \vdash A \otimes B, \Delta, \Delta'} \otimes_r$$

$$\frac{\Gamma, A \vdash \Delta}{\Gamma, A \& B \vdash \Delta} \&_l \qquad\qquad \frac{\Gamma \vdash A, \Delta \qquad \Gamma \vdash B, \Delta}{\Gamma \vdash A \& B, \Delta} \&_r$$

Both rules are just one in classical logic, due to weakening or contraction:

$$\frac{\dfrac{\Gamma, A \vdash \Delta}{\Gamma, A, B \vdash \Delta} \text{weak}_l}{\Gamma, A \wedge B \vdash \Delta} \wedge_l \qquad\qquad \frac{\dfrac{\Gamma \vdash A, \Delta \qquad \Gamma \vdash B, \Delta}{\Gamma, \Gamma \vdash A \wedge B, \Delta, \Delta} \wedge_r}{\Gamma \vdash A \wedge B, \Delta} \text{contr}_l, \text{contr}_r$$

For the same reason, there are two disjunctions in linear logic. The first disjunction, "par" ($\mathfrak{N}$), denotes some kind of causal dependance, without any chronological order, between two processes. The second disjunction, "plus" ($\oplus$), is a type of choice I am not responsible for, for example some choice left to somebody else.

$$\frac{\Gamma, A \vdash \Delta \qquad \Gamma', B \vdash \Delta'}{\Gamma, \Gamma', A \mathfrak{N} B \vdash \Delta, \Delta'} \mathfrak{N}_l \qquad\qquad \frac{\Gamma \vdash A, B, \Delta}{\Gamma \vdash A \mathfrak{N} B, \Delta} \mathfrak{N}_r$$

$$\frac{\Gamma, A \vdash \Delta \qquad \Gamma, B \vdash \Delta}{\Gamma, A \oplus B \vdash \Delta} \oplus_l \qquad\qquad \frac{\Gamma \vdash A, \Delta}{\Gamma \vdash A \oplus B, \Delta} \oplus_r$$

They also are the same disjunction in LK, because of weakening and contraction:

$$\frac{\dfrac{\Gamma, A \vdash \Delta \qquad \Gamma, B \vdash \Delta}{\Gamma, \Gamma, A \vee B \vdash \Delta, \Delta} \vee_l}{\Gamma, A \vee B \vdash \Delta} \text{contr}_l, \text{contr}_r \qquad\qquad \frac{\dfrac{\Gamma \vdash A, \Delta}{\Gamma \vdash A, B, \Delta} \text{weak}_r}{\Gamma \vdash A \vee B, \Delta} \vee_r$$

There are thus two conjunctions ($\otimes$ and &) and two disjunctions ($\mathfrak{N}$ and $\oplus$), one of each using a kind of weakening or contraction (& and $\oplus$), whereas the other one does not

($\otimes$ and $⅋$). "Times" ($\otimes$) and "par" ($⅋$) are called *multiplicative* connectives; in each of their rules, the number of formulae is exactly the same before and after using the rule (as an example, there are four letters on the top of $\otimes_l$, and the same number on the bottom of this rule). It is never the case with "with" (&) and "plus" ($\oplus$), which are called *additive*: there are always more, or less, letters on the bottom than on the top (for instance, there are three letters on the top of $\&_l$, and four on the bottom).

As a consequence, two fragments may be distinguished in linear logic: the multiplicative one (MLL, Multiplicative Linear Logic) and the additive one (MALL, Multiplicative Additive Linear Logic, from which the additive part will almost exclusively be used in this paper).

The linear negation $A^{\perp}$ behaves exactly like the classical one; in particular, it is involutive. The linear implication $A \multimap B$ is, like in classical logic, equivalent with $A^{\perp} ⅋ B$. The main linear connectives are thus:

	conjunction	disjunction	implication
multiplicative	times ($\otimes$)	par ($⅋$)	implication ($\multimap$)
additive	with (&)	plus ($\oplus$)	

4.2 The Multiplicative Fragment (MLL)

Let us first examine the multiplicative fragment of linear logic (see Appendix B). Only the following statements may be proved:

$$\vdash_{LL} (A^{\perp})^{\perp} \multimap A \qquad \leftarrow\text{-contrariety}$$
$$\vdash_{LL} (A \otimes B)^{\perp} \multimap A^{\perp} ⅋ B^{\perp} \qquad \searrow\text{-contradiction}$$
$$\vdash_{LL} (A ⅋ B)^{\perp} \multimap A^{\perp} \otimes B^{\perp} \qquad \nearrow\text{-contradiction}$$
$$\vdash_{LL} (A^{\perp} \otimes B^{\perp})^{\perp} \multimap A ⅋ B \qquad \diagup\text{-contradiction}$$
$$\vdash_{LL} (A^{\perp} ⅋ B^{\perp})^{\perp} \multimap A \otimes B \qquad \diagdown\text{-contradiction}$$

But some statements are not provable:

$$\nvdash_{LL} A \otimes B \multimap A ⅋ B \qquad \text{left-subalternation}$$
$$\nvdash_{LL} A^{\perp} \otimes B^{\perp} \multimap A^{\perp} ⅋ B^{\perp} \qquad \text{right-subalternation}$$
$$\nvdash_{LL} (A ⅋ B) ⅋ (A^{\perp} ⅋ B^{\perp}) \qquad \text{subcontrariety}$$

This implies that the square of multiplicative linear logic is even less complete than the intuitionistic one:

$$A \otimes B \Longleftrightarrow A^{\perp} \otimes B^{\perp}$$

$$A ⅋ B \qquad A^{\perp} ⅋ B^{\perp}$$

This incompleteness does not prevent the multiplicative connectives $\otimes$ and $⅋$ from respecting a perfect duality. This is a first example of a logical system where the lack of subcontrariety does not imply a gap in contradiction.

No more than in intuitionistic logic, there can be a group of oppositions for MLL. Not even a monoid: once the square is not complete, there can be no associativity.

4.3 The Additive Fragment (MALL)

What about the additive fragment of linear logic? As one can see, the square is incomplete, as well as linear quantifiers' square (see the proofs in Appendix C)

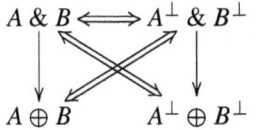 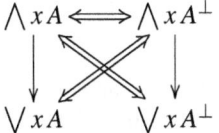

since the following sequents can be proved:

$\vdash_{LL} (A^{\perp})^{\perp} \multimap A$ ←-contrariety $\vdash_{LL} \bigwedge x (A^{\perp})^{\perp} \multimap \bigwedge x A$

$\vdash_{LL} A \& B \multimap A \oplus B$ left-subalternation $\vdash_{LL} \bigwedge x A \multimap \bigvee x A$

$\vdash_{LL} A^{\perp} \& B^{\perp} \multimap A^{\perp} \oplus B^{\perp}$ right-subalternation $\vdash_{LL} \bigwedge x A^{\perp} \multimap \bigvee x A^{\perp}$

$\vdash_{LL} (A \& B)^{\perp} \multimap A^{\perp} \oplus B^{\perp}$ ↘-contradiction $\vdash_{LL} (\bigwedge x A)^{\perp} \multimap \bigvee x A^{\perp}$

$\vdash_{LL} (A \oplus B)^{\perp} \multimap A^{\perp} \& B^{\perp}$ ↗-contradiction $\vdash_{LL} (\bigvee x A)^{\perp} \multimap \bigwedge x A^{\perp}$

$\vdash_{LL} (A^{\perp} \& B^{\perp})^{\perp} \multimap A \oplus B$ ↙-contradiction $\vdash_{LL} (\bigwedge x A^{\perp})^{\perp} \multimap \bigvee x A$

$\vdash_{LL} (A^{\perp} \oplus B^{\perp})^{\perp} \multimap A \& B$ ↖-contradiction $\vdash_{LL} (\bigvee x A^{\perp})^{\perp} \multimap \bigwedge x A$

Like in intuitionistic logic, there is a *lack of subcontrariety*:

$$\nvdash_{LL} (A \oplus B) \oplus (A^{\perp} \oplus B^{\perp}) \quad \text{subcontrariety} \quad \nvdash_{LL} \bigvee x A \oplus \bigvee A^{\perp}$$

But, unlike intuitionistic logic, there is no *gap in contradiction* any more, since we can prove all types of contradiction. This is therefore another example of a logical fragment where the lack of subcontrariety does not imply a gap in contradiction.

4.4 The Non-standard Linear Squares (MELL and MLL+mix)

However, the generalization of the multiplicative fragment, namely the Multiplicative Exponential Linear Logic (MELL, see Appendix B), where

$$!A = A \otimes A \otimes \cdots$$
$$?A = A \,\invamp\, A \,\invamp\, \cdots,$$

does have a full square, as well as classical logic:

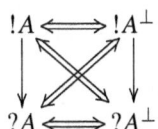

All following statements actually are provable:

$$\vdash_{LL} !(A^\perp)^\perp \multimap !A \qquad \leftarrow\text{-contrariety}$$
$$\vdash_{LL} (!A)^\perp \multimap ?A^\perp \qquad \searrow\text{-contradiction}$$
$$\vdash_{LL} (?A)^\perp \multimap !A^\perp \qquad \nearrow\text{-contradiction}$$
$$\vdash_{LL} (!A^\perp)^\perp \multimap ?A \qquad \swarrow\text{-contradiction}$$
$$\vdash_{LL} (?A^\perp)^\perp \multimap !A \qquad \nwarrow\text{-contradiction}$$
$$\vdash_{LL} !A \multimap ?A \qquad \text{left-subalternation}$$
$$\vdash_{LL} !A^\perp \multimap ?A^\perp \qquad \text{right-subalternation}$$
$$\vdash_{LL} ?A \,\bindnasrepma\, ?A^\perp \qquad \text{subcontrariety}$$

This complete square can not be found in MLL, except if we just add the "mix" rule to the set of linear rules, which "merges" different proofs:

$$\frac{\Gamma \vdash \Delta \qquad \Gamma' \vdash \Delta'}{\Gamma, \Gamma' \vdash \Delta, \Delta'} \; \text{mix}$$

Using this rule, we can prove

$$\vdash_{LL+mix} A \otimes B \multimap A \,\bindnasrepma\, B \qquad \text{left-subalternation}$$
$$\vdash_{LL+mix} A^\perp \otimes B^\perp \multimap A^\perp \,\bindnasrepma\, B^\perp \qquad \text{right-subalternation}$$
$$\vdash_{LL+mix} (A \,\bindnasrepma\, B) \,\bindnasrepma\, (A^\perp \,\bindnasrepma\, B^\perp) \qquad \text{subcontrariety}$$

as can be seen on the following proof-trees

$$\frac{\dfrac{\dfrac{\overline{A \vdash A}\;ax \quad \overline{B \vdash B}\;ax}{A, B \vdash A, B}\;mix}{\dfrac{A \otimes B \vdash A, B}{\dfrac{A \otimes B \vdash A \,\bindnasrepma\, B}{\vdash A \otimes B \multimap A \,\bindnasrepma\, B}\;\multimap_r}\;\bindnasrepma_r}}{}\;\otimes_l}{}$$

$$\frac{\overline{A \vdash A}\;ax \quad \overline{B \vdash B}\;ax}{\dfrac{A, B \vdash A, B}{\dfrac{A^\perp, B^\perp, A, B \vdash}{\dfrac{A^\perp, B^\perp \vdash A^\perp, B^\perp}{\dfrac{A^\perp \otimes B^\perp \vdash A^\perp, B^\perp}{\dfrac{A^\perp \otimes B^\perp \vdash A^\perp \,\bindnasrepma\, B^\perp}{\vdash A^\perp \otimes B^\perp \multimap A^\perp \,\bindnasrepma\, B^\perp}\;\multimap_r}\;\bindnasrepma_r}\;\otimes_l}\;\perp_r \times 2}\;\perp_l \times 2}\;mix}$$

$$\frac{\overline{A \vdash A}\;ax \quad \overline{B \vdash B}\;ax}{\dfrac{A, B \vdash A, B}{\dfrac{\vdash A, B, A^\perp, B^\perp}{\dfrac{\vdash A \,\bindnasrepma\, B, A^\perp \,\bindnasrepma\, B^\perp}{\vdash (A \,\bindnasrepma\, B) \,\bindnasrepma\, (A^\perp \,\bindnasrepma\, B^\perp)}\;\bindnasrepma_r}\;\bindnasrepma_r \times 2}\;\perp_r \times 2}\;mix}$$

and thus find a complete square for MLL+mix:

$$\begin{array}{ccc}
A \otimes B & \underset{}{\overset{}{\longleftrightarrow}} & A^\perp \otimes B^\perp \\
\Big\downarrow{\scriptstyle mix} & \diagdown\!\!\diagup & \Big\downarrow{\scriptstyle mix} \\
A \,\bindnasrepma\, B & \underset{mix}{\longleftrightarrow} & A^\perp \,\bindnasrepma\, B^\perp
\end{array}$$

Except in these last cases, the square of opposition in Linear Logic is incomplete. To become complete, it requires either the generalization of its multiplicative connectives (exponentials), or a non-standard rule ("mix"). In both cases, the completeness of the square does not belong to basic linear logic. Once linear logic is deprived of its "essentialist"[2] or strange rules, it respects the symmetry of contradiction (duality) but not the subcontrariety. Since the square is not complete, there can be no associativity; and without associativity, there can be no group of opposition in standard linear logic.

5 Conclusion

Let us summarize the different squares we have drawn in this paper:

Logical system	Square	Group conditions
Classical Logic (LK)		symmetrical subalternation
Intuitionistic Logic (LJ)		Gödel-translation
Linear Logic MLL		mix or exponentials
MALL Quantified MALL		
MLL+mix MELL		

However different they may be, all the linear squares show a perfect left-right symmetry, i.e. a perfect duality between connectives, which does not prevent this logic from being constructive. Consequently, even though constructivity is responsible for the lack of subcontrariety, it is absolutely innocent with respect to the gap in contradiction: weakening and contraction, more than constructivity, are responsible of it. Dropping subcontrariety and nothing else, linear logic exhibits what may be the minimal price to pay to save constructivity.

Moreover, a set of oppositions is a group if and only if every opposition is defined and involutive. Since it is not the case in constructive logics, there can be no group of opposition preserving their fundamental philosophical properties.

---

[2]Girard [4] calls "essentialism" the assumption that truth is eternal, and can be duplicated at will; in linear logic, a "truth" can not be used an infinite number of times without the help of exponentials.

Appendix A: Rules for Classical Linear Logic (CLL)

Axiom and Cut

$$\frac{}{A \vdash A}\ \text{ax} \qquad\qquad \frac{\Gamma \vdash A, \Delta \qquad \Gamma', A \vdash \Delta'}{\Gamma, \Gamma' \vdash \Delta, \Delta'}\ \text{cut}$$

Structural Rules

$$\frac{\Gamma \vdash \Delta}{\sigma(\Gamma) \vdash \Delta}\ \text{exchange}_l \qquad\qquad \frac{\Gamma \vdash \Delta}{\Gamma \vdash \sigma(\Delta)}\ \text{exchange}_r$$

Rules for Logical Connectives

$$\frac{\Gamma, A, B \vdash \Delta}{\Gamma, A \otimes B \vdash \Delta}\ \otimes_l \qquad\qquad \frac{\Gamma \vdash A, \Delta \qquad \Gamma' \vdash B, \Delta'}{\Gamma, \Gamma' \vdash A \otimes B, \Delta, \Delta'}\ \otimes_r$$

$$\frac{\Gamma, A \vdash \Delta}{\Gamma, A \& B \vdash \Delta}\ \&_l^l \quad \frac{\Gamma, B \vdash \Delta}{\Gamma, A \& B \vdash \Delta}\ \&_l^r \quad \frac{\Gamma \vdash A, \Delta \qquad \Gamma \vdash B, \Delta}{\Gamma \vdash A \& B, \Delta}\ \&_r$$

$$\frac{\Gamma, A \vdash \Delta \qquad \Gamma', B \vdash \Delta'}{\Gamma, \Gamma', A \,\invamp\, B \vdash \Delta, \Delta'}\ \invamp_l \qquad\qquad \frac{\Gamma \vdash A, B, \Delta}{\Gamma \vdash A \,\invamp\, B, \Delta}\ \invamp_r$$

$$\frac{\Gamma, A \vdash \Delta \qquad \Gamma, B \vdash \Delta}{\Gamma, A \oplus B \vdash \Delta}\ \oplus_l \quad \frac{\Gamma \vdash A, \Delta}{\Gamma \vdash A \oplus B, \Delta}\ \oplus_r^l \quad \frac{\Gamma \vdash B, \Delta}{\Gamma \vdash A \oplus B, \Delta}\ \oplus_r^r$$

$$\frac{\Gamma \vdash A, \Delta \qquad \Gamma', B \vdash \Delta'}{\Gamma, \Gamma', A \multimap B \vdash \Delta, \Delta'}\ \multimap_l \qquad\qquad \frac{\Gamma, A \vdash B, \Delta}{\Gamma \vdash A \multimap B, \Delta}\ \multimap_r$$

$$\frac{\Gamma \vdash A, \Delta}{\Gamma, A^\perp \vdash \Delta}\ \perp_l \qquad\qquad \frac{\Gamma, A \vdash \Delta}{\Gamma \vdash A^\perp, \Delta}\ \perp_r$$

Exponentials

$$\frac{\Gamma \vdash \Delta}{\Gamma, !A \vdash \Delta}\ !\text{weak} \qquad\qquad \frac{\Gamma \vdash \Delta}{\Gamma \vdash ?A, \Delta}\ ?\text{weak}$$

$$\frac{\Gamma, !A, !A \vdash \Delta}{\Gamma, !A \vdash \Delta}\ !\text{contr} \qquad\qquad \frac{\Gamma \vdash ?A, ?A, \Delta}{\Gamma \vdash ?A, \Delta}\ ?\text{contr}$$

$$\frac{\Gamma, A \vdash \Delta}{\Gamma, !A \vdash \Delta}\ !\text{der} \qquad\qquad \frac{\Gamma \vdash A, \Delta}{\Gamma \vdash ?A, \Delta}\ ?\text{der}$$

$$\frac{!\Gamma \vdash A, ?\Delta}{!\Gamma \vdash !A, ?\Delta}\ !\text{prom} \qquad\qquad \frac{!\Gamma, A \vdash ?\Delta}{!\Gamma, ?A \vdash ?\Delta}\ ?\text{prom}$$

Mix (in LL+mix)

$$\frac{\Gamma \vdash \Delta \qquad \Gamma' \vdash \Delta'}{\Gamma, \Gamma' \vdash \Delta, \Delta'}\ \text{mix}$$

Appendix B: Proofs for MLL±mix and MELL

Subalternation

$$
\cfrac{
  \cfrac{
    \cfrac{A \vdash A \;\; ax \qquad B \vdash B \;\; ax}{A, B \vdash A, B} \; mix
  }{
    \cfrac{A \otimes B \vdash A, B}{
      \cfrac{A \otimes B \vdash A \,\mathbin{\text{⅋}}\, B}{\vdash A \otimes B \multimap A \,\mathbin{\text{⅋}}\, B} \multimap_r
    } \mathbin{\text{⅋}}_r
  }{} \otimes_l
}{}
$$

$$
\cfrac{
  \cfrac{A \vdash A \;\; ax}{
    \cfrac{!A \vdash A}{
      \cfrac{!A \vdash ?A}{\vdash !A \multimap ?A} \multimap_r
    } \,?der
  } \,!der
}{}
$$

$$
\cfrac{A \vdash A \;\; ax \qquad B \vdash B \;\; ax}{
\cfrac{A, B \vdash A, B \; mix}{
\cfrac{A^{\perp}, B^{\perp}, A, B \vdash \; \perp_l \times 2}{
\cfrac{A^{\perp}, B^{\perp} \vdash A^{\perp}, B^{\perp} \; \perp_r \times 2}{
\cfrac{A^{\perp} \otimes B^{\perp} \vdash A^{\perp}, B^{\perp} \; \otimes_l}{
\cfrac{A^{\perp} \otimes B^{\perp} \vdash A^{\perp} \,\mathbin{\text{⅋}}\, B^{\perp} \; \mathbin{\text{⅋}}_r}{
\vdash A^{\perp} \otimes B^{\perp} \multimap A^{\perp} \,\mathbin{\text{⅋}}\, B^{\perp} \; \multimap_r}}}}}}
$$

$$
\cfrac{A \vdash A \;\; ax}{
\cfrac{A, A^{\perp} \vdash \; \perp_l}{
\cfrac{A^{\perp} \vdash A^{\perp} \; \perp_r}{
\cfrac{!A^{\perp} \vdash A^{\perp} \; !der}{
\cfrac{!A^{\perp} \vdash ?A^{\perp} \; ?der}{
\vdash !A^{\perp} \multimap ?A^{\perp} \; \multimap_r}}}}}
$$

Contradiction

$$
\cfrac{A \vdash A \;\; ax \qquad B \vdash B \;\; ax}{
\cfrac{A, B \vdash A \otimes B \; \otimes_r}{
\cfrac{(A \otimes B)^{\perp}, A, B \vdash \; \perp_l}{
\cfrac{(A \otimes B)^{\perp} \vdash A^{\perp}, B^{\perp} \; \perp_r \times 2}{
\cfrac{(A \otimes B)^{\perp} \vdash A^{\perp} \,\mathbin{\text{⅋}}\, B^{\perp} \; \mathbin{\text{⅋}}_r}{
\vdash (A \otimes B)^{\perp} \multimap A^{\perp} \,\mathbin{\text{⅋}}\, B^{\perp} \; \multimap_r}}}}}
$$

$$
\cfrac{A \vdash A \;\; ax}{
\cfrac{\vdash A, A^{\perp} \; \perp_r}{
\cfrac{\vdash A, ?A^{\perp} \; ?der}{
\cfrac{\vdash !A, ?A^{\perp} \; !prom}{
\cfrac{(!A)^{\perp} \vdash ?A^{\perp} \; \perp_l}{
\vdash (!A)^{\perp} \multimap ?A^{\perp} \; \multimap_r}}}}}
$$

$$
\cfrac{
\cfrac{A \vdash A \;\; ax}{\vdash A^{\perp}, A} \perp_r \qquad
\cfrac{B \vdash B \;\; ax}{\vdash B^{\perp}, B} \perp_r
}{
\cfrac{\vdash A^{\perp}, B^{\perp}, A \otimes B \; \otimes_r}{
\cfrac{\vdash A^{\perp} \,\mathbin{\text{⅋}}\, B^{\perp}, A \otimes B \; \mathbin{\text{⅋}}_r}{
\cfrac{(A^{\perp} \,\mathbin{\text{⅋}}\, B^{\perp})^{\perp} \vdash A \otimes B \; \perp_l}{
\vdash (A^{\perp} \,\mathbin{\text{⅋}}\, B^{\perp})^{\perp} \multimap A \otimes B \; \multimap_r}}}}
$$

$$
\cfrac{A \vdash A \;\; ax}{
\cfrac{\vdash A^{\perp}, A \; \perp_r}{
\cfrac{\vdash ?A^{\perp}, A \; ?der}{
\cfrac{\vdash ?A^{\perp}, !A \; !prom}{
\cfrac{(?A^{\perp})^{\perp} \vdash !A \; \perp_l}{
\vdash (?A^{\perp})^{\perp} \multimap !A \; \multimap_r}}}}}
$$

$$
\cfrac{
\cfrac{A \vdash A \;\; ax}{\vdash A, A^{\perp}} \perp_r \qquad
\cfrac{B \vdash B \;\; ax}{\vdash B, B^{\perp}} \perp_r
}{
\cfrac{\vdash A, B, A^{\perp} \otimes B^{\perp} \; \otimes_r}{
\cfrac{\vdash A \,\mathbin{\text{⅋}}\, B, A^{\perp} \otimes B^{\perp} \; \mathbin{\text{⅋}}_r}{
\cfrac{(A \,\mathbin{\text{⅋}}\, B)^{\perp} \vdash A^{\perp} \otimes B^{\perp} \; \perp_l}{
\vdash (A \,\mathbin{\text{⅋}}\, B)^{\perp} \multimap A^{\perp} \otimes B^{\perp} \; \multimap_r}}}}
$$

$$
\cfrac{A \vdash A \;\; ax}{
\cfrac{\vdash A, A^{\perp} \; \perp_r}{
\cfrac{\vdash ?A, A^{\perp} \; ?der}{
\cfrac{\vdash ?A, !A^{\perp} \; !prom}{
\cfrac{(?A)^{\perp} \vdash !A^{\perp} \; \perp_l}{
\vdash (?A)^{\perp} \multimap !A^{\perp} \; \multimap_r}}}}}
$$

$$\cfrac{\cfrac{\cfrac{\cfrac{\cfrac{\overline{A \vdash A}\ \text{ax}}{\vdash A^\perp, A}\ \perp_r \qquad \cfrac{\overline{B \vdash B}\ \text{ax}}{\vdash B^\perp, B}\ \perp_r}{\vdash A^\perp \otimes B^\perp, A, B}\ \otimes_r}{\vdash A^\perp \otimes B^\perp, A ⅋ B}\ ⅋_r}{(A^\perp \otimes B^\perp)^\perp \vdash A ⅋ B}\ \perp_l}{\vdash (A^\perp \otimes B^\perp)^\perp \multimap A ⅋ B}\ \multimap_r \qquad\qquad \cfrac{\cfrac{\cfrac{\cfrac{\cfrac{\overline{A \vdash A}\ \text{ax}}{\vdash A^\perp, A}\ \perp_r}{\vdash A^\perp, ?A}\ ?der}{\vdash {!}A^\perp, ?A}\ !prom}{({!}A^\perp)^\perp \vdash ?A}\ \perp_l}{\vdash ({!}A^\perp)^\perp \multimap ?A}\ \multimap_r$$

Subcontrariety

$$\cfrac{\cfrac{\cfrac{\cfrac{\overline{A \vdash A}\ \text{ax} \qquad \overline{B \vdash B}\ \text{ax}}{A, B \vdash A, B}\ \text{mix}}{\vdash A, B, A^\perp, B^\perp}\ \perp_r \times 2}{\vdash A ⅋ B, A^\perp ⅋ B^\perp}\ ⅋_r \times 2}{\vdash (A ⅋ B) ⅋ (A^\perp ⅋ B^\perp)}\ ⅋_r \qquad\qquad \cfrac{\cfrac{\cfrac{\cfrac{\overline{A \vdash A}\ \text{ax}}{\vdash A, A^\perp}\ \perp_r}{\vdash A, ?A^\perp}\ ?der}{\vdash {!}A, ?A^\perp}\ !prom}{\vdash {!}A ⅋ ?A^\perp}\ ⅋_r$$

Appendix C: Proofs for MALL and Quantified MALL

Subalternation

$$\cfrac{\cfrac{\cfrac{\overline{A \vdash A}\ \text{ax}}{A \,\&\, B \vdash A}\ \&_l}{A \,\&\, B \vdash A \oplus B}\ \oplus_r}{\vdash A \,\&\, B \multimap A \oplus B}\ \multimap_r \qquad\qquad \cfrac{\cfrac{\cfrac{\overline{A \vdash A}\ \text{ax}}{\bigwedge x A \vdash A}\ \textstyle\bigwedge_l}{\bigwedge x A \vdash \bigvee x A}\ \textstyle\bigvee_r}{\vdash \bigwedge x A \multimap \bigvee x A}\ \multimap_r$$

$$\cfrac{\cfrac{\cfrac{\cfrac{\cfrac{\overline{A \vdash A}\ \text{ax}}{\vdash A, A^\perp}\ \perp_r}{A^\perp \vdash A^\perp}\ \perp_l}{A^\perp \,\&\, B^\perp \vdash A^\perp}\ \&_l}{A^\perp \,\&\, B^\perp \vdash A^\perp \oplus B^\perp}\ \oplus_r}{\vdash A^\perp \,\&\, B^\perp \multimap A^\perp \oplus B^\perp}\ \multimap_r \qquad\qquad \cfrac{\cfrac{\cfrac{\cfrac{\cfrac{\overline{A \vdash A}\ \text{ax}}{\vdash A, A^\perp}\ \perp_r}{A^\perp \vdash A^\perp}\ \perp_l}{\bigwedge x A^\perp \vdash A^\perp}\ \textstyle\bigwedge_l}{\bigwedge x A^\perp \vdash \bigvee x A^\perp}\ \textstyle\bigvee_r}{\vdash \bigwedge x A^\perp \multimap \bigvee x A^\perp}\ \multimap_r$$

Contradiction

$$\cfrac{\cfrac{\cfrac{\cfrac{\cfrac{\overline{A \vdash A}\ \text{ax}}{\vdash A, A^\perp}\ \perp_r}{\vdash A, A^\perp \oplus B^\perp}\ \oplus_r \qquad \cfrac{\cfrac{\overline{B \vdash B}\ \text{ax}}{\vdash B, B^\perp}\ \perp_r}{\vdash B, A^\perp \oplus B^\perp}\ \oplus_r}{\vdash A \,\&\, B, A^\perp \oplus B^\perp}\ \&_r}{(A \,\&\, B)^\perp \vdash A^\perp \oplus B^\perp}\ \perp_l}{\vdash (A \,\&\, B)^\perp \multimap A^\perp \oplus B^\perp}\ \multimap_r \qquad\qquad \cfrac{\cfrac{\cfrac{\cfrac{\cfrac{\overline{A \vdash A}\ \text{ax}}{\vdash A, A^\perp}\ \perp_r}{\vdash A, \bigvee x A^\perp}\ \textstyle\bigvee_r}{\vdash \bigwedge x A, \bigvee x A^\perp}\ \textstyle\bigwedge_r}{(\bigwedge x A)^\perp \vdash \bigvee x A^\perp}\ \perp_l}{\vdash (\bigwedge x A)^\perp \multimap \bigvee x A^\perp}\ \multimap_r$$

$$\cfrac{\cfrac{\cfrac{\cfrac{\overline{A \vdash A}^{\ \text{ax}}}{\vdash A, A^\perp}\perp_r}{\vdash A \oplus B, A^\perp}\oplus_r \qquad \cfrac{\cfrac{\overline{B \vdash B}^{\ \text{ax}}}{\vdash B, B^\perp}\perp_r}{\vdash A \oplus B, B^\perp}\oplus_r}{\vdash A \oplus B, A^\perp \,\&\, B^\perp}\&_r}{\cfrac{(A \oplus B)^\perp \vdash A^\perp \,\&\, B^\perp}{\vdash (A \oplus B)^\perp \multimap A^\perp \,\&\, B^\perp}\multimap_r}\perp_l$$

$$\cfrac{\cfrac{\cfrac{\cfrac{\overline{A \vdash A}^{\ \text{ax}}}{\vdash A, A^\perp}\perp_r}{\vdash \bigvee xA, A^\perp}\bigvee_r}{\vdash \bigvee xA, \bigwedge xA^\perp}\bigwedge_r}{\cfrac{(\bigvee xA)^\perp \vdash \bigwedge xA^\perp}{\vdash (\bigvee xA)^\perp \multimap \bigwedge xA^\perp}\multimap_r}\perp_l$$

$$\cfrac{\cfrac{\cfrac{\cfrac{\overline{A \vdash A}^{\ \text{ax}}}{\vdash A^\perp, A}\perp_r}{\vdash A^\perp, A \oplus B}\oplus_r \qquad \cfrac{\cfrac{\overline{B \vdash B}^{\ \text{ax}}}{\vdash B^\perp, B}\perp_r}{\vdash B^\perp, A \oplus B}\oplus_r}{\vdash A^\perp \,\&\, B^\perp, A \oplus B}\&_r}{\cfrac{(A^\perp \,\&\, B^\perp)^\perp \vdash A \oplus B}{\vdash (A^\perp \,\&\, B^\perp)^\perp \multimap A \oplus B}\multimap_r}\perp_l$$

$$\cfrac{\cfrac{\cfrac{\cfrac{\overline{A \vdash A}^{\ \text{ax}}}{\vdash A^\perp, A}\perp_r}{\vdash A^\perp, \bigvee xA}\bigvee_r}{\vdash \bigwedge xA^\perp, \bigvee xA}\bigwedge_r}{\cfrac{(\bigwedge xA^\perp)^\perp \vdash \bigvee xA}{\vdash (\bigwedge xA^\perp)^\perp \multimap \bigvee xA}\multimap_r}\perp_l$$

$$\cfrac{\cfrac{\cfrac{\cfrac{\overline{A \vdash A}^{\ \text{ax}}}{\vdash A, A^\perp}\perp_r}{\vdash A, A^\perp \oplus B^\perp}\oplus_r \qquad \cfrac{\cfrac{\overline{B \vdash B}^{\ \text{ax}}}{\vdash B, B^\perp}\perp_r}{\vdash B, A^\perp \oplus B^\perp}\oplus_r}{\vdash A \,\&\, B, A^\perp \oplus B^\perp}\&_r}{\cfrac{(A^\perp \oplus B^\perp)^\perp \vdash A \,\&\, B}{\vdash (A^\perp \oplus B^\perp)^\perp \multimap A \,\&\, B}\multimap_r}\perp_l$$

$$\cfrac{\cfrac{\cfrac{\cfrac{\overline{A \vdash A}^{\ \text{ax}}}{\vdash A^\perp, A}\perp_r}{\vdash \bigvee xA^\perp, A}\bigvee_r}{\vdash \bigvee xA^\perp, \bigwedge xA}\bigwedge_r}{\cfrac{(\bigvee xA^\perp)^\perp \vdash \bigwedge xA}{\vdash (\bigvee xA^\perp)^\perp \multimap \bigwedge xA}\multimap_r}\perp_l$$

References

1. Gentzen, G.: Untersuchungen über das logische Schließen. I. Math. Z. **39**(1), 176–210 (1935)
2. Gentzen, G.: Untersuchungen über das logische Schließen. II. Math. Z. **39**(1), 405–431 (1935)
3. Gödel, K.: Eine Interpretation des intuitionistischen Aussagenkalküls. Ergeb. Math. Kolloqu. **4**, 39–40 (1933). English translation: An interpretation of the intuitionistic propositional logic. Reprinted in: Béziau, J.-Y. (ed.) Universal Logic: An Anthology, pp. 89–90. Springer, Basel (2012)
4. Girard, J.-Y.: Du pourquoi au comment: la théorie de la démonstration de 1950 à nos jours. In: Pier, J.-P. (ed.) Les mathématiques 1950–2000, pp. 515–545. Birkhäuser, Basel (2000)

B. Mélès (✉)
Université Blaise Pascal, Clermont-Ferrand, France
e-mail: baptiste.meles@normalesup.org

The Square of Opposition and Generalized Quantifiers

Duilio D'Alfonso

Abstract In this paper I propose a set-theoretical interpretation of the logical square of opposition, in the perspective opened by generalized quantifier theory. Generalized quantifiers allow us to account for the semantics of quantificational Noun Phrases, and of other natural language expressions, in a coherent and uniform way. I suggest that in the analysis of the meaning of Noun Phrases and Determiners the square of opposition may help representing some semantic features responsible to different logical properties of these expressions. I will conclude with some consideration on scope interactions between quantifiers.

Keywords Generalized quantifiers · Semantics · Logical duality

Mathematics Subject Classification Primary 03B65 · Secondary 68Q55

1 Noun Phrases as Generalized Quantifiers

Let M be a non-empty set, the universe, of a model $\mathcal{M}$ and Q_M any set of subsets of M. Then, in the Lindström terms [3], Q_M is a type $\langle 1 \rangle$ Generalized Quantifier (GQ, here and henceforth) whose meaning is given by:

$$Q_M(A) \text{ is true} \quad \text{iff} \quad A \in Q_M \tag{1}$$

where the Q_M on the left side of the iff-statement is meant to be a quantificational operator of the functional type $(et)t$ (a function mapping predicates onto truth-values, A being a first-order predicate, of type et), and the Q_M on the right side is to be interpreted as a subset of $\wp(M)$, the power set of M (Q_M is a second order unary predicate).

The type $\langle 1 \rangle$ generalized quantifier is a straightforward illustration of the "set-theoretical" conception of quantifiers in the GQ theoretical framework: a quantifier is conceived as a second order n-ary relation between sets (i.e. between first-order predicates). For instance, a type $\langle k_1 \ldots k_n \rangle$ quantifier, where $k_1 = k_2 = \cdots = k_n = m$, is a second order n-ary relation between m-ary predicates.

Generalized quantifiers can be written as a variable-binding operators. As a simple case, a type $\langle 1 \rangle$ quantifiers Q_M can be notated in the usual way as $Q(x)\phi(x)$. Let $\mathcal{M} \models \phi$ be the *satisfaction* relation between a model and a formula, and $[\![\phi(x)]\!]_{\mathcal{M}}$ the "extension" of $\phi(x)$ in $\mathcal{M}$ (the set of the elements satisfying ϕ). The following statement provides the semantics of a formula in which x is bound by the generalized quantifier Q_M, as variable-binding operator:

$$\mathcal{M} \models Qx\phi(x) \quad \text{iff} \quad [\![\phi(x)]\!]_{\mathcal{M}} \in Q_M \tag{2}$$

J.-Y. Béziau, D. Jacquette (eds.), *Around and Beyond the Square of Opposition*, 219–227
Studies in Universal Logic, DOI 10.1007/978-3-0348-0379-3_15, © Springer Basel 2012

This notation suggests a compositional account for the first-order logic formulas with quantifiers. It is well-known that standard universal and existential quantifiers do not have a denotation in the predicate calculus, and that, as a consequence, it is not possible to compositionally interpret formulas in which they occur. But, as type $\langle 1 \rangle$ quantifiers, standard universal and existential quantifiers denote sets of sets:

$$\forall_M = \{M\}$$
$$\exists_M = \{A \subseteq M : A \neq \emptyset\}$$

Now, we can easily express, in a "compositional fashion", the meaning of a formulas such as $\forall x \phi(x)$:

$$\mathcal{M} \vDash \forall x \phi(x) \quad \text{iff} \quad [\![\phi(x)]\!]_{\mathcal{M}} \in [\![\forall_M]\!] \quad \text{iff} \quad [\![\phi(x)]\!]_{\mathcal{M}} = M \tag{3}$$

The denotation of $\forall x \phi(x)$ is constructed out by applying the denotation of the quantifier to the denotation of the open formula. So conceived, standard quantifiers end up to show the same functional property of Noun Phrases in natural languages.

1.1 Noun Phrases as Type $\langle 1 \rangle$ Quantifiers

Noun Phrases (NP), typed $(et)t$, are uniformly interpreted (following Montague) as type $\langle 1 \rangle$ quantifiers, in their role of mapping the predicates expressed by the Verb Phrases onto $\{True, False\}$. The following 'parse tree' illustrates the point.

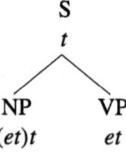

S
t

NP   VP
(et)t   et

This tree is the representation of the syntactic and semantic structure of the nuclear sentence[1] (in English, but also in other natural languages). As well as a sentence is constructed out by combining a NP with a VP, its meaning, a truth-value, is obtained by a step of functional application of the meaning of the NP to the meaning of the VP. In the following, is showed this process of meaning computation, for a simple sentence like "every sailor drink":

$$[\![[every\ sailor]_{NP}[drink]_{VP}]\!]_t$$
$$\text{iff} \quad [\![every\ sailor]\!]_{(et)t} ([\![drink]\!]_{et})$$
$$\text{iff} \quad drink \in \{A \subseteq M : sailor \subseteq A\} \Leftrightarrow drink \subseteq sailor \tag{4}$$

provided that the meaning of "every sailor" is conceived as a generalized quantifier:

$$[\![every\ sailor]\!]_{(et)t} = (every\ sailor)_M = \{A \subseteq M : sailor \subseteq A\} \tag{5}$$

The uniform treatment of NPs as type $\langle 1 \rangle$ quantifiers (Mostowski GQ, [5]) is assured by the "type-raising" of the meaning of proper names, as showed by the meaning computation for a sentence like "John drinks":

---

[1] S is the symbol for the syntactic category of 'sentence', VP for 'verb phrase'.

$$[[john]_{NP}[drinks]_{VP}]_t$$
$$\text{iff}\quad [john]_{(et)t}([drink]_{et})$$
$$\text{iff}\quad drink \in \{A \subseteq M : john \in A\} \Leftrightarrow john \in drink \tag{6}$$

provided that the meaning of "John" is conceived as a generalized quantifier:

$$[john]_{(et)t} = (john)_M = \{A \subseteq M : john \in A\} \tag{7}$$

1.2 Determiners as Type $\langle 1, 1 \rangle$ Quantifiers

As a natural consequence of Lindström generalization of type $\langle 1 \rangle$ quantifiers to type $\langle k_1 \ldots k_n \rangle$ quantifiers, it is possible to semantically construe NPs as the result of the combination of Determiners (DET) with common nouns (that are first-order predicates). DETs are interpreted as type $\langle 1, 1 \rangle$ quantifiers, i.e. functions mapping predicates into type $\langle 1 \rangle$ quantifiers. As for sentences, the meaning of NPs is the result of a compositional process, in which the meaning of a DET is applied to that of a common noun:

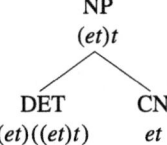

Type $\langle 1, 1 \rangle$ quantifiers can be defined along the same line of the previous definition for type $\langle 1 \rangle$ quantifiers. Let M be the universe of a model $\mathcal{M}$ and Q_M any set of pairs of subsets of M. Then Q_M is a type $\langle 1, 1 \rangle$ GQ whose meaning is given by:

$$Q_M(A, B) \text{ is true}\quad \text{iff}\quad \langle A, B \rangle \in Q_M \tag{8}$$

where $Q_M(A, B)$ on the left side of the iff-statement is a quantificational operator of type $(et)((et)t)$ and $Q_M(A, B)$ on the right side is a subset of $\wp(M) \times \wp(M)$.

Again, an example could help illustrating this definition at work. Let consider the meaning of a definite NP like "the boy", resulting from the application of the meaning of "the" to the meaning of "boy":

$$[the\ boy]_{(et)t} = [the]_{(et)((et)t)}([boy]_{et})$$

Since the meaning of the definite determiner is:

$$the(A, B) \text{ is true}\quad \text{iff}\quad A \subseteq B \wedge |A| = 1 \tag{9}$$

we have the following semantic computation for deriving the meaning of the definite NP:

$$[the]([boy])$$
$$\text{iff}\quad the(boy, B)$$
$$\text{iff}\quad boy \subseteq B \wedge |boy| = 1 \tag{10}$$

In the last line of this derivation there is a type $\langle 1 \rangle$ quantifier that can equivalently be written as: $the(boy)_M = \{A \subseteq M : boy \subseteq A \wedge |boy| = 1\}$. Moreover, note that (9) explicates the meaning of the singular *the*; for the plural case we have to modify the expression as follows:

$$the_{pl}(A, B) \quad \text{iff} \quad A \subseteq B \wedge |A| > 1 \tag{11}$$

In order to show the expressive power of GQ theory, and its wide applicability to the analysis of the semantics of natural language determiners, a list of DETs as type $\langle 1, 1 \rangle$ quantifiers is presented. The four quantifiers of the Aristotelian square of opposition are:

$$
\begin{aligned}
&all(A, B) \Leftrightarrow A \subseteq B \\
&no(A, B) \Leftrightarrow A \cap B = \emptyset \\
&some(A, B) \Leftrightarrow A \cap B \neq \emptyset \\
&not\ all(A, B) \Leftrightarrow A - B \neq \emptyset
\end{aligned}
\tag{12}
$$

Below some other well-known quantifiers are listed:

$$\left(Q^R\right)_M \Leftrightarrow |A| > \frac{|M|}{2} \quad \text{(the Rescher quantifier for finite universe)}$$

$$most(A, B) \Leftrightarrow |A \cap B| > |A - B|$$

$$ten(A, B) \Leftrightarrow |A \cap B| \geq 10$$

$$the\ ten(A, B) \Leftrightarrow |A| = 10 \wedge A \subseteq B$$

$$John's(A, B) \Leftrightarrow A \cap \{b : owner(j, b)\} \subseteq B \wedge \left| A \cap \{b : owner(j, b)\} \right| > 1$$

(in the last line is represented the case of possessive construction in plural NPs, like "John's bikes", while $|A \cap \{b : owner(j, b)\}| = 1$ holds in the case of singular NP, like "John's bike").

2 Squaring Generalized Quantifiers

As set-theoretical entities, generalized quantifiers can be the object of application of the boolean operations. In what follows I will focus on negative operations and on the structural and logical properties showed by quantifiers when they are undergoing those operations. The two negative operations on GQs are the outer negation and the inner negation.

Let Q be a GQ. We adopt the following notational convention:

i. $\neg Q$ is the outer negation of Q
ii. $Q\neg$ is the inner negation of Q
iii. $\neg Q\neg = Q^d$ is the dual of Q (outer negation of the inner negation).

A quantifier, together with its outer negation, inner negation and dual forms a natural unit, the *square*:

$$square(Q) = \left\{ Q, \neg Q, Q\neg, Q^d \right\} \tag{13}$$

Since a type $\langle 1 \rangle$ quantifier Q_M, on a domain M, is a subset of $\wp(M)$, the square of Q may straightforwardly be characterized in terms of set-theoretical formulas as follows:

i. $\neg Q_M(A) = \{\wp(M) - Q(A)\}$
ii. $Q_M\neg(A) = \{Q(M - A)\}$

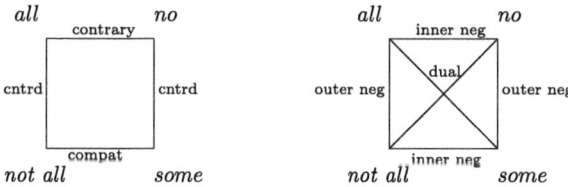

Fig. 1 On the *left* is showed the classical square of opposition, on the *right* the modern interpretation. The logical relations in the classical square involve the existential import

iii. $Q^d(A) = \{\wp(M) - Q(M - A)\}$

This square is naturally extensible for a type $\langle 1, 1 \rangle$ quantifier $Q(A, B)$:

  i. $\neg Q_M(A, B) = \{\wp(M)^2 - Q(A, B)\}$
 ii. $Q_M \neg(A, B) = \{Q(A, M - B)\}$
iii. $Q^d(A, B) = \{\wp(M)^2 - Q(A, M - B)\}$

The following are some obvious general facts about *square(Q)*:

- if Q is non-trivial ($Q_M \neq \wp(M)$ and $Q_M \neq \emptyset$), so are the other quantifiers in its square
- if $Q' \in square(Q)$ then $square(Q') = square(Q)$
- *square(Q)* has either two or four members
- if $Q' = Q^d$ then Q and Q' are inter-definable.

2.1 Classical vs Modern Square

The debate on the difference between classical and modern square of opposition (as showed in Fig. 1) is nowadays still on (see [7] and [6] for a wide discussion on the topic). The classical square represents logical relations among sentences in which a single quantificational determiners occurs in subject position, the so-called "categorical judgments". The modern version is intended to represent the "span" of a quantifier when undergoing the negative operations.

Clearly, the modern version is a generalization of the classical square of opposition. By interpreting the modern square in set-theoretical terms, the logical relations holding among quantifier of the classical square are easily reproducible, under some conditions on the set-theoretical objects involved. The main case is that of the existential import of *ALL* in the classical square. In the modern sense, *ALL* is conceived without existential import, but, given that $ALL_M(A, B) \Leftrightarrow A \subseteq B$ and $ALL_M \neg(A, B) \Leftrightarrow A \subseteq M - B$, to ensure that they can not both be true (for reproducing the contrary relation) is simply required that A is not empty.

Other consequence of the more generality of the modern square is that it is applicable to every quantifier and thus it displays more utility for investigating the semantics of NPs. Moreover, it is worth noting that, in many natural languages, there are pairs of quantificational expressions (DETs), but also of expressions belonging to other categories, like adverbials (*still/already*) or conjunctions (*because/although*), whose denotation exhibits the pattern of duality. This may be a sign that the square of logical duality codifies some deep and pervasive pattern of the semantics of natural languages expressions embodying some quantificational feature.

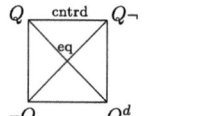 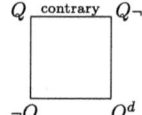 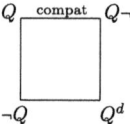

Fig. 2 Using the modern version of the square, we can easily visualize different patterns of logical relations in the *square(Q)* of a quantifier Q. From *left* to *right*, the *first square* sketches the logical pattern for ultrafilters (Q and $Q\neg$ are contradictory), the *second* for filters (Q and $Q\neg$ are contrary), the *third* for intersective quantifiers (Q and $Q\neg$ are compatible)

3 Squaring NPs Properties

The semantic homogeneity of NPs is due to their functional homogeneity as type $\langle 1 \rangle$ generalized quantifiers. But type $\langle 1 \rangle$ quantifiers have different logical properties, splitting them into three fundamental categories: ultrafilter, filter and "intersective" quantifiers, as outlined in the following table:

ultrafilter	Proper Names, definite NP_{sg}s, Pronouns *John, the boy, I*, etc.
filter	definite quantified NPs *every boy, no student*, etc.
intersective	indefinite quantified NPs *some boys, a boy, two boys*, etc.

Briefly, what distinguishes a filter Q from an intersective quantifier Q' is that Q and $Q\neg$ cannot be both true (they are contrary), while Q' and $Q'\neg$ do (they are compatibles). What distinguishes an ultrafilter Q from a filter Q' is that Q and $Q\neg$ cannot be neither both true nor both false (they are contradictory), while Q' and $Q'\neg$ can be both false.[2] As visualized in Fig. 2, this way of distinguishing filters from ultrafilters and intersective quantifiers suggests the use of the square in order to figure out logical properties of these classes of NPs. Ultrafilters are individuals, inner and outer negation having the same effect of determining contradiction. Filters and intersective quantifiers are "quantirelations": they have the so-called "isomorphism" (ISOM) property, whereas ultrafilters are non-ISOM. ISOM means that for holding the relation expressed by the quantifiers only the cardinality of sets involved is relevant.

The difference between filters and intersective quantifiers is a matter of different logical effect of inner negation, as a set-theoretical operation. For instance, *all* is a filter because $all_M(A, B) = |A - B| = 0$ and $all_M\neg(A, B) = |A - (M - B)| = |A \cap B| = 0$ cannot both be true, if A is not empty, while *some* is an intersective quantifier because $some_M(A, B) =$

---

[2]The distinction here introduced is slightly different from the original characterization of definite NPs as filters due to Barwise and Cooper [2] and revisited by Loebner [4]. For example, the classification here proposed has the concept of filter in common with the characterization of NPs in Loebner [4], but the method of "squaring" logical properties of quantifiers involves some not negligible differences. The effect of the inner negation in determining contradiction or contrariness is here the only parameter used in distinguishing quantifiers. As a consequence, three classes of quantifiers are proposed, while Loebner essentially distinguishes only between filters and ultrafilters, as his main concern, in the cited article, is the semantics of definite NPs, and not NPs overall considered as generalized quantifiers.

$|A \cap B| \neq 0$ and $some_M \neg (A, B) = |A \cap (M - B)| = |A - B| \neq 0$ can obviously be both true but not both false (if A is not empty). Moreover, we can notice that if Q is a filter, Q^d is an intersective quantifier. This is the case of a GQ like $most_M (A, B) = |A \cap B| > |A - B|$, as showed by its square:

   i. $most_M \neg (A, B) = |A \cap B| < |A - B|$
   ii. $most_M^d (A, B) = |A \cap B| \geq |A - B|$
   iii. $\neg most_M (A, B) = |A \cap B| \leq |A - B|$

By reading this square it is immediately verified that $most^d$ and its inner negation $most^d \neg = \neg most$ can be both true.

3.1 Nested Quantifiers

In natural language sentences is usual to have NPs both in subject and object position. And, given the formal under-determination of natural language grammar, each NP, as type $\langle 1 \rangle$ quantifier, can take scope over the other. If this scope commutativity between the two NPs produces a semantic ambiguity or not, is a matter of "relative logical properties" of the quantifiers involved.

Let briefly examine the "scope interaction" between quantifiers. If Q_1 and Q_2 are type $\langle 1 \rangle$ generalized quantifiers and R is a binary relation, we can abbreviate the formula $Q_1(x)Q_2(y)R(x, y)$ as $Q_1 Q_2 R$ and its inversion $Q_2(y)Q_1(x)R(y, x)$ as $Q_2 Q_1 R^{-1}$. We can now distinguish three fundamental cases (for each case an example is reported immediately below):

   i. *self-commuting* quantifiers:

$$Q(Q(R)) \Leftrightarrow Q(Q(R^{-1}))$$
$$all(all(R)) \Leftrightarrow all(all(R^{-1}))$$

   ii. *independent* quantifiers:

$$Q_1(Q_2(R)) \Leftrightarrow Q_2(Q_1(R^{-1}))$$
$$all(the(R)) \Leftrightarrow the(all(R^{-1}))$$

   iii. a quantifier is (scopally) *dominant* over the other:

$$Q_1(Q_2(R)) \Rightarrow Q_2(Q_1(R^{-1}))$$
$$some(all(R)) \Rightarrow all(some(R^{-1}))$$

Semantic ambiguity rises only in the third case, i.e. when Q_1 is dominant over Q_2. In order to illustrate this fact I will proceed with an example. Let consider the following two sentences:

All boys read the books   (14)

All boys read some book   (15)

For sentence (14), the *object narrow scope* (ons) reading and the *object wide scope* (ows) reading are equivalent, as a matter of logic of set-theoretical formulas:

$$(ons) \quad all(boy)\big[the_{pl}(book)[read]\big]$$
$$= boy \subseteq \{x : book \subseteq \{y : x\ read\ y\} \land |book| > 1\}$$

(16)

$$(ows) \quad the_{pl}(book)\big[all(boy)\big[read^{-1}\big]\big]$$
$$= book \subseteq \{y : boy \subseteq \{x : x\ read\ y\}\} \land |book| > 1$$

For sentence (15), (ons) and (ows) readings are not equivalent, given that (ows) ⇒ (ons), as a matter of logic:

$$(ons) \quad all(boy)\big[some(book)[read]\big]$$
$$= boy \subseteq \{x : book \cap \{y : x\ read\ y\} \neq \emptyset\}$$

(17)

$$(ows) \quad some(book)\big[all(boy)\big[read^{-1}\big]\big]$$
$$= book \cap \{y : boy \subseteq \{x : x\ read\ y\}\} \neq \emptyset$$

In literature [1], scope dominance is defined as follows: in finite domain Q_1 is dominant over Q_2 iff Q_1 is an "exist" quantifier or Q_2 is a "universal" quantifier. On the square, this condition could be expressed as follows: Q_1 is *dominant* over Q_2 iff Q_1 is an intersective quantifier or Q_2 is a filter. For what concerns independent quantifiers, retracing the definition in Westersthål [8], we can say that Q_1 and Q_2 are *independent* iff at least one of the following holds: (i) both Q_1 and Q_2 are intersective; (ii) both Q_1 and Q_2 are filters; (iii) one of the two is an ultrafilter. Actually, this definition should be refined, in order to account for cardinal quantifiers. In fact, cardinal quantifiers are intersective but it is easy to see that the bidirectional entailment is not valid:

$$n\big(m(R)\big) \not\Leftrightarrow m\big(n\big(R^{-1}\big)\big)$$

However, the undirectional entailment continues to hold:

$$n\big(m(R)\big) \Rightarrow m\big(n\big(R^{-1}\big)\big)$$

In other words, in the scope interaction of two cardinal quantifiers Q_1 and Q_2 it is the quantifier that takes scope over the other that is scopally dominant. Thus, each of two cardinal quantifier can be dominant over the other, according to the scope it takes relatively to the other. To accommodate this situation, the definition of independent quantifiers could be refined as follows: Q_1 and Q_2 are independent iff

$$Q_1\big(Q_2(R)\big) \Leftrightarrow Q_2\big(Q_1\big(R^{-1}\big)\big)$$

or

$$Q_1\big(Q_2(R)\big) \Rightarrow Q_2\big(Q_1\big(R^{-1}\big)\big) \land Q_2\big(Q_1(R)\big) \Rightarrow Q_1\big(Q_2\big(R^{-1}\big)\big)$$

The latter disjunct is required to account for the "reciprocity" of scope dominance characterizing cardinal quantifiers.

For filters (type $\langle 1, 1 \rangle$), we need a more subtle distinction between *co-intersective* (let $Q(A, B)$: only $|B - A|$ is relevant) and *proportional* (both $|A \cap B|$ and $|B - A|$ are relevant). Thus, although $all(A, B)$ and $most(A, B)$ are both filter, *most* is dominant over *all*, because of the relevance, in its meaning, of $|A \cap B|$. The (ons) interpretation of a sentence like "all boys read most books" implies the (ows) reading, but the vice-versa does not hold:

$$most(boy)\big[all(book)[read]\big] \Rightarrow all(book)\big[most(boy)\big[read^{-1}\big]\big]$$

but

$$most(boy)\big[all(book)[read]\big] \not\Rightarrow all(book)\big[most(boy)\big[read^{-1}\big]\big]$$

To sum up, in the paper I tried to apply the square of opposition, conceived as the explication of the logical dualities among quantifiers, to characterize some semantic features of natural language NPs. A strong hypothesis, to be validated, could be that all quantificational NPs fall into one of the three categories (ultrafilters, filters, intersective quantifiers) depending on the logical pattern they give rise when subjected to negative operations, i.e. depending on the "collocation" they find in the square.

Generally, the methodology of "squaring" logical relations of quantificational operators can reveal some utility, for various purposes, and this utility is undoubted in the analysis of the semantic and logical properties of natural language expressions such as NPs, in the Generalized Quantifier framework.

References

1. Altman, A., Peterzil, Y., Winter, Y.: Scope dominance with upward monotone quantifier. J. Log. Lang. Inf. **14**, 445 (2005)
2. Barwise, J., Cooper, R.: Generalized quantifiers and natural language. Linguist. Philos. **4**, 159–219 (1981)
3. Lindström, P.: First order predicate logic with generalized quantifiers. Theoria **32**, 186–195 (1966)
4. Loebner, S.: Natural language and generalized quantifier theory. In: Gärdenfors, P. (ed.) Generalized Quantifiers, pp. 181–202. Reidel, Dordrecht (1987)
5. Mostowsky, A.: On a generalization of quantifiers. Fundam. Math. **44**, 12–36 (1957)
6. Parsons, T.: The traditional square of opposition. In: Stanford Encyclopedia of Philosophy. http://plato.stanford.edu/entries/square (2006)
7. Peters, S., Westerståhl, D.: Quantifiers in Language and Logic. Oxford University Press, Oxford (2006)
8. Westerståhl, D.: Self-commuting quantifiers. J. Symb. Log. **61**, 212–224 (1996)

D. D'Alfonso (✉)
University of Calabria, via Pietro Bucci, 87036 Arcavacata di Rende (CS), Italy
e-mail: duilio.dalfonso@tin.it

Privations, Negations and the Square: Basic Elements of a Logic of Privations

Stamatios Gerogiorgakis

Abstract I try to explain the difference between three kinds of negation: external negation, negation of the predicate and privation. Further I use polygons of opposition as heuristic devices to show that a logic which contains all three mentioned kinds of negation must be a fragment of a Łukasiewicz-four-valued predicate logic. I show, further, that, this analysis can be elaborated so as to comprise additional kinds of privation. This would increase the truth-values in question and bring fragments of (more generally speaking) Łukasiewicz-n-valued predicate logics into the scene.

Keywords Privation · Negation · Łukasiewicz-n-valued predicate logic

Mathematics Subject Classification Primary 03B65 · Secondary 03B50

1 Motivating Arguments and Some History

Negation is what makes the square a square "of opposition" and a very intensively investigated logical constant.

Privation, a kind of truth-value inversion expressed by nouns like 'immorality', 'irresponsibility', 'unreadiness', 'anarchy', 'senselessness', 'blindness', 'deafness', 'ignorance', etc., is usually associated with negation in a very general sense of the word. Unlike "standard" negation however, in the recent decades privation has been neglected or suppressed in the discussion on the square and in logic altogether. This is one of these very usual sins, which are understandable but inexcusable.

It is inexcusable, since the very first hints to a logic of privations date already in the Middle Ages to be repeated now and then until the 19th century.

Greek and Arab Aristotelians dealt with three sorts of negation (cf. [16, pp. 173–184]):

1. the simple negation, which they understood in a way which roughly resembles our external negation—"external" because in our semi-formal English the particle "not" occurs, in this kind of negation, "externally" as part of the introductory expression: "It is not the case that ...";

For discussions concerning privation and amendments in the first drafts of this paper, I would like to thank Prof. Dr. Hans Burkhardt, Georgios Karageorgoudis (both Munich) and, last but not least, two anonymous referees.

J.-Y. Béziau, D. Jacquette (eds.), *Around and Beyond the Square of Opposition*, 229–239
Studies in Universal Logic, DOI 10.1007/978-3-0348-0379-3_16, © Springer Basel 2012

2. the negation of the predicate; and
3. the privation

Rough examples in English would be respectively: 'It is not the case that Ebenezer Scrooge is a generous person'; 'Ebenezer Scrooge is a non-generous person'; 'Ebenezer Scrooge is a stingy person'. NB being stingy is a privation of a virtue: generosity.

The negation of the predicate was known to the ancients and the medievals by two alternative names: infinite term (cf. Aristotle [3, 16a32]; Boethius [6, editio prima, I, chapter 2]) and negation by transposition (cf. Ammonius [2, pp. 161–162]; Alexander [1, p. 397]; John Italos [11, pp. 61–62, i.e. question 49]). The reason for giving it the name "negation by transposition" is the structure of the classical languages: in Greek and in Latin the negation of the predicate is formulated by taking a negative sentence with the usual syntax of these languages (i.e. with the negation particle just in front of the verb) and moving ("transposing") the negation particle ('ouk' or 'ou' in Greek, 'non' in Latin) to come immediately before the predicate. A standard Latin negation would be: 'Bucephalus non est asinus' (= It is not the case that (there is something like) Bucephalus (and that it) is a donkey) and the negation by transposition would be: 'Bucephalus est non asinus' (Bucephalus is no donkey). 'Non' was transposed from the first to the second place, to formulate the negation by transposition. The English syntax does not allow negation by transposition to be formulated by means of a transposition literally.

Some medieval Arab Aristotelians tended to confuse privations and negations of the predicate because of the linguistic production rules of Arabic: There is nothing like a monolectic privation with affixes in Arabic and the privation of the Indogermanic languages can only be expressed in this language by means of a periphrastic formulation analogous to the Indogermanic negation of the predicate—cf. [16, p. 178]. Regardless of the reasons which gave rise to it, there is something true in the idea of the Arab logic to regard the negation of the predicate as some kind of privation. In what follows we are going to see that the negation of the predicate is, like privation, a *strong* kind of negation.

John Italos [11, p. 62, i.e. question 49], an 11th-century Norman from Sicily who wrote in Greek, observed that the sentence 'Being just is predicated of humans' is true in just one case: obviously when the humans in question are just. Italos continues with the statement that the sentence: 'Being non-just is predicated of no humans' is true in three cases. He does not explicitly say so, however good candidates would be the following three cases: case (1) the humans in question are just (note that this case made also the sentence 'Being just is predicated of humans' true) or, case (2), there are no humans or, case (3), there is nothing which is non-just. Further, Italos observed that the sentence: 'Being unjust is predicated of no humans' is true in five cases. By this he rather meant the following five cases: the inclusive disjunction of the cases 1 till 3, or, case (4), injustice is a predicate which does not apply to the humans of the domain (like, for example, it does not apply to insects) or case (5), there is nothing unjust. Italos remarked that the humans' in question being unjust implies their being non-just, which implies their not being just. However, according to Italos these implications do not hold *vice versa*. And this is exactly what my interpretation of Italos does justice to.

Another example, not found in Italos but representative of what he had in mind, might be more illuminating: If Homer is blind, then Homer sees nothing. But it is not correct to say that if Homer sees nothing, then Homer is blind, since Homer may be unable to see only because the room is dark. Further, if Homer sees nothing, then seeing is not predicated of the reference of this name. But if seeing is not predicated of the reference of the name 'Homer', then from this does not follow that Homer is an existing individual and that other predicates of real people can be attached to him: like, say, being in a dark room, being blind and so on. In other words, if the only thing which we know about Homer is that it is not the case that he is there and watches around, then from this does not follow that he is there and watches nothing. He might not be there at all. Cf. also Wolfson [16, pp. 179–181].

After the 15th century, the distinction between simple (= external) negation and privation was criticized as unintelligible by authors who were anti-Aristotelians or nominalists or both; among many others by Valla [14, pp. 114–115], and Gassendi [7, lib. I, ex. 6, cap. 1]. Their arguments against privation reflect basic nominalistic beliefs which primarily concern *ontology* and affect the logic of privation only insofar one decides to *indulge in nominalist ontology*. It is interesting, however, to review these arguments, in order to understand how privation was dismissed from natural science, initially, and from logic subsequently. Since Aristotle, privation was defined with reference to natural properties. Privation-talk presupposes that an individual has properties, which are natural for a universal, to which it belongs, and others which are not natural for it. We say, for example, that Stevie Wonder is deprived of sight, because it is natural for humans to see, but we do *not* say that he is deprived of hearing ultrasounds. The nominalistic cluster of arguments against privation-talk denies the ontology which underlies properties natural for members of the one universal to have and unnatural for members of another universal to have. Based on this, nominalists stress that the distinction between simple negation and privation has to be dismissed. This view has had some very influential supporters not only in the early modern period but also in very recent times, above all Quine. Horn [10, p. 112], reviews briefly the contemporary discussions on this point.

These historical remarks concerning the topic: nominalist ontology and privation may suffice. The logical issue concerning privation is not directly linked to nominalist ontological concerns. The logic of privation formalizes the privation-talk of natural language. The ontological issue, which the traditional nominalists and their recent epigones raise, does not affect the way in which we express privation in natural language. While nominalist ontology rejects the idea of real universals with natural properties, the logical issue consists in taking privation-talk for granted and in attempting to model it in logic. In what follows, I will leave ontology aside and concentrate on the logical issue. In other words, I will take privation-talk seriously and by doing this I will engage in an ontology, which is not nominalistic.

This is not to say that I will put all issues raised by nominalism aside. My analysis does deal with some logical issues, which nominalism raises, though not with the ontological ones. A logical issue of nominalistic spirit, which my analysis will account for, was anticipated by Aristotle [4, 1022b32-33], who observed that no privation is meant like the next privation: an individual mole is not said to be *deprived* of sight in reference to creatures which have eyes (moles do have eyes). Therefore, it is *one kind* of blindness which is expressed in a mole's being blind and *another* kind of blindness which is expressed in a human's being blind as a result of an accident—in the sense that these individuals are deprived of one-and-the-same thing in different grades. The one is deprived of something she never had, the other is deprived of something she once had. Another, perhaps more appropriate or revealing example, would be the difference between blindness and half-blindness. Let me coin a new "-ism" for this feature of natural language: the pluralism of privations. At the end of this paper, I will attempt to show that the pluralism of privations does not dismiss privation.

Despite the criticism, to which the notion of privation was exposed in the 16th and 17th centuries, Wolff [15, §§109–110] and Kant [12, A70 = B95; A72-73 = B97-98] in the 18th century expressed the view that the sentences are according to their quality affirmative, negative or infinite. Kant's example was the difference between 'The soul is not mortal' (a negative sentence) and 'The soul is non-mortal' (an infinite sentence). We know what the case must be for the negative sentence 'The soul is not mortal' to be true: it has to fail to fulfill the sentence: 'The soul is mortal'. This is the case when the soul does not exist or is inorganic or is divine. But we do not know as exactly as this, what the case must be for the infinite sentence 'The soul is non-mortal' to be true. The things which are non-mortal form the complementary set of all mortal entities—and this set has infinite members. Since this set is infinite, we can never be certain whether there exist entities in it, perhaps the soul itself, which would surprise us in the sense that neither mortality nor non-mortality, or even both would be attributed to them. However, when we state that the soul is *not* mortal (instead of *non*-mortal), we claim that the soul fails to be an element of a finite set: the set of all mortal entities. And since we have defined this finite set, in principle we can be certain that mortality either is or is not attributed to the soul—for this line of argument cf. Kant [12, A72-73 = B97-98].

Subsequently there were Kantian scholars who expressed the view that it is one thing to negate the verb ('The soul is-not mortal') and another thing to negate the predicate ('The soul is non-mortal'); for a discussion cf. Hamilton [9, pp. 177–179]. But the issues, which Kantianism raised in this respect, got forgotten at the latest when this philosophical tradition declined in the second half of the 20th century.

Albeit inexcusable, the negligence of privation in logic is, as I said, understandable. It is understandable, since the syntactical means to express privation, unlike those which express external negation, do not seem to be propositional operators. I will argue that this is a false impression, that privations *are* propositional functions after all. I will show this by using an hexagon and a square of opposition.

2 Polygons of Opposition Edged by Privations, Negations of the Predicate, and External Negations

Take the following examples of a privation, a negation of the predicate, and an external negation. These are exactly the kinds of negation, which are known to Aristotelian logic.

Privation The philosopher's stone is immortal
Negation of the predicate The philosopher's stone is non-mortal
External negation[1] It is not the case that the philosopher's stone is mortal

---

[1] Some contemporary authors like Blau [5, p. 49] and Horn [10, pp. 122–132], call the negation of the predicate: 'presupposing' negation, because, unlike the external negation it presupposes the existence of the entity expressed in the subject. Especially Blau (loc. cit.) calls it 'strong negation' because it entails the external negation but is not entailed by it. Since in Blau's three-valued logic there are exactly two negations: (a) a formal construction, which formalizes the negation of the predicate/by transposition, and (b) a weak negation which formalizes the external negation of the natural language, the former is the only strong negation in Blau's three-valued logic. But, if I am allowed to anticipate one of my results, the negation of the predicate is *not* the strongest kind of negation in the logic of privation.

My example of an external negation is a true sentence. A stone cannot be said to be mortal. This seems to make my example of a negation of the predicate true (this conclusion is premature and inaccurate, but for now, let me continue). The philosopher's stone belongs to the entities which are neither mortal nor immortal. To say that a stone would belong to the entities which are immortal, is to commit a category mistake. The main difference between my example of a negation of the predicate and my example of a privation is that the privation invites a category mistake, the negation of the predicate does not. In fact, a negation of a predicate never involves category mistakes, privations sometimes do (cf. [10, pp. 110–122, i.e. chapter 2.3]). This is one of the reasons why the privation above is also not true: it is a category mistake to assign a stone immortality. To avoid the category mistake one must avoid to speak of immortality in this context, and say instead that the philosopher's stone belongs to the entities which are non-mortal. However effective for avoiding category mistakes, this trick, however, does not save from unjustified existential import.

Unjustified existential import is the other reason for which my example of a privation is not true (my example of a privation fails to be true because of two reasons). My example of a privation presupposes that the philosopher's stone exists. If the philosopher's stone existed, then it would be true to say that it is non-mortal, since it would be neither mortal nor immortal. But, alas, the philosopher's stone does not exist. To refer to the philosopher's stone is analogous to referring to entities like the present king of France, James Bond and Pegasus. Note that referring to the predicates which are not predicated of the present king of France is totally different than referring to the predicates which are not predicated of the present queen of the Netherlands. It is not true to say of Beatrix that she is bald, because she has plenty of hair on her head. And it is not true to say of the present king of France that he is bald (or that he is not bald alike), because he does not exist. To make a long argument short, my example of an external negation is true, my example of a negation of the predicate, however, is not true because it presents an unjustified existential import.

The hexagon in Fig. 1 may serve as a starting point to portray the subalternations between my examples.

The bottom sentences of Fig. 1 are true. Each of them contradicts to one of the top sentences, which, *eo ipso*, fail to be true. The sentences on the top right and at the bottom left, contradict each other in every respect. In the top right sentence an unjustified existential import and a category mistake are committed. The bottom left sentence discards the unjustified existential import and the category mistake. Analogous is the case with the bottom right sentence: it denies the unjustified existential import and the category mistake committed in the top left sentence.

I left the remaining contradiction between the middle right and the middle left sentence last, because it presents a disharmony compared to the other two contradictions of the hexagon. To begin with the problems involved, the middle right and the middle left sentences are a contradictory pair. Therefore they cannot be both false. But they both fail to be true for the following reasons: The middle right sentence fails to be true because of unjustified existential import. The middle left sentence appears syntactically to be an external negation of the middle right sentence. If classical logic is adequate to model the semantic nuances of this hexagon (this is a hypothesis, which we will have to drop in due course), then the middle left sentence is true, since it is an external negation of the middle right sentence, which is not true. But it cannot be true! The middle right sentence

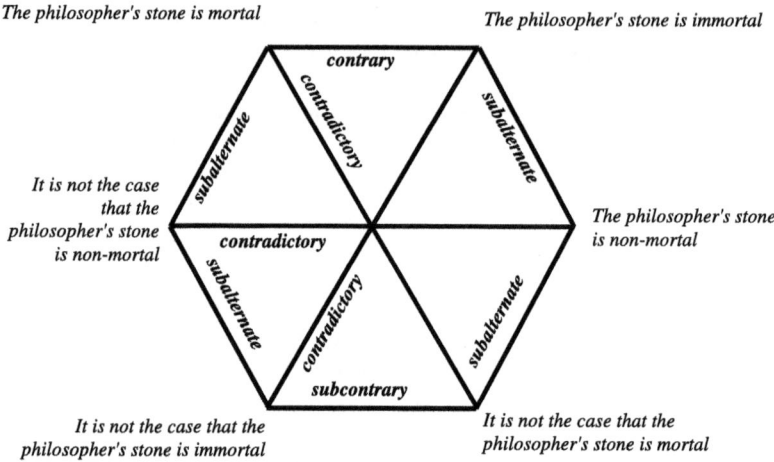

Fig. 1 An hexagon with three different kinds of negation

commits an unjustified existential import and no category mistake. It is not clear whether the middle left sentence says that the philosopher's stone is a non-existing non-mortal (in which case it is true) or a non-existing (maybe even an existing) mortal (in which cases it is not true). The relation between the middle right and the middle left sentence is not of the same kind as the contradictions between the two other contradictory pairs of the hexagon, where it was clear that the one sentence committed a category mistake, the other did not. The fact that the middle left sentence is susceptible to interpretations which fail to be true, although syntactically it contradicts a sentence which also fails to be true, is my main motivation to consider classical logic as inadequate for this context. This is also my reason to abandon the hexagon. An hexagon insinuates that all sentences connected by a diagonal, including the middle right and the middle left corners, contradict in the same way. But as we have just seen, the sentences of the middle right and the middle left corner are opposed to each other in a way different, than the way in which the other two contradictory pairs of the hexagon are opposed to each other. I would like to call the opposition between the middle right and the middle left corner: "relativization" (instead of contradiction). The reasons for this name are going to be clear once these sentences are assigned adequate values in a multivalued setting. Since the hexagon in Fig. 1 insinuates a misleading account of these "nuances", I suggest the good-old-square in Fig. 2 as a more fitting representation of the oppositions between my examples.

The top left and top right sentences of Fig. 2 are false for two reasons: they presuppose the existence of the philosopher's stone and they commit a category mistake. The middle right sentence is false for just one reason: unjustified existential import (but no category mistake). The middle left sentence is opposed to the middle right sentence in a way different than the way, in which the bottom right and the bottom left sentences are opposed to the top left and the top right sentences respectively.

I take it that a four-valued system is adequate to cope with the semantic "nuances" of my examples which edge the polygons of Fig. 1 and Fig. 2. In what follows, I will try to substantiate this claim.

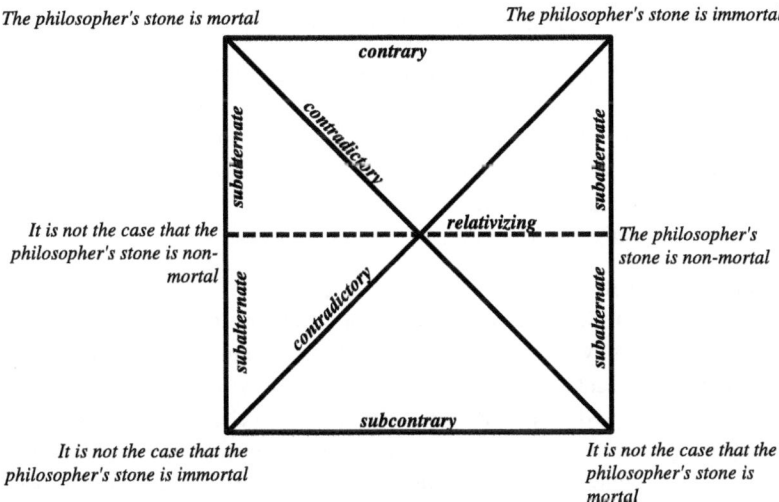

The philosopher's stone is mortal   *The philosopher's stone is immortal*

It is not the case that the philosopher's stone is non-mortal   *The philosopher's stone is non-mortal*

It is not the case that the philosopher's stone is immortal   *It is not the case that the philosopher's stone is mortal*

Fig. 2 A square with three different kinds of negation

3 What Is a Negation?

Privation is a kind of negation which is used in natural languages, but has never been used in a formal context. Drawing a semantics of privation is, for this reason, a task akin to redefining or reinterpreting negation in formal contexts. The following definition of negation is useful for my purposes.

Definition of Negation   A negation (comprising external negation, negation of the predicate and privation among others) is any monadic propositional function, such that it renders at least the maximum and the minimum truth values of a sentence which serves as the argument of the function, into other values.

Remark on the Definition of Negation   This definition implies that identity and the constant functions, which assign every sentence one single truth value, are not negations.

This captures, I think, the common element, which all sorts of negation, which are in use in logic, share. For, first, you need some truth-value change in a negation. This has to be at least the maximum and the minimum value. When a meteorologist predicts: 'It will rain tomorrow', then she might mean that this is true, or at least, that it is more true than false. To deny the accuracy of the meteorologist's prediction, is to say that, what she says is false or, at least, more false than true. If I take her prediction to be true granted that she means it to be true, but to be more false than true, granted that she means it to be more true than false, then, if anything, I cannot be said to deny the accuracy of her prediction. What I would do in such a case is maybe doubt the meteorologist's ability to maintain probabilities rather than necessities. But I would not *deny* the accuracy of the meteorologist's prediction *in every respect*. As one sees, a monadic propositional function, whose values deviates only from any intermediate value of the sentence which serves as its argument, but does not deviate from the maximum and the minimum values of this sentence, is *not* a negation.

In other words, for a monadic propositional function $f(p)$, if

- $v_i[p]$ is the intermediate value of the sentence p which serves as the argument of $f(p)$; and
- $v_x[f(p)]$ is the value of $f(p)$ if $v_i[p]$ is the value of p; and
- $v_y[f(p)]$ and $v_z[f(p)]$ are the values of $f(p)$ if p is assigned the values $v_{max}[p]$ and $v_{min}[p]$ respectively; and if finally
- $v_y[f(p)] = max$ and $v_z[f(p)] = min$

then $f(p)$ is *not* a negation, whatever the value $v_x[f(p)]$.

In order to have a negation, however, it is not necessary to have $v_y[f(p)] = min$ and $v_z[f(p)] = max$. Post's cyclical negation does not function in this way. Moreover, it is not necessary to change the intermediate values of the initial sentence. The strong negation in Blau's three valued logic does not. But, of course, one can change also some intermediate values along with the maximal and minimal values of the initial sentence to obtain a negation thereof. The symmetrical negation in Łukasiewicz's four-valued logic does exactly this.

Theorem *For any n-valued logic there are $n^{n-2} \times (n-1)^2 - (n-2)$ negations in the sense of the aforementioned Definition of Negation.*

Proof This number follows from the following considerations. The number of the rows of any truth-table which delivers the truth-functional definition of a monadic function in an n-valued logic is n. The number of all monadic functions of an n-valued logic is n^n. But since negations are only those monadic operations, in which the maximum and the minimum values of the initial sentence must be rendered into other values, in every possible truth table which delivers the truth-functional definition of a *negation in the sense of the aforementioned Definition of Negation*, there are two rows which cannot contain all possible values of the initial sentence. That is, there are n^{n-2} possible combinations of values in $n-2$ rows which are combined with, i.e. multiplied by $(n-1)^2$ possible combinations of values in 2 rows (the highest and the lowest). We have to subtract from this product the number of those monadic functions which are constant and give a value other than the maximum or the minimum. These are $n-2$. ◻

Remark on the Proof The two constant functions which give the maximum and the minimum values, are, of course, also not welcome as negations, but we have already excluded them by demanding the maximum and the minimum values $v_{max}[p]$ and $v_{min}[p]$ to result anything but $v_y[f(p)] = max$ and $v_z[f(p)] = min$ respectively. For the same reason we have already excluded identity from negations.

Corollary 1 *Since I am attempting to square the semantic "nuances" of my examples which edge the square in Fig. 2 by operating in a four-valued system, let me remark that in every four-valued logic there are 142 negations in the above sense.*

Among these kinds of negation, there are some awkward ones. One might doubt, for example, whether the function which turns true and false sentences to more-true-than-false and intermediate values to true, is a negation. I would say that it is. It is a considerably weak denial of maximal and minimal truth values.

4 The Semantics of the Logic of Privations

I attempt to model privation-talk in the logic of privations (LP), which I take to be a fragment of the Łukasiewicz-four-valued predicate logic; a fragment of an extension of the Łukasiewicz-four-valued propositional logic, that is, which is large enough to contain quantifiers, predicate symbols, individual variables and constants and indexical operators.[2]

LP is only a fragment of the Łukasiewicz-four-valued predicate logic because quantifiers and individual variables occur in LP only as parts of singular terms: definite descriptions or indexical/demonstrative expressions.

The reason for this proviso is the following: one basic idea of LP is that external negations are subalternate to privations. For example 'It is not the case that Stevie Wonder can see' is subalternate to 'Stevie Wonder is blind' (cf. the top left and the bottom left sentences of Fig. 2). Subalternations of this kind, however, *do not apply* in expressions like 'Some singers...', etc. Note, for example, that 'Some singers are blind' does *not* imply: 'It is not the case that some singers can see', since it is true that there are blind singers, it is not true however that all singers cannot see. In order to ensure, now, that external negations are subalternate to privations, by turning English contexts into LP-contexts, one has to restrict context to English sentences, whose grammatical subject is a proper name or, if it is any other (common) noun, it is governed by an indexical/demonstrative ('here', 'there', 'now', 'this', 'that', etc.) or by the definite article ('the').

LP contains the usual symbols for individual constants, individual variables and logical constants plus some symbols for alternative kinds of negation. Where we have to go into some detail, are the logical constants: $\rightarrow$, $\neg$, $-$, $\sim$, $\sqcap$.

The truth values are: 1 (true), 2 (more-true-than-false), 3 (more-false-than-true), 4 (false).

The arrow ($\rightarrow$) symbolizes implication with the meaning which this operation has in Łukasiewicz's four-valued propositional logic. The following table defines truth-functionally the propositional function $p \rightarrow q$:

$p \backslash q$	1	2	3	4
1	1	2	3	4
2	1	1	3	3
3	1	2	1	2
4	1	1	1	1

The other logical constants in question have to be understood as follows

- '$\neg$': weak negation;
- '$-$': presupposing negation (the formal counterpart of the negation of the predicate in English—note that it has been often held to be a privation in the history of philosophy);
- '$\sim$': symmetric negation (Łukasiewicz's four-valued negation);
- '$\sqcap$': privation.

---

[2]Gottwald [8, pp. 55–57] gives a brief exposition of the construction of the Łukasiewicz-four-valued *predicate* logic; he [8, pp. 45–47] also gives the axioms and the deduction rules of the Łukasiewicz-four-valued propositional logic. For a brief but full account of the semantics of the Łukasiewicz-four-valued propositional logic cf. [13, pp. 29–32].

The truth-functional definition of these kinds of negation is given by the following table:

p	$\neg p$	$-p$	$\sim p$	$\ulcorner p$
1	4	4	4	4
2	1	3	3	4
3	1	4	2	4
4	1	1	1	1

Now, let me assign the sentences of the square in Fig. 2 values between 1 and 4.

- Top right: v['The philosopher's stone is immortal'] $= 4$ (NB, a privation which is false for two reasons: unjustified existential import and category mistake);
- Top left: v['The philosopher's stone is mortal'] $= 4$ (an affirmative sentence which is false for the same reasons as its contrary—contrary sentences can be simultaneously false);
- Middle right: v['The philosopher's stone is non-mortal'] $= 3$ (NB, a presupposing negation, which is false for one reason: unjustified existential import);
- Middle left: v['It is not the case that the philosopher's stone is non-mortal'] $= 2$ (a *symmetric negation* (!) of the presupposing middle right negation—readers can very easily persuade themselves that, when a sentence has the value 3, as is here the case, then the symmetrical negation of this sentence gives the value 2);
- Bottom right: v['It is not the case that the philosopher's stone is mortal'] $= 1$ (a symmetric negation, forming a contradictory pair with the false top left affirmative sentence);
- Bottom left: v['It is not the case that the philosopher's stone is immortal'] $= 1$ (a *symmetric negation* of a privation, forming a contradictory pair with the false top right privation).

As one sees the external negation of English in the sentences which edge the square of Fig. 2 has to be formalized with a *symmetric*, *not* with a weak negation.

The peculiar name which I have chosen to express the opposition between the middle right and the middle left sentences ('relativizing' in Fig. 2 instead of 'contradictory' in Fig. 1) has to do with the truth-values 3 and 2 which I just assigned these sentences: the one says that the philosopher's stone is non-mortal—because it does not belong to the things which are either mortal or immortal. The other adds for consideration or, rather, relativizes, that it is not existing either, so that it is natural for it to fail being non-mortal.

As one sees, four values, some negations more and an understanding of subalternation as Łukasiewicz's four-valued implication are sufficient to solve the puzzle which arises out of a natural understanding of the middle-right and the middle-left sentences.

One last remark: In my outline of LP, I used only five negations (privation, presupposing negation, symmetric negation, symmetric negation of a presupposing negation, symmetric negation of a privation), although in the Corollary 1 spoke of 142 negations in a four-valued logic. Are the rest kinds irrelevant? The answer is no! The square could be edged by alternative kinds of negation, instead of those which I presented. And, what is much more, many additional kinds of negations and privations among them might be introduced. For example lacking magnificence, undoubtedly a privation, allows many intermediate grades of privation: lacking magnificence but not excellence; then lacking magnificence and excellence, but not goodness; moreover lacking magnificence, excellence, goodness, but still being mediocre; finally being bad. To be able to make a full account of the different sorts of truth, falsity and contradiction in a corresponding square, we might need to speak not of a four-valued logic, but in a more general manner of an n-valued logic. The number of the values would be dependent on the number of the negations and privations involved.

References

1. Alexander: In Priora Analytica. Wallies, M. (ed.). Reimer, Berlin (1883)
2. Ammonius: In Aristotelis De interpretatione commentarius. Busse, A. (ed.). Reimer, Berlin (1895)
3. Aristotle: De interpretatione. Cooke, H.P. (ed.). Harvard University Press, Cambridge (1938)
4. Aristotle: Metaphysica. Jaeger, W. (ed.). Clarendon, Oxford (1957)
5. Blau, U.: Die dreiwertige Logik der Sprache. de Gruyter, Berlin (1977)
6. Boethius: Commentarius in librum Aristotelis Peri hermeneias. Meiser, C. (ed.). Teubner, Leipzig (1877)
7. Gassendi, P.: Exercitationes paradoxicae adversus Aristoteleos. Amsterdam (1649)
8. Gottwald, S., et al.: Nichtklassische Logik. Eine Einführung. Akademie Verlag, Berlin (1988)
9. Hamilton, W.: Lectures on Metaphysics and Logic, vol. 2. Gould and Lincoln, Boston (1860)
10. Horn, L.R.: A Natural History of Negation. CSLI, Palo Alto (2001)
11. Italos, J.: Quaestiones quodlibetales. Joannou, P. (ed.). Studia Patristica et Byzantina, vol. 4. Buch-Kunstverlag, Ettal (1956)
12. Kant, I.: Kritik der reinen Vernunft. Suhrkamp, Frankfurt a. M. (1974)
13. Sinowjew, A.A.: Über mehrwertige Logik. Ein Abriss. VEB, Berlin (1968)
14. Valla, L.: Retractatio totius dialectice cum fundamentis universe philosophie (=Repastinatio dialectice et philosophie, vol. 1). Zippel, G. (ed.). Antenore, Padua (1982)
15. Wolff, Ch.: Philosophia rationalis sive logica, 2nd edn. Frankfurt a. M. and Leipzig (1732)
16. Wolfson, H.A.: Infinite and privative judgments in Aristotle, Averroes and Kant. Philos. Phenomenol. Res. **8**, 173–187 (1947)

S. Gerogiorgakis (✉)
University of Erfurt, Hauspostfach 64, PSF 900221, 99105 Erfurt, Germany
e-mail: stamatios.gerogiorgakis@uni-erfurt.de

Fuzzy Syllogisms, Numerical Square, Triangle of Contraries, Inter-bivalence

Ferdinando Cavaliere

Abstract A new Predicate Calculus, called "Distinctive" D, is presented in the form of the Hexagon of Opposition, in which the Particular Yba (= only some b are a) is the contradictory of the Universal Uba (= all or no b are a). Y is preferred, as a primitive, to I or O, because it is more "natural" than the others. Typical inferences of the systems are the obversions: Yba = Yba' (= only some b are not a), Uba = Uba' (= all or no b are not a).

D-Systems include traditional Syllogisms. Polygonal and Numerical developments are being developed, including intermediate quantifiers ("the majority of", …). Polygons are finally absorbed into the Numerical D-Square NDS. Isomorphisms are discovered between bivalent D-systems and some Non-Standard Logics by depriving the subject-class of the quantifier and transferring its (pre)numerical attribute to the truth value of the judgement. These '(poly-)inter-bivalent' logics admit intermediate values between true and false. In that way we get Fuzzy Logic, and present the Fuzzy Square.

Keywords Predicate logic · Quantifiers · Logical diagram · Square of opposition · Hexagon of opposition · Numerical syllogisms · Non-standard logic · Fuzzy syllogisms · Synonyms

Mathematics Subject Classification Primary 03A05 · 03B20 · 03B52 · Secondary 03B65 · 03B60 · 03B53 · 05C99

1 Introduction

A new, as yet unpublished Predicate Calculus, called "Distinctive" or "D", is presented here in the forms of the Hexagon of Opposition and other geometrical shapes. Its originality lies mainly with: (a) the use of quantifiers that are intermediate between the classical "all" and "none", (b) the semantic relationships with set-diagrams, (c) its non-standard interpretations. The aim is to develop a logic close to natural language and geometric intuition, and applicable to the technologies of knowledge organisation. The logical presentation is not the usual formalised description, but is suggested by meta-logical considerations.

Section 2 improves the Square of Opposition with new pre-numerical quantifiers and shows the new calculus. Section 3 enlarges the systems to numerical predications, thus bringing us into Sect. 4, where the above systems are interpreted in the light of non-standard logics, such as fuzzy and paraconsistent ones. Section 5 sketches possible future developments.

J.-Y. Béziau, D. Jacquette (eds.), *Around and Beyond the Square of Opposition*, 241–260
Studies in Universal Logic, DOI 10.1007/978-3-0348-0379-3_17, © Springer Basel 2012

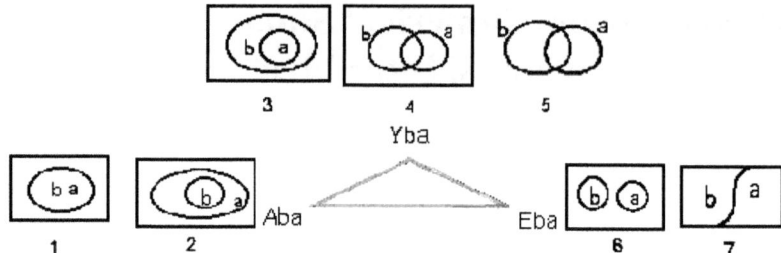

Fig. 1 Diagrammatic interpretation of the distinctive triangle

2 Distinctive Pre-numerical Calculus

The deductive systems are presented from the simplest to the most complex—the complexity depending on what kind of quantifier or predicate is used each time. From this point of view we start our analysis with the "pre-numerical" quantifier, considered as "generic" in traditional categorical logic.

2.1 Distinctive Simple Calculi

The D Simple Calculi consider reasoning with premises and conclusions consisting of single predicates, while the D Compound Calculi work with double predicates as premises or conclusions.

2.1.1 Categorical Triangular Calculus D3

Among the Distinctive Predicate Calculi presented here, the most primitive are the 'Triangular' ones based on the 'triangle' of contraries, which presents the following three categorical propositions: universal affirmative, universal negative, and, rather than the two Particulars of the classic *oppositive square*, their logical conjunction, which is called *Particular Distinctive* (see Fig. 1). The quantifier in the latter can be interpreted in natural language with expressions like "only some" in the sense of *excluding* universality, "at least one but not all", "all the … except some", "neither all, nor none", "only a part (among all the …)", etc.

Given an ordered pair of *classes*—a "b" subject class and an "a" predicate class—we may symbolise the three categoricals mentioned above in the following way: **Aba** (= every b is a); **Eba** (= no b is a); **Yba** (= only some b is a). The terminological choice of *Distinctive* (in contrast to *Affirmative* and *Negative*) is motivated by the need to place, next to judgements compatible with the totality of a whole, those judgements that produce a "distinction" (partition) between parts of a whole, affirming some of these parts in the negation of the rest.[1]

---

[1]For the purposes of this study we will utilise the terms "class" and "set" as synonyms.

In Fig. 1, next to the formulas marking the vertices of the triangle, one finds the seven corresponding Venn-diagrams for all possible relations between (contextually defined) pairs of sets that make the associated formula true. All sets in question are distinct from the *Universe of discourse* UD and from the *null set*, these restrictions, it seems, being presupposed in the Aristotelian-Boethian tradition as described in Seuren [20, 21]. In these diagrams, the boundary rectangle represents the UD, except in case 5, where the union of b and a covers the entire UD.[2] The seven diagrams are *exhaustive* of the relations between two pairs of complementary sets that are all *different from the Universal class and the Empty class* (see [2, chapter 3, section 27]). The *triangle of opposition* results in a partition of these seven cases, a circumstance that does not occur in the opposi- tive square. We can therefore propose the Axiom (**Aba**) $\underline{V}$ (**Eba**) $\underline{V}$ (**Yba**), where $\underline{V} =$ the exclusive *or*. Laws of Immediate inference: $a = a''$ (double negative); **Aba = Aa'b'** (contraposition); **Eba = Eab** (conversion); **Aba = Eba'** (obversion); **Eba = Aba'** (obversion); **Yba = Yba'** (obversion). The latter may also be proven within traditional Syllogistic (S) as follows: Yba = def: Iba * Oba: Iba * Oba = (Eba)' * (Aba)' = (Aba')' * (Eba')' = Oba' * Iba' = def: Yba'. For example: *Only some* scientists are logical if and only if only some scientists are not logical.

The obversion of the Particular Distinctive, with affirmative copula or predicates, re- sults in, without any change in the quality of the quantifier (as instead occurs for the four traditional ones) an equivalent Particular Distinctive, with negative copula or predicate.

According to these three categorical propositions, it is possible to construct a calculus that we will call *triangular* or D3, based on the classical syllogistic model. Among the 108 possible moods,[3] the 6 valid ones will be:

  (first) AAA (Barbara), EAE (Celarent);
 (second) EAE (Cesare), AEE (Camestres);
  (third) YAY (HydraLynx, new mood);
 (fourth) AEE (Camenes).

The YAY in the third figure may be taken as an axiom, but here we intend to demonstrate how it can be derived from traditional syllogistic.

Thesis (Yba * Abc) → Yca

  1. (Iba * Abc) → (Ica)   (Disamis)
  2. (Oba * Abc) → (Oca)   (Bocardo)
  (1. * 2.) → Ica * Oca   (Propositional Logic)
  (1. * 2.) → Yca   (Definition Yca = Ica * Oca)
  [(Iba * Abc) * (Oba * Abc)] → Yca
      1.   2.
  [(Iba * Oba) * Abc] → Yca   (Distributive property of the logical product)
  [Yba * Abc] → Yca   (Definition Yba)   □

---

[2]Examples: case 4: UD = quadrilaterals, b = rhombuses, a = rectangles; case 5: UD = polygons, b = polygons with fewer than five sides, a = polygons with more than three sides.

[3]This number result from the combination of the three quantifiers with the three predicates in the four figures.

Fig. 2 Exclusive square

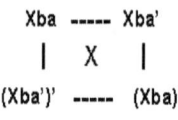

Fig. 3 Exclusive triangle

For example:

(Only some syllogisms are Aristotelian) * (Every syllogism is reasoning)

→ (*Only some* reasoning is not Aristotelian)

(Ysa * Asr → Yra′ substituting Yra with Yra′).

2.1.2 Exclusive Triangular Calculus D3x

In Scholastic Logic, "only b is/are a" (Tantum b est/sunt a) type predications were called *Exclusivae*. In the light of modern syllogistic interpretations they are = Ab′a′, or in our symbolism = Xba, standing for: "only b's are a's" or "no a is not b" (while "there are b's that are not a" is left undefined). Xba′ stands for: "only b is not a".

Therefore, Xba = Ab′a′ = Aab = Xa′b′, while Xba′ = Aa′b = Ab′a = Xab′. Oppositive squares were derived even from Exclusive propositions (Fig. 2). As for the particulars, we can once again generate an oppositive triangle (Fig. 3): Xba = only b is a; Wba = every complement of b is a = every not b ("b ad infinitum") is a; Jba = only some complement of b is a (only some not b is a).

For these predications, only rules of inference and axioms that are entirely analogous to those of the categoricals will be applied. It therefore gives rise to a distinctive calculus with the due replacements of the negative terms as required. We will see later its usefulness.

2.1.3 Hexagonal Calculus D6

At this point, triangular Categorical Propositions will be integrated with traditional Categorical Propositions enriched with negative terms into a Hexagonal Calculus D6.

From the axiom (Aba ⌄ Eba ⌄ Yba), we may derive the contradictions, or the negations, of the basic categorical propositions. Therefore, (Aba)′ ↔ (Yba ⌄ Eba); (Eba)′ ↔ (Aba ⌄ Yba); (Yba)′ ↔ (Aba ⌄ Eba). We recognise the traditional Iba = def. (Aba ⌄ Yba) and Oba = def. (Eba ⌄ Yba), while a new categorical proposition appears that we shall call Universal "*Distinctive*", which expresses the negation of the Particular Distinctive and is symbolised as **Uba** = def. Aba ⌄ Eba (or: "every or no b is a") from the modified formula: "Adf**I**rmo, n**Eg**O, in h**Y**bridis disting**U**o".

The obversion is also valid for the U-proposition: Uba = Uba′. Since Eba = Eab, (Eba)′ = (Eab)′; furthermore (Eba)′ = (Aba ⌄ Yba), as (Eab)′ = (Aab ⌄ Yab), therefore (Aab ⌄ Yab) = (Aba ⌄ Yba), i.e. Iab = Iba. In the resulting oppositive hexagons

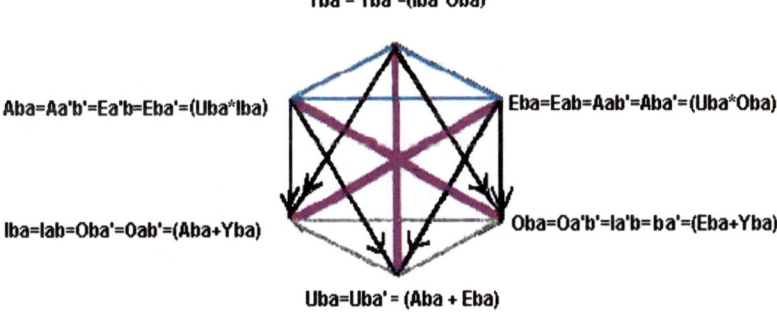

Fig. 4 Distinctive hexagon

Table 1 Distinctive basic propositions

$Aba = Aa'b' = Ea'b = Eba'$	contradictory of	$Oba = Oa'b' = Iba' = Ia'b$
$Eba = Eab = Aab' = Aba'$	,,	$Iba = Iab = Oba' = Oab'$
$Yba = Yba'$	,,	$Uba = Uba'$
$Ab'a' = Aab = Eab' = Eb'a$	,,	$Ob'a' = Oab = Iab' = Ib'a$
$Eb'a' = Ea'b' = Aa'b = Ab'a$	,,	$Ib'a' = Ia'b' = Oa'b = Ob'a$
$Yb'a' = Yb'a$	,,	$Ub'a' = Ub'a$
$Yab = Yab'$	,,	$Uab = Uab'$
$Ya'b' = Ya'b$	,,	$Ua'b' = Ua'b$

(see Fig. 4), the *contradictory categorical propositions* are arranged symmetrically around the centre, while the *contraries* (or *sub-contraries*) are arranged symmetrically around the vertical axis.

An ideal horizontal axis separates the primitives, located in the upper part of the hexagon, from the derived categorical proposition below. The possible sets that represent the derived categorical propositions Iba, Oba, Uba, can be obtained by disjunction.

Thus, a general law of the Calculus is identified. This law states that *for each categorical proposition, the same law of immediate inference is valid for its contradictory categorical proposition*: contraposition for A-O propositions, conversion for E-I propositions, obversion for Y-U propositions.[4]

By adding the two cases of obversion to the rules of immediate inference at hand and then systematically extending all of the transformations of equivalence to an unordered pair of terms taken as both negative and positive, 16 basic propositions with different meanings are obtained, as indicated in Table 1.[5]

The table in Appendix summarises all the deductive schemas admissible in the hexagonal system. By carrying out the necessary transformations (inverting the order of the

---

[4]Blanché [3] and Sesmat [19] already presented oppositive hexagons, but we have found no evidence of a calculus ever having been developed. As from 1910, Vasiliev [25] developed a 'triangular' syllogistic based on the "only some" quantifier, but different from ours in that it is similar to paraconsistent systems. See Béziau [1], Suchon [24] and Moretti [13].

[5]In the last three lines there are expressions not included in the initial hexagon, but rather generated from another 3 hexagons constructed on the pairs $b'a'$, ab, $a'b'$ (the remaining 4 possible hexagons [or triangles] are equivalent to the others by obversion).

Fig. 5 Quasi-numerical
polygon

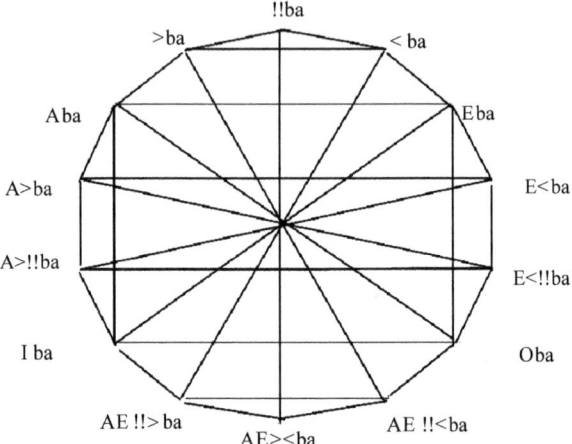

premises, conversions, etc.), the table encompasses all valid moods of the traditional syllogism (24 moods) and of the syllogism with negative terms, as well as adding new syllogistic forms. The new moods include: in the first figure, Uma*Acm → Uca (*UnaLux*), demonstrable by means of indirect reduction from the *HydraLynx* axiom; in the second figure, Uam*Ycm → Oca (*VultGyro*). Moreover, other deductive forms appear that are not strictly syllogistic, but are nonetheless valid (and derivable), such as: Uma*Ymc → Yca*Yc'a'.

The valid syllogisms that include a premise in Y are:

(first) AYI, EYO;
(second) AYO, EYO;
(third) AYI, YAY, YAI, YAO, EYO;
(fourth) YAI, EYO.

The logic of propositions allows us to derive the subordinate moods (in the third Figure) YAI and YAO (Y → I, Y → O) from *HydraLynx*.

The first of these, when the conclusion is converted and the premises switched, leads to AYI; the latter, by obversion of the conclusion and of the first premise, results in the EYO mood.[6]

2.1.4 Polygonal Distinctive Calculus—Quasi-numerical Calculus QD

The A-E side is broken by the Y-proposition or, if preferred, its hiatus is unambiguously closed. At this point, we can conceive further interruptions of the segment that are intermediary but not necessarily equidistant.

---

[6]If the first premise is converted, it is possible to reach EYO in the fourth Figure from EYO in the third Figure; this mood may result in YAI, still in the fourth Figure, by means of indirect reduction. From here, conversion of the conclusion and the exchanging of premises results in AYI in the first Figure, which by obversion of conclusion and the first premise generates EYO, which, by conversion of the first premise, generates EYO in the second Figure. AYO in the second Figure is generated from AYI in the first Figure as follows: contraposition of the first premise and obversion of second premise and conclusion.

Fig. 6 Quasi-numerical
square

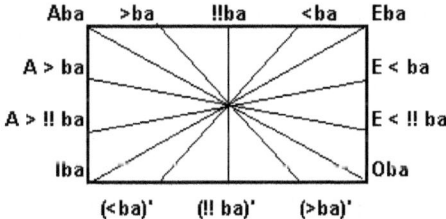

These new quantifiers, and corresponding predications, if chosen in such a way as to be incompatible and exhaustive among themselves, may give rise to new *Oppositive Polygons*, or bases for the construction of a *Polygonal Distinctive Calculus*. Figure 5 illustrates the oppositive relations among five quantifiers: the traditional A and E, and another three that are called Quasi-numerical, Qn, which we symbolise as >, !! , <, standing for "only a majority of", "the exact half of", "only a minority of", respectively.

Iba, is equivalent to A > !! < ba (every, or only a majority, or exact half, or only a minority of b is a) and Oba, to E < !! > ba (none or only a minority, or exact half, or only a majority of b is a). The expressions at the opposite vertices are reciprocally contradictory. The expressions that may be obtained through obversion with exchange of the quantifiers are arranged symmetrically around the central vertical axis. For example: >ba ↔ <ba′ as <ba ↔ >ba′, or A > !! ba ↔ E < !! ba′. !! is reciprocal to itself, like Y, which is equivalent to > !! <.

By simplifying we can compress the polygon into the classical Square, as in Fig. 6.

In order to develop a *Quasi-numerical Distinctive Calculus* QD, one must begin with Carnes and Peterson [4], who use the so-called 'intermediate' quantifiers, which we have positioned on the vertical sides of the Quadrilateral, while the Distinctive arrangement favours, as do primitives, the Qn positioned on the upper horizontal side.[7]

The *borderline* case reduces the figurations deal with up until now to simple segment (**collapsed polygon**): if b consists of only 1 element, it is either a or a′. This structure is analogous to that of the much discussed Singular Predications.[8]

The 5 Probabilistic modality operators may be considered analogous to Quasi-numerical ones: **c**ertain (or **t**rue), **p**robable, **e**qui-probable, im**p**robable, **i**mpossible (or **f**alse). Some possible disjunctions are: **verified** = (c + i), **uncertain** = (p + e + r), **possible** = (u + c).

2.2 Distinctive Compound Calculus

As seen with the three categorical propositions of the oppositive triangle, the seven topological situations in Fig. 1 have been partitioned. Each of these divisions requires a further predication in order to be distinguished from all the others.

---

[7] An example of QD may be: (Only a majority part of the animals is ill) * (no ill animal is prized) → (no, or only a minority part of animals is prized) (>ai * Eip) → (E < ap).

[8] In general, the validity of the 8 deductive formulas of modern Singular Syllogisms are confirmed, while the U-proposition schemes will be added. See [2, chapter 5, paragraph 45].

Table 2 Double triangle and 7 cases

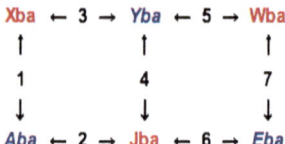

$$Xba \leftarrow 3 \rightarrow Yba \leftarrow 5 \rightarrow Wba$$
$$\uparrow \qquad\qquad \uparrow \qquad\qquad \uparrow$$
$$1 \qquad\qquad 4 \qquad\qquad 7$$
$$\downarrow \qquad\qquad \downarrow \qquad\qquad \downarrow$$
$$Aba \leftarrow 2 \rightarrow Jba \leftarrow 6 \rightarrow Eba$$

2.2.1 Distinctive Compound Calculus with Complementary D7c: Basic Expression, Immediate Inference, Mediated Deductive Rules

The additional information required to distinguish the seven cases may be provided by the Exclusive propositions.

In Table 2, we can see how the conjunction of pairs of predicates may do so by means of the two categorical and exclusive triangles (the blue one and the red one, with the latter being upside-down).

Table 3 presents equivalent forms in each line. In column II ('Scholastic'), the conjunctions of Categorical and Exclusive 'triangular' propositions of the same pair *ba* are listed.

These conjunctions identify the seven topological cases indicated in column I on a one-to-one basis. In column III (Explicit), each 'double' categorical proposition—or bicategorical proposition—may become a premise of a *Distinctive Compound Calculus* with *Complementaries*, D7c, involving three terms—one of which has middle function—as well as three corresponding complementary terms.

Column IV (Practice), in the light of calculation rules, simplifies the formula of each case by omitting the conjunction sign and the second pair of terms, indicated here by the double comma. We maintain the quantifier of the second pair, which is intended as the negated terms of the first.

An example of D7c reasoning: $(YscA,, * YctA,,) \rightarrow YstA,,$

Only some sound films are in colour and every mute film is in black and white;
Only some colour films are by Tornatore, and every film in black and white is not by him;
therefore: only some sound films are by Tornatore, and every mute film is not by him.

First of all, the *rules of immediate inference* must be established (with reference to column IV, 'Practice'):

- A..A inverts the sign of complementation of both terms. It is also immediately convertible.

Table 3 Equivalent formulations of the 7 cases

I Cases	II "Scholastic"	III Explicit	IV Practice	V Alternative	VI Synthetic
1	Aba * Xba	Aba * Ab'a'	AbaA,,	Aba * Eb'a	Abab'E
2	Aba * Jba	Aba * Yb'a'	AbaY,,	Aba * Yb'a	Abab'Y
3	Yba * Xba	Yba * Ab'a'	YbaA,,	Yba * Eb'a	Ybab'E
4	Yba * Jba	Yba * Yb'a'	YbaY,,	Yba * Yb'a	Ybab'Y
5	Yba * Wba	Yba' * Ab'a	Yba'A,,	Yba * Ab'a	Ybab'A
6	Eba * Jba	Aba' * Yb'a	Aba'Y,,	Eba * Yb'a	Ebab'Y
7	Eba * Wba	Aba' * Ab'a	Aba'A,,	Eba * Ab'a	Ebab'A

Table 4 Distinctive Compound deductions

	1 AbaA,,	2 AbaY,,	3 YbaA,,	4 YbaY,,	5 Yba′A,,	6 Aba′Y,,	7 Aba′A,,
1 AacA,,	AbcA,,	AbcY,,	YbcA,,	YbcY,,	Ybc′A,,	Abe′Y,,	Abc′A,,
2 AacY,,	AbcY,,	AbcY,,	Ibc	Ycb	Ybc′A,,	Ib′c	Ybc′A,,
3 YacA,,	YbcA,,	Ib′c′	YbcA,,	Yc′b	Ibc′	Abe′Y,,	Abc′Y,,
4 YacY,,	YbcY,,	Yb′c	Ybc		Ybc	Yb′c	YbcY,,
5 Yac′A,,	Ybc′A,,	Ib′c	Ybc′A,,	Ycb	Ibc	AbcY,,	AbcY,,
6 Aac′Y,,	Abc′Y,,	Abc′Y,,	Ibc′	Yc′b	YbcA,,	Ib′c′	YbcA,,
7 Aac′A,,	Abc′A,,	Abc′Y,,	Ybc′A,,	YbcY,,	YbcA,,	AbcY,,	AbcA,,

- A..Y or Y..A invert both the sign of complementation of each term and the succession of the quantifiers. The conversion of the terms must be accompanied by an analogous change in the position of the quantifiers (e.g.: YabA = AbaY = Λa′b′Y = Yb′a′A).
- Y..Y: the sign of complementation may be inverted, even of only one term. It is also immediately convertible (e.g.: YabY = YbaY = Yb′aY = Ya′b′Y …).

Table 4 illustrates the conclusions that may be extracted from 49 pairs of bi-premises with the exclusion of equivalent and subordinate moods.[9] The rules of practical deduction will be applied as follows. Commas will be ignored and eventually replaced upon conclusion of the calculation if necessary.

The *standard form* must be reached preliminarily: (1) the premises are arranged in sequence, as in the first figure of the traditional syllogism with inverted premises, (2) the middle term is directed to the same sign in both premises (always possible through the application of the laws of immediate inference). Therefore:

(A) When at least one of the premises has the form A..A, the non-middle term contained therein may substitute the middle term in the other premise, thereby producing the conclusion.

(B) If the premises have the form A..Y or Y..A, and are equally 'oriented' when in sequence (AY * AY, YA * YA), *the middle terms may be eliminated together with the nearby quantifiers*, leaving the conclusive sequence; if instead the 'orientation' is symmetrical (AY * YA, YA * AY), the result will be an *I-conclusion* with terms of *the same sign as in the premise* if the central quantifiers (near the middle terms) are A A, and of the opposite sign if the central quantifiers are Y Y.

(C) When an Y..Y type premise is combined with an A..Y or Y..A premise, the conclusion will be a Y-type whose subject will be the non-middle term that appears in the A..Y or Y..A premise and will have the same sign as in the premise if accompanied by a Y therein, and the opposite sign if accompanied by a A therein.

(D) Two Y..Y premises result in no conclusions as all seven relations between non-middle terms are possible in this case.

The rules may also be given an interpretation in terms of set theory, for example: (B) means: If the middle class is strictly included in the other two, then the intersection

---

[9]Conclusions are indicated by means of bi-categorical propositions, simple distinctive particulars, and traditional particulars, conjunctions of bi-categoricals being excluded for the purpose of conciseness.

between them is not empty; if the middle class includes the other two, the complementary of the latter will have a non-empty intersection. Distinctive Compound Calculi are more economical than their translation in equivalent traditional syllogisms with categorical propositions united by connectives.

2.2.2 Distinctive Compound Calculus with Subject "Ad Infinitum" D7s

Equivalent variations of D7c are possible. For example (see column V "Alternative" of Table 3), we can use two categorical propositions, of which the *second* presents only the subject of the first as negative, maintaining the predicate unchanged, also using the universal negative quantifier; in a 'synthetic' version (column VI), not to repeat the predicate "a", the second categorical proposition is read from right to left.

According to this system of notation, we will obtain a schema equivalent to the previous one in the following form: $(Y \, scs' \, E * Y \, ctc' \, E) \rightarrow Y \, sts' \, E$

> Only some sound films are in colour and no mute film is;
> only some colour films are by Tornatore, and no black and white is;
> therefore, only some sound films are by Tornatore, and no mute is.

2.3 Tri-hexagonal Extensions: Iconic Relational Calculus R7

Now Table 3 will be presented following its development with further interpretations (Table 5).

If the predicative structure of the Distinctive Compound Calculus places it within the framework of a natural and traditional language, the translation of column IV, through the equivalences of column VII, to column VIII makes this system belong to *Modern Symbolic Logic*, even if of a limited kind, in that it is *bi-argumentive* and *special* due to its type of *symbology* and the reference to seven cases.

In column VII there are seven cases described in a way 'equivalent' to those 'practice' column expressions already seen.

Now a relational notation system may be adopted (column VIII), according to the following translation code (the dots stand for the variables): A..A becomes ⊖; A..Y becomes)); Y..Y becomes)( )(.

Column IX calls attention to its *diagrammatic iconicity* and is determined by means of the *immediate inference* rules of the previous column: the simple rule of *specular* rotation of the *parenthesis* or other grapheme in conjunction with the inversion of the quality of the nearby term [e.g. b))a = b) (a′ = b′((a′ = b′( )a]. The "⊖" relation consists of two parts (upper and lower) each of which can refer to a term: if only one part rotates around its extremity, it results in a $\cup^{\cap}$ or $^{\cap}\cup$ relation, if they both rotate the "⊖" is restored. Instead, the rotation of)( )(is not affected by free changes in the quality of terms.

In column IX, a miniature scheme of the topological cases used as reference for each ordered pair of sets "b a" is graphically created. In column X, the seven relations are denoted by means of transitive verbs.

Table 5 Iconic and linguistic interpretations

Cases	VII	VIII Trirelational	IX Iconic	X Relations	XI Synonymity Antonymity
1	AbaA,,	b⊖a	b⊖a	Occupies	synonym
2	AbaY,,	b))a	b))a	Diminishes	hyponym
3	Ab'a'Y,,	b'))a'	b((a	Clothes	hyperonym
4	YbaY,,	b)( )(a	b)( )(a	EXtra-miXes	meronym
5	Ab'aY,,	b'))a	b( )a	Hyperintegrates	hypercomplement
6	Aba'Y,,	b))a'	b) (a	Keeps-out	hypocomplement
7	Aba'A,,	b⊖a'	b∪∩a	Stops	complement

3 Distinctive Numerical Calculus

As seen, the A-E side of the traditional square is broken by the Y and Qn-proposition. Numerical quantifiers and corresponding predications can make us conceive further interruptions of the segment, and the quantifier considered as 'generic' in traditional theories of the categoricals is given a precise numerical value.

We define a *Numerical Distinctive Proposition* as a relation of equivalence between two non-empty classes, a subject-class s and a predicate-class p, in a *UD* that is distinct from each of the two classes. This relation is defined by means of *numbers* (normally natural numbers) that make up the *Numerical Distinctive Quantifiers (NDq)* associated with the functors: $\geq$ (at least), $\leq$ (at most), $>$ (more than), $<$ (less than), $\neq$ (either more or less, or not equal), "$=$" (exactly or only); the latter is implied.

3.1 Distinctive Simple Numerical Calculus ND and Distinctive Numerical Square NDs

By definition, $\geq n$ is equal to $>n-1$, as $\leq n$ is to $<n+1$. The NDq calls for a reference to three numbers for each term: the first (Numerator N) indicates the *quantity* or cardinality (of the subset of the subject) involved in the predication, the second indicates the quantity within which (total or Maximum M) the first is distinguished, and the third, *from which* (Remainder R), still the first, is distinguished; given that the remainder can be defined by the difference between total and numerator, it is usually implied. In practice, often when we speak of 'quantifiers' or 'quantified', we intend to refer only to the numerator, e.g. "4 s are p" $= s^4 p$. All the types of traditional quantifiers may, in the final analysis, be reducible to the generalised numerical form of $sM^N p$. As is easily seen, we obtain Asp when $M = N$, Esp when $N = \emptyset$, Ysp when $(M \neq N) * (N \neq \emptyset)$, Isp when $N \neq \emptyset$, Osp when $M \neq N$, Usp when $(M = N$ aut $N = \emptyset)$. Even the QD may be reduced to the numerical schemes, with numerical variables subject to conditions: when $N > \frac{1}{2}M$ we may affirm that $>sp$, when $N = \frac{1}{2}M$, then !!sp, when $N < \frac{1}{2}M$, $<sp$. On this basis we can construct a *Simple Numerical Distinctive Calculus ND*: $[(b_6^4 r) * (r_9^0 c)] \rightarrow (b_6^{\leq 2} c)$ (of all our **6 b**icycles, only **4** are **r**usty) and (**none** of our **9 r**usty objects is **c**ostly), therefore:

Fig. 7 Numerical Distinctive square

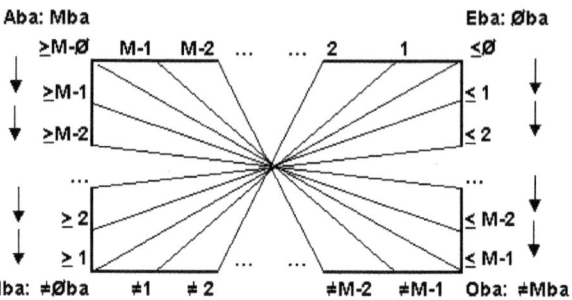

(at most 2 of our 6 bicycles are costly). As for the Qn propositions, we can use Oppositive Polygons even for the Numerical Distinctive Propositions; however, in order to *integrate* the consequential ND with other more Classic Numerical Syllogisms NS, we modify the Oppositive Polygon to adapt it to a Numerical Distinctive square NDs. We thereby obtain a representation of the intermediate quantifiers in an *orderly scale* between the two ends of segment Aba-Eba Each quantifier is diametrically opposed to its contradictory (Fig. 7: the dots leave the expansion to other intermediate quantifiers open).

The quantifiers may be interpreted in an *Exceptive* sense: "all except (or all but) Ø (zero), except 1, 2 and so on" to "all except all", i.e. none.[10] Each quantifier on the upper side (A-E) is arranged symmetrically to the median vertical line in correspondence to its own reciprocal; this may be defined by the equivalence $sM^N p = sM^{M-N} p'$. If N is equal to M (or to Ø), the equivalence above would become $sM^M p = sM^\emptyset p'$ (or with numeral M and Ø switched).[11] The interval between M − 1 and 1 illustrates the numerators which, when disjointed, compose Yba, while the complementary—or *external*—interval indicates Uba. Other types of internal or external intervals may be obtained from the schemes. Therefore, $s_6^{\geq 3 \leq 5} p$, will be read: *Among the s's, which number* 6 *in all, between* 3 *and* 5 *are among the p's, which number at least* 7 *in all*, while $s_6^{\leq 3 \geq 5} p$ will mean: *Among the s's, which number* 6 *in all, either at most* 3 *or at least* 5 *are among the p's.*

The non-Distinctive 'Numerical Syllogisms' by Hacker and Parry [8], Murphree [14] and others, ignore the upper side of the NDs. However, their deductive apparatus appears to be sound and founded on algebraic developments and diagrammatical tests (see Pfeifer [17], Pratt-Hartmann [18]). The deductive rules of these systems must be referred to in order to construct more economical NDs that maintain immediate intuition and the *principles of discrete mathematics* (inclusion-exclusion, Da Silva/Sylvester's formulas ...). Below two examples are given of NS in the terms of the above-mentioned authors, using our symbolism to give an idea of a potential fusion.

$$(m^{>60}p) * (m^{\leq 10}s') \rightarrow (s^{>50}p) \quad \text{(numerical Disamis)}$$

$$(m^{>16}p') * (m^{\leq 10}s') \rightarrow (s^{>6}p') \quad \text{(numerical Bocardo)}$$

---

[10] A long tradition, from the scholastic "excepta" to J.H. Lambert to A. De Morgan to the numerical propositions of Murphree [14], had made use of the numerical exceptive, which can be compared to the set operation of asymmetric difference: s\p (or s − p) = the s's that are not p (= the complement of p *in s*).

[11] If s has an even number of elements, a numerator will be placed exactly on the median axis and will be its own reciprocal.

As we know, the *HydraLynx* mode, typical of the Triangular Calculus, may be derived from the classic moods Disamis and Bocardo. Therefore, the examples given allow an analogous mood to arise in Numerical Distinctive Calculus:

$$\left(^{>16}m^{>60}p\right) * \left(m^{\le 10}s'\right) \rightarrow \left(^{>6}s^{>50}p\right) \quad \text{(numerical HydraLynx)}$$

3.2 Distinctive Numerical Compound Calculus DNC

The 'distinctive spirit' naturally leads to a 'numerical quantification of the predicate' and, in order to have full information, to a numerical quantification of the complement of the subject, or of the intersection of the complements of the subject and the predicate (bi-proposition). For example, $^{(2)}s8^6 9p$ means: "While exactly (2) remainders are not (are non-p) of 8s totals, exactly 6 are among the 9p totals".

The general formula of a complete numerical calculus proposition (also with a subject "ad infinitum") will be: $s_{...M}{}^{...Q...R}{}_{...G}p^{...T...L}{}_{...v}s'$ (the dots count as occurrences of the functors $\le$ or $\ge$) or, by ignoring functors, intervals and derivable data: s M^R G p Z s'. According to these bases we obtain a *Distinctive Numerical Compound Calculus DNC*.

Regarding immediate inferences, from the summed total of the subject and that of its complement (but the same can be said for the predicate), we obtain the total of the UD. The difference between total and numerator indicate the power of the intersection between the subject and the complement of the predicate. From the total of the subject and predicate, minus the intersection, in order to avoid counting twice, we obtain the total of the union set. The UD minus this union equals the total of the complement of the union. If zeros appear, some sectors will be absent.

One possible numerical development of the iconic relational system may lead to a simplification. In fact, we may attribute a number to each of the sectors separated by the parentheses or other icons.[12]

4 Non-standard Logical Interpretations

At the basis of standard logic, the Principle of Non-Contradiction or Pnc may be expressed faithfully in singular propositions and syllogisms where a specific attribute may be entirely attributed or entirely negated to a subject that is 'non-quantified' because it is unitary. Aristotle emphasised that "the same attribute cannot simultaneously belong and not belong to the same subject from the same point of view" (or aspect or interpretation). Let us then ask: What happens when we change point of view?

---

[12]For example:

b8⊖a 5	⇒	b's and a's are equivalent and 8 at all, the remaining are 5
b8)3)a 6	⇒	b's are 8, a's are 11, the not-b's not-a's are 6
6 b(3(8a	⇒	the not-b's not-a's are 6, the b's are 11, the a's are 8
b5)(3)(7a 9	⇒	the b's are 8, the a's are 10, not-a's not-b's are 9 [or b(5(3)7)a 9]
b 5(3)7 a	⇒	the b's are 8, the a's are 10, the not-a's are 5, the not-b's are 7
b8) 2 (11a	⇒	the b's are 8, the a's are 11, 2 are neither a nor b
b4∪∩3a	⇒	the b's are 4, the a's are 3.

4.1 The Transposition of Quantifiers to Truth Values: Inter-bivalence, Super-bivalence

Multiple points of view are introduced with the quantification of the subject, to meet the need to distinguish partitions or subsets within the subject-class, which is no longer seen as a monolithic entity. In the Particular Distinctive proposition, the *subject-class* (logic level 2), which is formally unitary, essentially consists of two subclasses (logic level 1) that are disjoint and exhaustive in relation to the mother-class, for which opposite *propositions* are defined that are, however, in conformity with their *elements* (logic level 0). In classical logic Pnc is associated with bivalence. "Are Europeans Greek?" As a rule, or generally speaking, they are not. But if what is true for one part is not true for the other part of the subject (level 1), how we can maintain that for the *whole* subject (level 2), only one of the two truth-values is true? A strict bivalence seems *inadequate* because it is asymmetrical. We know the properties of a class may be different to those of its own elements: why can this not be true for the kind of truth-values as well?

We can nullify this inadequacy by *weakening the Principle of Contradiction to level 2*, or in other words, partially violating it by means of a *flexible bivalence*. There will then be three possible and alternative answers to the question "Are x's y's?": **T** (True) = yes, it is True; **F** = no, it is False; **Γ** = so-so, **partially** = it is Limited (or Partial, Tolerant, Reductive, Moderate, Relative). If a proposition is partially true, it is also partially false, therefore partially not false and partially not true. The tripartition *each/none/only some* then shifts from the quantifier to the truth value, or even to the copula *is/is not/is partially*, if preferred.

It should be noted that the 'third' value option is nothing but a hybrid consisting exclusively of 'portions' of the other two, and therefore does not result as being 'alien' or 'heterogeneous', as required by an authentic third value, extra-bivalent (lacking in meaning, not well-formed, conventional, non-contextual, extra-categorical, inaccurate). This valence instead refers to all those cases ideally positioned in the interval between the two extremes (excluded) of "Every" and of "None". For this reason we have denoted it as *Inter-bivalence*, and *not Trivalence*.

Therefore, it is necessary to make a *conceptual distinction* between the two formulations that have always been believed to be equals: on the one hand the Aristotelian principle of the Excluded Middle ($\mu\varepsilon\tau\alpha\xi\upsilon$), and, on the other, its scholastic Latin version of "tertium non datur" (Excluded Third).

Abandoning the classic bivalence to make room for the Inter-bivalence, is due *not* to modal, temporal, metaphysical, or physical/probabilistic needs (free will, indeterminism of the future), as occurred historically for Aristotle, Lukasiewicz, and others, but to *multiple perspectives* and a *stratified way of observing* the *semantic-topological relations* between sets, subsets, and elements.

There is also a stronger way of remaining within the bivalence by violating the Pnc: that of Paraconsistent Logics or Dialectics, according to which something can be both true and false (**Ŧ**). We maintain that such Logics, which proponents themselves would save explicitly from trivial inconsistency, are in some way related to Inter-bivalence. In this case, the logic levels themselves (0, 1, 2, ...) will be translated in terms of valence, not the quantifier, with a 'metalogical' lateral leap. E.g.: all machines are useful (level 1A: if well used); all machines are harmful (level 1B: if badly used); thus all machines are both useful and harmful (level 2: without considering their use). Only some levels confirm them as being useful. It is true and false that machines are useful.

We can define this value (Ŧ) as a *Super-bivalence*.[13] The 'Super-bivalence' may interpret contradictory sets (and resolve the liar paradox) in that they are unfolded in overlapping dimensions or perspectives (spatial-temporal, topological, polysemy, metalinguistic, structural, etc.) with *synthetic judgements oscillating between the various planes*.

We can conclude that isomorphisms between the *bivalent D-systems* and many *Non-Standard Logics* may be identified in the following way: by depriving the subject class of its quantifier (or of its logic level) and transposing its (pre)numerical attribute to the truth value of the judgement. Their typical law is obversion: $Yba = Yba'$; $\Gamma ba = \Gamma ba'$; $Ŧba = Ŧba'$.

4.2 Scalar or Graduated (Discrete) Numerical Inter-bivalent Calculus

The *"natural" Numerical Calculus* discussed previously is translated here into a kind of *Poly-Inter-bivalence*. Or, to use a metaphor, if the bivalence uses black and white and the Inter-bivalence adds a grey tone, the Scalar logic increases the intermediate tones according to discrete unities organised in a scale, thereby ensuring a more precise judgement.

It is a first approximation of the *fuzzy propositions*, between term classes from confines that are still crisp (rough), where the belonging or not-belonging of any element to the given set is still clearly defined. Thus we assign a precise measure of truth of a subject-class predication towards a predicate-class. In reference to the numerical square, we can make a truth value correspond to a point or to an interval defined by points on the segment between the universal quantifiers.[14]

One of the results of this logic is that it can solve the *sorites paradox* (or heap).

The Inter-bivalence must be *distinguished from the probability calculus*, even if it uses an analogous mathematical method.[15]

Attention must be paid, however, to reductive interpretations: e.g. the NDq in s10[5], if interpreted as 'half of s's' (because arithmetically $\frac{1}{2} = 5/10$), leads to the loss of the absolute data of s, leaving us with only some relations that may be important in some contexts, but misleading in others.

4.3 Fuzzy (Continuous) Inter-bivalent Syllogisms

With the use of quantifiers with rational or real numbers, the infinite range of shades of grey is reached without an interruption.

---

[13] Many thinking systems adopt it, e.g: Imaginary logic of N. Vasiliev [25], Non-Monotonic Logics, Psychoanalysis (contrariety conscious-unconscious, bi-logic of I. Matte-Blanco) and, in general, the non-reductionist descriptions of reality, from biology to historical and sociological processes.

[14] Naturally, the measure of truth is relative to the totality of the subject-class (usable in various ways: absolute commensurable, correlative-statistical, relative-probabilistic, fractionary-percentages, etc.). Here, as in the following systems, all the deductive rules and laws for numerical calculus can be found.

[15] The 50 % uncertainty that the refrigerator contains an apple is everything but the certainty that there is half an apple (Kosko [9]).

Fig. 8 Fuzzy square

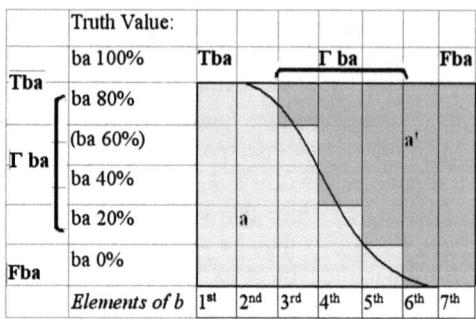

In order to complete this approach to fuzzy logic, there is still one last step: the attribution of the measure (or truth) of its membership to a class (thereby transforming it into a *fuzzy class*) as well as to each element (level 0) of the class itself and not only to the relation.

The fuzziness values can now be represented by means of a histogram in the form of a *Fuzzy Square* (FS), as in Fig. 8.

The gradability or scalarity thus turns from discrete into continuous or variegated ("fuzzy", or French "flou").

When we say "The herd is partially wet" (Γhw) we may intend to say that some sheep are and other sheep are not wet, or that each sheep is partially wet, or even a mixture of these readings. The model of FS shows at each level (elements, sets, predicative relations) how propositions such as the one just given are 'disambiguated' or made numerically specific.

It will then be possible to develop a *Fuzzy Simple and Compound Calculus* or *Fuzzy Syllogisms*.[16]

5 Potential for Development

In this section we draw some sketches and suggestions for developments that are premature at present.

5.1 The Oppositive Triangle in Propositional Logic

For the isomorphism that exists between Class Logic and Propositional Logic, we can export the tri-hexagonal scheme of distinctive calculus to the latter.

---

[16]Zadeh [26] proposed the term "Fuzzy syllogism" as a reasoning pattern based on vague quantifiers, like "most". Since then the term assumed different meanings in scientific literature. See for example the work of Dubois and Prade [6], Naddeo et al. [16] or, more recently, Kumova and Çakir [10]. The latter use the term for representing the truth degree of the traditional 256 moods. Their "fuzzy syllogistic system" is an original approach, whose objective is to model automated reasoning based on inductive knowledge. On the contrary, we use the same term to highlight that new quantifications (and new deductive moods) derivable from classical syllogistics are interpretable as fuzzy reasoning, without probability or statistical bases.

By beginning with the structural similarities between the Universal Aba and the implication b → a, or between Aba′ and b → a′, we may obtain a new *distinctive implication* or *semi-implication* by analogy with the Particular Distinctive Proposition, which can be expressed as b % → a and interpreted as **b only partially** (or only sometimes) **implies a**, equivalent by obversion to b % → a′.

5.2 The Oppositive Hexagon (and Other Polygons) in Modal Logics

The tri-hexagonal structure is perhaps capable of simplifying modal systems. If, within a UD of 'considerable worlds or cases', the *possible* worlds or cases p constitute a set *like any other* (complementary to that of the *impossible*, p′), we can express the relation with a second set, for example that of the worlds marked by the *c-characteristic*, as follows:

□   Apc = every possible case is c = c is (Necessary) Ascertained or Evident or Agreed
■   Epc = no possible case is c = c is (Impossible) Excluded or Rejected
◆   Ypc = only some possible cases are c = c is (Contingent) Variable Ambivalent Uneven Frayed Fuzzy Discussible Controversial
◆   Opc = at least 1 possible case is c′ = c is (Unnecessary) Excludable Rejectable
◇   Ipc = at least 1 possible case is c = c is (*Possible*) Ascertainable Agreeable Likely
❏   Upc = either every or no possible case is c = c is (either Necessary or Impossible) Prejudged Predetermined Dogmatic Absolute Clean Crisp

5.3 Other Interpretations

The seven cases may offer a complete panorama of possible reciprocal relations between truth evaluations of predicates.

Posing that the class b is that of all true predicates, and class a that of all false ones, we recognise in case 7 the classical logic where every predicate is true or false but not both; the class of true predicates is identical to that of not-false ones, and that of false predicates is identical to that of not-true ones (bivalent systems). In case 6 we can also see predicates that are neither true nor false (trivalence or gap systems). On the contrary, in case 5 predicates appear that are both true and false (paraconsistent or glut systems). Case 4 presents all four types of predicates described above. The cases 1, 2, and 3 are interesting because the 'gap' among the possible predicates regards at least one between the classical true or false predications (or both)—situations that can give rise to new kinds of logic.[17]

---

[17]For example, in case 3, all possible predicates are: those that are both true and false, those that are true but not false, those that are neither true nor false, the gap being between those that are false but not true. A model of this logic may fit into the idea that all falsehood hides a truth, as in the Freudian 'lapsus'. A model for case 2 is given by a philosophy that considers all truth relative (that is, false at the same time) but also admits the existence of pure (not-true) false predicates. As regard case 1, a model may be a sophistic position where all propositions are either contradictory or meaningless.

5.4 Limitations and Perspectives

The limitations of the D systems derive from the latter being founded on the subject-predicate structure (term logic), which is not suitable for expressing complex, multi-argument functions and relational logic.[18]

Sommers [22, 23] created a system (called "*Old New Logic*") which began with the subject-predicate structure as a base of the natural and syllogistic language and underwent progressive expansion (studies by Englebretsen [7] and Numerical Term Logic by Murphree [15]). Summers' type systems seem to have a deductive and expressive power that rivals first-order predicate calculus.

According to Lindell [12], these alternatives may actually be connected (and applicable in information technology and AI).

In *Lambert's* studies [11], expressions like "*all b's that are c's are not a's*" (which may be expanded iteratively to other terms: "that are ... that are ..."), are similar to *Boolean functors*.

All this suggests that an extension of the argumental and relational variables of term logic is possible without having to give up natural language and intuition.

Apart from the above limitative considerations, it must be recognised that the bi-argumental structure of language (subject-predicate) has a relevant role in thought and culture. One can found potential models of D-logic in many areas that apply the (bi-argumental) mechanisms of translation of terms: library indices, dictionaries, glossaries, technical jargons, and 'ontologies' for semantic search engines.

Column XI, Table 5, offers an interpretation of the seven relations in terms of synonymies; the first three and the last one are already used in linguistics and in dictionaries of synonyms and antonyms; the remaining three were coined by us to underline the importance of the *Conceptual Background*, or *Lexicographics*, in the definition of voices that plays a role analogous to that of the Universe of Discourse in the defining of the classes.

In various disciplines (e.g. Semiotics), the Definitions may make use of the Essential, Alien, Accidental triad (their contradictories are Unessential, Compatible, Discriminant or Crucial).

New semiotic polygons or square can be constructed.

We think that D-systems can give a contribution to the organisation of knowledge in these areas (see Cavaliere [5]).

Appendix: Deductive Conclusion in D6

See Table 6. The headings of the columns and lines function respectively as I and II premises. Each proposition inside the cells represents the 'stricter' or 'stronger' conclusion of the syllogism (subordinate moods are implied).

---

[18]However, the analyses of the distinctive propositions reveal a relation with four-five arguments (and even more in compound calculi) in which the first two (subsets) share the third (subject) and relate to a fourth and a fifth argument (the predicate and its negation). In the historical development of classical logic, the Exceptive Proposition ["Every b, except c, is a" or $A(b-c)a$] manifests a tri-argument structure.

Table 6 Distinctive simple deductions

	Ama	Am'a'	Ema	Em'a'	Ima	Im'a'	Oma	Om'a'	Yma	Ym'a'	Yam	Ya'm'	Uma	Um'a'	Uam	Ua'm'
Amc	Ica	Ac'a'	Oca	Ec'a'	Ica		Oca		Yca		Ica	Oca		Uc'a'	Oca+Ac'a'	Ica+Ec'a'
Am'c'	Aca	Ic'a'	Eca	Oc'a'	Aca	Ic'a'		Oc'a'		Yc'a'	Oc'a'	Ic'a'	Uca		Eca+Ic'a'	Aca+Oc'a'
Emc	Oc'a'	Eca	Ica	Aca	Oc'a'		Ic'a'		Yc'a'		Oc'a'	Ic'a'		Uca	Eca+Ic'a'	Aca+Oc'a'
Em'c'	Ec'a'	Oca	Ac'a'	Ica		Oca		Ica		Yca	Ica	Oca	Uc'a'		Oca+Ac'a'	Ica+Ec'a'
Imc	Ica	Ica	Oca			Oca	Ica								Oca+Ac'a'	Ica+Ec'a'
Im'c'	Oc'a'	Ic'a'		Oc'a'				Oc'a'							Eca+Ic'a'	Aca+Oc'a'
Omc	Oc'a'		Ic'a'			Ic'a'									Eca+Ic'a'	Aca+Oc'a'
Om'c'		Oca		Ica				Ica							Oca+Ac'a'	Ica+Ec'a'
Ymc	Yac	Ya'c	Ya'c	Yac									Yca+Yc'a'			
Ym'c'	Ya'c	Yac	Yac	Ya'c										Yca+Yc'a'		
Ycm	Ica	Oca	Oca												Oca	Ica
Yc'm'	Oc'a'	Ic'a'	Ic'a'	Oc'a'											Ic'a'	Oc'a'
Umc	Uac	Uac	Uac	Ua'c'										Uca+Uc'a'		
Um'c'	Ua'c'	Ua'c'											Uca+Uc'a'			
Ucm	Aca+Oc'a'	Eca+Ic'a'	Eca+Ic'a'	Aca+Oc'a'	Aca+Oc'a'	Eca+Ic'a'	Eca+Ic'a'	Aca+Oc'a'			Oc'a'	Ic'a'			Eca+Ic'a'	Aca+Oc'a'
Uc'm'	Ica+Ec'a'	Oca+Ac'a'	Oca+Ac'a'	Ica+Ec'a'	Ica+Ec'a'	Oca+Ac'a'	Oca+Ac'a'	Ica+Ec'a'			Ica	Oca			Oca+Ac'a'	Ica+Ec'a'

References

1. Béziau, J.-Y.: New light on the square of oppositions and its nameless corner. Log. Investig. **10**, 218–232 (2003)
2. Bird, O.: Syllogistic and Its Extensions. Prentice-Hall, Englewood Cliffs (1964)
3. Blanché, R.: Structures Intellectuelles: Essai sur l'Organisation Systematique des Concepts. Vrin, Paris (1966)
4. Carnes, R.D., Peterson, P.L.: Intermediate quantifiers versus percentages. Notre Dame J. Form. Log. **32**(2) (1991)
5. Cavaliere, F.: Motori di Ricerca Semantici con Antinomie Inedite e Sillogismi Sfumati. http://www.arrigoamadori.com/lezioni/AngoloDelFilosofo/AngoloDelFilosofo.htm
6. Dubois, D., Prade, H.: On fuzzy syllogism. Comput. Intell. **4**, 171–179 (1988)
7. Englebretsen, G.: The New Syllogistic. Peter Lang, New York (1987)
8. Hacker, E.A., Parry, W.T.: Pure numerical boolean syllogisms. Notre Dame J. Form. Log. **8**(4), (1967)
9. Kosko, B.: Fuzzy Thinking: The New Science of Fuzzy Logic. Hyperion, New York (1993)
10. Kumova, B.I., Çakir, H.: The fuzzy syllogistic system. In: Mexican International Conference on Artificial Intelligence (MICAI'10), Pachuca. LNAI. Springer, Berlin (2010)
11. Lambert, J.H.: Neues Organon [Liepzig, 1764]. Olms, Hildesheim (1965)
12. Lindell, S.: A term logic for physically realizable models of informations. In: The Old New Logic: Essays on the Philosophy of Fred Sommers, Chap. 8. MIT Press, Cambridge (2005)
13. Moretti, A.: The geometry of logical opposition. PhD Thesis, Université de Neuchâtel, Switzerland (2009)
14. Murphree, W.A.: The numerical syllogism and existential presupposition. Notre Dame J. Form. Log. **38**(1) (1997)
15. Murphree, W.A.: Numerical term logic. Notre Dame J. Form. Log. **39**(3) (1998)
16. Naddeo, A., Cappetti, N., Pappalardo, M., Donnarumma, A.: Fuzzy logic application in the structural optimisation of a support plate for electrical accumulator in a motor vehicle. J. Mater. Process. Technol. **120**(1–3), 303–309 (2002)
17. Pfeifer, N.: Contemporary syllogistics: comparative and quantitative syllogisms. In: Kreuzbauer, G., Dorn, G. (eds.) Argumentation in Theorie und Praxis: Philosophie und Didaktik des Argumentierens, pp. 57–71. Lit, Vienna (2006)
18. Pratt-Hartmann, I.: On the complexity of the numerically definite syllogistic and related fragments. Bull. Symb. Log. **14**(1), 1–28 (2008)
19. Sesmat, A.: Logique. Hermann, Paris (1951)
20. Seuren, P.A.M.: The natural logic of language and cognition. Pragmatics **16**(1), 103–138 (2006)
21. Seuren, P.A.M.: The Logic of Language. Language from Within, vol. 2. Oxford University Press, Oxford (2010)
22. Sommers, F.: The Logic of Natural Language. Oxford University Press, Oxford (1982)
23. Sommers, F.: Predication in the logic of terms. Notre Dame J. Form. Log. **31**(1) (1990)
24. Suchon, W.: Vasil'iev: what did he exactly do? Logic Logic. Philos. **7**, 131–141 (1999)
25. Vasiliev, N.A.: Imaginary (non Aristotelian) logic. In: Atti del V Congresso Internazionale di Filosofia, Napoli (1925)
26. Zadeh, Lofti A.: A computational approach to fuzzy quantifiers in natural languages. Comput. Math. Appl. **9**(1), 149–184 (1983)

F. Cavaliere (✉)
v. della Libertà 5, 47042 Cesenatico (FC), Italy
e-mail: cavaliere.ferdinando@gmail.com

Part IV
Constructions Generalizing the Square
of Opposition

General Patterns of Opposition Squares and 2n-gons

Ka-fat Chow

Abstract In the first part of this paper we formulate the General Pattern of Squares of Opposition (GPSO), which comes in two forms. The first form is based on trichotomies whereas the second form is based on unilateral entailments. We then apply the two forms of GPSO to construct some new squares of opposition (SOs) not known to traditional logicians. In the second part of this paper we discuss the hexagons of opposition (6Os) as an alternative representation of trichotomies. We then generalize GPSO to the General Pattern of 2n-gons of Opposition (GP2nO), which also comes in two forms. The first form is based on n-chotomies whereas the second form is based on co-antecedent unilateral entailments. We finally introduce the notion of perfection associated with 2n-gons of opposition (2nOs) and point out that the fundamental difference between a SO and a 6O is that the former is imperfect while the latter is perfect. We also discuss how imperfect 2nOs can be perfected at different fine-grainedness.

Keywords Trichotomy · Unilateral entailment · General Pattern of Squares of Opposition · General Pattern of 2n-gons of Opposition

Mathematics Subject Classification Primary 03B65

1 Introduction

The square of opposition (SO) has been an important entity studied in traditional logic since the ancient times. Yet, for centuries logicians seemed to know only one type of SO, i.e. the Boethian SO composed of the universal/particular affirmative/negative propositions arranged in the four corners named A, E, I and O. These four propositions are linked up by four types of binary opposition relations whose definitions are given in Table 1.[1]

In modern times, many scholars tried to unravel the underlying principles of opposition inferences with a view to achieving a new interpretation of SO. Such kind of studies includes Brown's (1984) classification of 4 main types and 34 subtypes of SOs [4], Jaspers' representation of SO in terms of two-dimensional Cartesian coordinate [6], Seuren's formulation of the improved square notation based on his Valuation Space Model [9]. In this paper, we will generalize the traditional Boethian SO to a general pattern of SOs. After

---

[1]This paper treats contrary, subcontrary and contradictory as three parallel relations, rather than treating contradictory as a special case of the other two relations, as some scholars do.

J.-Y. Béziau, D. Jacquette (eds.), *Around and Beyond the Square of Opposition*, 263–275
Studies in Universal Logic, DOI 10.1007/978-3-0348-0379-3_18, © Springer Basel 2012

Table 1 Opposition relations

Name	Definition
Subalternate	Unilateral entailment (i.e. In case the former proposition is true, the latter must be true, but not vice versa)
Contrary	Mutually exclusive but not collectively exhaustive (i.e. The propositions cannot be both true but can be both false)
Subcontrary	Collectively exhaustive but not mutually exclusive (i.e. The propositions cannot be both false but can be both true)
Contradictory	Both mutually exclusive and collectively exhaustive (i.e. The propositions cannot be both true nor both false)

doing so, we can then go further by generalizing the notion of "squares of opposition" to "2n-gons of opposition".

But before embarking, we have to clarify two points. First, in modern times some scholars (such as [5, 8, 11]) adopt a new definition of SOs based on the notions of negation and duality, rather than the traditional opposition relations as defined in Table 1. This paper will stick to the traditional definition of SOs and will not consider the new definition. Moreover, one should note that the general patterns discussed in this paper are derived from traditional SOs and are not generally applicable to the modern SOs (which should be more accurately called "squares of duality").[2]

Second, this paper adopts a graph-theoretic rather than geometrical view on the figures representing the logical relations. Instead of constructing all sorts of higher-dimensional geometrical figures as was done by the n-Opposition theorists such as [7] and [10], we will represent the logical relations by 2-dimensional labeled multidigraphs, with the vertices representing the propositions and the arcs representing the opposition relations. The unilateral subalternate relations will be represented by single arcs with arrow heads, whereas the bilateral contrary, subcontrary and contradictory relations will be represented by arcs without arrow heads which are understood to be double arcs going in opposite directions. Under this graph-theoretic view, the angles and lengths of the arcs have no significance.

2 General Pattern of Squares of Opposition

2.1 Some Preliminary Observations

We start from some preliminary observations of the Boethian SO. Consider the subalternate relation between the A (i.e. "All S are P") and I (i.e. "Some S are P") statements first. One crucial point is that if we rewrite the I statement as the disjunction "All S are P ∨ Part of the S are P",[3] then the subalternate relation automatically obtains as it is only a special

---

[2]Reference [8] defined a "square" function which may generate squares of duality composed of sentences with various types of determiners. This may be seen as a "general pattern of squares of duality".

[3]In this paper, "Part of the S are P" is to be interpreted as "Some but not all S are P".

Fig. 1 GPSO1

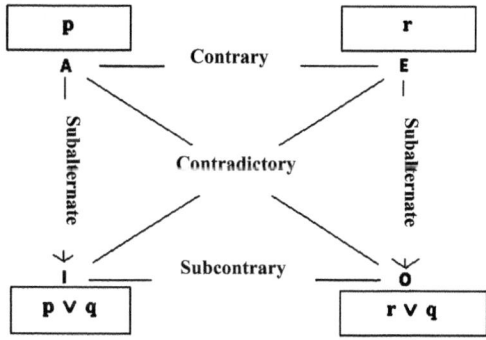

case of the unilateral entailment[4] $p_1 \Rightarrow_u (p_1 \vee p_2)$ in propositional logic. The same can be said of the subalternate relation between the E (i.e. "All S are not P", or equivalently "No S are P") and O (i.e. "Some S are not P", or equivalently "Not all S are P") statements if we rewrite the O statement as the disjunction "No S are P ∨ Part of the S are P". Moreover, we notice that the three statements "All S are P", "Part of the S are P" and "No S are P" constitute a trichotomy, i.e. these three statements are pairwise mutually exclusive and collectively exhaustive. Thus, we may say that the SO is derived from a trichotomy.

Second, we observe that by virtue of the two contradictory relations (i.e. $O \equiv \neg A$, $E \equiv \neg I$), the two subalternate relations (i.e. $A \Rightarrow_u I$ and $E \Rightarrow_u O$) are in fact contraposition of each other. Thus, we may also say that the SO is derived from a unilateral entailment (i.e. subalternate relation) and its contraposition.

2.2 Two Forms of the General Pattern of Squares of Opposition

We can generalize the above two observations to the General Pattern of Squares of Opposition (GPSO), which comes in two forms. The first form, henceforth GPSO1, is generalized from the first observation.

GPSO1 Given 3 non-trivial[5] propositions (or propositional functions) p, q and r that constitute a trichotomy, we can construct a SO as shown in Fig. 1.

In what follows we will show that the SO in Fig. 1 satisfies the definitions of the opposition relations. As mentioned above, the two subalternate relations are just special cases of $p_1 \Rightarrow_u (p_1 \vee p_2)$. The two contradictory relations follow from the fact that p, q and r constitute a trichotomy. The contrariety between p and r follows from the fact that any two members of a trichotomy are mutually exclusive but not collectively exhaustive. The subcontrariety between $(p \vee q)$ and $(r \vee q)$ follows from the fact that $(p \vee q) \vee (r \vee q) \equiv (p \vee q \vee r)$ and $(p \vee q) \wedge (r \vee q) \equiv q$. Thus, $(p \vee q)$ and $(r \vee q)$ are collectively exhaustive but not mutually exclusive.

---

[4]This paper uses "$\Rightarrow_u$" to represent "unilaterally entails", i.e. $(p \Rightarrow_u q) \Longleftrightarrow (p \Rightarrow q \wedge q \nRightarrow p)$.

[5]In this paper, non-trivial propositions refer to propositions that are neither tautologies nor contradictions. Similarly, non-trivial sets refer to sets that are neither empty nor equal to the universal set.

Fig. 2 GPSO2

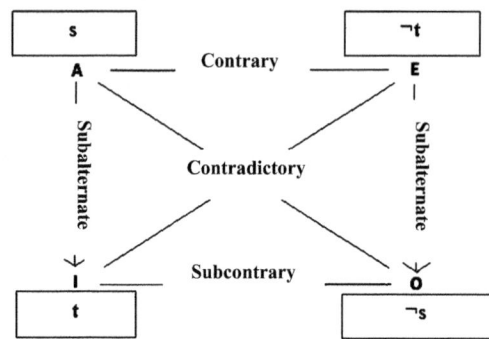

The second form, henceforth GPSO2, is generalized from the second observation.

GPSO2 Given 2 non-trivial propositions (or propositional functions) s and t such that (a) $s \not\equiv t$; (b) they constitute a unilateral entailment: $s \Rightarrow_u t$, we can construct a SO as shown in Fig. 2.

In what follows we will show that the SO in Fig. 2 also satisfies the definitions of the opposition relations. First, by definition of this SO, the two subalternate and contradictory relations are obviously satisfied. Next we consider the contrary relation. This can be proved by showing that two propositions p_1 and p_2 are contrary to each other if and only if $p_1 \Rightarrow_u \neg p_2$. This unilateral entailment is equivalent to $(p_1 \Rightarrow \neg p_2) \wedge (\neg p_2 \not\Rightarrow p_1)$. This can be rewritten as $(p_1 \wedge p_2 \equiv F) \wedge (p_1 \vee p_2 \not\equiv T)$,[6] which is equivalent to saying that p_1 and p_2 are mutually exclusive but not collectively exhaustive, and thus satisfy the contrary relation. Now since we have $s \Rightarrow_u t$, it thus follows that s and $\neg t$ are contrary to each other.

Finally we consider the subcontrary relation. This can be proved by showing that two propositions p_1 and p_2 are subcontrary to each other if and only if $\neg p_1$ and $\neg p_2$ are contrary to each other. Now the fact that $\neg p_1$ and $\neg p_2$ are contrary to each other can be expressed as $(\neg p_1 \wedge \neg p_2 \equiv F) \wedge (\neg p_1 \vee \neg p_2 \not\equiv T)$, which is equivalent to $(p_1 \vee p_2 \equiv T) \wedge (p_1 \wedge p_2 \not\equiv F)$. This shows that p_1 and p_2 are collectively exhaustive but not mutually exclusive, and thus satisfy the subcontrary relation. As we have previously shown that s and $\neg t$ are contrary to each other, it thus follows that $\neg s$ and t are subcontrary to each other.

2.3 Relations Between GPSO1 and GPSO2

Although GPSO1 and GPSO2 are based on a trichotomy and a unilateral entailment, respectively, the two forms are in fact two sides of the same coin. Given one form, we can always transform it into the other form, which we will show below.

First, suppose we are given a SO constructed from GPSO1, we immediately get a unilateral entailment, i.e. $p \Rightarrow_u (p \vee q)$ such that $p \not\equiv (p \vee q)$. With this unilateral entailment,

---

[6]"T" and "F" represent "true" and "false", respectively.

Fig. 3 Correspondence
between a unilateral
entailment and a trichotomy

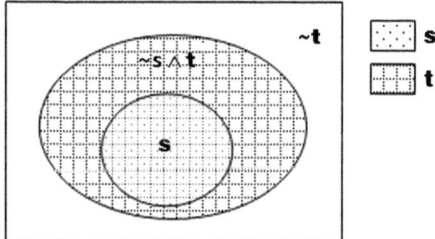

Fig. 4 SO of proportional
quantifiers

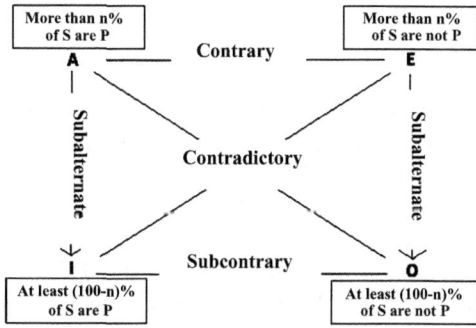

we can then construct another SO by invoking GPSO2. Please note that if either of p, q, r is trivial, then either p or $(p \vee q)$ is trivial, or $p \equiv (p \vee q)$.

On the other hand, suppose we are given an SO constructed from GPSO2, then s, $\neg t$ and $(\neg s \wedge t)$ constitute a trichotomy. To show that these three propositions are pairwise mutually exclusive, we first observe that $s \wedge \neg t \equiv F$ by the contrary relation between s and $\neg t$. Moreover, it is also obvious that $s \wedge (\neg s \wedge t) \equiv \neg t \wedge (\neg s \wedge t) \equiv F$. To show that the three propositions are collectively exhaustive, it suffices to show that $s \vee \neg t \vee (\neg s \wedge t) \equiv T$, which is obvious after we expand the left-hand side into $(s \vee \neg t \vee \neg s) \wedge (s \vee \neg t \vee t)$. The above fact can be illustrated by Fig. 3 in which s and t are depicted as sets and the unilateral entailment is depicted as proper set inclusion. From Fig. 3 we can see that if s is a proper subset of t, then s, $\neg t$ and $(\neg s \wedge t)$ constitute a partition of the universe.

The above discussion shows that a unilateral entailment is closely related to a trichotomy. With this trichotomy, we can then construct another SO by invoking GPSO1. Please note that if either s or t is trivial, or $s \equiv t$, then either of s, $\neg t$, $(\neg s \wedge t)$ is trivial.

2.4 Applications of GPSO1

By applying the two forms of GPSO, a great number of SOs not known to traditional logicians can be easily constructed. We first see some examples of the application of GPSO1. Let $50 < n < 100$, where n is a real number. Then the three intervals $[0, 100-n)$, $[100-n, n]$ and $(n, 100]$ constitute a tripartition of the interval $[0, 100]$. In other words, "Less than $(100-n)$ % of S are P", "Between $(100-n)$ % and n % of S are P" and "More than n % of S are P" constitute a trichotomy. With this trichotomy, we can construct the SO shown in Fig. 4.

Fig. 5 SO of pre-1789
French Estates General

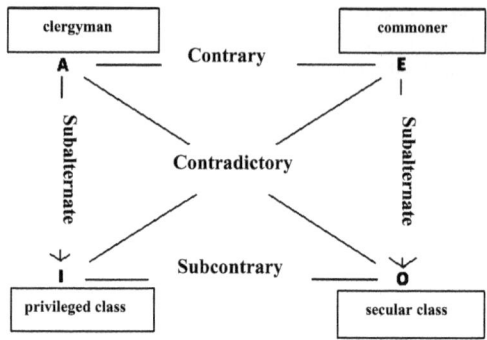

Please note that in constructing the SO in Fig. 4, we have also made use of the following equivalences: "Less than $(100 - n)$ % of S are P" $\equiv$ "More than n % of S are not P"; "At most n % of S are P" $\equiv$ "At least $(100 - n)$ % of S are not P".

The application of GPSO is not confined to quantified sentences, but may be extended to other types of sentences or even plain predicates. Suppose our domain of discourse consists of members of the pre-1789 French Estates General, which were divided into three estates: clergymen, noblemen and commoners, which constituted a trichotomy. If we now call "clergymen ∨ noblemen" the "privileged class" and "commoners ∨ noblemen" the "secular class", then we may construct the SO shown in Fig. 5.

Although the SO in Fig. 5 is composed of plain predicates rather than propositions, these plain predicates may be seen as short forms of propositional function with a variable x. For example, "clergymen" in the SO may be seen as short form of the propositional function "x was a clergyman".

2.5 Applications of GPSO2

According to GPSO2, a SO may be constructed from a unilateral entailment together with its contraposition. This is in fact the underlying principle of the "semiotic squares". Given a pair of contrary concepts, such as "happy" and "unhappy", we immediately obtain the unilateral entailment "happy $\Rightarrow_u$ not unhappy".[7] With this unilateral entailment, we can then construct the semiotic square shown in Fig. 6.

Unilateral entailments may also occur between sentences with more than one quantifier, such as "Every boy loves every girl $\Rightarrow_u$ Some boy loves some girl". One interesting subtype of this kind of entailments consists of unilateral entailments between the active and passive forms of a sentence with more than one quantifier. This is the subject area of "scope dominance" studied by [2] and [1].[8] These scholars have discovered a number

---

[7]It is essential that "happy" and "unhappy" constitute a pair of contrary rather than contradictory concepts. Otherwise, the entailment will be bilateral rather than unilateral.

[8]Scope dominance originally refers to unilateral entailments between the "direct scope" and "inverse scope" readings of a sentence with more than one quantifier. However, this phenomenon can also be reinterpreted as unilateral entailments between the active and passive forms of the sentence, assuming that both forms are interpreted under the direct scope reading.

Fig. 6 Semiotic square

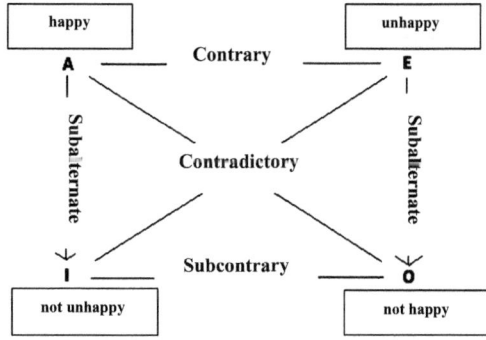

Fig. 7 SO of scope dominance

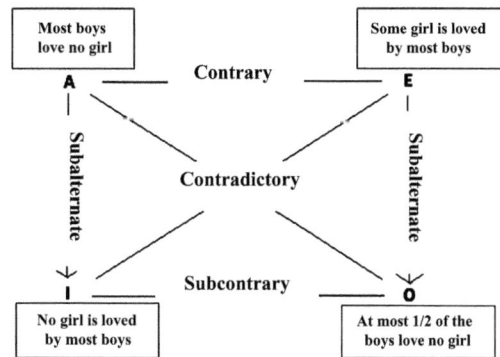

of valid unilateral entailments involving sentences with more than one quantifier. Here is one such example:

$$\text{Most boys love no girl} \Rightarrow_u \text{No girl is loved by most boys} \tag{2.1}$$

With this unilateral entailment, we can then construct the SO shown in Fig. 7.

We can thus see that by combining GPSO2 with the research findings on scope dominance, we can construct a great number of SOs composed of sentences with more than one quantifier.

3 General Pattern of 2n-gons of Opposition

3.1 Hexagons of Opposition

According to GPSO1, a SO can be derived from any trichotomy composed of 3 non-trivial propositions p, q and r. However, GPSO1 shows an asymmetry between these 3 propositions: while each of p and r appears as an independent proposition in the two upper corners (i.e. A and E), q only appears as parts of two disjunctions in the two lower corners (i.e. I and O).

To achieve symmetry, we need to add to the SO two new vertices denoted as Y and U corresponding to the propositions q and $p \vee r$, respectively, thus expanding the SO to a hexagon of opposition (6O) proposed by [3] as shown in Fig. 8.

Fig. 8 Hexagon of
opposition

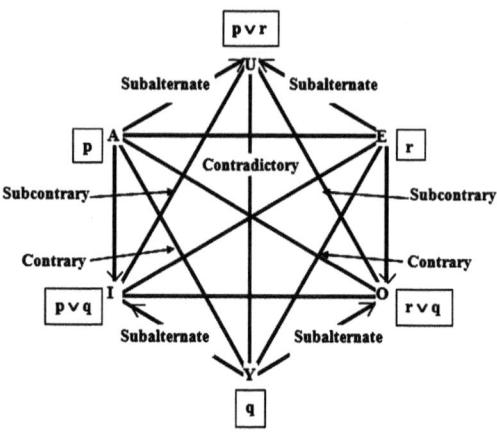

Fig. 9 6O constructed from a
trichotomy

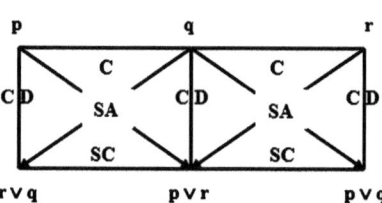

Please note that this figure contains the Boethian SO AEIO as a subpart. For this reason, only those arcs not being part of AEIO are labeled. This figure also contains the AYE triangle of contrariety, IOU triangle of subcontrariety as well as two more SOs: AYUO and YEIU, as its subparts.

For ease of comparison with other 2n-gons below, the components of the above 6O can be rearranged into the form shown in Fig. 9 which is, graph-theoretically speaking, isomorphic with the original form (in Fig. 9, SA = subalternate, CD = contradictory, C = contrary, SC = subcontrary).

To avoid messing up the figure, some arcs in Fig. 9 have been left out, e.g. the SA arc leading from p to $p \vee q$. One should view Fig. 9 on the understanding that there is a C arc between any two upper-row vertices, a SC arc between any two lower-row vertices and a SA arc leading from any upper-row vertex to any lower-row vertex not directly below it.

The above discussion shows that a trichotomy is best represented by a 6O. But as shown in Subsect. 2.3, trichotomies and unilateral entailments are closely related to each other. How can we construct a 6O from a unilateral entailment? If we review Fig. 3, then we can see that apart from the unilateral entailment $s \Rightarrow_u t$, there is in fact another less obvious one with s as antecedent, i.e. $s \Rightarrow_u (s \vee \neg t)$. By properly arranging the 3 propositions s, t, $(s \vee \neg t)$ and their contradictories $\neg s$, $\neg t$, $(\neg s \wedge t)$, we can then construct a 6O as shown in Fig. 10.

3.2 Two Forms of the General Pattern of 2n-gons of Opposition

The natural association between a trichotomy (or a unilateral entailment) and a 6O may be generalized to an association between a n-chotomy (or $n - 2$ unilateral entailments) and

Fig. 10 6O constructed from
a unilateral entailment

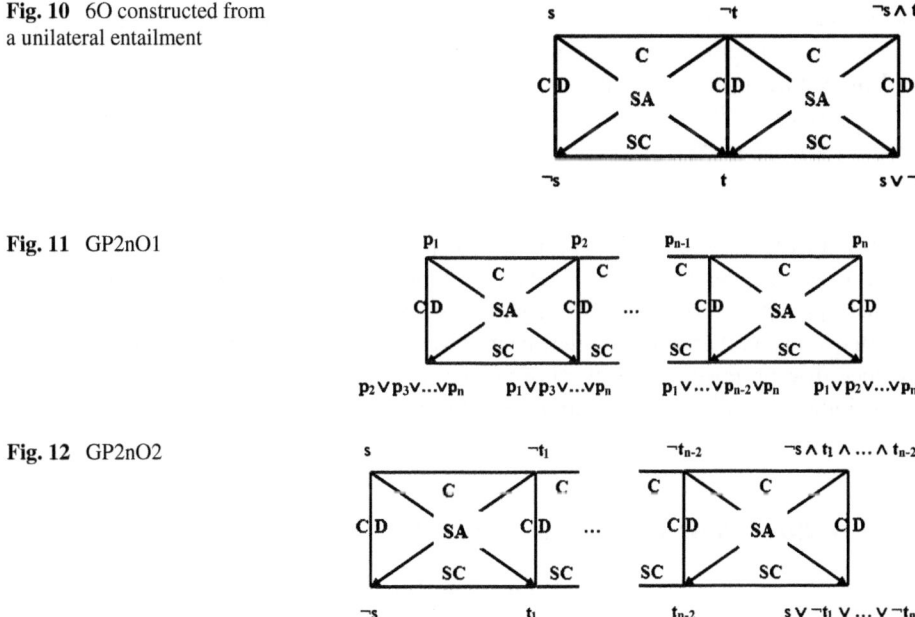

Fig. 11 GP2nO1

Fig. 12 GP2nO2

a 2n-gon of opposition (2nO), resulting in the General Pattern of 2n-gons of Opposition (GP2nO), which also comes in two forms, denoted GP2nO1 and GP2nO2 below.

GP2nO1 Given n $(n \geq 3)$ non-trivial propositions (or propositional functions) $p_1, p_2,$ $\ldots, p_n$ that constitute a n-chotomy (i.e. $p_1, p_2, \ldots, p_n$ are collectively exhaustive and pairwise mutually exclusive), we can construct a 2nO as shown in Fig. 11 (many arcs have been left out from this figure).[9]

It is easy to see that the 2nO in Fig. 11 satisfies the definitions of the opposition relations. For example, any two lower-row propositions are subcontrary to each other because they contain some common disjuncts (hence not mutually exclusive), and they collectively contain all the n propositions as disjuncts (hence collectively exhaustive).

GP2nO2 Given $n - 1$ $(n \geq 3)$ non-trivial propositions (or propositional functions) $s, t_1, \ldots, t_{n-2}$ such that (a) any two of $t_1, \ldots, t_{n-2}$ satisfy the subcontrary relation; (b) $s \neq (t_1 \wedge \cdots \wedge t_{n-2})$; (c) they constitute $(n - 2)$ co-antecedent unilateral entailments (i.e. unilateral entailments with the same antecedent): $s \Rightarrow_u t_1, \ldots, s \Rightarrow_u t_{n-2}$, then we have an additional unilateral entailment: $s \Rightarrow_u (s \vee \neg t_1 \vee \cdots \vee \neg t_{n-2})$ and we can construct a 2nO as shown in Fig. 12.

Next we show that the 2nO in Fig. 12 also satisfies the definitions of the opposition relations. The contradictory relations are obviously satisfied. So we consider the

---

[9]If $n = 2$, then the upper-row propositions are t_1 and t_2, while the lower-row propositions are t_2 and t_1. In this case, the subalternate relations will become equivalence relations, which are not among the opposition relations defined in Table 1. That is the reason why n must be at least 3.

subalternate relations. The subalternate relations between s and t_1, $\ldots$, s and t_{n-2}, as well as the subalternate relations between $\neg t_1$ and $\neg s$, $\ldots$, $\neg t_{n-2}$ and $\neg s$ are guaranteed by condition (c) above and its contraposition. We will now show that $\neg t_i \Rightarrow_u t_j$ $(1 \leq i, j \leq n - 2, i \neq j)$. By condition (a) above, any t_i and t_j satisfy the subcontrary relation, which can be expressed as $(t_i \vee t_j \equiv T) \wedge (t_i \wedge t_j \not\equiv F)$. This is equivalent to $(\neg t_i \Rightarrow t_j) \wedge (t_j \not\Rightarrow \neg t_i)$, which can be rewritten as $\neg t_i \Rightarrow_u t_j$. The remaining subalternate relations in Fig. 12 are just special cases of unilateral entailments in propositional logic. For example, $(\neg s \wedge t_1 \wedge \cdots \wedge \neg t_{n-2}) \Rightarrow_u t_1$ is just a special case of $(p_1 \wedge p_2) \Rightarrow_u p_1$.

Next we consider the contrary relations. Since it has been shown that for any two distinct upper-row propositions p_1 and p_2, we have $p_1 \Rightarrow_u \neg p_2$, thus p_1 and p_2 are contrary to each other. Finally, since each lower-row proposition is the contradictory of the upper-row proposition directly above it, the subcontrariety between any two lower-row propositions follows from the contrariety between any two upper-row propositions.

3.3 Relations Between GP2nO1 and GP2nO2

Just like GPSO, the two forms of GP2nO are also interdefinable. If we are given a 2nO constructed from GP2nO1, then this 2nO contains $n - 1$ propositions: p_1, $(p_1 \vee p_3 \vee \cdots \vee p_n)$, $\ldots$, $(p_1 \vee \cdots \vee p_{n-2} \vee p_n)$ that satisfy the three conditions of GP2nO2: (a) the subcontrariety between any two of $(p_1 \vee p_3 \vee \cdots \vee p_n)$, $\ldots$, $(p_1 \vee \cdots \vee p_{n-2} \vee p_n)$ is guaranteed by the subcontrariety between any two lower-row propositions in this 2nO; (b) $p_1 \not\equiv (p_1 \vee p_3 \vee \cdots \vee p_n) \wedge \cdots \wedge (p_1 \vee \cdots \vee p_{n-2} \vee p_n)$ because the right-hand side is equivalent to $(p_1 \vee p_n)$; (c) these propositions constitute $(n - 2)$ co-antecedent unilateral entailments: $p_1 \Rightarrow_u (p_1 \vee p_3 \vee \cdots \vee p_n)$, $\ldots$, $p_1 \Rightarrow_u (p_1 \vee \cdots \vee p_{n-2} \vee p_n)$.

This 2nO also contains an additional unilateral entailment: $p_1 \Rightarrow_u (p_1 \vee p_2 \vee \cdots \vee p_{n-1})$ whose antecedent is p_1 and whose consequent has the correct form as stipulated in GP2nO2, because $(p_1 \vee p_2 \vee \cdots \vee p_{n-1}) \equiv p_1 \vee \neg (p_1 \vee p_3 \vee \cdots \vee p_n) \vee \cdots \vee \neg (p_1 \vee \cdots \vee p_{n-2} \vee p_n)$. Moreover, the arrangement of the $2n$ propositions of this 2nO also satisfies GP2nO2. Please also note that if either of $p_1, \ldots, p_n$ is trivial, then either of the $n - 1$ propositions given above is trivial, or $p_1 \equiv (p_1 \vee p_3 \vee \cdots \vee p_n) \wedge \cdots \wedge (p_1 \vee \cdots \vee p_{n-2} \vee p_n)$. For example, if $p_2 \equiv F$, then $(p_1 \vee p_3 \vee \cdots \vee p_n) \equiv T$.

Conversely, if we are given a 2nO constructed from GP2nO2, then this 2nO contains n propositions s, $\neg t_1$, $\ldots$, $\neg t_{n-2}$, $(\neg s \wedge t_1 \wedge \cdots \wedge t_{n-2})$ that constitute a n-chotomy. That these n propositions are pairwise mutually exclusive is guaranteed by the contrary relations among these propositions. To show that these propositions are collectively exhaustive, it suffices to show that $s \vee \neg t_1 \vee \cdots \vee \neg t_{n-2} \vee (\neg s \wedge t_1 \wedge \cdots \wedge t_{n-2}) \equiv T$, which is obvious after we expand the left-hand side into $(s \vee \neg t_1 \vee \cdots \vee \neg t_{n-2} \vee \neg s) \wedge (s \vee \neg t_1 \vee \cdots \vee \neg t_{n-2} \vee t_1) \wedge \cdots \wedge (s \vee \neg t_1 \vee \cdots \vee \neg t_{n-2} \vee t_{n-2})$. The above fact can be illustrated by Fig. 13 for the case $n = 2$, which shows that the two co-antecedent unilateral entailments $s \Rightarrow_u t_1$ and $s \Rightarrow_u t_2$ give rise to the additional unilateral entailment: $s \Rightarrow_u (s \vee \neg t_1 \vee \neg t_2)$ and a 4-chotomy consisting of s, $\neg t_1$, $\neg t_2$ and $(\neg s \wedge t_1 \wedge t_2)$.

Moreover, the arrangement of the $2n$ propositions of this 2nO also satisfies GP2nO1. Please also note that if either of s, t_1, $\ldots$, t_{n-2} is trivial, or $s \equiv (t_1 \wedge \cdots \wedge t_{n-2})$, then either of s, $\neg t_1$, $\ldots$, $\neg t_{n-2}$, $(\neg s \wedge t_1 \wedge \cdots \wedge t_{n-2})$ is trivial.

Fig. 13 Correspondence between 2 co-antecedent unilateral entailments and a 4-chotomy

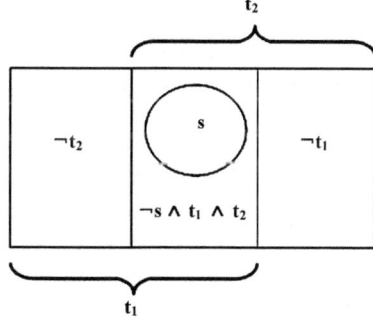

Fig. 14 An imperfect 4O

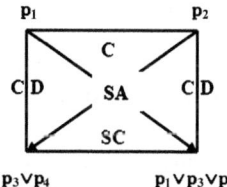

4 Perfection

Figures 11 and 12 show that every 2nO contains $C(n, m)$ 2mOs as its proper subparts,[10] for any m such that $m < n$ and $m \geq 2$.[11] To distinguish these 2mOs from the 2nO, we need to introduce the notion of perfection.

We say that a 2nO is perfect if the disjunction of all its upper-row propositions $\equiv$ the disjunction of all its lower-row propositions, otherwise it is imperfect. Given the pairwise subcontrariety of the lower-row propositions of any 2nO, this condition is equivalent to the fact that the disjunction of all its upper-row propositions $\equiv T$.

According to this definition, any 2nO that satisfies GP2nO1 or GP2nO2 is perfect, whereas a 2mO ($m < n$) which is a proper subpart of a perfect 2nO is imperfect. Thus, the 6O proposed by modern scholars as depicted in Fig. 9 above is perfect whereas the traditional SO (i.e. 4O) is imperfect. Moreover, the fundamental difference between the 6O and SO in terms of symmetry with respect to the three propositions in a trichotomy can now be more formally captured by the notion of perfection, i.e. the symmetric 6O is perfect and the asymmetric SO is imperfect.

Given an imperfect 2mO, we can always make it perfect. But the outcome of the perfection process is not unique because an imperfect 2mO can be perfected at different fine-grainedness by combining or splitting propositions.

Let's use an example to illustrate this point. Figure 14 shows an imperfect 4O because $(p_1 \vee p_2) \not\equiv (p_1 \vee p_2 \vee p_3 \vee p_4)$.

---

[10] $C(n, m) = n!/[(m!)(n - m)!]$ represents the number of possible ways of choosing m objects from a set of n objects.

[11] In fact, just as a 6O contains a triangle of contrariety and a triangle of subcontrariety as its proper subparts, every 2nO also contains a n-gon of contrariety (consisting of the upper-row propositions) and a n-gon of subcontrariety (consisting of the lower-row propositions) as its proper subparts. This paper will not discuss these n-gons any further.

The most straightforward way to perfect this 4O is to expand it into an 8O by adding two upper-row propositions: p_3 and p_4 (and corresponding lower-row propositions). But this is not the unique way. Another way is to expand this 4O into a 6O by adding just one upper-row proposition: $(p_3 \lor p_4)$ (and the corresponding lower-row proposition). This is tantamount to combining the two propositions p_3 and p_4 into one, hence transforming the original 4-chotomy to a trichotomy (comprising the three propositions p_1, p_2 and $(p_3 \lor p_4)$). Still another way is to expand this 4O into a 10O by rewriting p_4 as $p_{4a} \lor p_{4b}$ and adding three upper-row propositions: p_3, p_{4a} and p_{4b} (and corresponding lower-row propositions). This is tantamount to splitting the proposition p_4 into two non-trivial propositions p_{4a} and p_{4b} such that $p_4 \equiv p_{4a} \lor p_{4b}$, hence transforming the original 4-chotomy to a 5-chotomy (comprising the five propositions p_1, p_2, p_3, p_{4a} and p_{4b}).

5 Conclusion

In this paper we have shown how the Boethian SO is related to trichotomies or unilateral entailments and then formulated the two forms of GPSO, which enables us to construct a great number of new SOs not known to traditional logicians. We have also discussed 6O as an alternative representation of a trichotomies. Intuitively, a 6O differs from a SO in that the former is symmetric while the latter is asymmetric with respect to the three propositions in a trichotomy. This difference can be more formally captured by the notion of perfection in that a 6O is a perfect representation of trichotomies, whereas a SO, being a proper subpart of a 6O, is imperfect. By further generalizing trichotomies and unilateral entailments to n-chotomies and co-antecedent unilateral entailments, respectively, we come up with the notion of 2n-gons and the two forms of GP2nO. We finally discuss the distinction between perfect and imperfect 2nOs. Thus, GPSO and GP2nO have very different nature, in that any SOs constructed from the two forms of GPSO are imperfect whereas any 2nOs ($n \geq 3$) constructed from the two forms of GP2nO are perfect.

References

1. Altman, A., Peterzil, Y., Winter, Y.: Scope dominance with upward monotone quantifiers. J. Log. Lang. Inf. **14**(4), 445–455 (2005)
2. Ben-Avi, G., Winter, Y.: Scope dominance with monotone quantifiers over finite domains. J. Log. Lang. Inf. **13**(4), 385–402 (2004)
3. Blanché, R.: Sur l'opposition des concepts. Theoria **19** (1953)
4. Brown, M.: Generalized quantifiers and the square of opposition. Notre Dame J. Form. Log. **25**(4), 303–322 (1984)
5. Gottschalk, W.H.: The theory of quaternality. J. Symb. Log. **18**(3), 193–196 (1953)
6. Jaspers, D.: Operators in the lexicon: on the negative logic of natural language. PhD thesis, Leiden University (2005)
7. Moretti, A.: Geometry for modalities? Yes: through 'n-opposition theory'. In: Béziau, J.-Y., Costa-Leite, A., Facchini, A. (eds.) Aspects of Universal Logic, pp. 102–145. Centre de Recherches Sémiologiques, University of Neuchâtel, Neuchâtel (2004)
8. Peters, S., Westerståhl, D.: Quantifiers in Language and Logic. Clarendon, Oxford (2006)
9. Seuren, P.A.M.: The Logic of Language. Oxford University Press, Oxford (2010)
10. Smessaert, H.: On the 3D visualisation of logical relations. Logica Univers. **3**, 303–332 (2009)

11. van Eijck, J.: Generalized quantifiers and traditional logic. In: van Benthem, J., ter Meulen, A. (eds.) Generalized Quantifiers and Natural Language, pp. 1–19. Foris, Dordrecht (1984)

K.F. Chow (✉)
The Hong Kong Polytechnic University, Hong Kong, China
e-mail: kfzhouy@yahoo.com

The Cube Generalizing Aristotle's Square in Logic of Determination of Objects (LDO)

Jean-Pierre Desclés and Anca Pascu

Abstract In this paper we present a generalization of Aristotle's square to a cube, in the framework of an extended quantification theory defined within the Logic of Determination of Objects (LDO). The Aristotle's square is an image which makes the link between quantifier operators and negation in the First Order Predicate Language (FOPL). However, the FOPL quantification is not sufficient to capture the "meaning" of all quantified expressions in natural languages. There are some expressions in natural languages which encode a quantification on typical objects. This is the reason why we want to construct a logic of objects with typical and atypical objects. The Logic of Determination of Objects (LDO) (Desclés and Pascu, Logica Univers. 5(1):75–89, 2011) is a new logic defined within the framework of Combinatory Logic (Curry and Feys, Combinatory Logic I, North-Holland, Amsterdam, 1958) with functional types. LDO basically deals with two fundamental classes: a class of concepts ($\mathcal{F}$) and a class of objects ($\mathcal{O}$). LDO captures two kinds of objects: typical objects and atypical objects. They are defined by the means of other primitive notions: the intension of the concept f (Int f), the essence of a concept f (Ess f), the expanse of the concept f (Exp f), the extension of the concept f (Ext f). Typical objects in Exp f inherit all concepts of Int f; atypical objects in Exp f inherit only some concepts of Int f. LDO makes use of all the above notions and organizes them into a system which is a logic of objects (applicative typed system in Curry's sense—Curry and Feys, Combinatory Logic I, North-Holland, Amsterdam, 1958—with some specific operators). In LDO new quantifiers are introduced and studied. They are called star quantifiers: $\Pi^\star$ and $\Sigma^\star$ (Desclés and Guentcheva, in: M. Böttner, W. Thümmel (eds.), Variable-Free Semantics, pp. 210–233, Secolo, Osnabrück, 2000). They have a connection with classical quantifiers. They are considered as the determiners of objects of Exp f. They are different from the usual quantifiers Π and Σ expressed in the illative Curry version (Curry and Feys, Combinatory Logic I, North-Holland, Amsterdam, 1958) of Frege's quantifiers. They are defined inside the Combinatory Logical formalism (Curry and Feys, Combinatory Logic I, North-Holland, Amsterdam, 1958) starting from Π and Σ, by means of abstract operators of composition called combinators (Curry and Feys, Combinatory Logic I, North-Holland, Amsterdam, 1958). The system of four quantifiers Π, Σ, $\Pi^\star$ and $\Sigma^\star$ captures the extended quantification that means quantification on typical/atypical objects. The cube generalizing the Aristotle's square visualizes the relations between quantifiers Π, Σ, $\Pi^\star$ and $\Sigma^\star$ and the negation.

Keywords Typical/atypical instance · Determination · Indeterminate object · More or less determinate object · Intension · Expanse · Extension · Star quantifiers

Mathematics Subject Classification Primary 03B60 · Secondary 03B65

1 Introduction

Frege's famous quantification [15] theory with bound variables has obtained numerous and sound results by analyzing the language of mathematics. Linguists like Montague [16] used Frege's formalism to give representations of quantifiers in natural languages. However, questions concerning the adequation of this formalism should be raised. Some questions about Fregean quantification are:

1. Are there variables in natural languages?
2. Are there variables needed for analyzing quantified phrases? Can we conceive another logical analysis of natural language, different from Frege's and Montague's analyses? Are other non-classical logical systems useful for this purpose?
3. What do natural languages provide for quantificational expressions? What are the semiotic devices that are called upon to express quantification?
4. Is there a formalism which can adequately formalize the quantifications expressed by natural languages?
5. Is it possible to give logical representations without (bound) variables so that the structures of these representations are more adequate to the structures of sentences?

Our answer is based on Combinatory Logic [2] framework. The main concern will be to answer these questions and to show that:

1. Curry's Combinatory Logic is a "without (bound) variables logic" and it can be used for giving an analysis of noun phrases with quantifiers;
2. Curry's Combinatory Logic is related to a formalism "with bound variables", Church's λ-calculus, but its expressive power is more synthetic;
3. Both formalisms are applicative formalisms based on the primitive application of operators to operands;
4. Both formalisms (Church's λ-calculus and Combinatory Logic) can be typed;
5. Logical representations must explain how the analyzed sentences are built and how the logical representations are related to a syntactic analysis;
6. A new formalism answering questions above and based on Combinatory Logic can be built up. It is the Logic of Determination of Objects (LDO).

Considering the theory of quantification as part of logic reveals that there are two classes of theories:

- Fregean theories with bound variables:
    - Classical theory in First Order Predicate Language (FOPL);
    - Montague's quantification expressed in Church's λ-calculus.
- Fregean theory without bound variables:
    - Illative Theory expressed in Curry's Combinatory Logic.

We propose a new approach which is a non-Fregean theory without bound variables: the **Star theory** expressed in the Logic of Determination of Objects (LDO).

In this paper we present the Star theory which is the theory of quantification (star quantification) of LDO. The system of quantifiers of the Star theory and their relations with the negation can be represented by a cube, the **cube generalizing the Aristotle's square of quantifiers**.

Let us remember now some basic notions of LDO [10].

2 Basic Notions of LDO

The LDO is a non-classical logic of construction of objects. It contains a theory of typicality. It is described in [10]. LDO is defined within the framework of Combinatory Logic [2] with functional types.

LDO basically deals with two fundamental classes: a class of *concepts*, which is denoted by $\mathcal{F}$ and a class of *objects*, denoted by $\mathcal{O}$.

According to Frege [11, 14], concepts are functions from a set of objects into the set of truth values; objects are "saturated" and concepts are "unsaturated". In LDO, functions are considered as operators described by λ-expressions [1], that is, as processes constructing a result from an operand. Objects are always operands. An operator may be an operand of another operator.

Concepts are applied to objects. An object "falls under" a concept if the result of the application is the truth value $\top$. An object "does not fall under" a concept if the result of the application is the truth value $\bot$.

With every concept f the following are canonically associated [10]:

- an object called "typical object" τf, which represents the concept f as an object. This object is completely (fully) indeterminate;
- a determination operator δf applied to an object constructs an object which is more determinate than the first one;
- the intension of the concept f, Int(f), conceived as the class of all concepts that the concept f "includes", that is a semantic network of concepts structured by the relation "IS-A";
- the essence of a concept f, Ess f is the class of concepts such that they are inherited by all objects falling under the concept f;
- the expanse of the concept f, Exp f, which contains all "more or less determinate objects" which the concept f applies to;
- a part of the expanse is the extension Ext f of the concept f; it contains all fully (completely, totally) determinate objects such that the concept f applies to.

From the viewpoint of determination, in LDO, objects are of two kinds: "fully (completely, totally) determinate" objects and "more or less determinate" objects.

From the viewpoint of some of their properties, LDO captures two kinds of objects:

- Typical objects. Typical objects in Exp f inherit all concepts of Int f;
- Atypical objects. Atypical objects in Exp f inherit only some concepts of Int f.

2.1 Basic Operators of LDO

We recall some basic operators of LDO presented in [10]:

The constructor of "typical object": the operator τ This operator, denoted by τ and called *the constructor of typical object*, starting from a concept, builds a totally indeterminate object. Its type is **FFJHJ**; it canonically associates to each concept f, an indeterminate object τf, called "typical object". Its applicative scheme is:

$$\frac{\tau : \textbf{FFJHJ} \qquad\qquad f : \textbf{FJH}}{\tau f : \textbf{J}} \tag{2.1}$$

The object τf, is the "best representative" object of the concept f; it is totally indeterminate and abstractly represents the concept f in the form of an object which can then lead either to a typical object or to an atypical object more determinate.

The operator of determination: the operator δ The operator δ, called *the constructor of determination operators*, starting from a given concept, builds a *determination operator*. The operator δ canonically associates a determination operator of the type **FJJ** to each concept f. The type of operator δ is **FFJHFJJ**. Its applicative scheme is:

$$\frac{\delta : \textbf{FFJHFJJ} \qquad\qquad f : \textbf{FJH}}{\delta f : \textbf{FJJ}} \tag{2.2}$$

A determination operator δf is an operator which, being applied to an object x constructs another object y: $y = (\delta f)(x)$. The object y is more determinate than the object x, by means of the determination added by δf.

Definition 2.1 (The composition of determinations [10]) Let x be an object and δf, δg two determinations. In this case, we write:

$$(\delta g \circ \delta f)(x) = \big(\delta g\big((\delta f)x\big)\big)$$

The composition of determinations is associative and supposed to be commutative.

Definition 2.2 (Chain of determinations [10]) A *chain of determination* Δ is a finite string of determinations which can be composed of each other:

$$\Delta = \delta g_1 \circ \delta g_2 \circ \cdots \circ \delta g_n$$

Definition 2.3 (More or less determinate object [10]) A *more or less determinate object* is an object recursively obtained starting from the object τf by:

1. τf is a more or less determinate object;
2. If Δ is a chain of determinations, then $y = (\Delta\ x) = ((\delta g_1 \circ \delta g_2 \circ \cdots \circ \delta g_n)x) = (\delta g_1(\delta g_2 \ldots (\delta g_n x)))$ is a more or less determinate object;
3. Each more or less determinate object is obtained by the above rules.

Definition 2.4 (Fully determinate object [10]) An object x is *fully determinate* if and only if for each determination δg which can be applied to x, with $g \in \mathcal{F}$:

$$(\delta g\ x) = x$$

In LDO, objects are of two kinds:

- more or less determinate object: $x \in \mathcal{O}$;
- fully determinate object: $x \in \mathcal{O}_{det}$.

If x is a more or less determinate object obtained by a chain of determinations, its associated concept will be denoted by $\hat{x}$.

2.2 Theory of Typicality in LDO

We recall the typicality defined in [10]:

Definition 2.5 (Typical object of a concept f) An object $x \in \mathcal{O}$ is called *typical object* of the concept f if and only if:

- For each chain of determination $\Delta = (\delta g_1 \circ \delta g_2 \circ \cdots \circ g_i \circ \cdots \circ \delta g_n)$ such that $x = (\Delta \ \tau f)$;
- For each $i = 1, \ldots, n$, g_i is such that:
  - Either: if $g_i \in \text{Int} f$, then $N_1 g_i \notin \text{Int} f^1$ and $N_1 g_i \notin \text{Int} \hat{x}$ (a canary versus a bird);
  - Or: if $g_i \in \text{Int} f$, $N_1 g_i \in \text{Int} f$, then $g_i \in \text{Int-caract} \hat{x}$ (Julie versus an ostrich).

Definition 2.6 (Atypical object of a concept f) An object $x \in \mathcal{O}$ is called *atypical object* of the concept f if and only if:

- There is a chain of determination $\Delta = (\delta g_1 \circ \delta g_2 \circ \cdots \circ g_i \circ \cdots \circ \delta g_n)$ such that $x = (\Delta \ \tau f)$ and;
  - Either: there is g_i, $i = 1, \ldots, n$ such that $N_1 g_i \in \text{Int} f$ (an ostrich versus a bird);
  - Or: there is g_i, $i = 1, \ldots, n$ such that:
    * If $g_i \in \text{Int} f$ and $N_1 g_i \in \text{Int} f$, then $N_1 g_i \in \text{Int-caract} f$ and $g_i \in \text{Int-caract} \hat{x}$ (Rita versus an ostrich).
  - Or: there is g_i, $i = 1, \ldots, n$ such that:
    * If $g_i \in \text{Int} f$, then there is a y such that $x = (\Delta_1 \ y)$ and $y = (\Delta_2 \ \tau f)$ and y is an atypical f (Rita versus a bird considered to be an ostrich).

3 Star Quantification

In order to introduce the star quantification, first, we recall the quantification of FOPL presented by natural deduction i.e. by introduction and elimination rules.

3.1 First Order Predicate Logic (FOPL) Quantifiers

Quantifiers of FOPL are Π_1 for universal quantification and Σ_1 for existential quantification:

The universal quantifier All x (from the universe of discourse) is f.
Its introduction and elimination rules, following Gentzen [12], are:

$$
\begin{array}{|l}
x \\
\quad \begin{array}{|l} (f\ x) \\ \vdots \end{array} \\
\Pi_1 f
\end{array}
\qquad\qquad [\,i\text{-}\Pi_1\,] \qquad (3.1)
$$

---

[1] N_1 is the functional negation, the negation of a concept [10].

This rule intuitively expresses: if we have $(f\ x)$ for any object x, then $\Pi_1 f$.

$$\frac{\Pi_1 f}{(f\ x)} \qquad\qquad [\,e\text{-}\Pi_1\,] \qquad (3.2)$$

This rule intuitively expresses: if we have $\Pi_1 f$, then $(f\ x)$ for all x.

$\Pi_1\ f$ is read: All is f (all object of the universe of discourse is f).

The existential quantifier There is an x (of the universe of discourse) which is f.

Introduction and elimination rules for Σ_1 are:

$$
\begin{array}{|l|l}
a & (f\ a)\\
 & \vdots\\
\hline
\Sigma_1 f &
\end{array}
\qquad\qquad [\,i\text{-}\Sigma_1\,] \qquad (3.3)
$$

This rule expresses: if one has $(f\ a)$, then one has $\Sigma_1 f$.

$$\frac{\Sigma_1 f,\ (f\ a)\vdash B}{B} \qquad\qquad [\,e\text{-}\Sigma_1\,] \qquad (3.4)$$

This rule expresses the fact: if from $\Sigma_1 f$ and $(f\ a)$ one deduces a sentence B not depending on a, then one has B.

$\Sigma_1 f$ is read: There is a x such that $(f\ x)$.

Restricted Quantification in FOPL   Restricted quantification contains two operators: Π, Σ.

    Universal restricted quantification expresses the idea: All f is g. Its operator is Π:

$$
\begin{array}{|l|l}
x & \\
 & \begin{array}{|l} (f\ x)\\ \vdots\\ (g\ x)\end{array}\\
\hline
\Pi f g &
\end{array}
\qquad\qquad [\,i\text{-}\Pi\,] \qquad (3.5)
$$

This rule expresses: if x is an object such that from $(f\ x)$ one infers $(g\ x)$, then $\Pi f g$.

$$\frac{\Pi f g,(f\ x)}{(g\ x)} \qquad\qquad [\,e\text{-}\Pi\,] \qquad (3.6)$$

This rule expresses: if $\Pi f g$ and $(f\ x)$, then $(g\ x)$.

$\Pi f g$ is read: all object of f is an object of g or, briefly, all f is g.

    Restricted existential quantification expresses the idea: there is an f which is g. Its operator is Σ. The rules of this operator are:

$$\frac{(f\ a)\wedge(g\ a)}{\Sigma f g} \qquad\qquad [\,i\text{-}\Sigma\,] \qquad (3.7)$$

That is: if for an indeterminate object a one has $(f\ a)$ and $(g\ a)$, then one has $\Sigma f g$.

$$\frac{\Sigma f g,(f\ a)\vdash B,(g\ a)\vdash B}{B} \qquad\qquad [\,e\text{-}\Sigma\,] \qquad (3.8)$$

If one has $\Sigma f g$, from $(f\ a)$ one can deduce B (not depending on a) and from $(g\ a)$ one can deduce also B, then one has B.

$\Sigma f g$ is read: there is an f which is g.

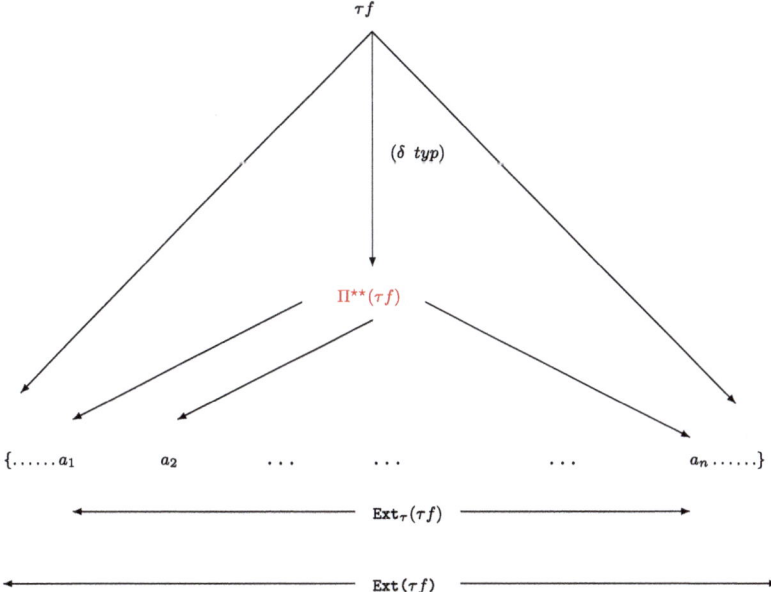

Fig. 1 The construction of the "typical arbitrary object"

3.2 Star Quantifiers

The star quantification (the system of quantification of LDO) is represented by two operators, Π^{**} and Σ^{**} and a system of four quantifiers, Π, Σ, Π^{*}, Σ^{*}. This system modelizes restricted quantification taking into account typical objects.

The Operator Constructor of the "Typical Arbitrary Object": Π^{**} (see Fig. 1) This operator is applied to the object τf constructing the object "the object whatever" $\Pi^{**}(\tau f)$. Its scheme with types is:

$$\frac{\Pi^{**} : \mathbf{FJJ} \qquad \qquad \tau f : \mathbf{J}}{\Pi^{**}(\tau f) : \mathbf{J}}$$

The object $\Pi^{**}(\tau f)$ is different from τf:

- τf is a potential generator of all object (typical or not) from $\mathrm{Exp}(\tau f)$;
- while $\Pi^{**}(\tau f)$ is a generator of all typical object, that is, all objects in $\mathrm{Exp}_\tau(\tau f)^2$;
- operator Π^{**} can be regarded as equivalent to the determination δ *typ* which is the determination by the concept *to-be-typical*:

$$\left(\Pi^{**}(\tau f)\right) \equiv \left((\delta\ typ)(\tau f)\right) \tag{3.9}$$

The Operator Constructor of the "Indeterminate Object": Σ^{**} This operator is applied to the object τf for building the "indeterminate object" $\Sigma^{**}(\tau f)$ (see Fig. 2). Its type is **FJJ**:

---

[2]$\mathrm{Exp}_\tau(\tau f)$ is the extension containing all typical objects generated from τf.

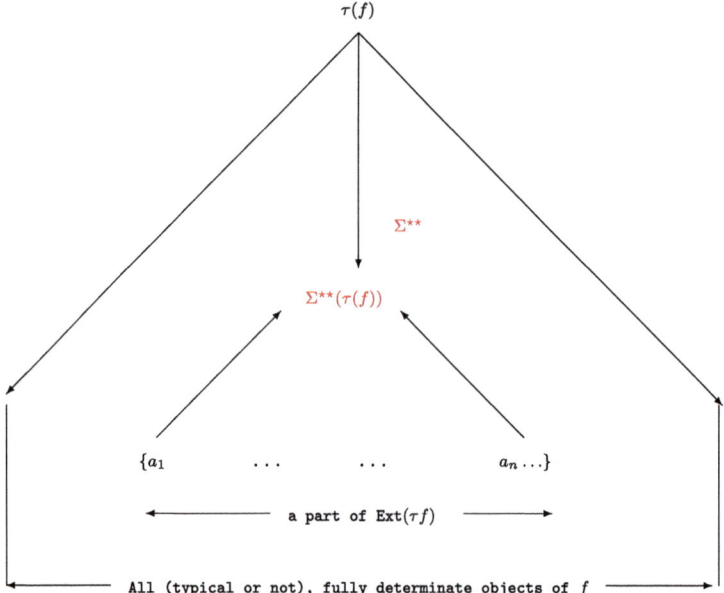

Fig. 2 The construction of the "indeterminate object"

$$\frac{\Sigma^{\star\star}\colon \text{FJJ} \qquad\qquad \tau f\colon \text{J}}{\Sigma^{\star\star}(\tau f)\colon \text{J}}$$

or

$$\frac{\Sigma^{\star\star}\colon \text{FJJ} \qquad\qquad\qquad \tau f\colon \text{a-man}}{\Sigma^{\star\star}(\tau f)\colon \text{an-indeterminate-man (typical or atypical)}}$$

Why do we need these two operators $\Pi^{\star\star}$, $\Sigma^{\star\star}$?

The answer is the following: the construction of the "typical arbitrary object" is always formally possible, that is:

$$f\left(\Pi^{\star\star}(\tau f)\right) = \top \tag{3.10}$$

As for the "indeterminate object", its existence is related to the axiom $A\tau\delta6^3$ (see [10]). This object represents a proof that $\text{Ext} f$ is not empty:

$$f\left(\Sigma^{\star\star}(\tau f)\right) = \top \quad \text{if and only if} \quad \text{Ext} f \neq \emptyset \tag{3.11}$$

Star Quantifiers   The reason to introduce star quantifiers comes from logical formalization of linguistic expressions. For example:

An Alsatian drinks beer. (*a typical Alsatian*)

Formalizations in LDO, FOPL and Curry's Logic are:

---

[3] $A\tau\delta6$, that is: $((\forall f), f \in \mathcal{F})[(f(\tau f)) = \top \text{ iff } \text{Ext}_\tau(\tau f) \neq \emptyset].$

- In LDO:

$$(\textit{to-drink-beer})\left(\Pi^{**}\left(\tau(\textit{to-be-an-Alsatian})\right)\right)$$

is true.
- In FOPL:

$$(\forall x)\left((\textit{Alsatian } x) \supset \left((\textit{to-drink-beer}) \, x\right)\right)$$

or equivalently,
- In Curry's Logic:

$$\Pi(\textit{to-be-Alsatian})(\textit{to-drink-beer})$$

are false.

The first one is consistent to the semantic context because of the fact that "A typical Alsatian drinks beer". The second and the third ones are not consistent because it is not true that "All Alsatians drink beer".

The system of star quantification is composed by four quantifiers: Π, Π^*, Σ^* and Σ. Quantifiers Π and Σ^* are Curry's quantifiers [2] but Π^* and Σ are new quantifiers taking into account the typicality.

The Operator of General Universal Quantification Π Expressing "All f Is g"   Its rules are these of (3.5) and (3.6).

The operator of typical universal quantification Π^* expresses: All typical f is g. Its introduction and elimination rules are:

$$
\left|
\begin{array}{l}
x = (\Pi^{**}(\tau f)) \\
\quad \left| \begin{array}{l} (f \; x) \\ \vdots \\ (g \; x) \end{array} \right. \\
g(\Pi^* f)
\end{array}
\right.
\qquad\qquad [\, i\text{-}\Pi^\star\,] \qquad (3.12)
$$

The introduction rule (3.12) captures the idea:

If x is a typical object generated from τf and from $(f \; x)$ one deduces $(g \; x)$, then $g(\Pi^* f)$.

$$\frac{g(\Pi^* f),(f \; x)}{x=(\Pi^{**}(\tau f)),(g \; x)} \qquad\qquad [\, e\text{-}\Pi^\star\,] \qquad (3.13)$$

For the elimination rule (3.13):

If $g(\Pi^\star f)$ and $(f \; x)$, then x is a typical object generated from τf and one has $(g \; x)$.

The operator of general existential quantification Σ^* expresses the idea: There is an f which is g.

Its introduction rule is:

$$\frac{(f \; (\Sigma^{**}(\tau f))) \wedge (g \; (\Sigma^{**}(\tau f)))}{g(\Sigma^* f)} \qquad\qquad [\, i\text{-}\Sigma^\star\,] \qquad (3.14)$$

This rule expresses the fact: if there is an indeterminate object (generated starting from τf, $(\Sigma^{**}(\tau f))$) which falls under f and g, then one has $g(\Sigma^* f)$.

The elimination rule of Σ^* is:

$$\frac{g(\Sigma^* f),(f \; (\Sigma^{**}(\tau f)))\vdash B,(g \; (\Sigma^{**}(\tau f)))\vdash B}{B} \qquad\qquad [\, e\text{-}\Sigma^\star\,] \qquad (3.15)$$

This rule says: if one has $g(\Sigma^\star f)$ and if from indeterminate object which falls under f and g one can deduce a sentence B depending on no object falling under f, then we have B.

The operator of typical existential quantifier Σ expresses: There exists a typical f which is g.

Its introduction and elimination rules are:

$$\frac{a=(\Delta\ \Pi^{**}(\tau f)),(f\ a)\wedge(g\ a)}{\Sigma fg} \qquad\qquad [\,\mathrm{i}\text{-}\Sigma\,] \qquad (3.16)$$

$$\frac{a=(\Delta\ \Pi^{**}(\tau f)),\Sigma fg,(f\ a)\vdash B,(g\ a)\vdash B}{B} \qquad\qquad [\,\mathrm{e}\text{-}\Sigma\,] \qquad (3.17)$$

In (3.16) and (3.17) the object a is a typical object, i.e. obtained starting from the object $\Pi^{**}(\tau f)$.

3.3 Quantifiers in the Axioms and Rules of LDO

The system of star quantifiers is related to the axioms of the LDO [10] by means of axiom 6 [10] and a new axiom formulated here, axiom 7. The axiom 6 [10]:

[$A\tau\delta 6$:] Existence and non-existence of occurrences of a concept

$$((\forall f),\ f\in\mathcal{F})\left[\text{If }\text{Ext}_\tau(\tau f)\neq\emptyset\text{ then }\big(f(\tau f)\big)=\top\right]$$

can be formulated equivalently as axiom 6*:

[$A\tau\delta 6^\star$:] $((\forall f),\ f\in\mathcal{F})$ [If $\text{Ext}_\tau(\tau f)\neq\emptyset$ then $f(\Pi^{**}(\tau f))=\top$]

On the other hand, one introduces the axiom 7. It is necessary in the case of a universe of discourse containing non-existent objects: "round square" or "monster" [18]:

[$A\tau\delta 7$:] The object $\Sigma^{**}(\tau f)$ exists iff $f(\tau f)=\top$

Then, the set of 6 axioms formulated in [10] is completed with the axiom 7 above.

The system of star quantifiers is related to the rules of the LDO [10] by the mean of following rules:

1. Rules relating quantifiers to other operators:
    (a) Typical universal quantifier related to operator Π^{**}:

$$\frac{g(\Pi^\star f)}{g(\Pi^{**}(\tau f))} \qquad (3.18)$$

$$\frac{g(\Pi^{**}(\tau f))}{g(\Pi^\star f)} \qquad (3.19)$$

These two rules are read: If all typical f is g, then the object "the typical arbitrary object" is g and conversely.

    (b) General existential quantifier related to operator Σ^{**}:

$$\frac{g(\Sigma^\star f)}{g(\Sigma^{**}(\tau f))} \qquad (3.20)$$

$$\frac{g(\Sigma^{**}(\tau f))}{g(\Sigma^\star f)} \qquad (3.21)$$

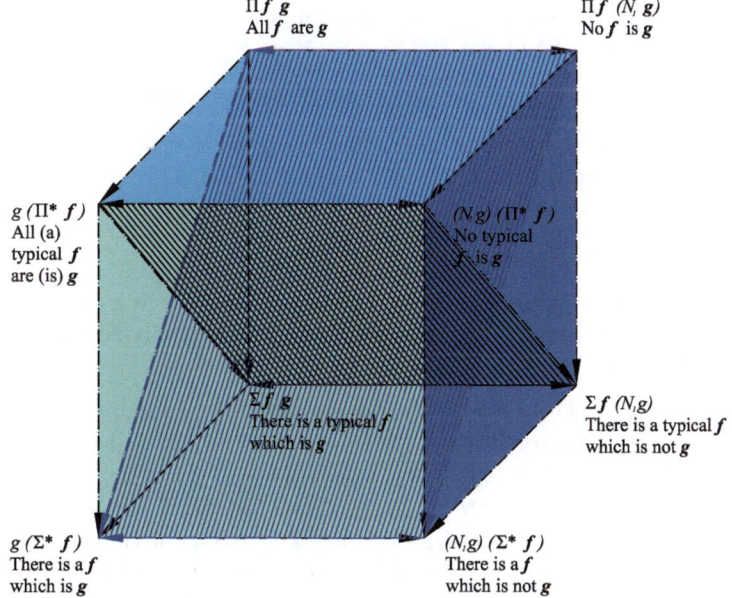

Fig. 3 The cube of star quantifiers

These two rules are read: If there is an f which is g, then the "indeterminate object" is g and conversely.

2. Atypicality and typical universal quantification (operator $\Pi^\star$):

$$\frac{(f\ x) = \top,\ ((N_1 g)\ x) = \top,\ g(\Pi^\star f)}{ATYP(f)(x)} \tag{3.22}$$

This rule is read: If x is an f, but it is not a g and all typical f falls under g, then x is not a typical f.

These rules are added to the rules formulated in [10].

3.4 The Cube of Quantifiers in LDO

The cube generalizing the Aristotle's square (see Fig. 3) visualizes the relations between quantifiers Π, Σ, $\Pi^\star$ and $\Sigma^\star$ and the negation.

1. The square $\Pi f g$, $\Pi f (N_1 g)$, $(N_1 g)(\Pi^\star f)$, $g(\Pi^\star f)$ captures:

   • Oppositions: $\Pi f g$ opposite to $\Pi f (N_1 g)$ and $g(\Pi^\star f)$ opposite to $(N_1 g)(\Pi^\star f)$ (they are denoted by double arrows ($\leftrightarrow$)).
   • Implications between general universal quantifier and typical universal one (they are denoted by dotted simple arrows): $\Pi f g \vdash g(\Pi^\star f)$ and $\Pi f (N_1 g) \vdash (N_1 g)(\Pi^\star f)$.

2. The square $\Sigma f g$, $\Sigma f (N_1 g)$, $(N_1 g)(\Sigma^\star f)$, $g(\Sigma^\star f)$ captures:

   • Oppositions: $\Sigma f g$ opposite to $\Sigma f (N_1 g)$ and $g(\Sigma^\star f)$ opposite to $(N_1 g)(\Sigma^\star f)$ (they are denoted by double arrows ($\leftrightarrow$)).

- Implications between typical existential quantification and general existential quantification: $\Sigma fg \vdash g(\Sigma^\star f)$ and $\Sigma f(N_1 g) \vdash (N_1 g)(\Sigma^\star f)$.

3. The square $g(\Pi^\star f)$, $(N_1 g)(\Pi^\star f)$, $(N_1 g)(\Sigma^\star f)$, $g(\Sigma^\star f)$ captures:

   - Oppositions: $g(\Pi^\star f)$ opposite to $(N_1 g)(\Pi^\star f)$; $g(\Sigma^\star f)$ opposite to $(N_1 g)(\Sigma^\star f)$;
   - Implications between typical universal quantification and general existential quantification: $g(\Pi^\star f) \vdash g(\Sigma^\star f)$ and $(N_1 g)(\Pi^\star f) \vdash (N_1 g)(\Sigma^\star f)$.

4. The square Πfg, $\Pi f(N_1 g)$, $\Sigma f(N_1 g)$, Σfg captures:

   - Oppositions: Πfg opposite to $\Pi f(N_1 g)$ and Σfg opposite to $\Sigma f(N_1 g)$;
   - Implications between general universal quantification and typical existential one: $\Pi fg \vdash \Sigma fg$ and $\Pi f(N_1 g) \vdash \Sigma f(N_1 g)$.

5. The square on the side $g(\Pi^\star f)$, Πfg, Σfg, $g(\Sigma^\star f)$ captures also implications between universal and existential for typical: $g(\Pi^\star f) \vdash \Sigma fg$ and for general quantification: $\Pi fg \vdash g(\Sigma^\star f)$, while the square $(N_1 g)(\Pi^\star f)$, $\Pi f(N_1 g)$, $\Sigma f(N_1 g)$, $(N_1 g)(\Sigma^\star f)$ the implications $(N_1 g)(\Pi^\star f) \vdash \Sigma f(N_1 g)$ and $\Pi f(N_1 g) \vdash (N_1 g)(\Sigma^\star f)$.

6. The rectangle Πfg, $\Pi f(N_1 g)$, $(N_1 g)(\Sigma^\star f)$, $g(\Sigma^\star f)$ represents implications between general universal and general existential quantifier: $\Pi fg \vdash g(\Sigma^\star f)$ and $\Pi f(N_1 g) \vdash (N_1 g)(\Sigma^\star f)$. It is the rectangle of classical quantifiers and it corresponds to the *Aristotle's square*.

7. The rectangle $g(\Pi^\star f)$, $(N_1 g)(\Pi^\star f)$, Σfg, $\Sigma f(N_1 g)$ represents implications between typical universal quantifier and typical existential quantifier: $g(\Pi^\star f) \vdash \Sigma fg$ and $(N_1 g)(\Pi^\star f) \vdash \Sigma f(N_1 g)$. It is the *rectangle of typical quantifiers*.

3.5 Star Theory

In [7] two relations between Π, $\Pi^\star$ and Σ, $\Sigma^\star$ respectively, are obtained. Let us denote by Q, Π or Σ and by $Q^\star$, $\Pi^\star$ or $\Sigma^\star$ respectively. The relations mentioned above are obtained by the following string on inferences using combinators and the relation of β-reduction (see [2]):

$$
\begin{array}{lll}
1. & g\,(Q^\star\, f) & \\
2. & \mathbf{C}^\star\,(Q^\star\, f)g & 1.,\text{i} - \mathbf{C}^\star \\
3. & \mathbf{B}\,\mathbf{C}^\star\,Q^\star\, f\, g & 2.,\text{i} - \mathbf{B} \\
4. & (\mathbf{B}\,\mathbf{C}^\star)Q^\star\, f\, g &
\end{array}
\tag{3.23}
$$

The two syntactic relations are:

$$\Pi \equiv \mathbf{BC}^\star\Pi^\star \tag{3.24}$$

$$\Sigma \equiv \mathbf{BC}^\star\Sigma^\star \tag{3.25}$$

They can be seen as "definitions" of Π starting from $\Pi^\star$ and of Σ starting from $\Sigma^\star$, respectively.

Why:

- For universal quantifiers: Π is the general one and $\Pi^\star$ is the typical one?
- For existential quantifiers: Σ is the typical one and $\Sigma^\star$ is the general one?

It is because one can prove the existence of an object by a constructive procedure. Generally, the result is a typical object. Moreover we have:

$$\Pi f g \vdash g(\Pi^* f) \tag{3.26}$$

If all f are g, then all typical f are g.

$$\Sigma f g \vdash g(\Sigma^* f) \tag{3.27}$$

If there is a typical f which is g, then there is an f which is g.
There are several differences between "classical" quantifiers and star quantifiers:

1. From the syntactic point of view, quantifiers Π and Σ are operators applied to two concepts f and g. Star quantifiers are applied to only one concept f. This idea is related to the fact that in linguistics the quantifier is applied to the nominal phrase;
2. There is an asymmetry in the definitions of Π and Π^* and, Σ and Σ^*, respectively with regard to the typicality: for the universal quantification, Π^* is the quantifier acting on typical objects, while for the existential quantification, Σ is acting on typical objects. The reason is given by relations (3.26) and (3.27);
3. Q^* et Q are of different types. The type $\mathbf{F(FJH)(F(FJH)H)}$ is the type of Curry's quantifiers Q.

$$\frac{\dfrac{Q : \mathbf{F(FJH)(F(FJH)H)} \qquad f : \mathbf{FJH}}{Qf : \mathbf{F(FJH)H}} \qquad g : \mathbf{FJH}}{Qfg : H} \tag{3.28}$$

The type of star quantifiers, Q^*, is $\mathbf{F(FJH)J}$

$$\frac{g : \mathbf{FJH} \qquad \dfrac{Q^* : \mathbf{F(FJH)J} \qquad f : \mathbf{FJH}}{Q^* f : \mathbf{J}}}{g(Q^* f) : H} \tag{3.29}$$

4. From the viewpoint of cognition, it results from relations (3.24) and (3.25) that Π^* and Σ^* are more primitive than Π and Σ.

Generally speaking, we can state that the construction of typical objects raises a set of cognitive operations which determine the quantification. Natural languages express this aspect emphasized by the Star Theory.

From technical viewpoint, the star quantification of LDO is represented by the rules (3.5), (3.6), (3.12), (3.13), (3.14), (3.15), (3.16), (3.17). Star Theory is composed of the same rules adding the rules (3.23), (3.24) and (3.25).

The Star Theory is an extension of the quantification of illative theory expressed in Curry's Combinatory Logic.

4 Some Examples

In this section we give examples of formalization in LDO of some sentences of natural languages:

1. general universal quantification:
   (a) *Man is mortal. All men are mortal.*
       Π(to-be-a-man)(to-be-mortal)

2. typical universal quantification:
   (a) *All (typical) men are biped. A (typical) man is biped.*
       (to-be-biped)(Π^*(to-be-a-man))
   (b) *All (typical) Alsatian drinks beer. Every (typical) Alsatian drink beer.*
       (to-drink-beer)(Π^*(to-be-an-Alsatian)
   (c) *All (typical) Generals (of the Army) die in their bed. Every (typical) General dies in his bed.*
       (to-die-in-his-bed)(Π^*(to-be-a-general))
   (d) *All (typical) Frenches are moaners. Every (typical) French is a moaner.*
       (to-be-a-moaner)(Π^*(to-be-a-French))
   (e) *A (typical) beaver builds dams.*
       (to-build-dams)(Π^*(to-be-a-beaver))
3. general existential quantification:
   (a) *There is a prime number between 2 and 4.*
       (to-be-between-2-and-4)(Σ^*(to-be-a-prime-number))
   (b) *There is a man in the garden.*
       (to-be-in-the-garden)(Σ^*(to-be-a-man))
4. typical existential quantification:
   (a) *There is a (typical) prime number between 1 and 10. (Here "typical" means "odd number".)*
       Σ(to-be-between-1-and-10)(to-be-a-prime-number)
   (b) *There is an (atypical) continuous function. (Here "atypical" means "everywhere continuous but nowhere differentiable".)*
       $\Sigma(N_1$(to-be-differentiable))(to-be-a-continuous-function)

5 Conclusions

The system of quantifiers of LDO is a new approach of quantification opening new ways for solving general problems of quantification in relation with: predication operations, determination operations, topicalization operations (theme–rheme; topic–comment) analyzed by linguistics, on one hand, and typical and atypical representations of a concept, categorization approach inherited from investigations in anthropology and cognitive psychology [17], on the other hand.

This system of quantifiers captures two aspects of natural languages, one semantic and the other syntactic:

- From the semantic viewpoint, it captures the typicality/atypicality whose models are in LDO;
- From the syntactic viewpoint, it captures the fact that the quantifier is an operator applied to operand (noun phrase) as taken into account by linguistics.

The cube generalizing Aristotle's square is a diagram visualizing the relations between the four star quantifiers and the negation.

Acknowledgements   Many thanks to our reviewers for their interesting remarks.

References

1. Church, A.: The Calculi of Lambda Conversion. Princeton University Press, Princeton (1941)
2. Curry, H.B., Feys, R.: Combinatory Logic I. North-Holland, Amsterdam (1958)
3. Desclés, J.-P.: Implication entre concepts, la notion de typicalité. In: Travaux de linguistique et de lit-térature, XXIV, 1, pp. 79–102 (1986)
4. Desclés, J.-P.: Langages applicatifs, langues naturelles et cognition. Hermès, Paris (1990)
5. Desclés, J.-P.: Dialogue sur les prototypes et la typicalité. In: Denis, M., Sabah, G. (eds.) Modèles et concepts pour la science cognitive, hommage J.-F. Le Ny, pp. 139–163. Presses de l'Université de Grenoble, Grenoble (1993)
6. Desclés, J.-P., Cheong, K.-S.: Analyse critique de la notion de variable. Math. Sci. Hum. **173**, 43–102 (2006)
7. Desclés, J.-P., Guentcheva, Z.: Quantification without bound variables. In: Bőttner, M., Thűmmel, W. (eds.) Variable-Free Semantics, pp. 210–233. Secolo, Osnabrück (2000)
8. Desclés, J.-P., Pascu, A.: Logic of determination of objects—the meaning of variable in quantifica-tion. Int. J. Artif. Intell. Tools **15**(6), 1041–1052 (2006)
9. Desclés, J.-P., Pascu, A.: Logique de la détermination des objets (LDO): une logique pour l'analyse des langues naturelles. Rev. Roum. Linguist./Rom. Rev. Linguist. **LII**(1–2), 55–96 (2007)
10. Desclés, J.-P., Pascu, A.: Logic of determination of objects (LDO): how to articulate "extension" with "intension" and "objects with concepts. Logica Univers. **5**(1), 75–89 (2011)
11. Geach, P., Black, M.: Translation from the Philosophical Writings of Gottlob Frege (1891), 2nd edn. (with corrections). Oxford Basil Blackwell, Oxford (1960)
12. Gentzen, G.: Recherches sur la déduction logique. Traduit de l'allemand par Feys, R., Ladrière, J. Presses Universitaires de France, Paris (1955)
13. Hilbert, D., Bernays, P.: Grundlagen der Mathematik II. Springer, Berlin (1939)
14. Frege, G.: Grundgesetze der Arithmetik, begriffsschriftlich abgeleitet (1893). Translated and edited with an introduction by Furth, M.: Basic Laws of Arithmetic (Exposition of the System). University of California Press, Los Angeles (1967)
15. Frege, G.: Begriffsschrift, eine der arithmetischen nachgebildete Formelsprache des reinen Denkens. Halle (1879). Version francaise par Besson, C.: l'Idéographie. Vrin, Paris (1999)
16. Montague, R.: The proper treatment of quantification in ordinary English. In: Thomason, R.H. (ed.) Formal Philosophy: Selected Papers of Richard Montague. Yale University Press, New Haven (1974)
17. Rosch, E.: Principles of categorization. In: Cognition and Categorization. Lawrence Erlbaum, Hills-dale (1978)
18. Pascu, A.: Logique de Détermination d'Objets: concepts de base et mathématisation en vue d'une modélisation objet. Thèse de doctorat, Université de Paris-Sorbonne, Paris (2001)
19. Pascu, A.: Les objets dans la représentation des connaissances. Application aux processus de caté-gorisation en informatique et sciences humaines. Habilitation à diriger des recherches, Université de Paris-Sorbonne, Paris (2006)

J.-P. Desclés (✉)
LaLIC (Langues, Logiques, Informatique et Cognition), Université de Paris Sorbonne,
28, rue Serpente, 75006 Paris, France
e-mail: Jean-Pierre.Descles@paris-sorbonne.fr

A. Pascu
LaLIC, Université de Bretagne Occidentale Brest, 20, rue Duquesne, 93837-29238,
Brest Cedex 03, France
e-mail: Anca.Pascu@univ-brest.fr

Hypercubes of Duality

Thierry Libert

Abstract We define hypercubes of duality—of which the modern square of opposition is an emblematic example—in proper mathematical terms, as orbits under some action of the additive group $\mathbb{Z}_2^m$, with $m \in \mathbb{N}$. We then introduce a notion of dimension for duality in classical logic and show, for example, how propositional expressions in at most three variables can be classified according to that notion. The paper ends with logical formulations of some representation theorem for Galois connections on powersets, where we see an underlying square of duality.

Keywords Square of opposition · Duality · Galois connections

Mathematics Subject Classification 03B05 · 03B10 · 03B80

The square of opposition and other related diagrams are meant to picture logical relationships that can exist between propositions with the same terms. Their geometrical shape, however, simply reflects the way these propositions can be obtained from each other by means of simple operations on the terms. So such diagrams classify propositions according to their symmetric properties with respect to those operations. In mathematical terms, one will say that such diagrams represent the orbits for the action generated by the operations. It is that geometric view on logical diagrams we want to emphasize here.

On the other hand, propositions may themselves have natural geometric representations from which logical properties can be inferred. This is the case for propositional expressions, as we shall see. So geometry can help to understand certain forms of symmetry in logic, and then forms of reasoning. Whether reasoning has any geometrical root is another interesting question.

An important form of symmetry in (classical) logic is duality. Since this is a very broad concept, we have opted for a presentation in which this is the central notion, though we shall mainly restrict ourselves to logical duality. Let us start with a brief description of the kind of duality we have in mind.

Many forms of duality (but not all) involve the action of an involution.[1] Roughly, we are given a class $\mathscr{C}$ of mathematical objects[2] together with a map

$$\iota : \mathscr{C} \longrightarrow \mathscr{C} : c \longmapsto c^\iota$$

---

[1] There are references in various areas of logic where the notion of duality is introduced in a more general way; see [1] for instance.

[2] These can be semantic objects (structures) as well as syntactical ones (formulas).

J.-Y. Béziau, D. Jacquette (eds.), *Around and Beyond the Square of Opposition*, 293–301
Studies in Universal Logic, DOI 10.1007/978-3-0348-0379-3_20, © Springer Basel 2012

such that $(c^\iota)^\iota = c$ for each $c \in \mathscr{C}$. The proof of the corresponding principle of duality, however formulated, will essentially rely on that property of ι.

Nevertheless, there are situations where the underlying involution can be regarded as a product of finitely many *independent* involutions, so a better analysis can be given in terms of the action of the additive group $\mathbb{Z}_2^n$, $n \in \mathbb{N}$.[3] That is, we are given a class $\mathscr{C}$ of mathematical objects together with a map

$$\iota : \mathbb{Z}_2^n \times \mathscr{C} \longrightarrow \mathscr{C} : \langle \zeta, c \rangle \longmapsto c^{\iota(\zeta)}$$

such that for every $c \in \mathscr{C}$ and all $\zeta, \eta \in \mathbb{Z}_2^n$,

$$\left(c^{\iota(\zeta)} \right)^{\iota(\eta)} = c^{\iota(\zeta + \eta)} \quad \text{and} \quad c^{\iota(\mathbf{0})} = c.$$

We note that for all $\zeta \in \mathbb{Z}_2^n$, the map

$$\iota(\zeta) : \mathscr{C} \longrightarrow \mathscr{C} : c \longmapsto c^{\iota(\zeta)}$$

is an involution, since $\zeta + \zeta = \mathbf{0}$, and that all these maps are generated by the $\iota(\mathbf{1}_k)$'s, where $\mathbf{1}_k$ is the n-uple with 1 in the k-th position, 0 elsewhere. We will let $\mathbf{1}$ stand for $\sum_{k=1}^n \mathbf{1}_k$ and consider $\iota(\mathbf{1})$ as the principal involution; so objects c, c' in $\mathscr{C}$ such that $c' = c^{\iota(\mathbf{1})}$ will be said to be *dual* of each other.

As was said above, our guiding examples will be related to logical duality.

Example 1 Let $\mathscr{C}$ be the set of all propositional expressions up to logical equivalence— this means that we essentially looking at $\mathscr{C}$ as the set of all truth functions. We then define $\iota : \mathbb{Z}_2^2 \times \mathscr{C} \longrightarrow \mathscr{C}$ as follows:

- $\iota(\mathbf{1}_1)$ is negation acting internally, on each propositional variable;
- $\iota(\mathbf{1}_2)$ is negation acting externally, on the whole propositional expression.

So $\iota(\mathbf{1})$ is just the involution associated with duality in propositional logic.

On the other hand, when the number of propositional variables is bounded, it is also natural to look at the following refinement of Example 1:

Example 2 Let $m \in \mathbb{N}$ and then let $\mathscr{C}$ be the set of all propositional expressions in (at most) m variables, say $p_1, \ldots, p_m$, up to logical equivalence. We now define $\iota : \mathbb{Z}_2^{m+1} \times \mathscr{C} \longrightarrow \mathscr{C}$ as follows:

- For each $k \in \{1, \ldots, m\}$, $\iota(\mathbf{1}_k)$ is negation acting on p_k only;
- $\iota(\mathbf{1}_{m+1})$ is negation acting on the whole propositional expression.

So again $\iota(\mathbf{1})$ is the involution associated with duality in propositional logic, but restricted here to propositional expressions in (at most) m variables.

These examples can easily be adapted to the case of classical predicate logic, or even modal logic, and they have obvious semantic counterparts as well.

---

[3]Throughout this paper, $\mathbb{Z}_2^n$ will be regarded as the set of all n-uples of 0 and 1's with addition performed component-wise and modulo 2.

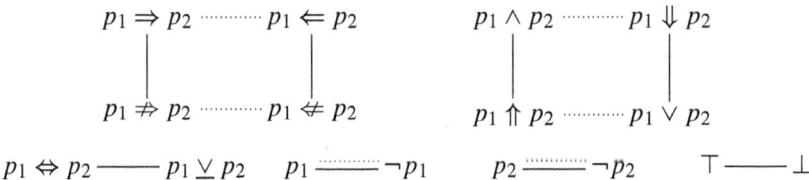

Fig. 1 Orbits for the action of Example 1

Now, given an action ι of $\mathbb{Z}_2^n$ on $\mathscr{C}$, one defines the *orbit* of $c \in \mathscr{C}$ by:

$$c^\iota := \left\{ c^{\iota(\zeta)} \mid \zeta \in \mathbb{Z}_2^n \right\}.$$

The cardinality of c^ι must be a power of 2, so it can be written as $2^{d(c)}$, where $d(c)$ is what we call the *dimension* of c w.r.t. ι. Then c^ι can be represented as a hypercube of $d(c)$ dimensions, which we refer to as a *hypercube of duality*. However, geometrically speaking, this is not a hypercube, for the symmetry group of c^ι is $\mathbb{Z}_2^{d(c)}$, by definition, and that is not the group of a hypercube unless $d(c) = 0$ or $d(c) = 1$. Typically, $\mathbb{Z}_2^2$ is known as the *Klein four-group*, and this is (isomorphic to) the symmetry group of a rectangle, not of a square!

In Fig. 1, we show all the orbits for the action of Example 1 when this is restricted to propositional variables in (at most) two variables. The actions of the generating maps $\iota(\mathbf{1}_1)$ and $\iota(\mathbf{1}_2)$, when not reduced to identity, have been depicted by ' $\cdots\cdots$ ' and ' $\underline{\hspace{1cm}}$ ' respectively. We have used (fairly) standard notations for the binary connectives, including the Sheffer stroke $\Uparrow$ ('NAND') and its dual, the Pierce arrow $\Downarrow$ ('NOR').[4] Given that there are exactly, up to logical equivalence, $2^{2^2} = 16$ propositional expressions in (at most) two variables, all possible orbits have been drawn. Obviously, the dimension of an object can never exceed 2 under the action of Example 1. That is why its orbits are referred to as "squares of quaternality" in [4].

On the other hand, given $m \geq 2$, one can find objects of dimension $m + 1$ by considering the action of Example 2. We show all possible orbits for $m = 2$ in Fig. 2, where the actions of the generating maps $\iota(\mathbf{1}_1)$, $\iota(\mathbf{1}_2)$ and $\iota(\mathbf{1}_3)$—again when not reduced to identity— have been depicted by ' $\cdots\cdots$ ', ' $---$ ' and ' $\underline{\hspace{1cm}}$ ' respectively. The reader will notice that there is no object of dimension 2 here. This fact, and all other features of Fig. 2, can easily be explained from a pure geometric viewpoint, by looking at propositional expressions as square-shaped objects.

Indeed, up to logical equivalence, each propositional expression in two variables can be identified with its truth function/table $\{0, 1\}^2 \longrightarrow \{0, 1\}$, which can then be regarded as a 2×2 matrix of 0's and 1's, or alternatively, as a square-shaped domino with white or black dots (white for 0, black for 1). In this view, the actions of $\iota(\mathbf{1}_1)$ and $\iota(\mathbf{1}_2)$ correspond to bilateral symmetries with vertical and horizontal axes respectively, whereas the action

---

[4]Curiously, it seems there is no natural *positive* notation in the literature for the dual connective of $\Rightarrow$, that is, the binary connective that is denoted here by $\nRightarrow$. Same for $\Leftarrow$.

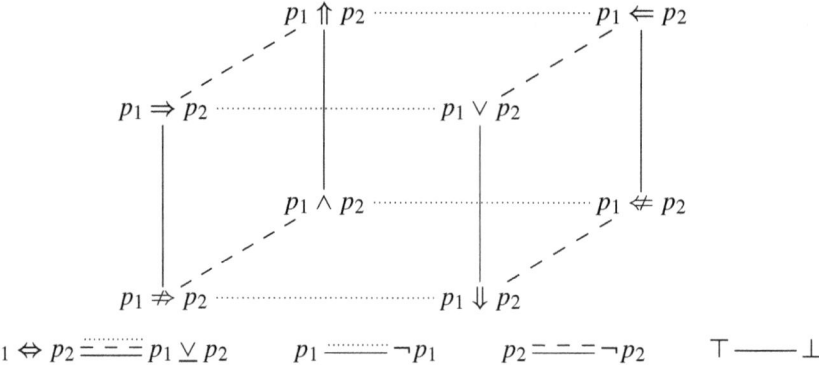

Fig. 2 Orbits for the action of Example 2, with $m = 2$

of $\iota(1_3)$ consists in taking the *negative* image, that is, reversing black and white. Thus, for instance, the orbit of $p_1 \Leftrightarrow p_2$ and $p_1 \veebar p_2$ in Fig. 2 will appear as

where we see that bilateral symmetries just interchange white and black dots, so negating one variable does correspond to negating the whole expression.

That geometric view of the action of Example 2 generalizes to any $m \in \mathbb{N}$, by thinking of (the truth function/table of) a propositional expression in m variables as a hypercube-shaped domino of dimension m—which remains intuitive as far as $m \leqslant 3$. For example, here are the cube-shaped representations of $p_1 \Rightarrow (p_2 \Rightarrow p_3)$, $(p_1 \Rightarrow p_2) \Rightarrow p_3$ and $(p_1 \Rightarrow p_2) \Leftrightarrow p_3$, respectively:

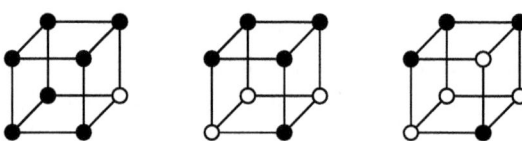

It is easy to see from these that $p_1 \Rightarrow (p_2 \Rightarrow p_3)$ and $(p_1 \Rightarrow p_2) \Rightarrow p_3$ are both of dimension 4 w.r.t. the action of Example 2, though they lie in different orbits, whereas $(p_1 \Rightarrow p_2) \Leftrightarrow p_3$ is of dimension 3.[5] In fact, with that view, one could easily list all the orbits for the action of Example 2 when $m = 3$ and prove, among other things, that there is again no square. See the Appendix.

There are objects of dimension 2 for the predicate and the modal versions of Example 2 (with $m \geqslant 1$). Natural examples are provided by the formulas $\forall x P_1(x)$, where P_1 is a predicate variable, and $\Box p_1$ in the modal case. Their orbits are traditionally represented

---

[5]Clearly, $\iota(1_3)$ & $\iota(1_4)$ have the same action on $(p_1 \Rightarrow p_2) \Leftrightarrow p_3$, not on $(p_1 \Rightarrow p_2) \Rightarrow p_3$.

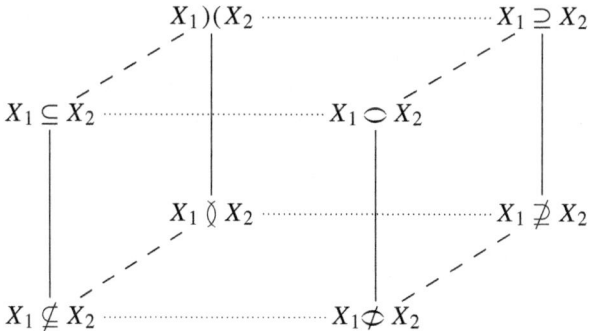

Fig. 3 The cube of Aristotelian relations

in the literature with *contradictory* statements in opposition to each other, rather than dual ones, as follows:

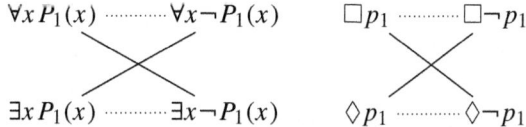

And these are referred to as "modern squares of opposition" in the literature.

The most famous example of modern square of duality/opposition is the Aristotelian one. This just appears as the orbit of the formula $\forall_P x\, P_1(x)$, where '$\forall_P x$' is some *restricted* quantifier, that is, $\forall_P x\, P_1(x)$ is the abbreviation for $\forall x (P(x) \Rightarrow P_1(x))$, where P is some predicate constant—which is tacitly assumed not to be identically false, otherwise the orbit will reduce to $\{\top, \bot\}$. It should be remarked too that the orbit will not be of dimension 2 either when $\{x \mid P(x)\}$ is a singleton, in which case $\forall_P x\, P_1(x)$ and $\exists_P x\, P_1(x)$ are obviously logically equivalent. Typically, in predicate logic with equality, this happens by taking $P(x)$ to be $x = a$, where a is any object constant.

That said, the orbit of $\forall x (P_1(x) \Rightarrow P_2(x))$ for the predicative version of the action of Example 2 (with $m \geqslant 2$) will naturally provide us with a 3-dimensional generalization of the Aristotelian square. This is referred to as "the cube of opposition" in [6]. We show a semantic version of it in Fig. 3, where X_1, X_2 are variables for subsets of some (nonempty) set U, and where the actions of $\iota(\mathbf{1}_1)$ ('⋯⋯') and $\iota(\mathbf{1}_2)$ ('− − −') correspond to complementation (w.r.t. U) acting on X_1 and X_2 respectively.

The binary relations symbolized at the eight corners of that cube will be referred to as the *Aristotelian relations*. These arise in a natural way from the Venn diagram for two subsets of U by writing either 0 or 1 in each of the four regions, to mean that this is either empty or non-empty. Some notations are quite standard in the literature, or natural, such as $X_1 \between X_2$ for $X_1 \cap X_2 \neq \emptyset$ and $X_1)(X_2$ for its negation.[6] However, notations for their duals seem hard to find; by default, we have used here $X_1 \subset\!\supset X_2$ for $X_1 \cup X_2 = U$ and $X_1 \not\subset\!\supset X_2$ for its negation.[7] The cube of Aristotelian relations is just the predicate counterpart of the cube of duality appearing in Fig. 2.

---

[6]The use of $\between$ can be found in [2] for instance, but there is no notation for its negation.

[7]Again, it seems there is no natural *positive* notation for the dual of the inclusion relation $\subseteq$, that is, the binary relation that is denoted here by $\not\supseteq$. Likewise for $\supseteq$.

These so-called Aristotelian relations, especially $\subseteq$, $\not\subseteq$, $\emptyset$ and $)($, are particularly helpful—we think—to express and to teach Aristotelian syllogisms. We shall now describe another situation where reference to Aristotelian relations can be illuminating, which was precisely the subject of the talk given by the author at the 2nd World Congress on the Square of Opposition.

1 The Square of Galois Connections

Recall that, given posets $\mathcal{A}$ and $\mathcal{B}$, a *Galois connection*—also called an *adjunction*—from $\mathcal{A}$ to $\mathcal{B}$ is defined as a pair $\langle \alpha, \beta \rangle$ of maps $\alpha : \mathcal{A} \longrightarrow \mathcal{B}$ and $\beta : \mathcal{B} \longrightarrow \mathcal{A}$ such that $x \leqslant_{\mathcal{A}} \beta(y)$ iff $\alpha(x) \leqslant_{\mathcal{B}} y$, for all x in $\mathcal{A}$ and y in $\mathcal{B}$. Then α is called the *lower* or *left adjoint*, and β the *upper* or *right adjoint*. That special kind of correspondence between posets pervades mathematics. There is a more general definition of a Galois connection in categorical terms, of which the one above is just a particular instance. However, we shall restrict ourselves in what follows to the case where the posets $\mathcal{A}$, $\mathcal{B}$ are powersets.

Then, given sets A and B, one can distinguish four types of Galois connections from $\mathcal{P}(A)$ to $\mathcal{P}(B)$, according to whether each of the powersets is equipped with $\subseteq$ or $\supseteq$. So the defining property of a Galois connection from $\mathcal{P}(A)$ to $\mathcal{P}(B)$ may take one of following four forms: for all $X \subseteq A$, $Y \subseteq B$,

1. $X \subseteq \beta(Y)$ iff $\alpha(X) \subseteq Y$.
2. $X \subseteq \beta(Y)$ iff $Y \subseteq \alpha(X)$.
3. $\beta(Y) \subseteq X$ iff $\alpha(X) \subseteq Y$.
4. $\beta(Y) \subseteq X$ iff $Y \subseteq \alpha(X)$.

In the pioneered work on the subject, as in [5], the definition of a Galois connection between powersets is based on 2, because the motivating example—in Galois theory precisely—is of type 2. As a matter of fact, there is a natural *generic* example for such Galois connections, as follows.

For any binary relation R and set X, let $X_R := \{y \mid \forall x \in X, \ x \, R \, y\}$.

Now, given $R \subseteq A \times B$, define $\alpha : \mathcal{P}(A) \longrightarrow \mathcal{P}(B)$, $\beta : \mathcal{P}(B) \longrightarrow \mathcal{P}(A)$ by

$$\alpha(X) := X_R, \qquad \beta(Y) := Y_{R^*},$$

for all $X \subseteq A$, $Y \subseteq B$, where R^* is the opposite relation of R. Then it is easy to see that $\langle \alpha, \beta \rangle$ is a Galois connection of type 2 from $\mathcal{P}(A)$ to $\mathcal{P}(B)$.

Furthermore, it is proved in [5] that *any* Galois connection of type 2 from $\mathcal{P}(A)$ to $\mathcal{P}(B)$ is of this form, for some binary relation $R \subseteq A \times B$, which gives a representation theorem for Galois connection of type 2. Representation theorems for the other types can be derived by noting that one form can be changed into another by complementation acting on $\mathcal{P}(A)$ or $\mathcal{P}(B)$—there is indeed a square of duality here! Surprisingly, little is said about that in the literature, except perhaps in [3], but not in simple terms. Here are formulations that—we think—best reflect the underlying logical structure.

Theorem v.1 *Suppose $\langle \alpha, \beta \rangle$ is a Galois connection from $\mathcal{P}(A)$ to $\mathcal{P}(B)$. There exists $R \subseteq A \times B$ such that, if $\langle \alpha, \beta \rangle$ is respectively of type* 1, 2, 3, 4, *then: for all $X \subseteq A$, $Y \subseteq B$,*

1. $\alpha(X) = \{y \mid \{y\}_{R^*} \not\supseteq X\}$, $\beta(Y) = \{x \mid \{x\}_R \subset Y\}$.
2. $\alpha(X) = \{y \mid \{y\}_{R^*} \supseteq X\}$, $\beta(Y) = \{x \mid \{x\}_R \supseteq Y\}$.
3. $\alpha(X) = \{y \mid \{y\}_{R^*} \not\subset X\}$, $\beta(Y) = \{x \mid \{x\}_R \not\subset Y\}$.
4. $\alpha(X) = \{y \mid \{y\}_{R^*} \subset X\}$, $\beta(Y) = \{x \mid \{x\}_R \not\supseteq Y\}$.

Proof First we observe that $X_R = \{y \mid \{y\}_{R^*} \supseteq X\}$, by definition of X_R, since $\{y\}_{R^*} = \{x \mid x \, R \, y\}$, again by definition of X_R, for all R and X.

Now suppose $\langle \alpha, \beta \rangle$ is a Galois connection of type 1 from $\mathcal{P}(A)$ to $\mathcal{P}(B)$. Then $\langle (\_)^c \circ \alpha, \beta \circ (\_)^c \rangle$, where $(\_)^c$ is complementation, is a Galois connection of type 2 from $\mathcal{P}(A)$ to $\mathcal{P}(B)$, since $X \subseteq \beta(Y^c)$ iff $\alpha(X) \subseteq Y^c$ iff $Y \subseteq \alpha(X)^c$, for all $X \subseteq A, Y \subseteq B$. Therefore, assuming the representation theorem for type 2, there exists $R \subseteq A \times B$ such that for all $X \subseteq A, Y \subseteq B$,

$$\alpha(X)^c = \{y \mid \{y\}_{R^*} \supseteq X\}, \qquad \beta(Y^c) = \{x \mid \{x\}_R \supseteq Y\},$$

and then, for all $X \subseteq A, Y \subseteq B$,

$$\alpha(X) = \{y \mid \{y\}_{R^*} \supseteq X\}^c, \qquad \beta(Y) = \{x \mid \{x\}_R \supseteq Y^c\},$$

that is,

$$\alpha(X) = \{y \mid \{y\}_{R^*} \not\supseteq X\}, \qquad \beta(Y) = \{x \mid \{x\}_R \subset Y\}.$$

Likewise for type 3 [resp. 4], seeing that if $\langle \alpha, \beta \rangle$ is of type 3 [resp. 4], then $\langle (\_)^c \circ \alpha \circ (\_)^c, (\_)^c \circ \beta \circ (\_)^c \rangle$ [resp. $\langle \alpha \circ (\_)^c, (\_)^c \circ \beta \rangle$] is of type 2. $\qquad \square$

Equally, there is another formulation involving $\subseteq, \not\subseteq, \emptyset$ and $)($, namely:

Theorem v.2 *Suppose $\langle \alpha, \beta \rangle$ is a Galois connection from $\mathcal{P}(A)$ to $\mathcal{P}(B)$. There exists $S \subseteq A \times B$ such that, if $\langle \alpha, \beta \rangle$ is respectively of type 1, 2, 3, 4, then: for all $X \subseteq A, Y \subseteq B$,*

1. $\alpha(X) = \{y \mid \{y\}_{S^*} \, \emptyset \, X\}$, $\beta(Y) = \{x \mid \{x\}_S \subseteq Y\}$.
2. $\alpha(X) = \{y \mid \{y\}_{S^*})(X\}$, $\beta(Y) = \{x \mid \{x\}_S)(Y\}$.
3. $\alpha(X) = \{y \mid \{y\}_{S^*} \not\subseteq X\}$, $\beta(Y) = \{x \mid \{x\}_S \not\subseteq Y\}$.
4. $\alpha(X) = \{y \mid \{y\}_{S^*} \subseteq X\}$, $\beta(Y) = \{x \mid \{x\}_S \, \emptyset \, Y\}$.

Proof Apply v.1 and take $S := R^c$, so $\{y\}_{R^*} = \{y\}^c_{S^*}$ and $\{x\}_R = \{x\}^c_S$. $\qquad \square$

Appendix

Here we list all the orbits for the action of Example 2 when $m = 3$. These are classified according to the number b of black dots in the cube-shape representation of a propositional expression, and their possible arrangements up to permutations of the variables p_1, p_2, p_3. For each configuration C_k^b, we give one generic propositional expression as example, with dimension $|C_k^b|$. The index k just serves here as a numbering for different configurations with the same b. We use multiplicative notation (namely a dot) for conjunction and write E_{ij} for $(p_i \Leftrightarrow p_j)$, with $i \neq j$. Note that the actions of $\iota(1_1), \iota(1_2), \iota(1_3)$ preserve the numbers of white dots and black dots, whereas the action of $\iota(\mathbf{1}_4)$ must exchange these numbers; so the values of b are restricted to 0, 1, 2, 3, 4.

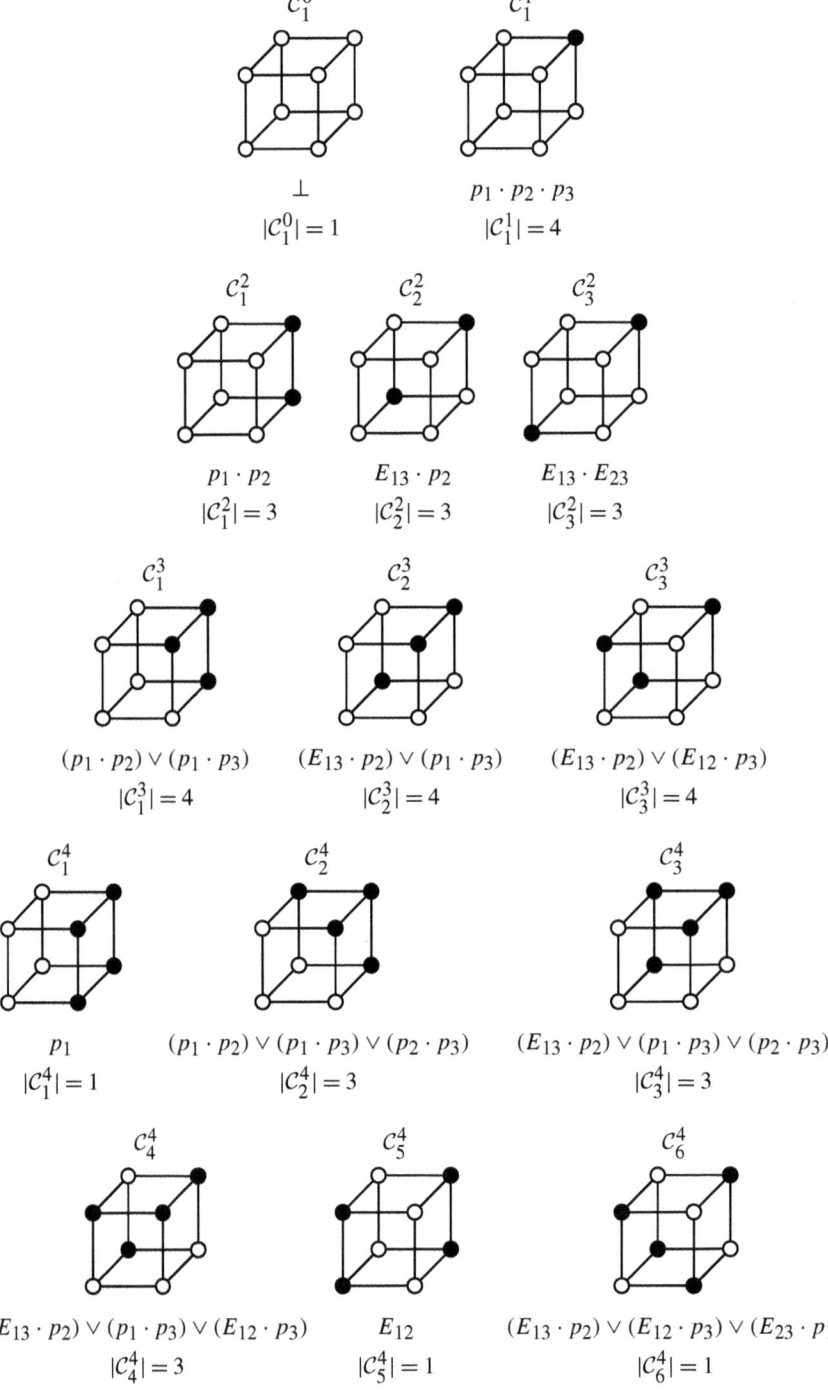

$$\mathcal{C}_1^0$$

$$\perp$$
$$|\mathcal{C}_1^0| = 1$$

$$\mathcal{C}_1^1$$

$$p_1 \cdot p_2 \cdot p_3$$
$$|\mathcal{C}_1^1| = 4$$

$$\mathcal{C}_1^2 \qquad\qquad \mathcal{C}_2^2 \qquad\qquad \mathcal{C}_3^2$$

$$p_1 \cdot p_2 \qquad\qquad E_{13} \cdot p_2 \qquad\qquad E_{13} \cdot E_{23}$$
$$|\mathcal{C}_1^2| = 3 \qquad\qquad |\mathcal{C}_2^2| = 3 \qquad\qquad |\mathcal{C}_3^2| = 3$$

$$\mathcal{C}_1^3 \qquad\qquad \mathcal{C}_2^3 \qquad\qquad \mathcal{C}_3^3$$

$$(p_1 \cdot p_2) \vee (p_1 \cdot p_3) \qquad (E_{13} \cdot p_2) \vee (p_1 \cdot p_3) \qquad (E_{13} \cdot p_2) \vee (E_{12} \cdot p_3)$$
$$|\mathcal{C}_1^3| = 4 \qquad\qquad |\mathcal{C}_2^3| = 4 \qquad\qquad |\mathcal{C}_3^3| = 4$$

$$\mathcal{C}_1^4 \qquad\qquad \mathcal{C}_2^4 \qquad\qquad \mathcal{C}_3^4$$

$$p_1 \qquad (p_1 \cdot p_2) \vee (p_1 \cdot p_3) \vee (p_2 \cdot p_3) \qquad (E_{13} \cdot p_2) \vee (p_1 \cdot p_3) \vee (p_2 \cdot p_3)$$
$$|\mathcal{C}_1^4| = 1 \qquad\qquad |\mathcal{C}_2^4| = 3 \qquad\qquad |\mathcal{C}_3^4| = 3$$

$$\mathcal{C}_4^4 \qquad\qquad \mathcal{C}_5^4 \qquad\qquad \mathcal{C}_6^4$$

$$(E_{13} \cdot p_2) \vee (p_1 \cdot p_3) \vee (E_{12} \cdot p_3) \qquad E_{12} \qquad (E_{13} \cdot p_2) \vee (E_{12} \cdot p_3) \vee (E_{23} \cdot p_1)$$
$$|\mathcal{C}_4^4| = 3 \qquad\qquad |\mathcal{C}_5^4| = 1 \qquad\qquad |\mathcal{C}_6^4| = 1$$

The generic propositional expression for $\mathcal{C}_6^4$ is in fact logically equivalent to $p_1 \Leftrightarrow p_2 \Leftrightarrow p_3$, which is not to be confused with $E_{12} \cdot E_{13} \cdot E_{23}$, that is, $\mathcal{C}_3^2$. Note also that the ternary connective 'If..., then...; else...', i.e. the propositional expression $(p_1 \Rightarrow p_2) \wedge (\neg p_1 \Rightarrow p_3)$, coincides with $(\mathcal{C}_3^4)^{\iota(\mathbf{1}_1)}$.

Now, in the next table, we count the number n_k^b of orbits one obtains by permuting the variables p_1, p_2, p_3 from each configuration $\mathcal{C}_k^b$, and then the number $m^b = \sum_k 2^{|\mathcal{C}_k^b|} \cdot n_k^b$ of propositional expressions whose orbits are generated by a cube-shaped domino with b black dots.

n_k^b	0	1	2	3	4
1	1	1	3	3	3
2	–	–	3	3	1
3	–	–	1	1	3
4	–	–	–	–	3
5	–	–	–	–	3
6	–	–	–	–	1
m^b	2	16	56	112	70

So the number of orbits is $\sum_{b,k} n_k^b = 30$, and we have $\sum_b m^b = 256$, which is all right equal to the number of propositional expressions in (at most) three variables—up to logical equivalence. For those familiar with finite group theory, we note that the total number of orbits under the action Example 2 can easily be obtained, for any m, by Burnside's counting theorem.

References

1. Adámek, J., Herrlich, H., Strecker, G.E.: Abstract and Concrete Categories: The Joy of Cats. Dover, Mineola (2004)
2. Dubreil, P.: Algèbre, Tome 1, 2ième édn. Cahiers scientifiques, Fascicule XX. Gauthier-Villars, Paris (1954)
3. Erné, M., Koslowski, J., Melton, A., Strecker, G.E.: A primer on Galois connections. In: Papers on General Topology and Applications. Annals of the New York Academy of Sciences, vol. 704, pp. 103–125 (1993)
4. Gottschalk, W.H.: The theory of quaternality. J. Symb. Log. **18**, 193–196 (1953)
5. Ore, O.: Galois connexions. Trans. Am. Math. Soc. **55**, 493–513 (1944)
6. Reichenbach, H.: The syllogism revised. Philos. Sci. **19**, 1–16 (1952)

T. Libert (✉)
Département de mathématiques, Université Libre de Bruxelles (U.L.B.), Bd. du triomphe, CP 211, 1050 Brussels, Belgium
e-mail: tlibert@ulb.ac.be

Part V
Applications of the Square of Opposition

How to Square Knowledge and Belief

Wolfgang Lenzen

Abstract Taken for granted that knowledge satisfies the "truth-requirement", $K(a, p) \rightarrow p$, it immediately follows that the elementary epistemic attitudes $K(a, p)$, $K(a, \neg p)$, together with their respective negations, $\neg K(a, p)$, $\neg K(a, \neg p)$, fit into a standard modal "square of knowledge". In contrast to $K(a, p)$, the operator of belief, $B(a, p)$, does not satisfy the truth-requirement. Even if 'belief' is interpreted in the strong sense of *certainty* (i.e. maximal subjective probability), it still doesn't follow that whenever subject *a* is firmly convinced that *p*, *p* must therefore be true—as we all know, humans are fallible. Nevertheless, if it is only assumed that the concept of belief satisfies the "consistency principle" $B(a, p) \rightarrow \neg B(a, \neg p)$, one obtains a standard "square of belief". In view of the so-called entailment-thesis, $K(a, p) \rightarrow B(a, p)$, both squares can be combined into a corresponding "double square" where, interestingly, given certain assumptions of epistemic/doxastic logic, the doxastic "inner square" may be shown to be *identical* to the following square of *complex* epistemic modalities:

$$\neg K\big(a, \neg K(a, p)\big) \qquad \neg K\big(a, \neg K(a, \neg p)\big)$$
$$K\big(a, \neg K(a, \neg p)\big) \qquad K\big(a, \neg K(a, p)\big).$$

Keywords Epistemic logic · Modal square of opposition · Knowledge · Weak and strong belief

Mathematics Subject Classification Primary 03B42 · Secondary 03B45

1 Modal Square of Opposition

Before investigating the issue of *epistemic* and *doxastic* squares of opposition, let me briefly summarize the basic idea of a *modal* square in general. Abstractly speaking (see Fig. 1), four propositions form a square of opposition if and only if

1. *A* and *B* are "contrary" to each other, i.e. *A* and *B* cannot both be true while *A* and *B* can both be false together;
2. *C* and *D* are "subaltern" to *A* and *B*, respectively, i.e. *A* logically entails *C* and *B* logically entails *D*;
3. *A* and *D* on the one hand, and *B* and *C* on the other hand, are "contradictory" to each other, i.e. $D = \neg A$ and $C = \neg B$;

J.-Y. Béziau, D. Jacquette (eds.), *Around and Beyond the Square of Opposition*, 305–311
Studies in Universal Logic, DOI 10.1007/978-3-0348-0379-3_21, © Springer Basel 2012

Fig. 1 Abstract square of
opposition

$$A \quad B$$
$$C \quad D$$

Fig. 2 Modal square of
opposition

$$\Box p \qquad \Box \neg p$$
$$\neg \Box \neg p \qquad \neg \Box p$$

4. C and D are "subcontrary", i.e. C and D cannot both be false while C and D can both
   be true together.

Note that it would suffice to postulate, e.g., conditions 1 and 3 because the other requirements 2, 4 logically follow from these.[1]

If, as usual, '$\Box$' denotes the operator of *necessity*, then the expressions $A = \Box p$, $B = \Box \neg p$, $C = \neg \Box \neg p$, and $D = \neg \Box p$ form a (modal) square of opposition provided only
that two conditions hold:

(i) Necessity entails possibility, $(\Box p \rightarrow \Diamond p)$,[2] but conversely
(ii) possibility does *not* entail necessity.

If the latter would not hold, all modal distinctions would collapse, i.e. necessity would
become provably equivalent to simple truth, $(\Box p \leftrightarrow p)$, and therefore also to possibility,
$(\Box p \leftrightarrow \Diamond p)$. But then the corresponding Fig. 2 would no longer constitute a square of
opposition because the subaltern formulas $\Diamond p$ and $\Diamond \neg p$ fail to satisfy the second part of
condition 4. These formulas, which are equivalent to p and $\neg p$, respectively, cannot both
be true together.

2 Epistemic Square of Opposition

In his contribution to the 1st International Congress on The Square of Opposition, Pascal Engel argued that the question "Can there be an Epistemic Square of Opposition?"
probably should be answered in the negative.

> It seems (more or less) easy to transpose the alethic square of opposition [i.e., Fig. 2] to deontic
> modalities, but what about epistemic modalities? It is tempting to put knowledge at the top left,
> ignorance at the top right, belief at the bottom left and disbelief at the bottom right.[3]

However, his attempt to construct a square of opposition out of a *mixture* of epistemic
and doxastic modalities along the quoted lines is bound to fail for several reasons. For the
sake of a closer discussion, let us use the standard abbreviations '$K(a, p)$' and '$B(a, p)$'

---

[1] Assume that condition 1 is satisfied and that A is true; then B must be false and hence, in view of
condition 3, C must be true, i.e. A logically entails C. Similarly, B logically entails D since, if B is true,
A must be false (by 1) and hence $\neg A$, i.e. (by 3) D, must be true. Furthermore, if A and B are contrary to
each other, then C and D cannot both be false, because otherwise $\neg C$ and $\neg D$ would both be true, i.e.,
in view of condition 3, A and B would both be true together.

[2] Of course, the possibility operator, $\Diamond p$, may be defined as $\neg \Box \neg p$.

[3] Engel [1], 5.

Fig. 3 Engel's (partial)
square of opposition

$$K(a, p) \quad B$$
$$B(a, p) \quad D$$

for the epistemic operators '(person) a knows that p' and 'a believes that p', respectively. According to Engel, these propositions are meant to occupy the A- and C-position in the left half of the square of opposition shown in Fig. 3.

Since the C-proposition is subaltern to the A-proposition, $B(a, p)$ has to follow logically from $K(a, p)$. This is exactly the content of the so-called "entailment-thesis" saying that knowing that p entails believing that p:

$$K(a, p) \rightarrow B(a, p). \hspace{3cm} \text{(ENTAILMENT 1)}$$

The other two propositions B and D are thought by Engel to represent the negative attitudes of "ignorance" and "disbelief", respectively. What does this, however, exactly mean? 'Disbelief' would normally be understood as the mere *negation* of belief, hence,

$$D_1(a, p) := \neg B(a, p). \hspace{3cm} \text{(DISBELIEF 1)}$$

However, also the stronger requirement

$$D_2(a, p) := B(a, \neg p) \hspace{3cm} \text{(DISBELIEF 2)}$$

appears reasonable. Similarly, 'ignorance' might be understood in either of the following two senses:

$$I_1(a, p) := \neg K(a, p) \hspace{3cm} \text{(IGNORANCE 1)}$$

$$I_2(a, p) := K(a, \neg p). \hspace{3cm} \text{(IGNORANCE 2)}$$

Now, no matter which particular interpretation of 'disbelief' and 'ignorance' is chosen, the resulting completion of Fig. 2—with either $\neg K(a, p)$ or $K(a, \neg p)$ substituted in place of the B-proposition and with either $\neg B(a, p)$ or $B(a, \neg p)$ substituted in place of the D-proposition—will *not* yield any square of opposition. In the first variant (see Fig. 4), none of the four conditions of a square of opposition would be satisfied. First, the A and B formulae $K(a, p)$ and $\neg K(a, p)$ are contradictory, but *not contrary* to each other, i.e. they cannot both be false together. Second, according to (ENTAILMENT 1), the A-formula $K(a, p)$ logically entails the C-formula $B(a, p)$, hence, by contraposition, the D-formula $\neg B(a, p)$ entails the B-formula $\neg K(a, p)$ while in a square of opposition the converse entailment $\neg K(a, p) \rightarrow \neg B(a, p)$ would have to hold. Third, since belief is not the same as knowledge, the D-proposition $\neg B(a, p)$ is not the negation of the A-proposition $K(a, p)$, and similarly the C-proposition $B(a, p)$ is not the negation of the B-proposition $\neg K(a, p)$. Fourth, the subaltern formulae $B(a, p)$ and $\neg B(a, p)$ are contradictory, but *not subcontrary* to each other, i.e. they cannot both be true together.

In the second variant (see Fig. 5), the A and B formulae, $K(a, p)$ and $K(a, \neg p)$, are contradictory to each other, as is required by condition 1 of the square of opposition. Furthermore, the law of subalternation is fully satisfied since $K(a, p)$ logically entails

Fig. 4 Complete square of
opposition à la Engel (1st
variant)

$$K(a, p) \quad \neg K(a, p)$$
$$B(a, p) \quad \neg B(a, p)$$

Fig. 5 Complete square of
opposition à la Engel (2nd
variant)

$$K(a, p) \qquad K(a, \neg p)$$

$$B(a, p) \qquad B(a, \neg p)$$

$B(a, p)$, and analogously $K(a, \neg p)$ entails $B(a, \neg p)$. However, the negation of the A-proposition, $\neg K(a, p)$, certainly is not *equivalent* to the D-proposition $B(a, \neg p)$. To be sure, if a believes that p is not true, it hence follows that a doesn't *know* that p; but conversely, the mere fact that a doesn't know that p fails to guarantee that a believes that not-p! In view of the so-called truth-condition (saying that a cannot know that p unless p in fact is true),

$$K(a, p) \rightarrow p \qquad \text{(TRUTH)}$$

it might well be the case that p is false, and that *therefore* $\neg K(a, p)$, while a nevertheless *believes* that p is true, $B(a, p)$, and hence a doesn't disbelieve that p![4] Finally, the formulae C and D are contrary rather than "subcontrary" because it is well nigh possible that both C and D are false together! If a is any normal human subject, there exist many propositions p such that a neither believes that p nor believes that not-p![5]

3 Squares of Opposition for Knowledge, Strong Belief and Weak Belief

In order to answer the question "how to square knowledge and belief" properly, it should be observed that since the knowledge-operator $K(a, p)$ satisfies the (TRUTH)-requirement, it also satisfied the epistemic analogue of the modal condition (i) mentioned above, i.e.:

$$K(a, p) \rightarrow \neg K(a, \neg p). \qquad \text{(K-CONSISTENCY)}$$

Since, furthermore, the converse implication $\neg K(a, \neg p) \rightarrow K(a, p)$ does not generally hold,[6] it immediately follows that the elementary epistemic attitudes $K(a, p)$, $K(a, \neg p)$, together with their respective negations, fit into the "square of knowledge" shown in Fig. 6.

Fig. 6 Square of knowledge

$$K(a, p) \qquad K(a, \neg p)$$

$$\neg K(a, \neg p) \qquad \neg K(a, p)$$

---

[4]In the same way, the C-proposition $B(a, p)$ entails, but is not itself entailed by, the negation of the B-proposition, $\neg K(a, \neg p)$.

[5]As the reader may verify, also the remaining two "inhomogeneous" variants

$$\begin{array}{ll} K(a, p) & \neg K(a, p) \\ B(a, p) & B(a, \neg p) \end{array} \quad \text{and} \quad \begin{array}{ll} K(a, p) & K(a, \neg p) \\ B(a, p) & \neg B(a, p) \end{array}$$

fail to satisfy several conditions for a square of opposition.

[6]Such a condition at best holds for a supernatural omniscient being like God.

Fig. 7 Square of "strong belief"

$$C(a, p) \qquad C(a, \neg p)$$
$$\neg C(a, \neg p) \qquad \neg C(a, p)$$

Fig. 8 Square of "weak belief"

$$B(a, p) \qquad B(a, \neg p)$$
$$\neg B(a, \neg p) \qquad \neg B(a, p)$$

In contrast to $K(a, p)$, the belief operator $B(a, p)$ does *not* satisfy the (TRUTH)-requirement, but it is quite reasonable to suppose that it satisfies at least the weaker requirement of

$$B(a, p) \to \neg B(a, \neg p). \qquad \text{(B-CONSISTENCY)}$$

For a closer discussion of this topic, it will be helpful to distinguish (at least) two different notions of belief. In the everyday use of the notion, to *believe* that something, p, is the case means as much as to think that p is *probable*, i.e. more precisely, that p is *more probable* than not-p. Let this concept of "*weak* belief" be symbolized by $B(a, p)$. Furthermore a new operator $C(a, p)$ shall be introduced to abbreviate the special case of "*strong* belief", or conviction, which holds if and only if the respective subject a is absolutely *certain* that p. With the help of a subjective probability function $\text{Prob}(a, p)$ which numerically defines the degree of likelihood that a assigns to (the event described by proposition) p, the two notions of belief might be defined as follows:

$$B(a, p) := \text{Prob}(a, p) > \text{Prob}(a, \neg p), \text{ i.e. } \text{Prob}(a, p) > 0.5 \qquad \text{(PROB 1)}$$

$$C(a, p) := \text{Prob}(a, p) = 1. \qquad \text{(PROB 2)}$$

These definitions immediately entail that "strong belief", $C(a, p)$, is just a special form of "weak belief", for if $\text{Prob}(a, p) = 1$, then a fortiori $\text{Prob}(a, p) > 0.5$; hence we obtain principle

$$C(a, p) \to B(a, p). \qquad \text{(ENTAILMENT 2)}$$

Now, even if 'belief' is taken in the strong sense of subjective *certainty*, $C(a, p)$, it still doesn't follow that whenever subject a is firmly *convinced* that p, p must therefore be *true*. As we all know, humans are *fallible*. However, the concept of "strong belief" satisfies the principle of

$$C(a, p) \to \neg C(a, \neg p), \qquad \text{(C-CONSISTENCY)}$$

which maintains that if subject a is firmly convinced that p (is true), a cannot simultaneously be convinced that $\neg p$, i.e. that p is false. This follows immediately from the probabilistic interpretation (PROB 2), for if $\text{Prob}(a, p) = 1$, then $\text{Prob}(a, \neg p) = 0$, i.e. a is not at all convinced that $\neg p$, but a is rather convinced that $\neg p$ is false!

Similarly, the weaker condition (PROB 1) guarantees that if a weakly believes that p, a will not also weakly believe that $\neg p$, for according to the standard laws of subjective probability, if $\text{Prob}(a, p) > 0.5$, then $\text{Prob}(a, \neg p) = (1 - \text{Prob}(a, p))$ and hence $\text{Prob}(a, \neg p) < 0.5$. Thus principle (B-CONSISTENCY) is satisfied. In view of the general results concerning modal squares, it follows that both notions of "strong" and "weak" belief give rise to the squares of opposition shown in Figs. 7 and 8.

Fig. 9 Doxastic square of
opposition

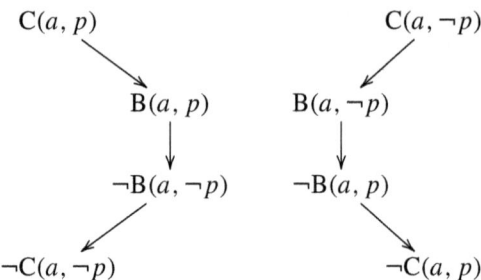

In view of principle (ENTAILMENT 2), these two squares can be combined into
the double square shown in Fig. 9 (where each arrow symbolizes a logical implication).

In the same way, the former principle (ENTAILMENT 1) allows us to join all three
squares of knowledge, of strong and of weak belief into the triple square shown in
Fig. 10.

Interestingly, given certain assumptions that have been defended elsewhere,[7] the doxastic square of strong belief turns out to be *identical* to the square of *complex* epistemic
modalities shown in Fig. 11.

All that has to be assumed is (α) that someone's probabilistic judgments are always
transparent (or epistemically accessible) to the subject a himself, and (β) that the extreme
case of complete certainty, $C(a, p)$, implies the complex, iterated attitude of a *believ-*

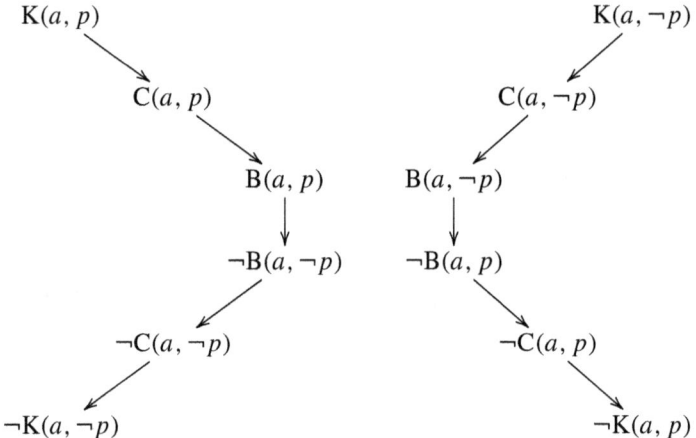

Fig. 10 Nested epistemic square of opposition

---

[7]Cf. especially Lenzen [4, 5] and the brief summaries in [6, 7].

Fig. 11 Square of complex
epistemic modalities

$$\neg K(a, \neg K(a, p)) \qquad \neg K(a, \neg K(a, \neg p))$$
$$K(a, \neg K(a, \neg p)) \qquad K(a, \neg K(a, p)).$$

ing that she knows that p. From the latter assumption which has been dubbed "Moore's principle"[8]

$$C(a, p) \rightarrow B\big(a, K(a, p)\big), \qquad \text{(MOORE)}$$

it follows (via the chain of implications in the lower left part of Fig. 10) that a fortiori

$$C(a, p) \rightarrow \neg K\big(a, \neg K(a, p)\big). \qquad \text{(LENZEN 1)}$$

Furthermore, in view of the transparency condition (α), whenever subject a is not entirely certain that p, a herself *knows* that she is not entirely certain that p:

$$\neg C(a, p) \rightarrow K\big(a, \neg C(a, p)\big). \qquad \text{(TRANSPARENCY 1)}$$

Hence by conjunction of (LENZEN 1) and (TRANSPARENCY 1), one obtains the important equivalence

$$C(a, p) \leftrightarrow \neg K\big(a, \neg K(a, p)\big) \qquad \text{(LENZEN 2)}$$

which shows that the A-proposition of the square of strong belief (Fig. 7) is provably equivalent to the A-proposition of the square of complex epistemic modalities (Fig. 11).

References

1. Engel, P.: Can there be an Epistemic Square of Opposition? In: Béziau, J.-Y., Payette, G. (eds.) Handbook of the First World Congress on the Square of Opposition, Montreux, 2007
2. Hendricks, V.F.: Hintikka on epistemological axiomatization. In: Kolak, D., Symons, J. (eds.) Questions, Quantifiers and Quantum Physics—Essays on the Philosophy of Jaakko Hintikka, pp. 3–32. Springer, Vienna (2004)
3. Lamarre, P., Shoham, Y.: Knowledge, belief, and conditionalization. In: Doyle, J., Sandewall, E., Torraso, P. (eds.) Principles of Knowledge Representation and Reasoning, pp. 415–424. Morgan Kaufmann, San Mateo (1994)
4. Lenzen, W.: Recent Work in Epistemic Logic. Acta Philosophica Fennica. North-Holland, Amsterdam (1978)
5. Lenzen, W.: Glauben, Wissen und Wahrscheinlichkeit. Springer, Vienna (1980)
6. Lenzen, W.: Knowledge, belief, and subjective probability: outlines of a unified system of epistemic/doxastic logic. In: Hendricks, V.F., Jørgensen, K.F., Pedersen, S.A. (eds.) Knowledge Contributors, pp. 17–31. Kluwer, Dordrecht (2003)
7. Lenzen, W.: Epistemic logic. In: Niiniluoto, I., Sintonen, M., Woleński, J. (eds.) Handbook of Epistemology, pp. 963–983. Kluwer, Dordrecht (2004)
8. Moore, G.E.: Certainty. In: Moore: Philosophical Papers, pp. 226–251. George, Allen & Unwin, London (1959)

W. Lenzen (✉)
Dept. of Philosophy, University of Osnabrueck, 49069 Osnabrueck, Germany
e-mail: lenzen@uos.de

---

[8]Cf. Hendricks [2] who refers in particular to Lamarre and Shoham [3]; however, Moore [8] at best put forward the ambiguous principle that believing that p entails believing that one knows that p; but Moore was not aware of the distinction between "weak" belief and conviction; nor did he therefore recognize that the crucial principle does not hold for the operator $B(a, p)$, but only for $C(a, p)$.

Structures of Oppositions
in Public Announcement Logic

Lorenz Demey

Abstract In this paper we study dynamic epistemic logic, and in particular public announcement logic, using the tools of n-opposition theory. Dynamic epistemic logic is a contemporary development of epistemic logic, which takes into account changes of knowledge through time. It studies, for example, public announcements and the influence they have on the agents' knowledge. n-Opposition theory is a systematic generalization of the traditional squares of oppositions. In the paper we provide introductions to the intuitions behind and the formal details of both of these disciplines. We construct a square and a hexagon of oppositions for the structural properties of public announcement. We then generalize this to opposition structures for *any* partially functional process, and show how this generalization can be used to support the structuralist philosophy surrounding n-opposition theory. Next, we focus on the epistemic properties of public announcement, and construct an octagon and a (three-dimensional) rhombic dodecahedron of oppositions. Many of the techniques applied in this paper were originally developed by Smessaert (Logica Univers. 3:303–332, 2009); they are thus shown to be very powerful and widely applicable. Summing up: we establish a strong connection between public announcement logic and n-opposition theory, and show that this connection has definite advantages for both of the disciplines involved.

Keywords Epistemic dynamics · Public announcement logic · n-Opposition theory · Square of oppositions · Structure of oppositions

Mathematics Subject Classification Primary 03B42 · 03A05 · Secondary 03G10 · 03B70 · 68T27

A preliminary version of this paper was presented at the *Second World Conference on the Square of Oppositions* (Corte, June 17–20, 2010). I would like to thank the audience of that talk for their valuable questions and comments. In particular I would like to thank Alexandru Marcoci, Eric Pacuit, Hans Smessaert and Margaux Smets for their detailed feedback on earlier versions of this paper. I would also like to thank four anonymous referees for their helpful suggestions. The early stages of this research were funded by the University of Leuven's Formal Epistemology Project and the Huygens Scholarship Programme; during the final stages I held a PhD fellowship of the Research Foundation—Flanders (FWO).

J.-Y. Béziau, D. Jacquette (eds.), *Around and Beyond the Square of Opposition*, 313–339
Studies in Universal Logic, DOI 10.1007/978-3-0348-0379-3_22, © Springer Basel 2012

1 Introduction

The square of oppositions is a historically rich and interesting subject in philosophical logic. It is a compact way of representing various logical relations between formulas, and thus serves as an illustration of the underlying logic's expressive and deductive powers. Therefore, throughout the history of logic many authors have found it worthwhile to show that the logics they were studying gave rise to square-like structures. Typical examples include the construction of squares for deontic logic [21] and modal logic [12, ch. 1].

In this paper, however, we will be concerned with yet another branch of logic, viz. (*dynamic*) *epistemic logic*. Lenzen, one of the main epistemic logicians of the 20th century [18, 19], has shown how squares can be constructed for notions such as knowledge and belief [22, p. 98].[1] Lenzen thus connected 'classical' epistemic logic and 'classical' squares of oppositions. In recent years, however, both topics have been developing rapidly. On the one hand, squares of oppositions were generalized (mainly by Moretti [22]) to higher-order structures of oppositions, leading to *n*-opposition theory. On the other hand, epistemic logic has been dynamified (mainly by the 'Amsterdam school', led by van Benthem [31]), leading to various versions of dynamic epistemic logic.

The main purpose of this paper is to study these dynamic epistemic logics using the tools of *n*-opposition theory. In other words, we will establish a connection between 'contemporary' epistemic logic and 'contemporary' generalizations of the square of oppositions.[2]

Establishing such a connection has many advantages for both of the topics involved. On the one hand, dynamic epistemic logic in general, and public announcement logic in particular, will benefit from the representational powers of *n*-opposition theory. Epistemic dynamic phenomena (such as public announcements) have various structural and epistemic properties. Because of the expressivity of public announcement logic, especially the latter tend to become quite subtle. The tools of *n*-opposition theory allow us to (re)present this vast amount of information in a compact way. They can thus help us to understand the subject more thoroughly, and even gain new insights.[3] On the other hand, there are also several advantages for *n*-opposition theory itself. In the first place, public announcement logic serves as a new example of the use of *n*-opposition theory, thus widening its scope of applicability. In the second place, this paper shows that not only classical static modalities, but also dynamic modalities give rise to oppositional phenomena, thus broadening the scope of *n*-opposition theory in a more fundamental way. Finally, based on the technical properties of such dynamic modalities, we will corroborate the structuralist philosophy surrounding *n*-opposition theory.

---

[1] The most comprehensive account of this work was presented by Lenzen during the *Second World Conference on the Square of Oppositions* (Corte, June 17–20, 2010).

[2] I would like to emphasize that Moretti has already shown how Lenzen's original squares for epistemic logic can be generalized to higher-order structures of oppositions (in particular, tetraicosahedra) [22, pp. 306–308]. In other words, he has used the 'contemporary' machinery of *n*-opposition theory to look at 'classical' epistemic logic.

[3] This heuristic role is not often acknowledged. A notable exception are Davey and Priestley, who write that "Diagram-drawing is as much an art as a science [...] good diagrams can be a real asset to understanding and to theorem-proving" [8, p. 11].

The remainder of this paper is organized as follows. In Sect. 2 we provide a brief introduction to the theme of epistemic dynamics in general, and public announcements in particular. We also present some of the technical details of public announcement logic,[4] and discuss some structural and epistemic properties of public announcements. In Sect. 3 we begin our investigation of opposition structures in public announcement logic, focusing on the *structural* properties of public announcements. We construct a square and a hexagon of oppositions, and use them to show that dynamic modalities fit well in the structuralist philosophy of *n*-opposition theory. Section 4 continues our investigation, now shifting attention to the *epistemic* properties of public announcements. We construct an octagon and also a three-dimensional structure: a rhombic dodecahedron. We then compare our results with those of Smessaert [26]. Section 5, finally, summarizes the results obtained in this paper, and mentions some open questions.

2 An Overview of Public Announcement Logic

In this section we provide a brief overview of public announcement logic (PAL), focusing on those aspects that are most important for the further development of the paper. Subsection 2.1 first sketches the topic of epistemic dynamics in general, and then focuses on one particular type, viz. public announcements. Subsection 2.2 introduces the syntax and semantics of PAL. Finally, in Subsect. 2.3 we discuss some important structural and epistemic properties of public announcements.

2.1 Epistemic Dynamics in Nature

Classical epistemic logic is *static*: it aims to describe the knowledge of one (or several) agent(s) at one point in time.[5] In real life, however, an agent's knowledge can change over time. Consider the following three examples:

Example 1 Ann does not know that Paris is the capital of France. She is chatting with her best friend, Bob. During their conversation, Bob tells Ann that Paris is the capital of France. After this dialog, Ann does know that Paris is the capital of France.

Example 2 Cath does not know that Paris is the capital of France. She is chatting with Bob (who knows that Paris is the capital of France). During their conversation, Bob tells Cath that Brussels is the capital of France. After this dialog, Cath thinks that she knows that Brussels is the capital of France, but actually she does not know this.

Example 3 Fred and Tom are competing treasure hunters, looking for a particular treasure. Fred finds a map, thereby learning the treasure's exact location. A few miles away,

---

[4]The aim is to keep this paper more or less self-contained. We *do* presuppose, however, a (very) basic knowledge of classical, 'static' epistemic logic, i.e. modal S5 and its Kripke semantics.

[5]Hintikka, for example, rules out occasions "on which people are engaged in gathering new factual information. Uttered on such an occasion, the sentences 'I don't know whether *p*' and [later] 'I know that *p*' need not be inconsistent" [16, pp. 7–8]. Contrast this with our Example 1.

Tom has a conversation with an old magician, who tells him the treasure's exact location. The day before they want to go dig up the treasure, Fred and Tom accidentally meet each other. There is a strange tension between them...

Example 1 is a straightforward case of a learning process (i.e. a transition from not-knowing to knowing) through communication. In Example 2 Bob is *deceiving* Cath: he knows a certain proposition to be false, yet still he communicates it to Cath. This deception leads Cath to think that she has acquired new knowledge (i.e. that she has gone through a learning process), but actually she hasn't. Finally, Example 3 illustrates the difference between *private* and *public* communication: Fred and Tom have both, through independent, private events, learned about the treasure's location. Hence they both know the treasure's location, but they do not know of each other that they know it. When they meet each other, they do their best not to share their newly acquired knowledge (although both of them actually already possess it).

These examples show the variety and subtlety of dynamic epistemic phenomena. All of these phenomena (including deception, private communication, etc.) can be studied in the general framework of dynamic epistemic logic [1, 2, 10, 30, 31]. In this paper, however, we will restrict ourselves to one particular type of epistemic dynamics (and the logic which studies it), viz. *public announcements*.[6] Here is a typical example:

Example 4 Ann, Bob and Cath are all sitting together in the living room, watching television. The news starts, and the newsreader announces that there will be a train strike tomorrow. Ann, Bob and Cath now know that there will be a train strike tomorrow, and they start discussing the consequences this strike will have for them.

In this scenario the newsreader's message is a public announcement. This particular type of dynamic phenomenon has at least three distinct features. First of all, the announcement is made by an *outside source* (which will not be explicitly represented in the logic). In the scenario, this outside source is the newsreader on television, not Ann, Bob or Cath. (Contrast this with Examples 1 and 2.) Second, the announcement is *truthful*: it is indeed true that there will be a train strike tomorrow. Only truths can be publicly announced. (Contrast this with Example 2.) Third, the announcement is *public*: afterwards, Ann, Bob and Cath not only know that there will be a train strike tomorrow, but they also know of each other that they know it. (Contrast this with Example 3.) This is clear from the fact that without any further ado, they start discussing the consequences of the strike. Ann, for example, does not need to tell Bob and Cath that there will be a strike; she knows that Bob and Cath were also watching television, and have thus heard the newsreader's announcement.

---

[6]The restriction from epistemic dynamics in general to public announcements (and thus also from 'full' dynamic epistemic logic to public announcement logic) is mainly for expository reasons, and because public announcement logic was, historically speaking, the first system of dynamic epistemic logic to be developed [13, 23]. Most of the results in subsequent sections can easily be generalized from public announcement logic to full dynamic epistemic logic. For the beginnings of such a generalization, see Subsect. 3.3.

2.2 Technical Background

Our discussion of public announcements so far has been informal. We will now formally introduce public announcement logic.[7] The main idea behind PAL (and other systems of dynamic epistemic logic) is that the models on which the object language is interpreted may be changed ('updated'), and that the object language can talk about these changes.

The models we will be using are simply multi-agent Kripke models as in static epistemic logic. We fix a finite set I of agents and a countably infinite set $Prop$ of proposition letters.

Definition 2.1 A *Kripke model* is a structure $\mathbb{M} = \langle W, \{R_i\}_{i \in I}, V \rangle$, where W is a non-empty set of states, R_i is an equivalence relation on W (for each $i \in I$),[8] and $V : Prop \to \wp(W)$ is a valuation which specifies for each proposition letter $p \in Prop$ the states in which it is true.

The formal language $\mathcal{L}$ of PAL is recursively defined as follows:

$$\varphi ::= p \,|\, \neg\varphi \,|\, \varphi \wedge \varphi \,|\, K_i\varphi \,|\, [!\varphi]\varphi$$

The other connectives are defined as usual; in particular we put $\varphi \vee \psi := \neg(\neg\varphi \wedge \neg\psi)$, and $\varphi \to \psi := \neg\varphi \vee \psi$. Also the dual of the K_i-operator is defined as usual: $\hat{K}_i\varphi := \neg K_i \neg\varphi$. The intuitive interpretation of the language will be discussed later. First, however, we define its formal semantics. Consider an arbitrary Kripke model $\mathbb{M}$ and state w; then we will put:

$$
\begin{array}{lll}
\mathbb{M}, w \models p & \text{iff} & w \in V(p) \\
\mathbb{M}, w \models \neg\varphi & \text{iff} & \mathbb{M}, w \not\models \varphi \\
\mathbb{M}, w \models \varphi \wedge \psi & \text{iff} & \mathbb{M}, w \models \varphi \text{ and } \mathbb{M}, w \models \psi \\
\mathbb{M}, w \models K_i\varphi & \text{iff} & \text{for all } v \text{ such that } (w, v) \in R_i : \mathbb{M}, v \models \varphi \\
\mathbb{M}, w \models [!\varphi]\psi & \text{iff} & \text{if } \mathbb{M}, w \models \varphi \text{ then } \mathbb{M}|\varphi, w \models \psi
\end{array}
$$

As usual, we will write $\mathbb{M} \models \varphi$ iff $\mathbb{M}, w \models \varphi$ for all states w of $\mathbb{M}$. We will also write $\models \varphi$ iff $\mathbb{M} \models \varphi$ for all Kripke models $\mathbb{M}$. If $\models \varphi$, φ is said to be *(PAL-)valid*.[9]

Note that the final clause of the formal semantics, for $[!\varphi]\psi$, mentions not only the model $\mathbb{M}$, but also the updated model $\mathbb{M}|\varphi$, which is defined as follows:

---

[7]This subsection is not meant to be an exhaustive overview of PAL. For example, we do not discuss axiom systems for PAL, their soundness and completeness, or the interesting phenomenon of 'non-successful formulas'. For an in-depth introduction to PAL, the reader is referred to [33, ch. 4].

[8]For any state $w \in W$, we define $R_i[w] := \{v \in W \mid (w, v) \in R_i\}$.

[9]PAL also has a stronger notion of validity. A formula φ is said to be *schematically valid* iff all of its substitution instances $\varphi[\psi/p]$ are valid (where $\varphi[\psi/p]$ is the formula that results from uniformly replacing every occurrence of p in φ with an occurrence of ψ). Obviously, schematic validity implies validity. In logics which have the *uniform substitution* property (if φ is valid, then every substitution instance $\varphi[\psi/p]$ is valid), validity also implies schematic validity, and thus the two notions coincide. PAL, however, does not have the uniform substitution property, and hence in PAL schematic validity is strictly stronger than validity [28, 29]. Also see Footnote 15.

Definition 2.2 Consider an arbitrary Kripke model $\mathbb{M} = \langle W, \{R_i\}_{i \in I}, V \rangle$ and a formula $\varphi \in \mathcal{L}$. We define the *updated model* $\mathbb{M}|\varphi := \langle W^\varphi, \{R_i^\varphi\}_{i \in I}, V^\varphi \rangle$ by putting:

1. $W^\varphi := [\![\varphi]\!]^{\mathbb{M}} = \{w \in W \mid \mathbb{M}, w \models \varphi\}$
2. $R_i^\varphi := R_i \cap (W^\varphi \times W^\varphi)$, for all $i \in I$
3. $V^\varphi(p) := V(p) \cap W^\varphi$, for all $p \in Prop$

The updated model $\mathbb{M}|\varphi$ is the formal representation of the public announcement of the formula φ in the model $\mathbb{M}$.[10] The main idea is that a public announcement of φ *deletes* all $\neg\varphi$-states from the model. The relations and valuation are changed accordingly. For example, if two states w and v were related by R_i before the announcement (i.e. $(w, v) \in R_i$), and they both 'survive' the announcement (because they make φ true), then they will still be related after the announcement (i.e. $(w, v) \in R_i^\varphi$). Also note that public announcements do not change the 'ground facts': the valuation V^φ is the same as the old valuation V (except, of course, that the $\neg\varphi$-states have been deleted). Public announcements can cause changes (e.g. going from not-knowing that φ to knowing that φ; cf. Example 4), but these changes are always on the epistemic level, never on the ontic level. Informally: public announcements (can) change our *knowledge about* the world, but they do not change the world *itself*.

As promised, we now return to the intuitive interpretation of the language $\mathcal{L}$. Formulas such as $K_i\varphi$ should be read as: "agent i knows that φ is the case". (Since R_i is an equivalence relation, K_i is an S5-type modal operator.)[11] More importantly, $[!\varphi]\psi$ should be read as: "after any public announcement of φ, it will be the case that ψ". Note that this does not imply that φ is true (i.e. that there actually *are* any public announcements of φ). If φ is false, then it cannot be announced at all (because of the truthfulness condition of public announcements, cf. supra), and $[!\varphi]\psi$ itself will thus be vacuously true.

The operator $[!\varphi]$ (for any formula $\varphi \in \mathcal{L}$) has a dual, $\langle!\varphi\rangle$, which is defined by putting $\langle!\varphi\rangle\psi := \neg[!\varphi]\neg\psi$. The formal semantics of this dual operator is:

$$\mathbb{M}, w \models \langle!\varphi\rangle\psi \quad \text{iff} \quad \mathbb{M}, w \models \varphi \text{ and } \mathbb{M}|\varphi, w \models \psi$$

Informally, the formula $\langle!\varphi\rangle\psi$ should be read as: "φ *can* be announced (i.e. there exists an announcement of φ), and after this announcement it will be the case that ψ". Hence, this formula *does* imply that φ is true (and thus that there actually are public announcements of φ).

A notion which will often be used later in the paper is that of a contingency, i.e. a formula which can be true and which can be false:

Definition 2.3 An $\mathcal{L}$-formula φ is *PAL-contingent* iff $\not\models \varphi$ and $\not\models \neg\varphi$.

---

[10] Note that if φ is false in all states in $\mathbb{M}$, then the updated model $\mathbb{M}|\varphi$ is empty, and thus not well-defined. However, the formal semantics defined above guarantees that this will never have any practical consequences, because the updated model $\mathbb{M}|\varphi$ is used only when $\mathbb{M}, w \models \varphi$ (in which case $\mathbb{M}|\varphi$ is well-defined).

[11] Of course, this gives rise to many philosophical difficulties, e.g. concerning *positive introspection*: $K_i\varphi \to K_iK_i\varphi$. These questions are not specific to epistemic dynamics (they already arise for static epistemic logic), and will therefore not be discussed here.

We end this subsection by emphasizing once more the *dynamic* nature of the public announcement operators.[12] Considering the formal semantic clauses for $[!\varphi]\psi$ and $\langle !\varphi \rangle \psi$, it might be tempting to think that these (dynamic) formulas are equivalent to the (static) formulas $\varphi \rightarrow \psi$ and $\varphi \wedge \psi$, respectively. However, in the static formulas, φ and ψ are being interpreted on the *same* model (viz. $\mathbb{M}$), whereas in the dynamic formulas, they are being interpreted on *different* models: φ is interpreted on $\mathbb{M}$, but ψ is interpreted on the updated model $\mathbb{M}|\varphi$. Another way to see that $[!\varphi]\psi$ is not in general equivalent to $\varphi \rightarrow \psi$ is by noting that $\models \varphi \rightarrow \varphi$, whereas $\not\models [!\varphi]\varphi$. For example, taking $\varphi := p \wedge \neg K_i p$, one can easily construct a Kripke model $\mathbb{M}$ and state w such that $\mathbb{M}, w \not\models [!\varphi]\varphi$.[13]

2.3 Structural and Epistemic Properties of Public Announcements

We now discuss some important formal properties of public announcements, which will be used in Sects. 3 and 4. We distinguish between two types of properties: *structural* properties, which describe the general structure of public announcement as a dynamic phenomenon, and *epistemic* properties, which describe the interaction between public announcement and knowledge.

The structural properties of public announcements are summarized by the following lemma:[14]

Lemma 2.4 *Consider arbitrary formulas $\varphi, \psi \in \mathcal{L}$. The following hold:*

1. $\models [!\varphi](\psi \rightarrow \chi) \rightarrow ([!\varphi]\psi \rightarrow [!\varphi]\chi)$
2. *if* $\models \psi$ *then* $\models [!\varphi]\psi$
3. $\models \varphi \leftrightarrow \langle !\varphi \rangle \top$
4. *if φ is PAL-contingent, then* $\not\models \langle !\varphi \rangle \top$
5. *if φ is PAL-contingent, then* $\not\models [!\varphi]\psi \rightarrow \langle !\varphi \rangle \psi$
6. $\models \langle !\varphi \rangle \psi \rightarrow [!\varphi]\psi$

Items 1 and 2 say that the public announcement operator $[!\varphi]$ (for any $\varphi \in \mathcal{L}$) satisfies distributivity and the 'necessitation' rule, which is often summarized by saying that it is a *normal* modal operator.[15] Item 3 says that announceability equals truth: a formula can be announced iff it is true. Since not all formulas are true, it follows that not all formulas can be announced, which is what item 4 says. This property is called the *partiality* of public announcements. Item 5 says that $[!\varphi]\psi$ does not require that φ can be announced, whereas $\langle !\varphi \rangle \psi$ does (cf. supra). Finally, item 6 says that public announcement is *functional*: if a formula φ is announced in identical states (in which it *can* be announced, i.e. in which it

---

[12] Thanks to Fabien Schang for advising me to be more explicit about this.

[13] This means that $\mathbb{M}, w \models \varphi$ and $\mathbb{M}|\varphi, w \not\models \varphi$, i.e. φ is initially true, but after announcing it, it becomes false. Such announcements are called 'non-successful'; cf. Footnote 7.

[14] For proofs of Lemmas 2.4 and 2.5, the reader is referred to [33, ch. 4].

[15] However, it should be emphasized that PAL, as a logical system, is not a normal modal logic, because it does not have the uniform substitution property. For example, it holds that $\models [!p]p$, but $\not\models [!(p \wedge \neg K_i p)](p \wedge \neg K_i p)$ [33, p. 106].

is true), this will always lead to identical 'outcome states' (formally: the updated model $\mathbb{M}|\varphi$, if defined at all, is uniquely determined by $\mathbb{M}$ and φ; cf. Definition 2.2 and Footnote 10).

We now turn to the epistemic properties of public announcements:

Lemma 2.5 *Consider arbitrary formulas* $\varphi, \psi \in \mathcal{L}$. *The following hold*:

1. $\models \langle!\varphi\rangle K_i \psi \rightarrow K_i[!\varphi]\psi$
2. $\models K_i[!\varphi]\psi \rightarrow [!\varphi]K_i\psi$
3. $\models \langle!\varphi\rangle\neg K_i\psi \rightarrow \neg K_i[!\varphi]\psi$
4. $\models \neg K_i[!\varphi]\psi \rightarrow [!\varphi]\neg K_i\psi$

Items 3 and 4 are merely the contrapositives of items 2 and 1, respectively, so they will not be discussed separately. Item 1 says that if φ can be announced and after that announcement agent i knows that ψ is the case, then she already knows 'now' (i.e. before any announcements) that ψ will be the case after any announcement of φ. Similarly, item 2 says that if i knows 'now' that ψ will be the case after any announcement of φ, then after any announcement of φ she will know that ψ is the case.

The general idea behind these principles is quite clear: they all state sufficient or necessary conditions for an agent to know something *after* a public announcement in terms of what she knows *before* the announcement. The more fine-grained distinctions between them (in particular, whether $[!\varphi]$ or $\langle!\varphi\rangle$ is used) show the subtlety of the interactions between public announcement and knowledge, and thus illustrate the expressivity of PAL.

3 Structural Oppositions for Public Announcements

In this section we begin our investigation of opposition structures in PAL. For now, we will focus on the *structural* properties of public announcements, and thus consider only 'structural oppositions'. Subsection 3.1 introduces some key notions from n-opposition theory, loosely based on [26]. Subsection 3.2 then uses these notions to construct a square and a hexagon of oppositions. Finally, in Subsect. 3.3 we make some remarks about the role of partial functionality, and connect this with the structuralist philosophy that seems to surround n-opposition theory.

3.1 Some Notions from n-Opposition Theory

The fundamental building blocks of the traditional square of oppositions, and any higher-order opposition structure, are the Aristotelian relations:[16]

Definition 3.1 The *Aristotelian relations* are binary relations on $\mathcal{L}$, defined as follows:

1. φ and ψ are *contradictory* iff $\models \neg(\varphi \wedge \psi)$ and $\models \varphi \vee \psi$
   (i.e. they cannot be true together and they cannot be false together)

---

[16]For expository purposes, all definitions in this subsection are specific to the logic PAL and its language $\mathcal{L}$. Generalizations to any other (sufficiently expressive) logic and language are straightforward.

2. φ and ψ are *contrary* iff $\models \neg(\varphi \wedge \psi)$ and $\not\models \varphi \vee \psi$
   (i.e. they cannot be true together, but they can be false together)
3. φ and ψ are *subcontrary* iff $\not\models \neg(\varphi \wedge \psi)$ and $\models \varphi \vee \psi$
   (i.e. they can be true together, but they cannot be false together)
4. φ and ψ are in *subalternation* iff $\models \varphi \rightarrow \psi$ and $\not\models \psi \rightarrow \varphi$
   (ψ is called the *superaltern* and ψ the *subaltern*)

It is easy to check that the first three of these relations are symmetrical, and that the fourth one is (by definition) asymmetrical. Another interesting property is that these four relations are mutually exclusive (at least when one restricts to contingent formulas):[17,18]

Lemma 3.2 *The Aristotelian relations are mutually exclusive for PAL-contingent formulas; i.e. if two PAL-contingent formulas φ and ψ stand in one of these four relations, then they cannot stand in any of the other three.*

Proof First of all, note that it follows trivially from the definitions that the three symmetrical Aristotelian relations are mutually exclusive (one does not even need to restrict to contingent formulas). It thus suffices to show that two contingent formulas cannot simultaneously be in subalternation and in one of the three other Aristotelian relations. For this purpose, let $\varphi, \psi \in \mathcal{L}$ be contingent formulas, and suppose that φ and ψ are in subalternation (and hence $\models \varphi \rightarrow \psi$). Then:

1. Suppose, towards a contradiction, that φ and ψ are contradictories. Hence $\models \neg(\varphi \wedge \psi)$. From $\models \varphi \rightarrow \psi$ it follows that $\models \varphi \leftrightarrow (\varphi \wedge \psi)$, and hence $\models \neg\varphi$, which contradicts the contingency of φ.
2. Suppose, towards a contradiction, that φ and ψ are contraries. By an analogous argument it follows that $\models \neg\varphi$, which contradicts the contingency of φ.
3. Suppose, towards a contradiction, that φ and ψ are subcontraries. Hence $\models \varphi \vee \psi$. Combining this with $\models \varphi \rightarrow \psi$ it follows that $\models \psi$, which contradicts the contingency of ψ. □

We are now ready to define the general notion of an opposition structure:

Definition 3.3 A *structure of oppositions (for PAL)* is a graph, which has as vertices PAL-contingent $\mathcal{L}$-formulas, and as edges the Aristotelian relations between these formulas

---

[17]To the best of our knowledge, this observation (which is of course more general than the version for PAL given here) has never appeared in print before.

[18]The Aristotelian relations generate an *Aristotelian geometry*, in which the opposition structures of Definition 3.3 exist. The Aristotelian relations are mutually exclusive only under the restriction to contingent formulas; furthermore, they are not exhaustive: there are pairs of formulas that stand in no Aristotelian relation at all. (For example, a formula never stands in any Aristotelian relation with itself.) In ongoing work with Hans Smessaert [11, 27], we define an *oppositional geometry* and an *implicational geometry*, each of which is generated by four relations which *are* (i) exhaustive and (ii) mutually exclusive (without restricting to contingent formulas). The Aristotelian geometry turns out to be a *hybrid* system: it shares the relations of contradiction, contrariety and subcontrariety with the oppositional geometry, and that of subalternation with the implicational geometry.

Fig. 1 Code for visually
representing the Aristotelian
relations

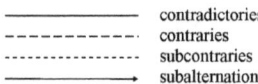

(according to the code in Fig. 1).[19] Finally, it is required that for any φ, ψ in the graph, it holds that $\not\models \varphi \leftrightarrow \psi$.

Note that this definition allows only *contingent* formulas to appear in opposition structures. The first reason for this restriction is of a historical nature: the classical (ancient and medieval) squares of oppositions only contained contingencies, and no tautologies or contradictions. The second reason is based on the informational use of opposition structures. Non-contingent formulas are not very informative and can thus be left out of the structures. Furthermore, non-contingent formulas enter 'vacuously' into Aristotelian relations with contingencies;[20] representing those vacuous Aristotelian relations would only mess up the diagrams.

Note, furthermore, that Definition 3.3 does not mention visual simplicity, two- or three-dimensionality, compactness, elegance, etc. Although the great practical use of *n*-opposition theory is largely based on these notions, they cannot easily be captured in a formally precise definition. Perhaps Definition 3.3 should be thought of as the best mathematical approximation of the 'real' notion of opposition structure. (Recall Footnote 3 about diagram-drawing as a science versus an art.)

Finally, the condition that $\not\models \varphi \leftrightarrow \psi$ for any φ, ψ in an opposition structure says that opposition structures are essentially *semantic* entities: they never contain formulas that are merely syntactically different forms expressing the same proposition. (For example, we do not allow an opposition structure to contain p and $\neg\neg p$ as separate formulas.)

A powerful technique for generating new opposition structures from already existing ones is that of taking *Boolean closures*. We first define this notion for sets of formulas, and then for opposition structures.

Definition 3.4 A set S of PAL-contingent formulas is called *Boolean closed* iff for any φ, $\psi \in S$, we have:

1. if $\not\models \neg(\varphi \wedge \psi)$ then there is a $\theta \in S$ such that $\models \theta \leftrightarrow (\varphi \wedge \psi)$
2. there is a $\theta \in S$ such that $\models \theta \leftrightarrow \neg\varphi$

Definition 3.5 Let S be a set of contingent formulas. Then the *Boolean closure of S* is the set B such that (i) $S \subseteq B$, (ii) B is Boolean closed, and (iii) for any set B' such that $S \subseteq B'$ and B' is Boolean closed, it holds that $B \subseteq B'$.

Definition 3.4 is *semantic* in nature: if $\varphi \in S$, then this definition does not require that S contains $\neg\varphi$ *itself*, but at least a formula that is *equivalent* to $\neg\varphi$ (similarly for $\wedge$).[21] Note,

---

[19]Note that the three symmetrical relations are represented by *non-directed* edges, whereas the asymmetrical subalternation relation is represented by *directed* edges going *from* the superaltern *to* the subaltern.

[20]Contradictions are superaltern and contrary to any contingent formula. Tautologies are subaltern and subcontrary to any contingent formula. (Note that these statements do not contradict Lemma 3.2, since in both cases a non-contingent formula is involved.)

[21]It suffices to consider the connectives $\wedge$ and $\neg$, since these two are functionally complete, i.e. any other truth-functional connective (of any arity) can be written in terms of them [32, pp. 24–25].

furthermore, that the $\wedge$-closure condition is also subject to a consistency requirement;[22] for example, if a set S contains the contingencies p and $\neg p$, then it is not required to contain also $p \wedge \neg p$ (or any other equivalent formula), since the latter is not consistent. Finally, note that it is clear from Definition 3.5 that if S is already Boolean closed, then it is its own Boolean closure.

We now lift these notions from sets of formulas to opposition structures:

Definition 3.6 An opposition structure S is called *Boolean closed* iff the set of formulas appearing as vertices in S is Boolean closed (according to Definition 3.4).

Definition 3.7 Let S be an opposition structure. Let F be the set of formulas appearing as vertices in S, and let F' be the Boolean closure (according to Definition 3.5) of this set. Then the *Boolean closure of S* is the opposition structure which has as its vertices the formulas of F'.

Definition 3.6 is the natural generalization of Definition 3.4. Note that in Definition 3.7, the set F is a set containing only contingent formulas (cf. Definition 3.3), so that it indeed makes sense to take its Boolean closure F' (cf. Definition 3.5). Finally, note that these definitions yield some easy consequences. First of all, the Boolean closure of an opposition structure is itself always Boolean closed (as an opposition structure). Secondly, if an opposition structure is already Boolean closed, then it is its own Boolean closure.

3.2 From Square to Hexagon

We will now start putting the tools of n-opposition theory to use, by constructing opposition structures for the structural properties of public announcement (which describe the general structure of public announcement as a dynamic phenomenon).

Despite its dynamic nature, the public announcement operator $[!\varphi]$ (for any $\varphi \in \mathcal{L}$) is essentially a *modal* operator; cf. items 1 and 2 of Lemma 2.4. Modal operators come in dual pairs; for example in alethic modal logic (necessary/possible), epistemic logic (knowing/holding possible), deontic logic (obligatory/permissible), and temporal logic (always/sometimes). The first item of each of these dual pairs is given a universal reading, and the second one an existential reading (for example: 'in *all* possible worlds'/'in *some* possible worlds'). Furthermore, in the squares of oppositions the universal notions occupy the two upper corners (so that they are superaltern to the existential notions), and the existential notions occupy the two lower corners (so that they are subaltern to the universal notions).

The public announcement operator comes in a dual pair as well ($[!\varphi]/\langle!\varphi\rangle$). Furthermore, the first item of this pair can (informally) be read as a universal notion, and the second one as an existential notion ('after *all* public announcements of φ'/'after *some* public announcement of φ'); cf. Subsect. 2.2. Hence it might be tempting to construct a square of oppositions for PAL with $[!\varphi]\psi$ and $[!\varphi]\neg\psi$ in the upper corners, and $\langle!\varphi\rangle\psi$

Fig. 2 Square of oppositions
for the structural properties of
public announcement

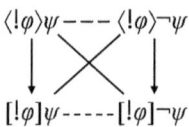

and $\langle!\varphi\rangle\neg\psi$ in the lower corners. However, $[!\varphi]\psi$ and $\langle!\varphi\rangle\psi$ (and similarly $[!\varphi]\neg\psi$ and $\langle!\varphi\rangle\neg\psi$) do not form a superaltern/subaltern pair (cf. item 5 of Lemma 2.4), and thus the proposed square cannot be constructed.

Because of the special structure of public announcement (as a dynamic phenomenon), however, it is still possible to obtain a square of oppositions. The key idea is to *reverse* the direction of the subalternation relation. The following theorem says this reversing operation 'respects' the other Aristotelian relations, and thus that the reversed square in Fig. 2 is a valid opposition structure.

Theorem 3.8 *Consider arbitrary formulas $\varphi, \psi \in \mathcal{L}$. Suppose that the formulas in Fig. 2 are contingent and pairwise non-equivalent. Then Fig. 2 shows an opposition structure.*

Proof It suffices to check that the Aristotelian relations represented in Fig. 2 (according to the code of Fig. 1) do indeed hold (Lemma 3.2 then guarantees that there are no other Aristotelian relations that have been 'left out'):

contradictions Trivial, given the definition of $\langle!\varphi\rangle\psi$ as $\neg[!\varphi]\neg\psi$.

subalternations By item 6 of Lemma 2.4 we have $\models \langle!\varphi\rangle\psi \rightarrow [!\varphi]\psi$. If also $\models [!\varphi]\psi \rightarrow \langle!\varphi\rangle\psi$, we would get $\models [!\varphi]\psi \leftrightarrow \langle!\varphi\rangle\psi$, which contradicts the non-equivalence of these formulas. Hence $\not\models [!\varphi]\psi \rightarrow \langle!\varphi\rangle\psi$, and thus there is a subalternation from $\langle!\varphi\rangle\psi$ to $[!\varphi]\psi$. Completely analogously we show that there is a subalternation from $\langle!\varphi\rangle\neg\psi$ to $[!\varphi]\neg\psi$.

contrariety We need to check that $\models \neg(\langle!\varphi\rangle\psi \wedge \langle!\varphi\rangle\neg\psi)$ and $\not\models \langle!\varphi\rangle\psi \vee \langle!\varphi\rangle\neg\psi$ (cf. Definition 3.1). Consider an arbitrary model and state $\mathbb{M}, w$. If $\mathbb{M}, w \models \langle!\varphi\rangle\psi \wedge \langle!\varphi\rangle\neg\psi$, then $\mathbb{M}, w \models \varphi$, and $\mathbb{M}|\varphi, w \models \psi$ and $\mathbb{M}|\varphi, w \models \neg\psi$, contradiction. Hence $\models \neg(\langle!\varphi\rangle\psi \wedge \langle!\varphi\rangle\neg\psi)$. Suppose, for a contradiction, that $\models \langle!\varphi\rangle\psi \vee \langle!\varphi\rangle\neg\psi$. This is equivalent to $\models [!\varphi]\psi \rightarrow \langle!\varphi\rangle\psi$, which together with item 6 of Lemma 2.4 yields that $\models [!\varphi]\psi \leftrightarrow \langle!\varphi\rangle\psi$. This contradicts the non-equivalence of these formulas, and hence $\not\models \langle!\varphi\rangle\psi \vee \langle!\varphi\rangle\neg\psi$.

subcontrariety We need to check that $\not\models \neg([!\varphi]\psi \wedge [!\varphi]\neg\psi)$ and $\models [!\varphi]\psi \vee [!\varphi]\neg\psi$ (cf. Definition 3.1). Suppose, for a contradiction, that $\models \neg([!\varphi]\psi \wedge [!\varphi]\neg\psi)$. This is equivalent to $\models [!\varphi]\psi \rightarrow \langle!\varphi\rangle\psi$, which together with item 6 of Lemma 2.4 yields that $\models [!\varphi]\psi \leftrightarrow \langle!\varphi\rangle\psi$. This contradicts the non-equivalence of these formulas, and hence $\not\models \neg([!\varphi]\psi \wedge [!\varphi]\neg\psi)$. To show that $\models [!\varphi]\psi \vee [!\varphi]\neg\psi$, consider an arbitrary model and state $\mathbb{M}, w$. On the one hand, if $\mathbb{M}, w \not\models \varphi$, then trivially $\mathbb{M}, w \models [!\varphi]\psi \vee [!\varphi]\neg\psi$. On the other hand, if $\mathbb{M}, w \models \varphi$, then either $\mathbb{M}|\varphi, w \models \psi$ (and thus $\mathbb{M}, w \models [!\varphi]\psi$) or $\mathbb{M}|\varphi, w \not\models \psi$ (and thus $\mathbb{M}, w \models [!\varphi]\neg\psi$), so in all cases $\mathbb{M}, w \models [!\varphi]\psi \vee [!\varphi]\neg\psi$. □

The reversed square in Fig. 2 is thus an opposition structure. It is, however, not Boolean closed:

Lemma 3.9 *Consider arbitrary formulas $\varphi, \psi \in \mathcal{L}$. Suppose that the formulas in Fig. 2 are contingent and pairwise non-equivalent. Then the opposition structure in Fig. 2 is not Boolean closed.*

Fig. 3 Hexagon of
oppositions for the structural
properties of public
announcement

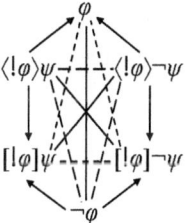

Proof Consider, for example, the conjunction of the formulas in the two lower corners, $[!\varphi]\psi \wedge [!\varphi]\neg\psi$ (we will abbreviate this conjunction with χ).

First of all, we show that χ is not equivalent to any of the four formulas already present in the square. Since there is a subalternation from $\langle!\varphi\rangle\psi$ to $[!\varphi]\psi$, it holds that $\not\models [!\varphi]\psi \rightarrow \langle!\varphi\rangle\psi$, and hence there exist a model and state $\mathbb{M}_1, w_1$ such that $\mathbb{M}_1, w_1 \models [!\varphi]\psi$ and $\mathbb{M}_1, w_1 \not\models \langle!\varphi\rangle\psi$. It follows that $\mathbb{M}_1, w_1 \models \neg\varphi$, and thus $\mathbb{M}_1, w_1 \models \chi$. Hence $\not\models \chi \leftrightarrow \langle!\varphi\rangle\psi$. The same model and state can also be used to show that $\not\models \chi \leftrightarrow \langle!\varphi\rangle\neg\psi$. Since $\langle!\varphi\rangle\psi$ is contingent, there exists a model and state $\mathbb{M}_2, w_2$ such that $\mathbb{M}_2, w_2 \models \langle!\varphi\rangle\psi$. Then $\mathbb{M}_2, w_2 \models [!\varphi]\psi$ while $\mathbb{M}_2, w_2 \not\models \chi$, and thus $\not\models \chi \leftrightarrow [!\varphi]\psi$. Since $\langle!\varphi\rangle\neg\psi$ is contingent, there exists a model and state $\mathbb{M}_3, w_3$ such that $\mathbb{M}_3, w_3 \models \langle!\varphi\rangle\neg\psi$. Then $\mathbb{M}_3, w_3 \models [!\varphi]\neg\psi$ while $\mathbb{M}_3, w_3 \not\models \chi$, and thus $\not\models \chi \leftrightarrow [!\varphi]\psi$.

Also note that because $\mathbb{M}_1, w_1 \models \chi$ and $\mathbb{M}_2, w_2 \not\models \chi$, it follows that χ is contingent. Hence χ is contingent and yet not equivalent to any of the formulas already present in the square, and thus by Definitions 3.4 and 3.6 the square is not Boolean closed.   □

Because the square in Fig. 2 is not Boolean closed, we can extend it to a new opposition structure by constructing its Boolean closure. This turns out to be the hexagon in Fig. 3.

Theorem 3.10 *Consider arbitrary formulas $\varphi, \psi \in \mathcal{L}$. Suppose that the formulas in Fig. 3 are contingent and pairwise non-equivalent. Then the hexagon in Fig. 3 is the Boolean closure of the square in Fig. 2.*

Proof We will construct the Boolean closure of the square in Fig. 2 'step by step', i.e. adding formulas as we go along, and show that in this way we arrive at the hexagon in Fig. 3.

The square is closed under $\neg$ (consider the contradiction relations). It is, however, not closed under $\wedge$: in the proof of Lemma 3.9 we showed that the conjunction of $[!\varphi]\psi$ and $[!\varphi]\neg\psi$ is contingent, and yet not equivalent to any of the four formulas appearing in the square. Hence we will simply *add* this conjunction.

Note, however, that $[!\varphi]\psi \wedge [!\varphi]\neg\psi$ is equivalent to $\neg\varphi$. To see this, consider an arbitrary model and state $\mathbb{M}, w$. Obviously, if $\mathbb{M}, w \models \neg\varphi$, then $\mathbb{M}, w \models [!\varphi]\psi \wedge [!\varphi]\neg\psi$. For the other direction: suppose $\mathbb{M}, w \models [!\varphi]\psi \wedge [!\varphi]\neg\psi$ and yet $\mathbb{M}, w \not\models \neg\varphi$; then $\mathbb{M}|\varphi, w \models \psi$ and $\mathbb{M}|\varphi, w \models \neg\psi$, contradiction; hence we conclude that $\mathbb{M}, w \models \neg\varphi$.

Recall that we wanted to add the conjunction $[!\varphi]\psi \wedge [!\varphi]\neg\psi$, which we have shown to be equivalent to $\neg\varphi$. Since opposition structures are semantic entities, it thus suffices to add $\neg\varphi$ (rather than the conjunction). We now have a set containing five formulas. This set is not closed under $\neg$ anymore: using an argument as in the proof of Lemma 3.9, we can show that the negation of the newly added $\neg\varphi$ is not equivalent to any of the five formulas in the set. Hence we simply *add* this negation $\neg\neg\varphi$, i.e. we add φ. We thus arrive at a set containing six formulas. This set is obviously closed under $\neg$ (consider the contradiction relations). To see that it is also closed under $\wedge$, consider the following table:[23]

$\wedge$	$\langle!\varphi\rangle\psi$	$\langle!\varphi\rangle\neg\psi$	$[!\varphi]\psi$	$[!\varphi]\neg\psi$	φ	$\neg\varphi$
$\langle!\varphi\rangle\psi$	$\langle!\varphi\rangle\psi$	$\bot$	$\langle!\varphi\rangle\psi$	$\bot$	$\langle!\varphi\rangle\psi$	$\bot$
$\langle!\varphi\rangle\neg\psi$		$\langle!\varphi\rangle\neg\psi$	$\bot$	$\langle!\varphi\rangle\neg\psi$	$\langle!\varphi\rangle\neg\psi$	$\bot$
$[!\varphi]\psi$			$[!\varphi]\psi$	$\neg\varphi$	$\langle!\varphi\rangle\psi$	$\neg\varphi$
$[!\varphi]\neg\psi$				$[!\varphi]\neg\psi$	$\langle!\varphi\rangle\neg\psi$	$\neg\varphi$
φ					φ	$\bot$
$\neg\varphi$						$\neg\varphi$

Hence, the set of six formulas appearing in the hexagon is Boolean closed. (Because of the construction process, it is also the smallest such set.) It thus suffices to show that the hexagon in Fig. 3 is indeed an opposition structure, i.e. that the Aristotelian relations represented in Fig. 3 do indeed hold (Lemma 3.2 then guarantees that there are no other Aristotelian relations that have been 'left out').

This is done exactly as in the proof of Theorem 3.8; we will therefore treat only one case explicitly. We show that $\langle!\varphi\rangle\psi$ and $\neg\varphi$ are contraries, i.e. that $\models\neg(\langle!\varphi\rangle\psi\wedge\neg\varphi)$ and $\not\models\langle!\varphi\rangle\psi\vee\neg\varphi$. For any model and state $\mathbb{M}, w$, if $\mathbb{M}, w\models\langle!\varphi\rangle\psi$ then $\mathbb{M}, w\not\models\neg\varphi$, and thus $\models\neg(\langle!\varphi\rangle\psi\wedge\neg\varphi)$. Furthermore, since $\langle!\varphi\rangle\neg\psi$ is contingent, there exists a model and state $\mathbb{M}, w$ such that $\mathbb{M}, w\models\langle!\varphi\rangle\neg\psi$. Then $\mathbb{M}, w\not\models\langle!\varphi\rangle\psi\vee\neg\varphi$, and thus $\not\models\langle!\varphi\rangle\psi\vee\neg\varphi$.   □

We have obtained the hexagon of oppositions in Fig. 3 as the Boolean closure of the square of oppositions in Fig. 2. Readers who are familiar with (the historical precursors of) n-opposition theory, however, might recognize this structure as the 'Sesmat-Blanché hexagon' [4–7, 25]—especially if we recall that the lowest formula, $\neg\varphi$, can be thought of as (is equivalent to) the conjunction of the two formulas in the lower square corners ($[!\varphi]\psi\wedge[!\varphi]\neg\psi$), and similarly, that the highest formula, φ, can be thought of as (is equivalent to) the disjunction of the two formulas in the upper square corners ($\langle!\varphi\rangle\psi\vee\langle!\varphi\rangle\neg\psi$). This observation is more general than the framework of PAL for which it was proved here: *any* of the traditional squares of oppositions gives rise to a Sesmat-Blanché hexagon *via* its Boolean closure.[24]

We finish this subsection by some remarks on how the hexagon of oppositions in Fig. 3 represents the structural properties of public announcements. First of all, it shows that public announcement operators are modal operators and thus, like any other modal operator, give rise to opposition structures. More importantly, however, the subalternations from $\langle!\varphi\rangle(\neg)\psi$ to $[!\varphi](\neg)\psi$ capture the *functionality* of public announcements, while the subalternations involving $(\neg)\varphi$ capture the *partiality* of public announcements

---

[23] $\dfrac{\wedge\;|\;\beta}{\alpha\;|\;\gamma}$ means that $\models\gamma\leftrightarrow(\alpha\wedge\beta)$. Because of the commutativity of $\wedge$, it suffices to give only the upper right half of the table.

[24] In the specific framework of PAL, the hexagon is a bit more elegant than in other frameworks, because the conjunction and disjunction 'reduce to' the simpler formulas $(\neg)\varphi$. (The formula φ is 'simpler' than the disjunction $\langle!\varphi\rangle\psi\vee\langle!\varphi\rangle\neg\psi$ in an obvious sense; however it should be emphasized that φ itself can be a formula of any complexity—as long as it is contingent; for example $\varphi = K_i(p\wedge q)\wedge\neg K_i r$.) However, when we turn to opposition structures for the epistemic properties of public announcements in Sect. 4, we will not be able to maintain this simplicity, and we *will* have to work with 'irreducibly complex' formulas.

(the precondition for a public announcement to be executable is the truth of the announced formula). We will elaborate on the role of partial functionality in the next subsection.

3.3 The Role of Partial Functionality

We have just seen that partial functionality (cf. Lemma 2.4) plays a key role in the opposition structures for the structural properties of public announcements. This property is, however, not specific to public announcements. We will now examine its role from a more abstract perspective.

For *any* 'process' or 'operation' (epistemic or otherwise), it makes sense to ask whether it is partially functional. Consider an arbitrary process π (π can be an epistemic process, such as the public announcement of some formula, but it can also be a non-epistemic process, such as the execution of a computer program). Then the following questions arise:

1. Is π functional?[25] I.e. if π is executed in identical states (in which it *can* be executed), will it always lead to identical 'outcome states'?
2. Is π partial? I.e. are there states where π cannot be executed at all? (And can the states where π *is* executable be characterized by means of a 'precondition' $pre(\pi)$?)

The logical properties of such arbitrary processes are studied by *propositional dynamic logic* (PDL) [15, 17]. PDL represents a process π by means of dynamic modal operators $[\pi]$ and $\langle\pi\rangle$. As is to be expected, $[\pi]\psi$ means that after *any* execution of π it will be the case that ψ, whereas $\langle\pi\rangle\psi$ means that after *some* execution of π it will be the case that ψ.

It is easy to see that a process π is functional iff $\models \langle\pi\rangle\psi \rightarrow [\pi]\psi$, and partial iff $\not\models [\pi]\psi \rightarrow \langle\pi\rangle\psi$.[26] Furthermore, the partiality of π can be further specified by means of its precondition: π can be executed iff (i.e. in exactly those states where) $pre(\pi)$ is true; formally: $\models pre(\pi) \leftrightarrow \langle\pi\rangle\top$.

With the same reasoning as in Subsect. 3.2, we are now able to show that any partially functional process π (whose executability can be captured by means of a precondition $pre(\pi)$) gives rise to a hexagon of oppositions as in Fig. 4. In fact, if we take π to be the concrete process $!\varphi$ (public announcement of φ), and recall that the precondition of the public announcement of φ is the truth of the announced formula φ ($pre(!\varphi) = \varphi$), then the opposition structure in Fig. 3 turns out to be merely a particular instance of that in Fig. 4.

A question that arises naturally at this point is what happens when we move from *partial* functionality to *total* functionality, i.e. when we focus on processes that are still functional, but that can be executed in *all* states. (Note that in this context the notion of a precondition is meaningless, since a totally functional process has as its precondition always $\top$.) For totally functional processes π, we do not only have $\models \langle\pi\rangle\psi \rightarrow [\pi]\psi$, but also the stronger $\models \langle\pi\rangle\psi \leftrightarrow [\pi]\psi$.

---

[25]Computer scientists prefer the term 'deterministic'.

[26]In this subsection (and in this subsection *only*), we use $\models$ to denote PDL-validity, rather than PAL-validity.

Fig. 4 Hexagon of
oppositions for a partially
functional process π

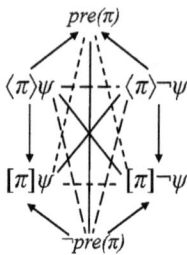

Given the semantic nature of opposition structures, this equivalence implies, however, that the two left formulas of (the square part of) Fig. 4 ($\langle\pi\rangle\psi$ and $[\pi]\psi$) collapse, and similarly that the two right formulas ($\langle\pi\rangle\neg\psi$ and $[\pi]\neg\psi$) collapse. Hence for totally functional processes the square of oppositions collapses into a single binary opposition (viz. a contradiction relation), cf. Fig. 5. Furthermore, it is easy to show that this opposition structure is already Boolean closed, and is thus its own Boolean closure.

We conclude that from the perspective of n-opposition theory, partially functional processes are more interesting than totally functional ones, because the latter only give rise to degenerate opposition structures (Fig. 5), whereas the former give rise to rich, non-degenerate opposition structures (Fig. 4).

We finish this subsection by showing how the partial functionality of dynamic modal operators nicely corroborates the structuralist philosophy surrounding n-opposition theory [22, in particular ch. 6]. According to structuralism, opposition structures are in the first place determined by their constituent *relations*, rather than by their constituent *formulas*. The identity and properties of the concrete formulas only matter insofar as they stand in one of the Aristotelian relations.[27]

An important argument for structuralism stems from the fact that opposition structures can be constructed for a wide variety of logics. For ease of exposition, let us focus on *squares* of oppositions. The structuralist says that there are squares for Aristotelian syllogistics, alethic modal logic, epistemic logic, deontic logic, temporal logic, etc. All of these logics have formulas of very different kinds ('all As are Bs', 'it is necessary that φ', 'agent i knows that φ', 'it is obligatory that φ', 'it is always the case that φ', etc.), yet all of these formulas enter into the same kinds of Aristotelian relations (*in casu*: a contradiction relation, a contrariety relation, and (being the superaltern in) a subalternation relation). Hence the 'relational' properties of these formulas are more important than the 'subject matter' that they speak about (knowledge/time/etc.).

Opponents of structuralism grant all of this, but they maintain that at a sufficiently high level of abstraction, the formulas still have 'independent (non-relational)' properties that determine their relational properties. For example, de Pater and Vergauwen [9] *define* the contrariety relation as holding between universal formulas of different quality, and the subcontrariety relation as holding between existential formulas of different quality.[28]

---

[27]Note that we implicitly subscribed to this structuralist philosophy when we claimed in Subsect. 3.1 that the 'fundamental building blocks' of opposition structures are the Aristotelian relations, rather than the concrete formulas appearing in the structures.

[28]The original Dutch text: "Contrair zijn universele uitspraken die in kwaliteit verschillen. [...] Subcontrair zijn particuliere proposities die in kwaliteit verschillen." [9, p. 100]. De Pater and Vergauwen *do* state the characterizations of contrariety and subcontrariety in terms of 'being able to be true/false together' (cf. our Definition 3.1), but only as further *lemmas* about these relations, not as their *definitions*.

$$\langle \pi \rangle \psi \text{ --------- } \langle \pi \rangle \neg \psi$$

Fig. 5 Degenerate opposition structure for a totally functional process π

Hence the formulas' independent properties (universality/existentiality) are still essential to determine their relational properties.

De Pater and Vergauwen are right that in all traditional squares of oppositions, the contrariety relation holds between 'universal' formulas of different quality (for example, between $\Box \varphi$ and $\Box \neg \varphi$) and the subcontrariety relation holds between 'existential' formulas of different quality (for example, between $\Diamond \varphi$ and $\Diamond \neg \varphi$); see also our remarks at the beginning of Subsect. 3.2. The squares of oppositions obtained from partially functional dynamic modal operators, however, are counterexamples to de Pater and Vergauwen's claims: in these structures the formulas in the contrariety relation are *existential* ($\langle \pi \rangle \psi$ and $\langle \pi \rangle \neg \psi$), and the formulas in the subcontrariety relation are *universal* ($[\pi]\psi$ and $[\pi]\neg \psi$).

This clearly shows that the independent properties of the formulas (universality/existentiality) do *not* suffice to determine their relational properties. In other words, the Aristotelian relations cannot be reduced to the properties of independent formulas, and thus still need to be taken as primitive—i.e., we have arrived back at the original structuralist position.

4 Epistemic Oppositions for Public Announcements

This section continues our investigation of opposition structures in PAL. We shift our attention from the structural to the epistemic properties of public announcements, thus arriving at 'epistemic oppositions'. In Subsect. 4.1 we show that this naturally leads to the construction of an octagon. In Subsect. 4.2 this octagon is further generalized to a three-dimensional opposition structure, viz. a rhombic dodecahedron. Subsection 4.3 is a comparison between our results and those of Smessaert [26].

4.1 Adding Knowledge: From Hexagon to Octagon

In Subsect. 3.3 we showed that the essence of the hexagon of oppositions (Fig. 3) was the partial functionality of public announcement, and that this can be generalized to *any* partially functional process π (Fig. 4). In Sects. 1 and 2, however, we stated that this paper would be concerned with (a branch of) *epistemic* logic.

The main difference between public announcements and other (partially functional) processes (for example, the process of watering the flowers) is that public announcements are *epistemic* processes: they interact with the agents' knowledge. For example, after the public announcement of a formula φ, it (usually)[29] becomes common knowledge between the agents that φ is the case. The most important of these interactions were stated

---

[29] Modulo unsuccessful formulas; cf. Footnotes 7 and 13.

Fig. 6 Octagon of
oppositions for the epistemic
properties of public
announcement

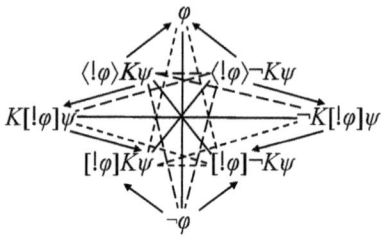

in Lemma 2.5. We will now use this lemma to build opposition structures for the epistemic properties of public announcements (i.e. for public announcements *qua* epistemic processes, not simply *qua* partially functional processes).

We will first informally describe how to build these new, 'epistemic' opposition structures as conservative extensions of the original hexagon (Fig. 3). The first step is to replace the formula ψ (which describes what is the case after the public announcement) with the epistemic formula $K\psi$.[30] For example, the formula in the upper left corner is now no longer $\langle!\varphi\rangle\psi$, but rather $\langle!\varphi\rangle K\psi$ (for some concrete $\varphi, \psi \in \mathcal{L}$). On the left side, we thus obtain the subalternation $\langle!\varphi\rangle K\psi \rightarrow [!\varphi]K\psi$. By inserting the formula $K[!\varphi]\psi$, this subalternation can be broken up into *two* subalternations. (Note that this new formula no longer speaks about what will be the case *after* any/some public announcement of φ; rather it says something about (what is known in) the *present*.) Similar remarks apply to the subalternation on the right side (inserting the formula $\neg K[!\varphi]\psi$).

These two new formulas also enter into Aristotelian relations with each other, and with all but two of the other formulas already present in the original hexagon. This leads to the octagon in Fig. 6.

Theorem 4.1 *Consider arbitrary formulas $\varphi, \psi \in \mathcal{L}$. Suppose that the formulas in Fig. 6 are contingent and pairwise non-equivalent. Then Fig. 6 shows an opposition structure.*

Proof Note that Fig. 6 represents no Aristotelian relations between $(\neg)K[!\varphi]\psi$ and $(\neg)\varphi$. Hence we need to check that there are indeed no Aristotelian relations between these formulas. We only treat $K[!\varphi]\psi$ and φ explicitly. Since $\langle!\varphi\rangle K\psi$ is contingent, there exist a model and state $\mathbb{M}, w$ such that $\mathbb{M}, w \models \langle!\varphi\rangle K\psi$. Then $\mathbb{M}, w \models K[!\varphi]\psi \wedge \varphi$, so $K[!\varphi]\psi$ and φ cannot be contradictories, nor contraries. For a reductio, suppose that there is a subalternation from φ to $K[!\varphi]\psi$. Then $\models \varphi \rightarrow K[!\varphi]\psi$, which together with item 2 of Lemma 2.5 yields that $\models \varphi \rightarrow [!\varphi]K\psi$. This is equivalent to $\models [!\varphi]K\psi$. But this contradicts the contingency of $[!\varphi]K\psi$; hence there is a no subalternation from φ to $K[!\varphi]\psi$. Finally, for a reductio, suppose that there is a subalternation from $K[!\varphi]\psi$ to φ. Then $\models K[!\varphi]\psi \rightarrow \varphi$, which together with item 2 of Lemma 2.5 yields that $\models K[!\varphi]\psi \rightarrow (\varphi \wedge [!\varphi]K\psi)$. This is equivalent to $\models K[!\varphi]\psi \rightarrow \langle!\varphi\rangle K\psi$. Together with item 1 of Lemma 2.5, this yields that $\models \langle!\varphi\rangle K\psi \leftrightarrow K[!\varphi]\psi$. But this contradicts the non-equivalence of these two formulas; hence there is a no subalternation from $K[!\varphi]\psi$ to φ.

---

[30]In the remainder of this section we will drop agent indices, since they are not crucial. We will thus use K instead of K_i to denote the knowledge operator in the formal language $\mathcal{L}$, and R instead of R_i to denote the equivalence relation in a Kripke model $\mathbb{M}$.

Fig. 7 Octagon of
oppositions for the S5 modal
operators

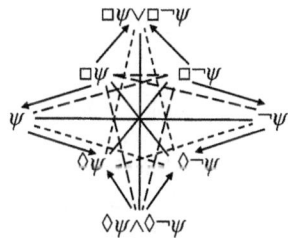

It now suffices to check that the Aristotelian relations that *are* represented in Fig. 6 indeed hold (Lemma 3.2 then guarantees that there are no other Aristotelian relations that have been 'left out').

We first check the subalternations. Item 1 of Lemma 2.5 states that $\models \langle !\varphi \rangle K \psi \rightarrow K[!\varphi]\psi$. If also $\models K[!\varphi]\psi \rightarrow \langle !\varphi \rangle K \psi$, then $\models \langle !\varphi \rangle K \psi \leftrightarrow K[!\varphi]\psi$, which contradicts the non-equivalence of these formulas. Hence $\not\models K[!\varphi]\psi \rightarrow \langle !\varphi \rangle K \psi$, and thus there is a subalternation from $\langle !\varphi \rangle K \psi$ to $K[!\varphi]\psi$. With analogous arguments, we can use items 2, 3 and 4 of Lemma 2.5 to show that there are subalternations from $K[!\varphi]\psi$ to $[!\varphi]K\psi$, from $\langle !\varphi \rangle \neg K \psi$ to $\neg K[\varphi]\psi$, and from $\neg K[!\varphi]\psi$ to $[!\varphi]\neg K \psi$, respectively. The remaining four subalternations (involving $(\neg)\varphi$) are established as in Sect. 3.

We now check the other Aristotelian relations; we treat only one case explicitly. We show that $\langle !\varphi \rangle K \psi$ and $\neg K[!\varphi]\psi$ are contraries, i.e. that $\models \neg(\langle !\varphi \rangle K \psi \wedge \neg K[!\varphi]\psi)$ and $\not\models \langle !\varphi \rangle K \psi \vee \neg K[!\varphi]\psi$. Suppose, towards a contradiction, that $\models \langle !\varphi \rangle K \psi \vee \neg K[!\varphi]\psi$. This is equivalent to $\models K[!\varphi]\psi \rightarrow \langle !\varphi \rangle K \psi$, which together with item 1 of Lemma 2.5 yields that $\models \langle !\varphi \rangle K \psi \leftrightarrow K[!\varphi]\psi$. This contradicts the non-equivalence of these formulas. Hence $\not\models \langle !\varphi \rangle K \psi \vee \neg K[!\varphi]\psi$. To show that $\models \neg(\langle !\varphi \rangle K \psi \wedge \neg K[!\varphi]\psi)$, consider an arbitrary model and state $\mathbb{M}, w$ and suppose, towards a contradiction, that $\mathbb{M}, w \models \langle !\varphi \rangle K \psi \wedge \neg K[!\varphi]\psi$. It then follows that $\mathbb{M}, w \models \varphi$, $\mathbb{M}|\varphi, w \models K \psi$, and there exists a state v such that $(w, v) \in R$ and $\mathbb{M}, v \not\models [!\varphi]\psi$. Hence $\mathbb{M}, v \models \varphi$ and $\mathbb{M}|\varphi, v \not\models \psi$. By Definition 2.2 we get $(w, v) \in R^\varphi$, so from $\mathbb{M}|\varphi, w \models K \psi$ it follows that $\mathbb{M}|\varphi, v \models \psi$, contradiction.   □

The key idea to obtain this octagon of oppositions was to 'split' the subalternation $\langle !\varphi \rangle (\neg) K \psi \rightarrow [!\varphi](\neg) K \psi$ into two subalternations, by inserting the new formula $(\neg) K[!\varphi]\psi$. Béziau [3] applied the same idea to the hexagon of oppositions for the S5 modal operators: the subalternation $\Box(\neg)\psi \rightarrow \Diamond(\neg)\psi$ is then split into two subalternations, by inserting the new formula $(\neg)\psi$. The additional Aristotelian relations that arise in this way are exactly the same as those we obtained in Theorem 4.1. In other words, our octagon of oppositions for the epistemic properties of public announcement (Fig. 6) is essentially the same as Béziau's octagon of oppositions for the S5 modal operators (Fig. 7).[31]

---

[31] Note that we are here again applying the structuralist philosophy mentioned in Subsect. 3.3: we are calling two opposition structures 'essentially the same', because they have the same configuration of Aristotelian relations. The formulas in both structures are concerned with different topics (public announcements/necessity), and even on a higher level of abstraction they differ significantly: in Fig. 6 the formulas in the upper corners of the square are *existential* ($\langle !\varphi \rangle (\neg) K \psi$), whereas in Fig. 7 they are *universal* ($\Box(\neg)\psi$).

4.2 A Three-Dimensional Opposition Structure

We have just established an analogy between our octagon for the epistemic properties
of public announcement and Béziau's octagon for modal S5. In this subsection we will
further exploit this analogy. Béziau's initial results on S5 [3] have been generalized by
Moretti [22], Smessaert [26], and others. We will now show that these generalizations can
perfectly be transferred from their original context (S5) to this context (PAL).

The first thing to note is that the octagon of oppositions we obtained in the previous
subsection is not Boolean closed.

Lemma 4.2 *Consider arbitrary formulas $\varphi, \psi \in \mathcal{L}$. Suppose that the formulas in Fig. 6
are contingent and pairwise non-equivalent. Then the opposition structure in Fig. 6 is not
Boolean closed.*

Proof Consider, for example, the conjunction of $\neg\varphi$ and $K[!\varphi]\psi$. Just like in the proof
of Lemma 3.9, one can show that this conjunction is not equivalent to any of the formulas
appearing in the octagon in Fig. 6.   □

In Sect. 3 we showed that the square of oppositions for PAL (Fig. 2) is not Boolean
closed (Lemma 3.9), and then immediately went on to construct its Boolean closure (Theorem 3.10). Now, however, we will proceed in a more indirect way. First, we construct
the Boolean closure of the *set of formulas* appearing in the octagon (so not of the octagon
itself, *qua* opposition structure).

Theorem 4.3 *Consider arbitrary formulas $\varphi, \psi \in \mathcal{L}$. Suppose that the formulas in Fig. 6
are contingent and pairwise non-equivalent. Let F be the set containing the eight formulas
appearing in the octagon in Fig. 6. The Boolean closure of F is the fourteen-element
set F', which has the following elements:*[32]

1. φ
2. $\langle!\varphi\rangle K\psi$
3. $K[!\varphi]\psi$
4. $[!\varphi]K\psi$
5. $\neg\varphi$
6. $\langle!\varphi\rangle\neg K\psi$
7. $\neg K[!\varphi]\psi$
8. $[!\varphi]\neg K\psi$
9. $\neg\varphi \wedge K[!\varphi]\psi$
10. $\neg\varphi \wedge \neg K[!\varphi]\psi$
11. $\varphi \vee \neg K[!\varphi]\psi$
12. $\varphi \vee K[!\varphi]\psi$
13. $[!\varphi]K\psi \wedge (\varphi \vee \neg K[!\varphi]\psi)$
14. $[!\varphi]\neg K\psi \wedge (\varphi \vee K[!\varphi]\psi)$

---

[32]In the remainder of this subsection, we will often refer to the formulas in F' using the numbers they are
given here. For example, we will refer to the formula $\neg\varphi \wedge \neg K[!\varphi]\psi$ using the number 10.

Proof First of all, note that F contains only contingent formulas, so that it indeed makes sense to construct its Boolean closure (cf. Definition 3.5). We will construct the Boolean closure of F 'step by step', i.e. adding formulas as we go along, and show that in this way we arrive at the fourteen-element set F' whose members are listed above. We will simply 'describe' this procedure; the details of each of the single steps are left to the reader.

Note that formulas 1–8 above are the elements of F itself. The set of these eight formulas is closed under $\neg$, but not under $\wedge$: it 'misses' the conjunctions $3 \wedge 5, 3 \wedge 8, 4 \wedge 7$ and $5 \wedge 7$. Note that $3 \wedge 5$ and $3 \wedge 8$ are equivalent, and also that $4 \wedge 7$ and $5 \wedge 7$ are equivalent. Hence we need to add only one of each of these pairs: we add $3 \wedge 5$ (this is formula 9) and $5 \wedge 7$ (this is formula 10).

We thus arrive at the set of formulas 1–10. By construction, this set is closed under $\wedge$. However, it is not closed under $\neg$: it 'misses' the negations of the newly added formulas, 9 and 10. Hence we add $\neg 9$ (this is formula 11) and $\neg 10$ (this is formula 12).

We thus arrive at the set of formulas 1–12. By construction, this set is closed under $\neg$. However, it is not closed under $\wedge$: it 'misses' the conjunctions $4 \wedge 11$ and $8 \wedge 12$. These conjunctions are not equivalent, so we need to add them both: we add $4 \wedge 11$ (this is formula 13) and $8 \wedge 12$ (this is formula 14).

We thus arrive at the set F' of formulas 1–14. By construction, this set is closed under $\wedge$. Furthermore, it is also closed under $\neg$, because the two formulas that were added at the last step, are each other's negations (up to equivalence): $\models \neg 13 \leftrightarrow 14$. By construction, this set F' is the Boolean closure of the original set F.

The following table shows the full details of the $\neg$-closedness of F':

	1	2	3	4	5	6	7	8	9	10	11	12	13	14
$\neg$	5	8	7	6	1	4	3	2	11	12	9	10	14	13

The following table shows the full details of the $\wedge$-closedness of F':

$\wedge$	1	2	3	4	5	6	7	8	9	10	11	12	13	14
1	1	2	2	2	$\bot$	6	6	6	$\bot$	$\bot$	1	1	2	3
2		2	2	2	$\bot$	$\bot$	$\bot$	$\bot$	$\bot$	$\bot$	2	2	2	$\bot$
3			3	3	9	$\bot$	$\bot$	9	9	$\bot$	2	3	2	9
4				4	5	$\bot$	10	5	9	10	13	3	13	9
5					5	$\bot$	10	5	9	10	10	9	10	9
6						6	6	6	$\bot$	$\bot$	6	6	$\bot$	6
7							7	7	$\bot$	10	7	6	10	6
8								8	9	10	7	14	10	14
9									9	$\bot$	$\bot$	9	$\bot$	9
10										10	10	$\bot$	10	$\bot$
11											11	1	13	6
12												12	13	14
13													13	$\bot$
14														14

We have constructed the Boolean closure F' of the set F of formulas appearing in the octagon in Fig. 6. By Definition 3.7 the fourteen formulas in F' are the formulas that

Fig. 8 Hasse diagram for the fourteen formulas in F'

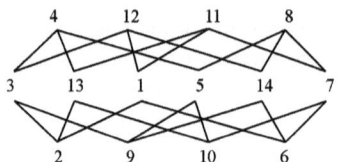

appear in the Boolean closure of this octagon (*qua* opposition structure). The question now arises how to 'organize' these formulas into an opposition structure.

Recall that we are only interested in the properties of the formulas in F' up to logical equivalence. In other words, we can identify any formula $\varphi \in F'$ with its equivalence class $[\varphi] := \{\psi \in \mathcal{L} \mid \models \varphi \leftrightarrow \psi\}$. From this perspective, it is a natural idea to construct a Hasse diagram for F'; cf. Fig. 8.

This Hasse diagram, however, is itself not an opposition structure, because it does not display all of the Aristotelian relations between the fourteen formulas in F' (it only represents the subalternation relations). A tedious but easy exercise leads to the following table, which represents the 79 Aristotelian relations that hold between any of the fourteen formulas in F'.[33]

	2	3	4	5	6	7	8	9	10	11	12	13	14
1	↙	NO	SC	CD	↙	NO	SC	C	C	↗	↗	NO	NO
2		↗	↗	C	C	C	CD	C	C	↗	↗	↗	C
3			↗	NO	C	CD	SC	↙	C	SC	↗	NO	NO
4				↙	CD	SC	SC	↙	↙	SC	SC	↙	SC
5					C	NO	↗	↙	↙	SC	SC	NO	NO
6						↗	↗	C	C	↗	↗	C	↗
7							↗	C	↙	↗	SC	NO	NO
8								↙	↙	SC	SC	SC	↙
9									C	CD	↗	C	↗
10										↗	CD	↗	C
11											SC	↙	SC
12												SC	↙
13													CD

Let us take stock. The aim is to construct the Boolean closure of the octagon of oppositions in Fig. 6. We already know the formulas of this Boolean closure: the fourteen elements of F'. We also already know the 79 Aristotelian relations which hold between these fourteen formulas: cf. the table above.

---

[33] 'NO' stands for no Aristotelian relation at all, 'CD' stands for contradictories, 'C' stands for contraries, 'SC' stands for subcontraries, and '↙ / ↗' stand for subalternation. In particular, '↙' says that there is a subalternation from the column-formula to the row-formula, and vice versa for '↗'. The table has 13×13 cells in total. Because only the upper right half is filled in, the number of entries is

$$\sum_{n=1}^{13}(14 - n) = \sum_{n=1}^{13}14 - \sum_{n=1}^{13}n = (14 \times 13) - \frac{14 \times 13}{2} = \frac{14 \times 13}{2} = 91.$$

Of these 91 entries, 12 are 'NO', so in total we arrive at $91 - 12 = 79$ Aristotelian relations.

Fig. 9 Rhombic
dodecahedron of oppositions
for the epistemic properties of
public announcement

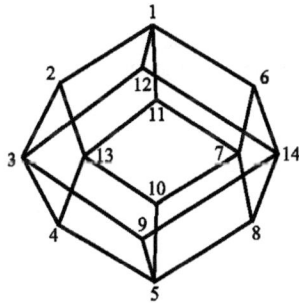

We are now ready to apply Smessaert's [26] results (which Moretti calls "simply *bril-liant*" [22, p. 217]). Reconsider the Hasse diagram for F' (Fig. 8). Smessaert's results say that this Hasse diagram can be transformed into a *rhombic dodecahedron*. This is a three-dimensional structure, which has twelve rhombus-shaped faces, and (of course) fourteen vertices. Figure 9 shows this dodecahedron; however, it does *not* show the dodecahedron *qua* opposition structure, because it does not show the 79 Aristotelian relations holding between the fourteen vertices (for reasons of visual clarity). Nevertheless, we will say that this dodecahedron is the (long-awaited) Boolean closure of the octagon in Fig. 6 (*qua* opposition structure).

Theorem 4.4 *The rhombic dodecahedron in Fig. 9 is the Boolean closure of the octagon in Fig. 6.*

Proof The formulas appearing in the dodecahedron are the elements of F', and so they are certainly the right ones by Theorem 4.3. That this dodecahedron completely represents the Aristotelian relations between these fourteen formulas follows from the fact that hexagons *i–vi* in Fig. 10 can be embedded in the dodecahedron: each of the 79 Aristotelian relations appears in at least one of these hexagons, and thus in the dodecahedron. □

The six hexagons in Fig. 10 that were used in the proof are the PAL-analogues of six 'original' hexagons for S5. These six original hexagons jointly illustrate the progress *n*-opposition theory has been making over the past few years. Béziau [3] already constructed hexagons *i*, *ii* and *iii* (for S5). Subsequently, hexagon *iv* (for S5) was constructed independently (and through different ways of reasoning) by Moretti and Smessaert. Finally, hexagons *v* and *vi* (for S5) were constructed by Smessaert [26].[34] In this paper, we constructed (the PAL-analogues of) these six hexagons in one swoop. This illustrates how we have been able to exploit the 'initial' analogy between our octagon of oppositions for the epistemic properties of public announcement (Fig. 6) and Béziau's octagon of oppositions for the S5 modal operators (Fig. 7).

---

[34]Furthermore, Pellissier [24] distinguishes between *strong* and *weak* hexagons. A hexagon is strong iff the disjunction of its three contrary formulas is tautologous; a hexagon is weak iff the disjunction of its three contrary formulas is not tautologous. The six hexagons discussed here (displayed in Fig. 10) are all strong. For example, for hexagon *i*, note that the disjunction of its three contrary formulas (2, 5 and 6) is equivalent to $\varphi \vee \neg\varphi$.

Fig. 10 Six hexagons that
are embeddable in the
rhombic dodecahedron

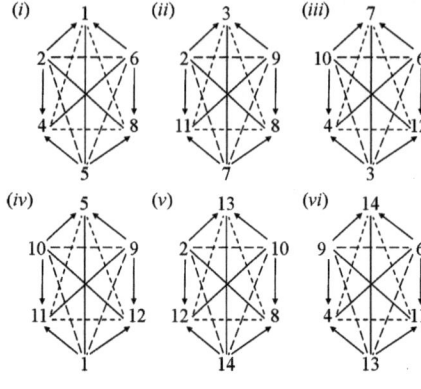

Fig. 11 Modal graph for S5

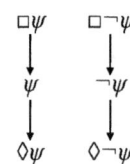

4.3 A Comparison with Smessaert's Results

As we emphasized throughout the previous subsection, our construction of the rhombic
dodecahedron of oppositions for the epistemic properties of public announcement (Fig. 9)
is essentially an application of the techniques developed by Smessaert [26] in the context
of S5. It should thus not be surprising that we obtain the same 'end results', viz. the rhombic dodecahedra for S5 and PAL. This also means that Smessaert, Pellissier and Moretti's
additional comments on the rhombic dodecahedron for S5 (it contains six strong and
four weak hexagons, it contains a cube and an octahedron, it is very close to Pellissier's
tetraicosahedron [24], etc.) also apply to our dodecahedron for PAL.

Despite these similarities, there are also some differences between Smessaert's original results on S5 and our results on PAL. The first difference is that the dodecahedron for
PAL does *not* arise from a modal graph. Smessaert's dodecahedron for S5 arises from the
modal graph for S5 (Fig. 11), which represents the six non-equivalent modalities expressible in S5 (including the 'naked' modalities). Furthermore, also the hexagon for the structural properties of public announcement (Fig. 3) can be seen as arising from the modal
graph for the public announcement operators.[35] The dodecahedron for PAL, however, is
based on Lemma 2.5, which talks about the interaction between public announcement and
knowledge, not about the non-equivalent public announcement modalities expressible in
PAL. This shows that there are opposition structures that do not arise from a modal graph,
and thus loosens the connection between *n*-opposition theory and modal graphs.[36]

---

[35]More generally: the hexagon of oppositions for any partially functional process π (Fig. 4) can be seen
as arising from the modal graph for the dynamic operators representing π.

[36]I conjecture that this observation will have (negative) consequences for Pellissier's view on opposition
structures as 'interpolations of consistency degrees' ([22, p. 211], [24, Section 2]).

The most important difference between Smessaert's results and ours, however, is the difference in *complexity*.[37] Smessaert constructed a dodecahedron for S5 and is thus concerned with *one* modality ($\square \cdot / \lozenge \cdot$), which is *one-placed* (unary). Our dodecahedron for the epistemic properties of public announcements, however, is concerned with *two* modalities: the knowledge operator ($K \cdot / \hat{K} \cdot$) and the public announcement operator ($[!\cdot] \cdot / \langle !\cdot \rangle \cdot$). The former is also one-placed, but the latter is *two-placed* (binary): if one has a public announcement operator, then one needs to supply two formulas to obtain a well-formed formula of $\mathcal{L}$. Hence according to both parameters (number of modalities/arity of the modalities), PAL is more complex than S5.

These differences exhibit the power and wide applicability of Smessaert's techniques. Although Smessaert initially applied them to obtain a dodecahedron for the 'simple' modal logic S5 (based on the modal graph for that logic), they can also be used to obtain opposition structures (which are not necessarily based on modal graphs) for more complex logics, such as PAL.

5 Conclusion

In this paper we have studied dynamic epistemic logic, and in particular public announcement logic, using the tools of n-opposition theory [22]. After giving a brief introduction to PAL, we constructed a square and a hexagon of oppositions for the structural properties of public announcements, and showed how to generalize them to opposition structures for *any* partially functional process. This can be used in support of the structuralist philosophy surrounding n-opposition theory. Finally, we focused on the epistemic properties of public announcements, and constructed an octagon and (using techniques developed by Smessaert) a (three-dimensional) rhombic dodecahedron of oppositions.

Some questions are left for further research. First of all, in this paper we have said nothing about *bitstrings*, one of the central topics in Smessaert [26]. It would certainly be nice to obtain a bitstring semantics for PAL as well. Secondly, although the dodecahedron in Fig. 9 is Boolean closed, other formulas can still be added. For example, because K is factive and $[!\varphi]$ is a normal modal operator, we have $\models [!\varphi]K\psi \rightarrow [!\varphi]\psi$. The antecedent of this implication is (equivalent to) one of the fourteen formulas in the dodecahedron, but the consequent is not. One may ask which formulas can still be added to the fourteen in the dodecahedron, which additional Aristotelian relations thus arise, and, most importantly, which opposition structures can be constructed for them. These questions, however, will be addressed in another paper.

References

1. Baltag, A., Moss, L.: Logics for epistemic programs. Synthese **139**, 165–224 (2004)

---

[37] The term 'complexity' is used in an intuitive sense here (as is explained in the main text). It is well-known that from the perspective of *computational complexity*, PAL is equally complex as S5: the satisfiability problem of both S5 and PAL is NP-complete in the single-agent case, PSPACE-complete in the multi-agent case without common knowledge, and EXPTIME-complete in the multi-agent case with common knowledge [14, 20].

2. Baltag, A., Smets, S.: A qualitative theory of dynamic interactive belief revision. In: Bonanno, G., van der Hoek, W., Woolridge, M. (eds.) Logic and the Foundations of Game and Decision Theory (LOFT 7). Texts in Logic and Games, vol. 3, pp. 9–58. Amsterdam University Press, Amsterdam (2008)

3. Béziau, J.-Y.: New light on the square of oppositions and its nameless corner. Log. Investig. **10**, 218–232 (2003)

4. Blanché, R.: Quantity, modality, and other kindred systems of categories. Mind **61**, 369–375 (1952)

5. Blanché, R.: Sur l'opposition des concepts. Theoria **19**, 89–130 (1953)

6. Blanché, R.: Opposition et négation. Rev. Philos. Fr. étrang. **167**, 187–216 (1957)

7. Blanché, R.: Structures intellectuelles. Essai sur l'organisation systématique des concepts. Vrin, Paris (1966)

8. Davey, B.A., Priestley, H.A.: Introduction to Lattices and Order, 2nd edn. Cambridge University Press, Cambridge (2002)

9. de Pater, W.A., Vergauwen, R.: Logica: Formeel en Informeel. Leuven University Press, Leuven (2005)

10. Demey, L.: Some remarks on the model theory of epistemic plausibility models. J. Appl. Non-Class. Log. **21**, 375–395 (2011)

11. Demey, L., Smessaert, H.: Logical geometries emanating from the traditional square of oppositions. Manuscript (2011)

12. Fitting, M., Mendelsohn, R.L.: First-Order Modal Logic. Kluwer, Dordrecht (1998)

13. Gerbrandy, J., Groeneveld, W.: Reasoning about information change. J. Log. Lang. Inf. **6**, 147–169 (1997)

14. Halpern, J.Y., Moses, Y.: A guide to completeness and complexity for modal logics of knowledge and belief. Artif. Intell. **54**, 319–379 (1992)

15. Harel, D., Kozen, D., Tiuryn, J.: Dynamic Logic. MIT Press, Cambridge (2000)

16. Hintikka, J.: Knowledge and Belief. An Introduction to the Logic of the Two Notions. Cornell University Press, Ithaca (1962)

17. Kozen, D., Parikh, R.: An elementary proof of the completeness of PDL. Theor. Comput. Sci. **14**, 113–118 (1981)

18. Lenzen, W.: Recent Work in Epistemic Logic. North-Holland, Amsterdam (1978)

19. Lenzen, W.: Glauben, Wissen und Wahrscheinlichkeit: Systeme der epistemischen Logik. Springer, Berlin (1980)

20. Lutz, C.: Complexity and succinctness of public announcement logic. In: Stone, P., Weiss, G. (eds.) AAMAS '06: Proceedings of the Fifth International Joint Conference on Autonomous Agents and Multiagent Systems, pp. 137–143. Association for Computing Machinery, New York (2006)

21. McNamara, P.: Deontic Logic. Stanford Encyclopedia of Philosophy (2010)

22. Moretti, A.: The geometry of logical opposition. PhD thesis defended at the University of Neuchâtel, Switzerland (2009)

23. Plaza, J.: Logics of public communications. In: Emrich, M.L., Pfeifer, M.S., Hadzikadic, M., Ras, Z.W. (eds.) Proceedings of the Fourth International Symposium on Methodologies for Intelligent Systems: Poster Session Program, pp. 201–216. Oak Ridge National Laboratory, Oak Ridge (1989). Reprinted in: Synthese **158**, 165–179 (2007)

24. Pellissier, R.: Setting n-opposition. Logica Univers. **2**, 235–263 (2008)

25. Sesmat, A.: Logique II. Les raisonnements, la logistique. Hermann, Paris (1951)

26. Smessaert, S.: On the 3D visualisation of logical relations. Logica Univers. **3**, 303–332 (2009)

27. Smessaert, H., Demey, L.: On the fourth Aristotelian relation: subalternation versus non-implication. Manuscript (2011)

28. van Benthem, J.: One is a lonely number: logic and communication. In: Chatzidakis, Z., Koepke, P., Pohlers, W. (eds.) Logic Colloquium '02. Lecture Notes in Logic, vol. 27, pp. 95–128. Association for Symbolic Logic & AK Peters, Wellesley (2006)

29. van Benthem, J.: Open problems in logical dynamics. In: Gabbay, D., Goncharov, S., Zakharyashev, M. (eds.) Mathematical Problems from Applied Logic I. International Mathematical Series, vol. 4, pp. 137–192. Springer, Berlin (2006)

30. van Benthem, J.: Dynamic logic for belief revision. J. Appl. Non-Class. Log. **17**, 129–155 (2007)

31. van Benthem, J.: Logical Dynamics of Information and Interaction. Cambridge University Press, Cambridge (2011)

32. van Dalen, D.: Logic and Structure, 4th edn. Springer, Berlin (2004)
33. van Ditmarsch, D., van der Hoek, W., Kooi, B.: Dynamic Epistemic Logic. Springer, Berlin (2007)

L. Demey (✉)
Center for Logic and Analytical Philosophy, Institute of Philosophy, KU Leuven – University of Leuven, Kardinaal Mercierplein 2, 3000 Leuven, Belgium
e-mail: lorenz.demey@hiw.kuleuven.be

Logical Opposition and Collective Decisions

Srećko Kovač

Abstract The square of opposition (as part of a lattice) is used as a natural way to represent different and opposite ways of who makes decisions, and in what way, in/for a group or a society. Majority logic is characterized by multiple logical squares (one for each possible majority), with the "discursive dilemma" as a consequence. Three-valued logics of majority decisions with discursive dilemma undecided, of veto, consensus, and sequential voting are analyzed from the semantic point of view. For instance, the paraconsistent and paracomplete logics M_3, M_{3veto} and C_3 are described. The distinction of designated and non-designated values is not used, and instead, the consequence relation is defined as a preservation of the minimum truth-value of the implying set of sentences. The consequence and opposition relations of the logics described are compared, and an ordering of the logics with respect to their opposition relations is established.

Keywords Collective decision · Discursive dilemma · Hexagon of opposition · Majority · Three-valued logic

Mathematics Subject Classification Primary 03B80 · Secondary 03B50

1 Introduction

In the first part of this paper, we represent opposition in collective decisions by a logical square as part of a lattice. It is shown that the "discursive dilemma" (see, for instance, [12]) arises in collective decisions because each possible majority has its own logical square within the common lattice. In the second part, we analyze some logics of collective decisions in a three-valued setting without a fixed distinction between designated and non-designated values and with a consequence relation defined by means of the relative preservation of truth values (cf. a similar idea in [9]). We semantically analyze majority decisions, right of veto, consensus, and sequential voting, and compare the consequence relation and opposition of the respective logics.

I am very grateful to the referees for valuable comments, corrections and suggestions.

J.-Y. Béziau, D. Jacquette (eds.), *Around and Beyond the Square of Opposition*, 341–356
Studies in Universal Logic, DOI 10.1007/978-3-0348-0379-3_23, © Springer Basel 2012

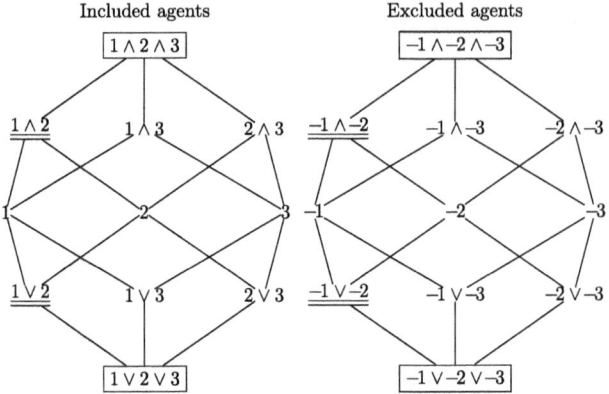

Fig. 1 Participation of the members of a group in decision-making (*boxed* and *double underlined numerals* indicate the vertices of two squares, respectively)

2 Collective Decisions in Squares

We analyze some aspects of the logics of collective decisions with particular respect to logical opposition as contained in the traditional square of opposition. We will conceive the square of opposition as part of a combination of two lattices presented by inverted Hasse diagrams (with meet up and join down in the diagram, where, consequently, what is down a line in the lattice follows from what is up). In the left-side lattice we can recognize the positive part of the square of opposition, and in the right-side lattice we can recognize the negative part of the square (see, for example, Fig. 1).

2.1 Who and in What Way Takes Collective Decisions?

The first question we address by means of a "latticed square" is *who* participates in collective decision-making in or for a group (e.g., society) of agents. As an example, we present the possibilities for a three-agent group (see Fig. 1). In the positive lattice, we can see that participation in decision-making ranges from the inclusion of each agent (in the form of consensual, majority or some other decision-making), to the inclusion of only one agent (one determinate agent decides in the name of the group), and to the inclusion of any one of the agents (i.e., whatever agent can decide in the name of the group). In the negative lattice, we can see that possibilities range from the exclusion of all agents from decision-making (a group governed entirely from the outside), the exclusion of this or that part of the group, to the exclusion of any one of the agents.

Once it is decided who participates in collective decision-making for a group, a natural question arises about *how* collective decisions are made? We focus on the class of cases where a group of agents participates in decision-making. The possibilities of how participants make decisions range from the consensual decision-making to individual decision-making for the whole group, simply by deciding for or against a proposal (some proposed proposition) (see Fig. 2). We can *combine* a positive and a negative lattice in one lattice, with a common meet and join at the top and at the bottom of the lattice, respectively. As an

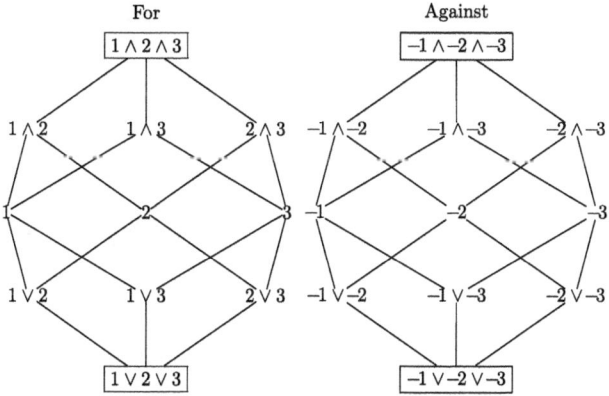

Fig. 2 Attitudes of the members of a group towards a proposal ($-n$ indicates that the agent n is against the proposal)

Fig. 3 Two-agent group

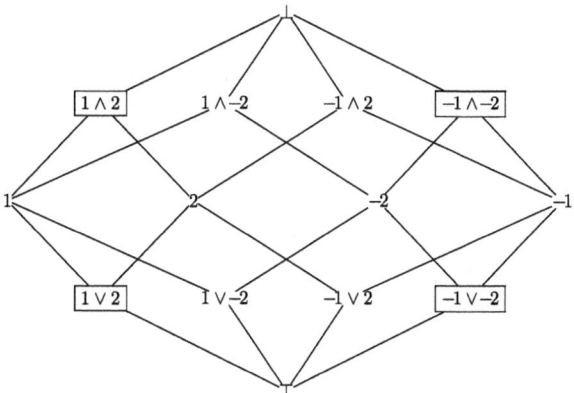

example, we show a combined lattice for a two-agent group (Fig. 3). In the example, we easily recognize the corners of the traditional square of opposition ($1 \wedge 2$, $-1 \wedge -2$, $1 \vee 2$, $-1 \vee -2$). We can also recognize another square in the middle of the lattice always with one positive and one negative member ($1 \wedge -2$, $-1 \wedge 2$, $1 \vee -2$, $-1 \vee 2$). Contradictories in the "latticed square" are points (nodes) that have the top of the lattice as their meet and the bottom of the lattice as their join. In addition, contraries are nodes that have the top of the lattice as their meet, and their join is always above the bottom. Subcontraries are nodes that have their meet always below the top of the lattice, and the bottom of the lattice is their join.

In the lattice, we can also insert the conditions for consensus, as well as for dissensus (disagreement) among agents. As we can see (Fig. 4), both nodes are near the center of the lattice.

Fig. 4 Consensus and dissensus in a two-agent group

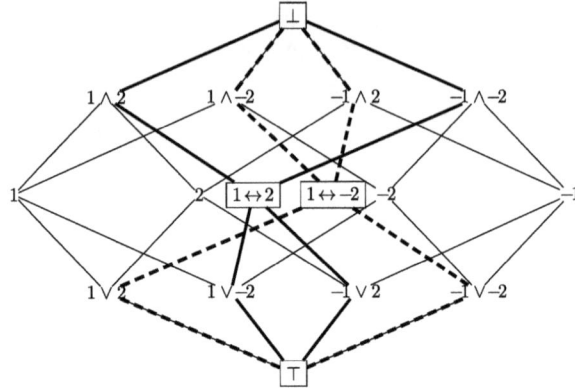

2.2 Majority Decision-Making and the Discursive Dilemma

Let us focus on the case where collective decisions are being made by a majority of the agents. Characteristically, each possible majority has its own square of opposition. Hence, majority decisions are characterized by a multiply squared opposition. As an example, we give a figure with multiple squares for a three-agent group (see Fig. 5). We notice that we can also construct a square consisting only of majorities in each of the corners, as in Fig. 6, where each corner is indicated by a dashed ellipse. Three elements within an ellipse should be conceived disjunctively, for instance $(1 \wedge 2) \vee (1 \wedge 3) \vee (2 \wedge 3)$ as the i-corner of the square.

The presence of multiply squared opposition in majority decisions is in fact a condition that makes the so-called "discursive dilemma" [6, 12] possible. In majority decision-making we could simultaneously have $P \rightarrow Q$, P as well as $\neg Q$ as decisions, despite the obvious contradiction (violation of modus ponens). But it is essential to see that each of these decisions is being made by another majority, each of which has its own square of opposition (see Fig. 7).

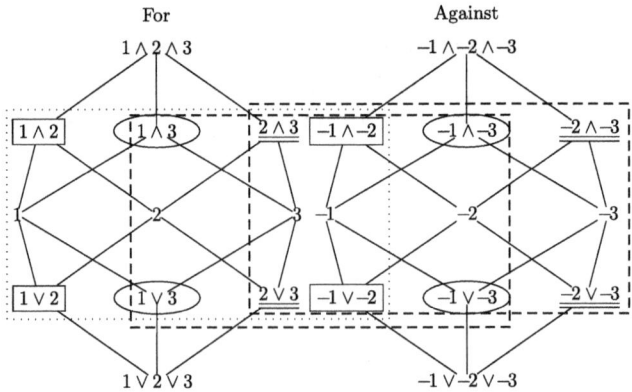

Fig. 5 Three majorities and three squares

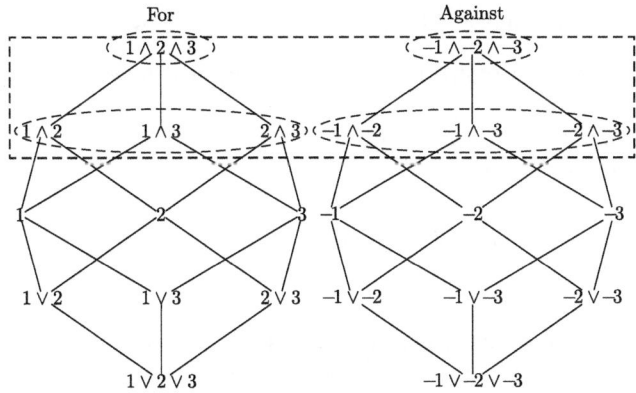

Fig. 6 Square of majorities

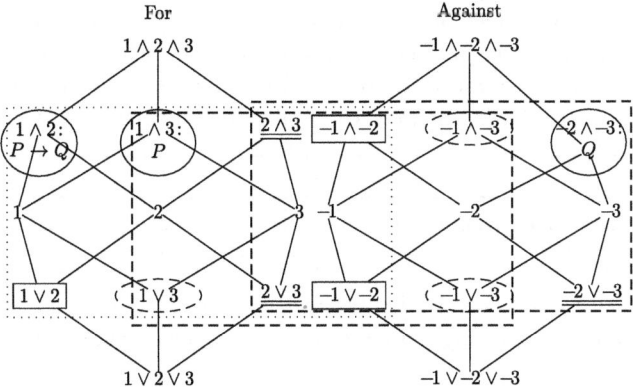

Fig. 7 Discursive dilemma

3 A Many-Valued Approach to Majority and Consensual Logic

In this and the following sections, we will describe several logics that in a somewhat sim-
plified and idealized manner formalize some characteristic ways of reasoning in making
collective decisions. The logics will be described from the semantic point of view in a
three-valued setting, and will be compared especially with respect to the opposition in a
logical hexagon.[1]

3.1 Language and Models

For each of the logics we will describe, we use the same propositional language $\mathcal{L}_m$ built
in the familiar way with $\neg, \wedge, \vee, \rightarrow$ as connectives. Let *Form* be the set of all formulas
of $\mathcal{L}_m$, and *At* the set of all atomic formulas of $\mathcal{L}_m$.

---

[1]On the aggregation of judgments in a many-valued setting, see [11].

Definition 1 (Model) A model $\mathfrak{M}$ is a 3-tuple $\langle V, \{f_*\}, \mathfrak{I}_0 \rangle$, where

1. V is an ordered set of truth values: $\mathbf{f} < \mathbf{u} < \mathbf{t}$ ($\mathbf{u}$, 'undefined', is short for '$\mathbf{t}$ or $\mathbf{f}$ or neither'),
2. f_* is the truth function of the connective $*$,
3. $\mathfrak{I}_0$ is a mapping $At \longrightarrow V$.

Definition 2 (Interpretation, $\mathfrak{I}$) $\mathfrak{I}$ is a mapping $Form \longrightarrow V$ inductively defined from $\mathfrak{I}_0$ by means of $\{f_*\}$.

We imagine that an atomic sentence is true ($\mathbf{t}$) or false ($\mathbf{f}$) in accordance with its acceptance or rejection, respectively, by voting. Further, an atomic sentence ϕ has the value "undefined" ($\mathbf{u}$) if, according to the voting results, it is not clear whether it should be accepted or not, or if it was simply not mentioned in any voting proposal. We notice that we do not have a set of designated values. The reason for this will become clear when we come to the definition of the consequence relation. For better understanding of the truth functions we will define, we give a tabular overview of the values with informal explanations. In the overview, we informally distinguish values $\mathbf{t}_F$ and $\mathbf{t}_A$ ('F' for 'for', 'A' for 'against'), as well as values $\mathbf{f}_F$ and $\mathbf{f}_A$. The distinctions are clear from the table.

	ϕ	$\neg\phi$
more than half for ϕ (simple majority) = less than half against ϕ	$\mathbf{t}_F = \mathbf{t}$	$\mathbf{f}_A = \mathbf{f}$
at most one half for ϕ (simple minority) = at least one half against ϕ	$\mathbf{f}_F = \mathbf{f}$	$\mathbf{t}_A = \mathbf{t}$

We also note that we treat withheld votes as not being counted (except for quorum). Finally, an ideal proviso in the logics to be proposed is that each agent reasons in a classical way.

3.2 Majority Logic M$_3$

We pause, first, for the case where decisions are being made by a majority answering the questions as, for example, at an referendum. A referendum can include a plurality of questions (for example, referendums with ten or fifteen question are known, but ideally we can imagine any finite number of questions posed at such an ideal referendum). We obtain a special 3-valued logic, differing in truth functions from Kleene's (strong) 3-valued logic (see [7]) in three places. We call this logic M$_3$. It is essential to note that in this kind of decision-making agents do not know the results of voting on any proposal (question) until after the whole voting has ended. We also assume that the discursive dilemma is not resolved.

Special cases as distinguished from Kleene's strong 3-valued logic are the following:

$$f_\wedge(\mathbf{t}, \mathbf{t}) = \mathbf{u} \qquad f_\vee(\mathbf{f}, \mathbf{f}) = \mathbf{u} \qquad f_\rightarrow(\mathbf{t}, \mathbf{f}) = \mathbf{u}$$

These differences are due to leaving the discursive dilemma undecided—which is probably the best option if we do not want to undertake an additional analysis of ballot papers or an additional voting.[2] If we resolve the dilemma classically (for example, conjunction entailing its conjuncts), we obtain Kleene's 3-valued logic (K_3). Here are the truth tables for majority logic M_3:

$f_\neg$		$f_\wedge$	t	u	f		$f_\vee$	t	u	f		$f_\to$	t	u	f
t	f	t	$\boxed{u}$	u	f		t	t	t	t		t	t	u	$\boxed{u}$
u	u	u	u	u	f		u	t	u	u		u	t	u	u
f	t	f	f	f	f		f	t	u	$\boxed{u}$		f	t	t	t

and the (well-known) truth tables for Kleene's K_3:

$f_\neg$		$f_\wedge$	t	u	f		$f_\vee$	t	u	f		$f_\to$	t	u	f
t	f	t	$\boxed{t}$	u	f		t	t	t	t		t	t	u	$\boxed{f}$
u	u	u	u	u	f		u	t	u	u		u	t	u	u
f	t	f	f	f	f		f	t	u	$\boxed{f}$		f	t	t	t

As we see from the truth tables for M_3, no truth values of ϕ and ψ can yield a true $\phi \wedge \psi$, nor false $\phi \vee \psi$, nor false $\phi \to \psi$.

Example 1 In the following three cases of agents' opinions we illustrate situations with a false conjunction of true conjuncts, a true disjunction of false disjuncts, and a true conditional of a true antecedent and false consequent. The possibility of situations with the classical valuation of conjunction, disjunction and conditional is obvious. Therefore, **u** is assigned in each case as an M_3 value.

$$f_\wedge(\mathbf{t}, \mathbf{t}) = \mathbf{u} \quad \mathbf{f} \text{ possibility: 3 agents, 1,2 for } \phi, \text{ 2,3 for } \psi, \text{ 2 for } \phi \wedge \psi,$$
$$f_\vee(\mathbf{f}, \mathbf{f}) = \mathbf{u} \quad \mathbf{t} \text{ possibility: 2 or 3 agents, 1 for } \phi, \text{ 2 for } \psi, \text{ 1,2 for } \phi \vee \psi,$$
$$f_\to(\mathbf{t}, \mathbf{f}) = \mathbf{u} \quad \mathbf{t} \text{ possibility: 3 agents, 1,3 for } \phi, \text{ 1 for } \psi, \text{ 1,2 for } \phi \to \psi.$$

For instance, a majority may be for sending person A in a shuttle to a space station, and a majority may be for sending person B in the same shuttle, but there may be no majority for sending both A and B in the same shuttle to the space station (say, because of some known personal incompatibilities between A and B). In real life, $f_\wedge(\mathbf{t}, \mathbf{t}) = \mathbf{u}$ means that after two positive voting results, we will still (if relevant) have to vote on the conjunction of the results. Formally, with that additional voting on the conjunction we generate in M_3 a new majority model, where informal conjunction is treated as an atom with a directly assigned truth value.

It is also interesting to compare M_3 with the other three-valued logics that already have an interpretation in a collective reasoning context (see [4] and [5]). In the following truth tables, the differences of P^1 and I^1 to M_3 are indicated by rectangles, and the differences between P^1 and I^1 are indicated by '!'.

---

[2]In ongoing research, we are investigating a majority logic with formal compound sentences (not only atomic sentences) directly subject to voting (making, for example, conjuncts true if the conjunction is true). This leads to a non-truth-functional approach where the logic is not generated by truth functions on the basis of atomic valuations (see [2, 10]).

P¹ (see [4, 5, 14]):

$f_\neg$	
t	f
t*	t!
f	t

$f_\wedge$	t	t*	f
t	t	t!	f
t*	t!	t!	f
f	f	f	f

$f_\vee$	t	t*	f
t	t	t	t
t*	t	t!	t!
f	t	t!	f

$f_\rightarrow$	t	t*	f
t	t	t!	f
t*	t	t!	f
f	t	t	t

I¹ (see [3–5]):

$f_\neg$	
t	f
f*	f!
f	t

$f_\wedge$	t	f*	f
t	t	f!	f
f*	f!	f!	f
f	f	f	f

$f_\vee$	t	f*	f
t	t	t	t
f*	t	f!	f!
f	t	f!	f

$f_\rightarrow$	t	f*	f
t	t	f!	f
f*	t	t!	t!
f	t	t	t

M₃:

$f_\neg$	
t	f
u	u
f	t

$f_\wedge$	t	u	f
t	t	u	f
u	u	u	f
f	f	f	f

$f_\vee$	t	u	f
t	t	t	t
u	t	u	u
f	t	u	f

$f_\rightarrow$	t	u	f
t	t	u	u
u	t	u	u
f	t	t	t

Note that whenever a conjunction in **P¹** is true, it is undefined in **M₃**, and whenever a disjunction in **I¹** is false, it is also undefined in **M₃**. In the next subsection, the opposition relation in **P¹** as well as in **I¹** will be represented by the corresponding logical hexagons.

3.3 Consequence, Equivalence, and Opposition

As already mentioned, in a majority logic model we do not have a designated set of values. Accordingly, we will define the consequence relation in a somewhat restricted way that does not allow any decrease of truth values in the transition from the set of assumptions to the consequence. This will make a distinction, for example, from Kleene's **K₃**, which allows a decrease of values from "undefined" to "false" (both being non-designated values). It will also make a distinction from Asenjo's/Priest's **LP** (see [1] and [13]), which allows a decrease of values from "true" to "undefined", both being designated values. In simple words, we say that in a majority logic ϕ is a consequence of a set Γ of sentences iff ϕ has at least the minimum value of the values of the members of Γ, or the highest possible value if $\Gamma = \varnothing$. But we define first what it is for sentence χ to have the minimum value in a model $\mathfrak{M}$ with respect to a set Γ of sentences. The following two definitions will be given, first, for logic **M₃** and later will be used for other logics of collective reasoning.

Definition 3 (Minimum value of a sentence with respect to Γ and $\mathfrak{M}$)

$$Min^{\Gamma,\mathfrak{M}}(\chi) \quad \text{iff} \quad \begin{cases} \chi \in \Gamma \wedge (\forall \psi \in \Gamma)\mathfrak{I}(\chi) \leq \mathfrak{I}(\psi) & \text{if } \Gamma \neq \varnothing \\ \forall \psi \mathfrak{I}(\psi) \leq \mathfrak{I}(\chi) & \text{otherwise} \end{cases}$$

where $Min^{\Gamma,\mathfrak{M}}(\chi)$ abbreviates 'χ has the minimum value in a model $\mathfrak{M}$ with respect to a set Γ', and $\mathfrak{M}$ is an **M₃** model.

Definition 4 (Consequence relation)

$$\Gamma \models_{\mathbf{M_3}} \phi \quad \text{iff} \quad \text{for each model } \mathfrak{M}, \, \forall \chi \left(Min^{\Gamma, \mathfrak{M}}(\chi) \to \mathfrak{I}(\chi) \leq \mathfrak{I}(\phi)\right)$$

where $\mathfrak{M}$ is an $\mathbf{M_3}$ model.

This definition of the consequence relation is very general. To say that the consequence ϕ should be true in each model where each member of Γ is true is simply a special case of Definition 4 above. The definition could be generalized even further (to include the consequence relation defined as the preservation of designated values in general) if we replace the *minimum* value (of sentences in Γ) concept with a *minimal* value concept. If we have a designated set of values in a model and if we define that each non-designated value strictly precedes each designated value, then each designated value is a minimal value of Γ if there are no members of Γ with a non-designated value, or, otherwise, each non-designated value is a minimal value of Γ.

By means of Definition 4 we exclude any possible inference from a **t**-valued sentence to a **u**-valued sentence, as well as any inference from a **u**-valued sentence to an **f**-valued sentence. The idea is to block such inferences since **u** includes the possibility of **t** as well as of **f**, and thus the inference from a **t**-valued sentence to a **u**-valued sentence and the inference from a **u**-valued sentence to an **f**-valued sentence would include the possibility of inference from a **t**-valued sentence to an **f**-valued sentence. Besides, in many cases there is a possibility to decide a question (a **u**-valued sentence) by voting. For example, instead of allowing an **f**-valued sentence to be a conclusion from a **u**-valued sentence, we should simply vote on the **u**-valued sentence to see whether it would be 'true' or 'false'.

According to Definition 4, a *counterexample* to $\Gamma \models_{\mathbf{M_3}} \phi$ is any model where ϕ has a value less than the minimum value of Γ.

In the following table of consequences, we compare $\mathbf{M_3}$ with $\mathbf{K_3}$ and $\mathbf{LP}$.[3]

Proposition 1 (Consequences)

$\mathbf{M_3}$	$\mathbf{K_3}$	$\mathbf{LP}$
$\{\phi \wedge \neg\phi\} \not\models_{\mathbf{M_3}} \psi$	$\models$	$\not\models$
$\{\phi, \psi\} \not\models_{\mathbf{M_3}} \phi \wedge \psi$	$\models$	$\models$
$\{\phi \wedge \psi\} \models_{\mathbf{M_3}} \phi, \{\phi \wedge \psi\} \models_{\mathbf{M_3}} \psi$	$\models$	$\models$
$\{\phi\} \models_{\mathbf{M_3}} \phi \vee \psi$	$\models$	$\models$
$\{\phi \vee \psi, \phi \to \chi, \psi \to \chi\} \not\models_{\mathbf{M_3}} \chi$	$\models$	$\not\models$
$\{\phi \to \psi, \psi \to \chi\} \models_{\mathbf{M_3}} \phi \to \chi$	$\models$	$\not\models$
$\{\phi \vee \psi, \neg\psi\} \not\models_{\mathbf{M_3}} \phi$	$\models$	$\not\models$
$\{\phi \to \psi, \phi\} \not\models_{\mathbf{M_3}} \psi$	$\models$	$\not\models$
$\{\phi \vee \psi\} \not\models_{\mathbf{M_3}} (\phi \to \psi) \to \psi$	$\not\models$	$\models$
$\not\models_{\mathbf{M_3}} \phi \vee \neg\phi$	$\not\models$	$\models$

The first line of Proposition 1 shows that $\mathbf{M_3}$ is *paraconsistent* (in contrast to $\mathbf{K_3}$, and like $\mathbf{LP}$). Let us also note that in $\mathbf{M_3}$ (as in $\mathbf{K_3}$) there are no valid sentences (for instance, take for each atomic sentence Φ, $\mathfrak{I}(\Phi) = \mathbf{u}$). Thus, $\phi \vee \neg\phi$ is not valid and hence $\mathbf{M_3}$ (like $\mathbf{K_3}$, and in contrast to $\mathbf{LP}$) is *paracomplete*.

---

[3] **LP** has the same truth tables as $\mathbf{K_3}$ (with **b** instead of **u**, **b** meaning 'true and false') but has $\{\mathbf{t}, \mathbf{b}\}$ as the designated set, as distinguished from $\mathbf{K_3}$, where **t** is the only designated value.

Fig. 8 Classical square of opposition

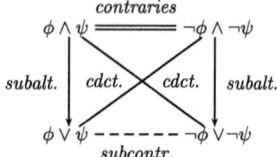

Fig. 9 Hassean classical hexagon of opposition

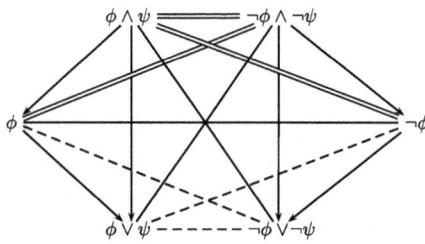

The following definitions of the equivalence and opposition relations hold for each logic analyzed in this paper.

Definition 5 (Equivalence) ϕ and ψ are equivalent ($\phi \simeq \psi$) iff $\{\phi\} \models \psi$ and $\{\psi\} \models \phi$.

Proposition 2 (Equivalences)

$$\phi \simeq_{\mathbf{M_3}} \neg\neg\phi$$

$$\neg(\phi \wedge \psi) \simeq_{\mathbf{M_3}} \neg\phi \vee \neg\psi$$

$$\neg(\phi \vee \psi) \simeq_{\mathbf{M_3}} \neg\phi \wedge \neg\psi$$

$$\phi \rightarrow \psi \simeq_{\mathbf{M_3}} \neg\phi \vee \psi$$

$$\phi \rightarrow \psi \simeq_{\mathbf{M_3}} \neg(\phi \wedge \neg\psi)$$

$$\phi \rightarrow \psi \simeq_{\mathbf{M_3}} \neg\psi \rightarrow \neg\phi$$

Each of the equivalences above holds in $\mathbf{K_3}$ *as well as in* $\mathbf{LP}$.

Definition 6 (Opposition) Two sentences are *contradictories* iff one of them is true iff the other is false. Two sentences are *contraries* iff, if one of them is true, the other is false, and in some models both are false. Two sentences are *subcontraries* iff, if one of them is false, the other is true, and in some models both are true. *Subalternation* between two sentences holds iff there is one of them that is true whenever the other is true. (Cf. similar definition for classical propositional logic in [8].)

In classical propositional logic the square of opposition as in Fig. 8 holds (if we exclude tautological and unsatisfiable ϕ and ψ). According to Definition 6, all forms of the opposition of the square hold in $\mathbf{M_3}$ (as well as in $\mathbf{K_3}$ and in $\mathbf{LP}$). However, in $\mathbf{P^1}$ contradiction and contrariety do not hold (counterexample: $\mathfrak{I}(\phi) = \mathfrak{I}(\psi) = \mathbf{t^*}$). In $\mathbf{I^1}$, contradiction and subcontrariety do not hold (counterexample: $\mathfrak{I}(\phi) = \mathfrak{I}(\psi) = \mathbf{f^*}$).

The square can be extended to the hexagon shown in Fig. 9, taking ϕ as an exemplary conjunct/disjunct. We have termed this hexagon a "Hassean classical hexagon" as distinguished from a Sesmat-Blanché (classical) hexagon, which extends the square to the top

Fig. 10 Sesmat-Blanché
diagram transformed

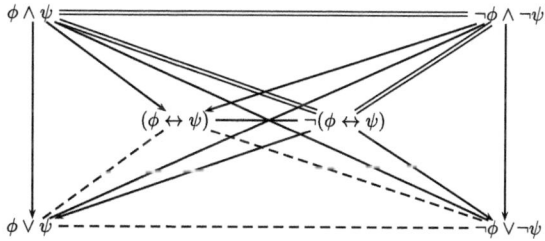

Fig. 11 The hexagon of
opposition in $\mathbf{P}^1$

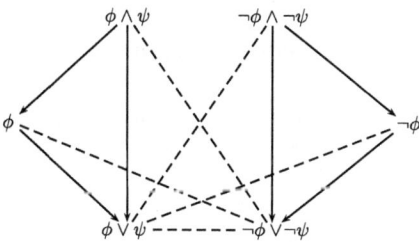

Fig. 12 The hexagon of
opposition in $\mathbf{I}^1$

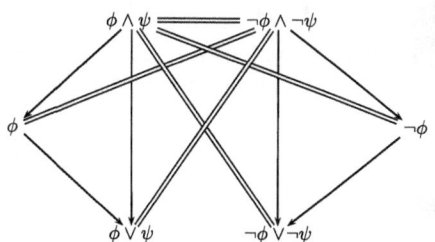

and the bottom instead of to the left and right. Whereas in a Sesmat-Blanché hexagon $(\phi \wedge \psi) \vee (\neg\phi \wedge \neg\psi)$ (equivalent to $\phi \leftrightarrow \psi$) comes to the top and $(\phi \vee \psi) \wedge (\neg\phi \vee \neg\psi)$ (equivalent to $\neg(\phi \leftrightarrow \psi)$) goes to the bottom of the diagram, in a Hassean diagram $(\phi \wedge \psi) \vee (\neg\phi \wedge \neg\psi)$ and $(\phi \vee \psi) \wedge (\neg\phi \vee \neg\psi)$ would come to the middle of the diagram (Fig. 10, see also Fig. 4).

Another way of extending the classical square of opposition in a Hassean way would be to extend it to the top by $\perp$ and to the bottom by $\top$. For simplicity, we treat $\phi \wedge \neg\phi$ and $\phi \vee \neg\phi$ separately from the diagram.

Let us return to Fig. 9. All forms of the opposition of the hexagon hold in $\mathbf{M}_3$ (as well as in $\mathbf{K}_3$). However, new non-opposition occurs in $\mathbf{P}^1$ and in $\mathbf{I}^1$: ϕ and $\neg\phi$ are not contradictories in either of the two logics, ϕ and $\neg\phi$ are in $\mathbf{P}^1$ generally not contraries to $\neg\phi \wedge \neg\psi$ and $\phi \wedge \psi$ respectively (counterexample: $\mathfrak{I}(\phi) = \mathfrak{I}(\psi) = \mathbf{t}^*$), and in $\mathbf{I}^1$ they are generally not subcontraries to $\neg\phi \vee \neg\psi$ and $\phi \vee \psi$ respectively (counterexample: $\mathfrak{I}(\phi) = \mathfrak{I}(\psi) = \mathbf{f}^*$). The hexagon for $\mathbf{P}^1$ is shown in Fig. 11 and the hexagon for $\mathbf{I}^1$ in Fig. 12.

Note that contradictions of the square of opposition become subcontrarieties in $\mathbf{P}^1$ and contrarieties in $\mathbf{I}^1$.

Fig. 13 The hexagon of opposition in $\mathbf{M_{3veto}}$

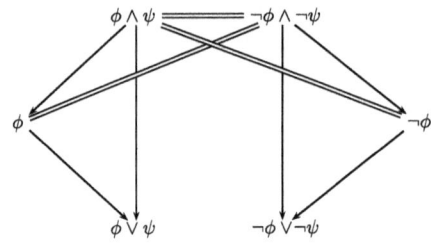

4 Veto and Consensus

4.1 Majority and Veto

In voting with the right of veto, every positive decision as being made in $\mathbf{M_3}$ can be canceled. Thus, the resulting logic $\mathbf{M_{3veto}}$ is very rudimentary, collapsing each **t** of $\mathbf{M_3}$ to **u**:

$f_\neg$		$f_\wedge$	t	u	f		$f_\vee$	t	u	f		$f_\rightarrow$	t	u	f
t	f	t	u	u	f		t	u	u	u		t	u	u	u
u	u	u	u	u	f		u	u	u	u		u	u	u	u
f	u	f	f	f	f		f	u	u	u		f	u	u	u

Truth tables with merely **u**s and **f**s clearly reflect a possible conservativism in decision-making with a right of veto. The consequence relation is defined as for $\mathbf{M_3}$ (see Definition 4).

The differences between $\mathbf{M_{3veto}}$ and $\mathbf{M_3}$ with respect to Proposition 1 are the following:

$$\{\phi\} \not\models_{\mathbf{M_{3veto}}} \phi \vee \psi$$

$$\{\phi \vee \psi\} \models_{\mathbf{M_{3veto}}} (\phi \rightarrow \psi) \rightarrow \psi$$

In addition, the Law of Double Negation and De Morgan's Law for $\neg(\phi \vee \psi)$ do not hold in $\mathbf{M_{3veto}}$.

With respect to the hexagon of opposition (see Fig. 13), it can easily be seen that contradiction and subcontrariety do not hold in this basic majority logic (since 'false' does occur and 'true' never occurs as a value of a connective function). Moreover, contrariety and subalternation hold only vacuously (since conjunction is never true in the truth tables).

4.2 Consensus

In consensual decision-making, no agent disagrees with a positively accepted proposal (i.e., accepted in voting for the proposal). We give the truth tables for consensual logic $\mathbf{C_3}$ ('!' marks differences from $\mathbf{M_3}$).

$f_\neg$		$f_\wedge$	t	u	f		$f_\vee$	t	u	f		$f_\rightarrow$	t	u	f
t	f	t	t!	u	f		t	t	t	t		t	t	u	f!
u	u	u	u	u	f		u	t	u	u		u	t	u	u
f	t	f	f	f	f		f	t	u	u		f	t	t	t

The consequence relation for $\mathbf{C_3}$ corresponds Definition 4.

Fig. 14 Hexagon of
opposition in C_3

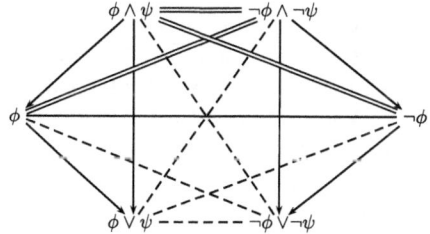

The differences between C_3 and M_3 regarding Proposition 1 are the following:

$$\{\phi, \psi\} \models_{C_3} \phi \wedge \psi$$

$$\{\phi \rightarrow \psi, \psi \rightarrow \chi\} \not\models_{C_3} \phi \rightarrow \chi$$

The Law of Double Negation holds, De Morgan's Laws and the equivalence of $\phi \rightarrow \psi$ and $\neg \phi \vee \psi$ do not hold.

The hexagon of opposition for C_3 is shown in Fig. 14. In comparison with M_{3veto}, the opposition in C_3 has grown, but still does not include the contradictions of the square.

5 Sequential Voting

M_3 has an essential characteristic—voting is simultaneous. The other possibility is **sequential** voting. In sequential voting (e.g., a consecutive show of hands), agents know (have common knowledge of) the results of the previous voting, and in voting on new questions they take this knowledge into account. That is, agents are aware of the consequences of their votes with respect to the previous voting results, and thus can change their opinions in the course of voting in order to support their best preferences.

In this kind of decision-making, the discursive dilemma disappears.

Example 2

1. For instance, agent 2 is for ϕ and against ψ.
2. $\phi \rightarrow \psi$ is accepted on the ground of the previous voting.
3. New voting: Who is for ϕ? ϕ automatically entails ψ according to common (classical) logic.
4. Agent 2 reasons: if (s)he chooses ϕ, (s)he automatically chooses ψ; if 2 wants $\neg\psi$, (s)he must vote against ϕ.
5. Agent 2 makes a decision whether to prefer ϕ or $\neg\psi$.
6. In either case, the result of voting will be in accord with the majority preference.

Example 3 Voting on ψ after ϕ has been accepted is at the same time voting on $\phi \wedge \psi$.

The resulting truth tables are the same as in K_3 with **u** for sentences not yet voted on (all classical valuation cases are preserved). The difference is in the concept of the consequence relation, which is defined corresponding to Definition 4. We will call this logic M_{3seq}. In fact, due to the difference from K_3 in the definition of the consequence relation, M_{3seq} is equivalent to Belnap's four-valued logic (B_4). As pointed out earlier, and according to Definition 4 of the consequence relation, the logic M_{3seq} of sequential voting differs from Kleene's $\mathbf{K_3}$ in not allowing inferences from $\mathbf{u}$-valued sentences to an $\mathbf{f}$-valued sentence (counting those cases as counterexamples to the consequence relation). These additionally blocked inferences are precisely those blocked by $\mathbf{LP}$. On the other hand, $\mathbf{M_{3seq}}$ differs from $\mathbf{LP}$ in counting the transition from $\mathbf{t}$-valued sentences to a $\mathbf{u}$-valued sentence as a counterexample to the consequence relation, that is, in blocking precisely those inferences that are blocked by $\mathbf{K_3}$. Hence, $\mathbf{M_{3seq}}$ is equivalent to Belnap's $\mathbf{B_4}$, where $\mathbf{t}$ and $\mathbf{b}$, or $\mathbf{t}$ and $\mathbf{u}$, can alternatively be taken as designated values without change in the consequence relation.

Since $\mathbf{M_{3seq}}$ is equivalent to $\mathbf{B_4}$, we recall some examples where the consequence relation does not hold: $\{\phi \wedge \neg\phi\} \not\models_{\mathbf{M_{3seq}}} \psi$, $\{\phi \vee \psi, \phi \to \chi, \psi \to \chi\} \not\models_{\mathbf{M_{3seq}}} \chi$, $\{\phi \to \psi, \psi \to \chi\} \not\models_{\mathbf{M_{3seq}}} \phi \to \chi$, $\{\phi \vee \psi, \neg\psi\} \not\models_{\mathbf{M_{3seq}}} \phi$, $\{\phi \to \psi, \phi\} \not\models_{\mathbf{M_{3seq}}} \psi$, $\{\phi \vee \psi\} \not\models_{\mathbf{M_{3seq}}} (\phi \to \psi) \to \psi$, $\not\models_{\mathbf{M_{3seq}}} \phi \vee \neg\phi$. On the other hand, in the following examples the consequence relation does hold: $\{\phi, \psi\} \models_{\mathbf{M_{3seq}}} \phi \wedge \psi$, $\{\phi \wedge \psi\} \models_{\mathbf{M_{3seq}}} \phi$ and $\{\phi \wedge \psi\} \models_{\mathbf{M_{3seq}}} \psi$, $\{\phi\} \models_{\mathbf{M_{3seq}}} \phi \vee \psi$. Further, De Morgan's Laws and the Law of Double Negation hold. In addition, $\phi \to \psi$ is equivalent to $\neg\phi \vee \psi$. $\mathbf{M_{3seq}}$ is paraconsistent (like $\mathbf{LP}$), paracomplete (like $\mathbf{K_3}$), and (like $\mathbf{K_3}$) has no valid sentences.

We note that consensual reasoning in sequential voting should have the same logic as the majority sequential reasoning. That is, in consensual reasoning, $f_\vee(\mathbf{f}, \mathbf{f})$ becomes $\mathbf{f}$, since after there was no consensus on ϕ, there will be consensus on ψ iff $\phi \vee \psi$, too, is consensually intended. Therefore, majority and consensual decisions in sequential voting can differ only when the valuations (models) are actualized in practice, the underlying logic being the same.

Since the opposition is defined only with respect to the values $\mathbf{t}$ and $\mathbf{f}$, disregarding the differences in the definition of the consequence relation (and disregarding the definition of the designated set of values), the opposition in $\mathbf{M_{seq}}$ is the same as in $\mathbf{K_3}$ and $\mathbf{M_3}$.

6 To Sum up

Here is a review of the consequence relations and equivalences in the logics considered above:

$\mathbf{M_3}$	$\mathbf{K_3}$	$\mathbf{LP}$	$\mathbf{M_{3veto}}$	$\mathbf{C_3}$	$\mathbf{M_{3seq}(B_4)}$
$\{\phi \wedge \neg\phi\} \not\models_{\mathbf{M_3}} \psi$	$\models$	$\not\models$	$\not\models$	$\not\models$	$\not\models$
$\{\phi, \psi\} \not\models_{\mathbf{M_3}} \phi \wedge \psi$	$\models$	$\models$	$\not\models$	$\models$	$\models$
$\{\phi \wedge \psi\} \models_{\mathbf{M_3}} \phi, \{\phi \wedge \psi\} \models_{\mathbf{M_3}} \psi$	$\models$	$\models$	$\models$	$\models$	$\models$
$\{\phi\} \models_{\mathbf{M_3}} \phi \vee \psi$	$\models$	$\models$	$\not\models$	$\models$	$\models$
$\{\phi \vee \psi, \phi \to \chi, \psi \to \chi\} \not\models_{\mathbf{M_3}} \chi$	$\models$	$\not\models$	$\not\models$	$\not\models$	$\not\models$
$\{\phi \to \psi, \psi \to \chi\} \models_{\mathbf{M_3}} \phi \to \chi$	$\models$	$\not\models$	$\models$	$\not\models$	$\not\models$
$\{\phi \vee \psi, \neg\psi\} \not\models_{\mathbf{M_3}} \phi$	$\models$	$\not\models$	$\not\models$	$\not\models$	$\not\models$
$\{\phi \to \psi, \phi\} \not\models_{\mathbf{M_3}} \psi$	$\models$	$\not\models$	$\not\models$	$\not\models$	$\not\models$
$\{\phi \vee \psi\} \not\models (\phi \to \psi) \to \psi$	$\not\models$	$\models$	$\models$	$\not\models$	$\not\models$
$\not\models \phi \vee \neg\phi$	$\not\models$	$\models$	$\not\models$	$\not\models$	$\not\models$
$\phi \simeq_{\mathbf{M_3}} \neg\neg\phi$	$\simeq$	$\simeq$	$\not\simeq$	$\simeq$	$\simeq$
$\neg(\phi \wedge \psi) \simeq_{\mathbf{M_3}} \neg\phi \vee \neg\psi$	$\simeq$	$\simeq$	$\simeq$	$\not\simeq$	$\simeq$
$\neg(\phi \vee \psi) \simeq_{\mathbf{M_3}} \neg\phi \wedge \neg\psi$	$\simeq$	$\simeq$	$\not\simeq$	$\not\simeq$	$\simeq$
$\phi \to \psi \simeq_{\mathbf{M_3}} \neg\phi \vee \psi$	$\simeq$	$\simeq$	$\simeq$	$\not\simeq$	$\simeq$
$\phi \to \psi \simeq_{\mathbf{M_3}} \neg(\phi \wedge \neg\psi)$	$\simeq$	$\simeq$	$\simeq$	$\simeq$	$\simeq$
$\phi \to \psi \simeq_{\mathbf{M_3}} \neg\psi \to \neg\phi$	$\simeq$	$\simeq$	$\simeq$	$\simeq$	$\simeq$

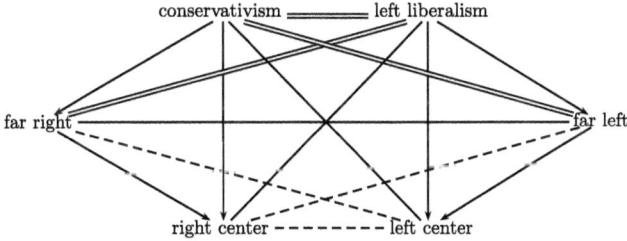

Fig. 15 Opposition in a parliament

Let $Opp(\mathbf{L})$ be the opposition of the hexagon in some logic $\mathbf{L}$, and $Contradict^{\phi,\neg\phi}$ the contradictory opposition between ϕ and $\neg\phi$. The following is a result of the previous analysis:

Corollary 1

1. $Opp(\mathbf{M_{3veto}}) \subset Opp(\mathbf{C_3}) \subset Opp(\mathbf{M_3}) = Opp(\mathbf{M_{3seq}})$
2. $Opp(\mathbf{I^1}) \cap Opp(\mathbf{C_3}) = Opp(\mathbf{M_{3veto}})$
3. $Opp(\mathbf{P^1}) \cup Opp(\mathbf{M_{3veto}}) \cup Contradict^{\phi,\neg\phi} = Opp(\mathbf{C_3})$

Proof Can easily be read from the hexagon diagrams of the respective logics. □

Informally, the opposition in a society or a group of agents increases with the transition from veto to consensus to majority decisions.

Besides, one negative result is that the increase of opposition is not proportional to the increase or decrease in the consequence relation. For example, $\mathbf{M_{3veto}}$ and $\mathbf{C_3}$ do not stand in a subset relation regarding their respective consequence relations, although $Opp(\mathbf{M_{3veto}}) \subset Opp(\mathbf{C_3})$.

Appendix

Let us illustrate the results of our analysis with the example of the opposition in a parliament. The maximal opposition is present when majority voting takes place, and this opposition can be schematized by the logical hexagon in Fig. 15.

We have put right-wing parties on the positive, left side of the diagram to indicate that left-wing parties in general oppose some already established, traditional values represented by right-wing parties (and somewhat corresponding to the seating arrangement in a parliament as seen from the viewpoint of the representatives). With each ideology label in the hexagon, we refer in fact to the representatives of the respective ideology. Further, by conservativism and left liberalism, we mean (relatively) complete and closed ideologies with a defined view on each politically relevant question. By 'far right', we mean particularistic options, such as nationalistic, religious, etc. fundamentalism. By 'far left', we mean particularistic options such as green parties, radical liberalism, orthodox communism, etc. We could extend the opposition to the octagon with 'anarchism' as $\bot$ on the top, and 'opportunism' as $\top$ on the bottom (but such parties usually do not occur in parliaments in a formal way).

References

1. Asenjo, F.G.: A calculus of antinomies. Notre Dame J. Form. Log. **7**, 103–105 (1966)
2. Béziau, J.-Y.: Non truth-functional many-valuedness. In: Béziau, J.-Y., Costa-Leite, A., Facchini, A. (eds.) Aspects of Universal Logic, pp. 199–218. Université de Neuchâtel, Neuchâtel (2004)
3. Bočvar, D.A.: Ob odnom trehznačnom isčislenii i ego primenenii k analizu paradoksov klassičeskogo rasširennogo funkcional'nogo isčisleniâ. Mat. Sb. **4**(46), 287–308 (1938)
4. Carnielli, W.A., Lima-Marques, M.: Society semantics and multiple-valued logics. In: Carnielli, W.A., D'Ottaviano, I.M.L. (eds.) Advances in Contemporary Logic and Computer Science, pp. 33–52. American Mathematical Society, Providence (1999)
5. Fernández, V.L., Coniglio, M.E.: Combining valuations with society semantics. J. Appl. Non-class. Log. **13**, 21–46 (2003)
6. Eckert, D., Pigozzi, G.: Belief merging, judgment aggregation and some links with social choice theory. In: Delgrande, J., Lang, J., Rott, H., Tallon, J.-M. (eds.) Belief Change in Rational Agents: Perspectives from Artificial Intelligence, Philosophy, and Economics. Internationales Begegnungs- und Forschungszentrum für Informatik (IBFI), Schloss Dagstuhl (Germany), Dagstuhl (2005). http://drops.dagstuhl.de/opus/volltexte/2005/333
7. Kleene, S.C.: On notation for ordinal numbers. J. Symb. Log. **3**, 150–155 (1938)
8. Kovač, S., Žarnić, B.: Logička pitanja i postupci. Kruzak, Zagreb (2008)
9. Machina, K.: Truth, belief, and vagueness. J. Philos. Log. **5**, 47–78 (1976)
10. Marcos, J.: What is a non-truth-functional logic? Stud. Log. **92**, 215–240 (2009)
11. Pauly, M., van Hees, M.: Logical constraints on judgement aggregation. J. Philos. Log. **35**, 569–585 (2006)
12. Pigozzi, G.: Belief merging and the discursive dilemma: an argument-based account to paradoxes of judgment aggregation. Synthese **152**, 285–298 (2006)
13. Priest, G.: Logic of paradox. J. Philos. Log. **8**, 219–249 (1979)
14. Sette, A.M.: On the propositional calculus P^1. Math. Jpn. **16**, 173–180 (1973)

S. Kovač (✉)
Institute of Philosophy, Ul. grada Vukovara 54, 10000 Zagreb, Croatia
e-mail: skovac@ifzg.hr

A Metamathematical Model for A/O Opposition in Scientific Inquiry

Mark Weinstein

Abstract The paper looks at the logic of scientific inquiry through the lens of physical chemistry in order to offer an account of the function of counter-examples to generalizations in scientific inquiry. A metamathematical model of emerging truth (MET) offers an image of truthlikeness in terms of model progressiveness, which supports an analysis of warrant strength that is connected with adaptive logic. The MET is exemplified by the history of the Periodic Table and examples from the history of chemistry are included by way of illustration. The discussion supports a rethinking of the relation of counter-examples to generalizations in terms of relative warrant strength challenging the standard view that calls for the rejection of a generalization in light of available counter-evidence.

Keywords Square of Opposition · Truth · Science · Inquiry · Chemistry · Periodic Table · Adaptive logic · Metamathematics · Argument

Mathematics Subject Classification Primary 03A10 · Secondary 80A50

1 Opposition in Scientific Inquiry

Of all of the relations on the square of opposition, contradiction is the most durable and apparently vital in any dialectical account of the power of logical relations. The force of refutation by contradiction is recognized as fundamental in theories of argument of all sorts. Nevertheless, an examination of the dialogical use of contradictions in the history of science presents a more nuanced picture. Successful processes of inquiry include long periods when contradictions are disregarded in the interest of continuing inquiry. This is readily seen in the history of, perhaps, the most successful inquiry project in human experience, that is, the development of a coherent and far-reaching image of the material world as exemplified by the Periodic Table of Elements. Throughout the history of the introduction, elaboration and extension of the basic insight of periodicity, generalizations drawn from empirical fact and theory were confronted by large bodies of inconsistent evidence. This, contrary to the doctrine of falsification, did not result in the abandonment of such generalizations; rather the generalizations were used as the basis for additional experiments and theoretic elaborations, as often as not, leading to continuing empirical and theoretic success. If such inquiry procedures are to be seen as reasonable, a rather different account of contradiction than the standard must be attempted.

The key to such an account of contradiction is to see truth as an emergent property across inquiry, rather than through a simple correspondence relation. In addition, the locus of argument in inquiry needs to be moved from micro-arguments, seen as a recursive basis for more elaborate arguments, to a notion of arguments as embedded in fields, networks of generalizations. Truth, rather than being directly assigned to elements, is seen as a property of elements in relation to their place in the field. Truth is a field property defined on emergent relations across a network, as in Quine's metaphor of the 'web of belief' (Weinstein [17]).

After a discussion of opposition in science, I will present the salient characteristics of a metamathematical model for emerging truth (Weinstein [19] has an elaborated version) and connect it with the notion of warrant borrowed from Stephen Toulmin's [15] classic model of argument. I will then turn to some recent work in adaptive logic by Straßer and Seselja [14] and make a connection with my approach.

The model of emerging truth (MET) furnishes an intuitive account of warrant strength, where warrants are seen as covering generalizations that support acceptable inference within a field of inquiry. Inferences are non-classical in that they are adaptive (possibly non-monotonic) and open to approximations defined in terms of neighborhood relations on models that serve as the interpretations of warrants. Most important, the levels of acceptable approximation are subject to the constraints applied within the field of inquiry. I take this to be central to Toulmin's [15] notion of backing and it is indicative of my concerns, which owe as much to argumentation theory as they do to logic (Weinstein [18]). Like much of current argument theory my work is essentially rooted in a pragmatic perspective. My concern is with how actual arguments are rationally grounded in a coherent logical structure. And I take as a paradigm for such a structure inquiry in chemistry. Chemistry is relatively non-controversial, its reach is profound and the practical applications point to the epistemic adequacy of its procedures (Weinstein [20]). My starting place is not with argument in the abstract sense of entailment familiar from standard first-order quantification theory but with a limited notion of warranted entailments that begins with concepts and procedures in use (Weinstein [19] offers the details of the construction). This move permits me to disregard the classic arguments against earlier formalist accounts in philosophy of science based on construing scientific inferences (explanations and reductions) as open to the entire range of logical equivalents and moves the discussion from simple logical models towards more complex constructions.

The core of this paper is an account of how counter-examples function dialectically within the context of the warrants that they contradict, hence the title 'A/O Opposition.' The section that follows gives an overview of the issue as seen within chemistry. We then look at the essential aspects of the MET that yield a hierarchical theory of warrant strength, which supports principled criteria for evaluating the dialectical acceptance or rejection of counter-examples to generalizations, I then move to a sketch of a possible logic that includes the hierarchy of warrants and indicate the connection with the square of opposition.

2 Universal Propositions and Science

The use of 'all' in describing physical reality is odd on its face and its oddity both requires and enables a significant rethinking of the dialectic of hypothesis and evidence. In terms of the square of opposition, the crucial example is A/O opposition, that is, the confrontation of a universal generalization and counter-evidence. My work in this area has focused on the clearest and least controversial of all of the areas of physical science, chemistry (Weinstein [20]). Within chemistry I look at the most deeply entrenched and accepted structural item in chemical theory, the Periodical Table of Elements. The history of the developing Periodic Table over the last one hundred years offers both insights and challenges for a logical account (Scerri [12] offers a detailed account of the history of the table). Take as an example Prout's hypothesis, a corner stone of the Periodic Table and, in 1817, and a bold and ultimately fruitful conjecture. In syllogistics terms: All elements are composed of hydrogen atoms. The evidence that prompted the conjecture was an outgrowth of a deep exploratory principle in early atomic theory, that is, that molecules could be described in terms of whole number ratios among their constituent parts. This was itself based on both the underlying intuition of atomic theory in its original form (that atoms, being indivisible, would only enter into combination as discrete individuals) and a growing body of evidence showing whole number ratios among the experimentally ascertained weights of naturally occurring substances after chemical decomposition. This led to the correlative theoretic notion of atomic weight as an overlay of the empirical results of measuring weights on increasingly sensitive balances. Unfortunately, in 1825, the noted chemist Jacob Berzelius compiled a set of improved atomic weights the disproved Prout's hypothesis (Scerri [12, p. 40]). Prout's hypothesis remained inconsistent with the evidence for at least a century. Nevertheless, Prout was correct in seeing hydrogen as the basis of the elements, since hydrogen with one proton serves as the basis as we move across the Periodic Table, each element adding protons in whole number ratios based on hydrogen with one proton. The core insight remained at the center of later work that strove to develop coherent chemical models based on multiples of fundamental elements. Problems with anomalies persisted despite the fact that the number of protons yielded the final organizing principle of the table. These were finally resolved, once atomic number, distinguished from atomic weight, which includes the contribution from neutrons unknown until the mid-20th century, finally vindicated Prout's bold conjecture. All of this was based on warrants that supported inferences, and appear to function inferentially as universal generalizations. Generalizations sustained in the face of counter-evidence, forming the basis for a sustained and successful research program. Scerri [12] offers a detailed and philosophically sensitive account against which my perspective may be evaluated.

What is the meaning of the universal quantifier in such a context of inquiry? In the traditional logic of categorical propositions the meaning of 'all' is seemingly non-controversial, tied to the underlying ontology. Given Aristotle's views that essential definitions of natural kinds were there to be found, there was some sense of defining 'all' in relation to the philosophical ideal (Weinstein [17]). 'All men are mortal' takes 'men' to be definable in principle, so it is not much of a stretch to think that 'all' can be sensibly seen to range over the class. In recent times the centrality of arithmetic in the foundational work in logic, and eventually the Skolem-Lowenheim theory (every consistent model had a model in the natural numbers) offered an even firmer basis for universal quantification. The definition of the natural numbers was clear in terms of Peano's axioms and the definition precisely characterized the domain. Quantifying over all of the numbers was both intuitive and furnished powerful logical tools e.g. mathematical induction. But whether we focus on predicates, as is natural in the syllogism, or focus on the domain, as in first-order logic, what it means to universally quantify over expressions in even as stable an inquiry as physical chemistry is none to clear. At no stage in the inquiry can the range of concern be precisely stated, since the task of inquiry is to find out, among other things, what is there and how and to what extent, the available theory comports with what is to be found. Since in physical chemistry there is no way to delimit the domain there is no extensional definition of the domain, so the arithmetic paradigm, as the guiding intuition on the semantics of first order languages, has no purchase what so ever.

Less obvious perhaps, is the fact that there are no clear intentional definitions. In the early stages of the history of chemistry, when the fundamental principles that were to permeate the developing science were laid down, the extent of the elements was unknown. The prevailing notion of decomposition into elements through chemical means was in its childhood, if not its infancy, and there was no reason to believe in the adequacy of any available conception of what the elements were. Progress was expected and changes in basic concepts welcomed in the face of empirical advance. Philosophical hand waving about all possible elements, or about where we would end up in the long run gives small comfort. And so, to put the point boldly, if the domain is not determinable, the universal quantifier is not logically defined. In such a situation it seems reasonable to say that 'all' is a vague statement of hope and intention and leave it at that. But that leaves us nowhere, for without some idea as to how generalizations comport with their evidence, no logic of science is possible. And yet throughout the history of physical chemistry, generalizations were made and functioned logically in terms of confirmation and contrary evidence. To understand the logic of all of this is essential to understanding argumentation in inquiry, and perhaps this may offer some connection with the new possibilities that modern logic provides.

Science when sufficiently mature and theoretical, as in physical chemistry, requires clear and often mathematical theories. In such formal theoretic contexts, as in mature scientific theories, predicates often refer to an unbounded, yet definable domain as specified by the theory. That is to say that within a theory predicates are given clear and explicit definitions in respect of an equally explicit domain of theoretic entities, and so 'all' makes clear sense as applying to the entire unbounded range of possible instances. That is, science, like mathematics, takes 'all' seriously. But when the theory reaches out to reality, to the models of data, which stand as its confirmatory basis, such clarity is often obscured by the empirical facts. For although experimental data is interpretable in terms of the theory and its predicates, the world has something to say about the specifics. This is required for the theory to be empirical, that is, falsifiable.

The history of theoretic generalizations in relation to their empirical database, however, is not the simple one of refutation by experimental counter-example as in the classic view of Karl Popper [10]. Theories often resist anomalous data in light of the robustness of the theoretic contexts within which the interpretation of empirical data occurs. Theories are subject to modification or even disconfirmation in light of recalcitrant facts, interpreted within the domain and predicates of the theory, which fail to support its theoretic generalizations. But empirical data may also be resisted in the name of the power of the theory, measured by its over-all empirical basis, and as important, its place in a network of other theories, each of which is supported by its own empirical basis and its place in the network of related theories. It is this give and take between theoretic embeddedness on the one hand, and risk of modification or falsification in light of recalcitrant empirical data that the discussion that follows attempts to illuminate. Such an account should explain how counter examples are to be considered on their merit, rather than serving as automatic refutations as in the traditional square or in the theories of those who followed Popper in philosophy of science.

The claim derived from my work is that an empirical generalization in a theoretical science such as chemistry is a promissory note about continuing advance. A generalization put forward based on empirical evidence has the following four essential properties:

1. The domain of its application is to be successively redefined in relationship to the ontological commitment of supporting theories.
2. It supports contrary to fact conditionals as a function of the generalization being reducible to a theory that explains the generalization in terms of these underlying theories.
3. Inferences (including resistance to counter examples via modes tollens) are adaptive, that is they respond to new information (monotonic if the new information is a consequence of the original set of proposition; ampliative adaptive logics are non-monotonic if they respond to additions to the set of propositions).
4. Estimates of truth likeness of the generalization can be rationally supported in light of evidence over time (I say 'truth likeness' rather than 'likelihood' since I don't want to commit to a probabilistic metric, although my tendency is towards a Bayesian approach with prior probabilities derived from the available underlying theoretic structure).

What is both relevant and clear is that generalizations, as in the nomic generalizations or other universal tending operations are, contrary to the traditional square, reasonably insensitive to refutation by a particular counter examples. And so my goal is to articulate what it is about the structure that underlies such generalizations that accounts for their resistance to counter example in perfectly appropriate ways. That is, I propose to devise a logical structure that weights the embeddedness of a generalization as against the force of the counter example. I focus on the connection between universal affirmative and particular negative, A/O opposition, since that is the most essential relationship in inquiry. The paradigmatic case for our purposes is where a theory T is confronted with a counterexample, a specific model of a data set inconsistent with T. The interesting case is where T has prima facie credibility, that is, is increasingly confirmed over time.

3 The Underlying Model of Truth

The basis of my proposal is a model of emerging truth (MET). My purpose is one of rational reconstruction and so even though the logical clarity of the presentation trumps its descriptive adequacy, descriptive adequacy is still at the heart of the intuition. Further, my intended audience includes argumentation theorists and informal logicians interested in inquiry, so I use the intuitive language of elementary set theory and offer constructions that lay out the details so that the component parts are easily distinguished.

The basic construction is a set of functions defined on a theory, T, which includes at least some generalizations. The theory is closed under a restricted entailment relation and is seen in relation to sequences of models, including models generated from other theories. Entailment is engineered to eliminate manifest irrelevancies based on the professional judgments of those working in the filed of inquiry (Weinstein [19]). This points to a central pragmatic component in the construction. Inferences are determined by actual argumentation put forward by members of the field and so do not include all logically possible consequences, only consequences actually drawn in accordance with acceptable inference procedures in the field. These include, of course, inferences drawn by opponents of the theory in light of counter evidence. Manifestly irrelevant premises are discarded as their irrelevance is ascertained by the discussants. In addition, discussants work within a discourse frame that establishes standards of rigor as the argumentation proceeds. This is part of what the Amsterdam theorists call the opening stage and forms part of the backing in Toulmin's [15] sense. Inference is non-monotonic and dynamic. Arguments are reconsidered in the light of new evidence and the status of a claim is based on the history of argumentation. An inference is evaluated within a network of associated inferences and the history of the network.

This is a basis for a reconsideration of the nature of warrants. The clear limitations on the scope and rigor of warrants in actual cases of use, what Toulmin [15] calls rebuttals, indicate warrants of varying strength. A number of clear structural distinctions can be made in based on the architecture of the MET to indicate dimensions of strength: strong warrants are housed within a body of generalizations, super strong warrants connect such bodies of generalization into progressive networks and the strongest warrants are super strong warrants of exceptional breadth and depth. It these last that offer candidate domains of ontological significance, in terms of which a Tarski-like definition of truth may be defined (Weinstein [17]).

Before I offer the MET a word of clarification as to what it is not. The formal tradition in philosophy of science included attempts to axiomatize significant portions of physics. These attempts were met with significant criticisms and I will not attempt to enter into the debate about the fruitfulness of such endeavors. For the MET is not such an attempt. The desire to axiomatize was based on the success of similar attempts in respect of portions of mathematics, and part of my purpose is to move away from the model of formal mathematics to present a rather different role for metamathematical accounts of science and especially scientific argument. The MET attempts to offer reasonable, transparent and rigorous metaphors for essential aspects of scientific inquiry. In particular, in this context, I want to substitute a more nuanced approach to counter-examples in their dialectical role as challenges to generalizations. The use of physical chemistry as a paradigm for the reevaluation of the role of counter-examples to generalizations is based on its effectiveness as a research program. My hope is to offer an alternative image of the underlying logic of generalization and counter-examples, what I call A/O opposition. Chemistry offers a salient indicator of how this functions in a rigorous domain. The MET is no substitute for the formal structures found in chemical theories, rather it attempts to offer a noetically transparent image of how inquiry in chemistry advances in the face of evidence that supports and challenges generalizations put forward.

4 The MET

The formalism in the MET is characterized by two essential logical concepts, that of modeling, the models that support various restricted entailment relationships (1–1.3 below) and a deeper sense of entailment that relies on the connection between discrete sets of principles and even vocabulary, what in the philosophy of science is called reduction (2–2.4, below). Reductions are not unlike the analytic entailments of logical empiricism, more rigid then weaker warrants seen as empirical or even nomic generalizations; warrants that support reductions are resistant to anomalies and form the foundation of inter-theoretic connections that fuel conceptual advance and deeper understanding.

Although the second of the two functions has the most novelty the first one permits a radical shift in perspective to be immediately seen: that is, it is the dynamics, not the statics of argument that furnishes the essential logical concepts. It is not truth in a model that is at issue, but truth seen across models. Truth is a relation among models and ultimately a criterion for selecting models. The basic logical structure of the MET is as follows:

1. A *scientific structure*, $TT = \langle T, FF, RR \rangle$ where T is a set of sentences that constitute the linguistic statement of TT closed under some appropriate consequence relation and where FF is a set of functions F, such that for each F in FF, there is a map f in F, such that $f(T) = m$, for some model or near model of T. And where RR is a set of sets of representing functions, R, such that for all R in RR and every r in R, there is some theory T^* and r represents T in T^*, in respect of some subset of T.

A scientific structure is first of all, a set of nomic generalizations, the theoretic commitments of the members of the field in respect of a given body of inquiry. These support inferences and a restricted consequence relation that rules out manifest irrelevancies of the sort that bedeviled the classic discussion of the D-N model of explanation. Details of my version of such a restricted entailment relation are found in Weinstein [19].

We then include distinguishable sets of possible models (or appropriately approximate models) and a set of reducing theories (or near reducers). The richness of such possible approximation relations was first indicated to me by the work of Apostel [2]. I rely heavily on the fact that such approximation relations are both mathematically definable and can be reasonably supported in light of the normative constraints found in successful disciplinary practices, for example, standards for measurement.

What we will be interested in here is a *realization* of TT, that is to say a triple $\langle T, F, R \rangle$ where F and R represent choices from FF and RR, respectively. What we look at is the history of realizations, that is an ordered n-tuple: $\langle \langle T, F_1, R_1 \rangle, \ldots, \langle T, F_n, R_n \rangle \rangle$ ordered in time. The claim is that the adequacy of TT as a scientific structure is a complex function of the set of realizations.

The notion of a realization is the deepest pragmatic element in the construction of MET, we are not interested in all possible models, or even all actual models. We are only interested in the models that are put forward in inquiry, that is to say, the bodies of evidence, both empirical and theoretical that are put forward in support and attack by those engaged in the inquiry. This point of view is less familiar in formal philosophy of science based as it was in the past on a logical model drawn from the study of foundations of mathematics. In mathematics one deals with necessities and so all possible models are of concern. In science one attempts to construct selected models that reflect the empirical realities as known as well as conjectures about yet to be found empirical models, and more essentially in physical chemistry, theoretic models that through their fecundity prove adequate to the emerging realties the theoretic models both anticipate and explain. It is such intended models that are the logical core of the constructions that follow.

1.1. Let T' be a subtheory of T in the sense that T' is the restriction of the relational symbols of T to some subset of these. Let f' be subset of some f in F, in some realization of TT. Let $\langle T'_1, \ldots, T'_n \rangle$ be an ordered n-tuple such that for each i, j $(i < j)$, T'_i reflects a subset of T modeled under some f' at some time earlier than T'_j. We say the T is *model progressive under* f' iff:
   (a) T'_k is identical to T for all indices k, or

(b) the ordered n-tuple $\langle T_1', \ldots, T_n' \rangle$ is well ordered in time by the subset relation. That is to say, for each T_i', T_j' in $\langle T_1', \ldots, T_n' \rangle$ $(i < j \geq 2)$, if T_i' is earlier in time than T_j', T_i' is a proper subset of T_j'.

I use elementary set theoretic constructions and rely on the subset relation as an ideal type although the realities undoubtedly require much more complex constructions. We cannot assume that all theoretic advances are progressive. Frequently, theories move backwards without being, thereby, rejected. We are looking for a preponderance of evidence or where possible, a statistic. Nor can we define this a priori. What counts as an advance is a judgment in respect of a particular enterprise over time in light of the implicit normative criteria as evidenced by the standards of practice of those engaged with the theory, an essential pragmatic turn if we are to understand argumentation in inquiry.

1.2. We define a *model chain* C, for theory, T, as an ordered n-tuple $\langle m_1, \ldots, m_n \rangle$, such that for each m_i in the chain $m_i = \langle d_i, f_i \rangle$ for some domain d_i, and assignment function f_i, and where for each i and j $(i < j)$, m_i is an earlier realization (in time) of T then m_j.

Let m^* be an intended model of T, making sure that $f(T) = m^*$ for some f in F (for some realization $\langle T, F, R \rangle$) and T is model progressive under f. We then say that C is a *progressive model chain* iff either:
(a) for every m_i in C, m_i is isomorphic to m^*, or
(b) there is an ordering of models in C such that for pairs m_i, m_j $(j > i)$ in C, m_j is a nearer isomorphs to m^* than m_i.

Condition (a) reflects the standard idea that, for example, all models of data are entailed by a theory. Condition (b) is the essential case that drives the model, where models proffered and accepted as confirming models in an inquiry are appropriately close approximations given the standards of the inquiry. Such a constraint permits inquiry to be logical and yet not driven by ideals of consistency drawn from mathematics. Again, this is an idealization, in practice a preponderance of closer approximations would suffice.

1.2.1. By 'nearer isomorph' we intend the following. Let $P = \{p_1, \ldots, p_n\}$ of predicates and relational symbols of T. Let v_i be an assignment of elements of m_i to P for a model m_i of model chain $C = \langle m_1, \ldots, m_n \rangle$ of T, where the nth member is the intended model, m^*. Let v_j $(j = i + 1)$ be as assignment of elements of m_j of P for a model m_j in model chain C of T, that is to say, $m_{i,j}$, are adjacent in the ordering and so constitute a minimal subchain.

Let v^* be the overlap $v_i \wedge v_j$ of assignment functions where for a graph of $v_i \wedge v_j$ (an ordered n-dimensional assignment of data points to predicates and relations of T) each element of v_i is identical to some element of v_j or is in a near neighborhood relation to it. And where the width of the near neighborhood is defined by current methodological practice in the field in which T is proposed, another aspect of the pragmatic turn.

The intuition is that v^* identifies the elements in a pair of models that are identical or close enough to be deemed functionally equivalent in terms of inferences. Since by assumption model chain C orders the models it contains, we can order the various function v^* accordingly, that is, construct a sequence $\langle v_1^*, \ldots, v_n^* \rangle$ where each v_i^* is the overlap of a pair-wise assignment, to the members of C. Since the v_i^* are sets we can order them by the subset relation. Since v_n^* is the overlap of m_{n-1}

and m_n and m_n is the intended model, m^*, for each minimal sub-chain, $m_{i,j}$, m_j is a nearer isomorph to m^* than m_i.

1.3. Let $\langle C_1, \ldots, C_n \rangle$ be a well ordering of the progressive model chains of TT, such that for all i, j ($i > j$), C_i is a later model chain than C_j. TT is *model chain progressive* iff $\langle C_1, \ldots, C_n \rangle$ is well ordered in time. That is to say each later model includes and extends the models antecedent to it in time.

As before, this is an idealization indicating a theory that is uniformly progressive. Much more complex judgments are required in the face of the vagaries of evidence and instrumentation, and the most that we can expect of a theory is that progress is predominant, that is, most model chains are mainly progressive. The history of the Periodic Table is full of such contentious episodes, where evidence for a theoretic model is partial and difficulties remain unresolved. An example is the attempt to organize known elements into triples with simple ratios, so called triads, where the values of key properties of a 'middle' element were seen as intermediate between two flanking elements (Scerri [12, pp. 42ff]). Decades of different theoretic attempts to organize the known elements in such triples, with ratios that approached whole number values were, driven by the basic intuition of the atomic theory, substances seen as composed of a basic element. This idea, ultimately proving to be the basic principle of the Periodic Table, was confronted with empirical approximations of various sorts as theorists looked at such empirical properties as weight, volume and eventually atomic weight and as measurements became more precise with improvements of instrumentation. It was only after the discovery of isotopes and the movement from atomic weight to atomic number that the available empirical models uniformly supported the theoretic model. Before that unification different strands of evidence with their own history of consistency with evidence offered individuals lines of support for the basic concepts of chemical analysis (Scerri [12]).

2. We now turn our attention to the members of some R in RR. The construction is based on the seminal work of Eberle [6] on representing functions, functions that map elements of one theory onto another. The crucial result is, for functions f^* in F^* for reducing theory, T^*, and reduced theory, T, f^* *reduces* T to T^* is equivalent to, for every model m^* of T^* there exists a model of m of T such that, for every sentence s of T, s is true in m if and only if $f^*(s)$ is true in m^*. And similarly for formulas and assignments in m and corresponding formulas in m^* for T^*.

2.1. Representing function can be limited to subsets of T, $k(T)$. This yields the following core construction.

The members of RR represent T in T^* in respect of some subset of T, $k(T)$. Let $\langle k_1(T), \ldots, k_n(T) \rangle$ be an n-tuple of representations of T over time, that is if $i > j$, then $k_i(T)$ is a partial representation of T in T^* at a time later that $k_j(T)$. We say that TT is *reduction progressive* iff either,
(a) $k(T)$ is identical to T for all indices, or
(b) the n-tuple is well ordered by the subset relation.

As before, condition (a) represents the best-case scenario, (b) represents a weaker but still acceptable condition. The intuition behind the construction reflects the historic fact that reductions (reinterpretation of a theory in terms of the concepts of a higher order, more abstract theory) rarely capture the entire scope of the reduced theory. A theory is reduction progressive if more of the reduced theory is represented in the reducing theory

of time. As before, the construction is an idealization, as new aspects of a theory are reduced, frequently earlier reduction need to be reconsidered, so in reality we are faced with
determinations based on trends rather than with a univocal picture of success. Examples
from the history of the physical science abound. The classic example of the reduction of
the Boyles-Charles Law via statistical mechanics could not be easily transferred outside
of the realm of its initial application to gases without significant modification (Nagel [9,
pp. 359ff]). Yet the reduction of all sorts of phenomena to physical principles ultimately
resulted in a wealth of physical explanations that shared a common conceptual core in
statistical mechanics, including fluid dynamics, rigid body dynamics, electromagnetism
and ultimately atomic theory.

2.1.1. We call an n-tuple of theories $RC = \langle T_1, \ldots, T_n \rangle$ a *reduction chain*.

2.1.2. $RC = \langle T_1, \ldots, T_j, \ldots, T_n \rangle$ is a *deeper reduction chain* than j-tuple $RC' = \langle T_1', \ldots, T_j' \rangle$, iff $n > j$ and for all $i < j$ there is a r_i in R_i such that r_i represents T_i in T_{i+1} and similarly for T_i' and for all T_k ($k \leq j$).

2.2. We call a theory, T, *reduction chain progressive* iff for an n-tuple of reduction chains $\langle RC_1, \ldots, RC_n \rangle$ and for each RC_i ($i < 1$), RC_{i+1} is a deeper reduction chain than RC_i.

The point of 2.2 is to call attention to the fact that there may be more than one chain
of reductions, that is, different aspects of physical reality are independently reduced to
some powerful reducing theory. The best example is atomic theory that although general
in its scope is first applied to gases and then only later is successfully applied to fluids
and solids. The historical phenomena of a deep theory, such as atomic theory, that had
as its impetus a desire to incorporate vast areas of actual and potential empirical knowledge, is one of large ambitions faced with small advances. Atomic theory was originally
formulated to deal with gases, it was than expanded laterally to apply to other chemical reactions, to liquids, crystals, solids, electrical phenomena and ultimately yielded to
particle physics which transformed atomic theory in unimaginable ways. This idea of a
theory branching out to include new areas of concern is the focus of the next series of
constructions.

Before we move on, a word of clarification is required. The MET might appear to
support a tree-structure, in which many of the distinctions appear to collapse. But since
there is no unifying relation across the MET, the tree structure is merely apparent. The
MET may include trees governed by a uniform relation, but to see the MET as a tree
waits upon the ultimate unification of physical science through a theory of everything.
An aspect of the MET that is irrelevant to our discussion here, attempts to capture that
deep ontological potential of physical theory through the notion of an ontic set, the set of
models that ultimate captures the entire field of theories by offering a univocal ontology
(Weinstein [17]). In such a total unification some of the distinctions here may prove to be
formally redundant, although they remain epistemologically essential for understanding
the process of inquiry.

2.3. T is a *branching reducer* iff there is a pair (at least) T' and T^* in set of realization of TT such that there is some r' and r^* in R' and R^*, respectively, such that r' represents T' in T and r^* represents T^* in T and neither T' is represented in T^* nor conversely.

We now turn to sequences of realizations, TT_1, TT_2, etc.

2.3.1. $B = \langle TT_1, TT_2, \ldots, TT_n \rangle = \langle \langle T_1, F_1, R_1 \rangle, \langle T_2, F_2, R_2 \rangle, \ldots, \langle T_n, F_n, R_n \rangle \rangle$ is a *reduction branch* of TT_1 iff T_1 is a branching reducer in respect of all T_i, and T_j ($i \geq 2$; $j \geq 3$ for $i, j \leq n$).

Yet again, this is an idealization; in the history of reductions there are often gaps, but to limit the construction to some T_i would be much too weak and would represent a period of theoretic conjecture. MET seeks to represent theoretic consolidation, since as we shall we below, the degree of consolidation is the basic logical element against which counterexamples must be weighed.

2.4. Since each branch, B, is an n-tuple the construction for a progressively branching reducer is more complex. Construct the set BB of all branches; BB is non-empty and has n-elements. We then proceed to 'count' the n-order subsets of BB, in the standard way.

For each B_i in BB we construct an indicator function, f, such that for all integers, n, B_i, $f(n) = 1$, if B_i is in BB and $f(n) = 0$ is B_i is not in BB. This gives us the n-tuple, $BB^* = \langle (f(0), f(1), \ldots, f(n-1) \rangle$, an index of BB. We say that a branching reducer, T, is a *progressively branching reducer* iff the n-tuple BB^* is well ordered in time that is, for each pair if indices, $f(i), f(j), i, j$ ($j > i$) B_j is a later branch than B_i, that is, the number of branching reducers has been increasing in breadth as inquiry persists.

5 A/O Opposition and the Theory of Warrants

Given this much we move to the central focus of the paper, A/O opposition and an account of the dialectics of generalization and counter-example that the MET affords. The core construction is where a theory T is confronted with a counterexample, a specific model of a data set inconsistent with T. The interesting case is where T has prima facie credibility, that is, where T is at least model progressive under some assignment function, that is, is increasingly confirmed over time (MET, 1.1).

A. The basic notion is that a model, cm, is a *confirming model* of theory T in TT, a model of data, of some experimental set-up or a set of systematic observations interpreted in light of the prevailing theory that warrants the data being used. And where:
(1) cm is either a model of T or
(2) cm is an approximation to a model of T and is the nth member of a sequence of models ordered in time and T is model progressive (MET, 1.2).
B. A model interpretable in T, but not a confirming model of T is an *anomalous model*.

The notion of confirming model is a modification of the standard idea that models consistent with a theory (models of the theory) confirm the theory. The modification of course is the core second condition (b) of the various constructions of the MET, approximate models defined in terms of near isomorphisms that are progressive. That is, a chain of models that approach the intended model over time. In terms of measurement theory, this can be thought of as 'goodness of fit,' a locution we will use below for its familiarity and succinctness. The notion of a 'model interpretable in T' is also fairly standard. Such a model shares the domain and the partition of the domain via assignment functions f in F, that is, it shares the conceptual structure to T, via assignment of predicates and relations in the domain. Anomalies can be thought of as recalcitrant empirical data that contradict implications of T although our construction accommodates theoretic anomalies via their empirical consequences.

With warrants construed as the nomic generalizations that characterize T we build a hierarchy of warrant strength that reflects a natural hierarchy of theoretic embeddedness derived from the MET: model progressive (MET, 1.2), model chain progressive (MET, 1.3) reduction progressive (MET, 2.1), reduction chain progressive (MET, 2.2), branching reducers (MET, 2.3) and progressively branching reducers (MET, 2.4).

We first look at the relation of anomalies to the theories they challenge. A/O opposition, at this level of analysis, varies with the strength of the theory. So, roughly, if T is merely model progressive, an anomalous model is type-1 anomalous, if in addition, model chain progressive, type-2 anomalous etc. up to type-6 anomalous for theories that are progressively branching reducers. A more precise characterization moves from theories to their determining warrants, those nomic generalizations that serve as the core of the warranting apparatus of explanations and other inferences in TT.

A warrant, W1 of type-1 is model progressive (MET, 1.2) if it satisfies its intended model or the set of models shows an increasing goodness of fit with the intended model.

A warrant, W2 of type-2 is model chain progressive (MET, 1.3) if it is of type-1 and the number of number of model chains that satisfies its intended model or show an increasing goodness of fit with the intended model are themselves increasing.

A warrant, W3 of type-3 is reduction progressive (MET, 2.1), if it is of type-2 and it satisfies an intended model drawn from a reducing theory or the set of models shows an increasing goodness of fit with the intended model drawn from the reducing theory.

A warrant, W4 of type-4 is reduction chain progressive (MET, 2.2), if it is of type-3 and the reducing theory includes reduction chains that are increasing over time.

A warrant, W5 of type-5 is of type-4 and is reduced by a theory that is a branching reducer (MET, 2.3).

A warrant, W6 of type-6 is of type-5 and is reduced by a theory that is a progressively branching reducer (MET, 2.4).

Warrants of type-1 and type-2 may be thought of as strong warrants. Warrants of type-3 through type-6 may be thought of as super strong warrants (Weinstein [17]). We first characterize an anomalous model in relation to the power of the warrant, such that an anomaly was seen in relation to the power of the theory for which it was an anomaly. The ordering of warrants is essential for evaluating dialectical resistance to the force of the anomaly.

Given the hierarchy of warrants we can define a plausible dialectic of counter-examples to generalization, opposition based on anomalous models. Opposition varies with the strength of the theory comprised by a warrant and its consequences under initial conditions. So, for a warrant in T, if T is merely model progressive, an anomalous model is type-1 anomalous, if in addition, model chain progressive, type-2 anomalous etc. up to type-6 anomalous for theories that are progressively branching reducers. This characterization sets the burden that the anomaly must bear. Type 1 anomalies have more dialectical power, which decreases across the types. That is to say, an anomaly of a higher type must bear a larger burden of epistemic force to be an effective challenge to a deeply embedded warrant. This constitutes its prima-facie strength.

The ordering of anomalies is essential for evaluating dialectical resistance to the force of the anomaly. This yields the first of two dialectical principles.

P1: The prima-facie strength of the anomaly is inversely proportional to dialectical resistance, that is, counter-evidence afforded by an anomaly will be considered as a
refutation of T as a function of the strength of T in relation to TT.

In terms of dialectical obligation, a claimant is dialectically responsible to account for
type 1 anomalies or reject T and less so as the type of the anomalies increases.

To make the logic simpler I assume that reductions involve theories with strong warrants. That is, the super-strong warrants that are supported by theoretical structures of exceptional breadth and depth are themselves strong that is, increasingly confirmed in their
initial applications, finding increasing success in being applied to additional experimental or naturally occurring configurations. That enables us to develop a linear hierarchy of
warrant strength along with the correlative type of anomaly. This, of course, is an idealization, a logical metaphor for a historical messy process, yet as a metaphor it seems to me to
capture the underlying logic of the history of theoretic development in physical sciences.
The movement from warrants of type-1 to type-2 is a natural interpretation that reflects
the history of chemistry. Success in applying a theory to a particular set of elements is
more often then not followed by increasing the sort of elements to which the theory can
be applied. So, for example, in the early experiments with isolating chemical components
of naturally occurring substances and comparing their constituent parts by volume, not
alone did the empirical results show an increasingly goodness fit with theoretic expec-
tations as techniques for isolating and comparing physical substances improved, but the
techniques of isolating substances and comparing them isolates by volume were applied
to additional elements, in our terms, resulting in new model chains. The issue is less clear
with the shift from models of the theory and models imposed on the theory by reductions
(warrants of type-3 through type-6), since reductions are often partial and deep theories
frequently exist that are not applicable to current empirical generalizations in a unique
way (Cartwright [4]).

It is certainly plausible given the power of higher order physical theories that a weak
warrant, in terms of its own confirmatory range of application, will fall under a reducing
theory before its power to explain is experimentally obvious. This seems true of developments in materials science where a theoretic construct results in a substance heretofore
unknown and so without clear evidence as to its empirical properties. But, although at
a given point the weight of the reduction may count more than the empirical evidence
that supports generalizations about its empirical properties, this seems only temporary,
since if the reductive basis is sound, the empirical properties will quickly manifest themselves in a host of applications yielding supported empirical generalizations about the new
substance's properties and relations. Whatever the temporary difficulties in applying the
basic structure of periodicity to new elements, as chemistry develops not only are new
substances found to conform to the requirements of the Periodic Table, but the ability to
articulate the chemical properties of these substances based on underlying theoretic analysis has resulted in the understanding of the physical world around us at a level of precision
that was barely imaginable by the early chemists. The result is an increase of goodness
of fit with theory as the empirical properties of new elements are ascertained through
new instrumentation and more adequate measurements. A recent example is the theoretic
understanding of semi-conductors, first envisioned in the 1920s, but only experimentally
applied in the 1950s, leading to both theoretical and practical advances including the enormous range of applications of transistor technology and the unification of chemistry and
electronics at a micro-level (Riordan and Hoddeson [11]).

Our principle P1 above sets the dialectical burden an anomaly must bear in terms of the warrant that they challenge. To be taken seriously as a challenge, anomalies require support based on their relation to the theories in which they are embedded. The second principle, P2, below, is based on the epistemic force of the anomalous model itself. If the model is a purely empirical model, that is, a set of data that is free of theoretic support it offers a minimal challenge, we call it a model of type-0. We then apply our hierarchy, to yield models of increasing epistemic power due to increasing theoretic warrant. This yields analogous characterizations of models type-1 through type-6. Challenges by anomalous models of type-0 are rare in modern science, where even models of data are theoretically supported, if only through the theory of instruments and measurement that they reflect. An interesting example a model of type-0 is the statistical analysis of John Snow who in the 1850s offered a paradigm of the use of statistical arguments, founding the scientific discipline of epidemiology in his investigation of the causes of a cholera epidemic. It was not, however, until the germ theory of disease was accepted that his work could be integrated into an acceptable explanation (Goldstein and Goldstein [7, pp. 36ff]). An anomaly, a non-confirming model is merely suggestive when seen in relation to a strong warrant, when it is free of theoretic support.

We now build correlative notions of models of type 1 through type-6 in an obvious fashion in terms of the theoretic support for the anomalous model in an explanatory framework other than T. So, an anomaly of type-n is a model that is supported by a warrant of type-n. So, in particular, if an anomalous model (a model of type-1) is explained by explained by an acceptable generalization (a type-1 warrant) it has more dialectical power and the supporting generalization must be addressed if the anomalous model is to be disregarded. This increases with the next level, where the supporting generalization has reasonable wide application (a model of type-2), becomes severe when the supporting generalizations is itself theoretically warranted by a reducing theory and increasingly so when the reducing theory is deeply explanatory and broadly applicable (models of type-3 through type-6). This yields our second principle.

P2: The strength of an anomaly is directly proportional to dialectical advantage, that is, the anomalous evidence will be considered as refuting as a function of the power of the explanatory structure within which it sits.

This yields the core intuition:

P*: The dialectical use of refutation is rational to the extent that it is an additive function of P1 and P2

Applying our distinction to A/O opposition, at the one extreme, type-1 anomalies approach classical opposition, the anomalous model falsifies the theory or must itself be rejected at the other extreme type-6 anomalies promote scientific revolutions, that is, require a thoroughgoing replacement of an entire scientific structure otherwise the anomalous model is rejected. The replacement of Aristotelian by Newtonian physics is a classic example of a response to type-6 anomalies; foundational concepts are replaced (Toulmin and Goodfield [16, pp. 210ff]). Type-2 anomalies require that there be a modification of the theory that preserves most model chains and the sequence of type-2 anomalies over time is model progressive within the modified theory, otherwise the anomalous model is rejected. The discovery of isotopes is based on type-2 anomalies, the basic structure of chemical knowledge is retained, but significant problems in the details of, e.g. atomic weight are reconciled when it is realized that substances that have isotopes vary in atomic structure and so generate alternative measurements (Scerri [12, pp. 73ff]). Type-3 anomalies require that there be an alternative reducing theory or the anomaly is rejected. That is, the anomalous model can be seen within a reinterpretation of the theory without substantial modifications of the theory itself. The kinetic theory of heat is an example of a reinterpretation based on type-3 anomalies, rather then seeing heat as a substance, the kinetic theory sees heat as a property of substances. Rumford's experiments that appeared to show that heat as 'inexhaustible' could not be subsumed under theories of heat as a substance and were disregarded until the kinetic theory was able to support a well-established body of empirical evidence (Goldstein and Goldstein [7, pp. 119ff]). Type-4 anomalies require the reorganization of interconnected theories or the anomalies are rejected. The development of the Periodic Table of Elements can be seen as a response to type-4 anomalies. The Periodic Table was a response to the need to reorganize available chemical knowledge in a new structure after decades of attempt to organize fragments of chemical knowledge based on their chemical properties. The result was not merely additional theoretic support in the form of the oft-cited predications of Mendeleyev, but more importantly resulted in a deep reorganization of chemical knowledge with the concomitant enormous advance of chemical understanding characteristic of the 20th century (Scerri [12, pp. 123ff]). Type-5 anomalies require major theoretic advances that reorganize and modify existing theories deeper reduction chains (2.1) and potentially branching reducers. Black box radiation is an example of a type-5 anomaly that, if not rejected, prompts theoretic advance of deep ontological significance, in this case, quantum mechanics [12, p. 189].

In addition to the degree of anomaly, P1, degree of adequacy for anomalous models is required as well, P2. At the basic level of empirical evidence put forward, empirical claims need to sit within a model chain of available corroborative data that demonstrates epistemic adequacy (MET, 1). The empirical claim must be replicable, or increasingly supported by newly available evidence. That is to say, anomalous models of type-0 have little purchase in the dialectic of theoretic advance. Anomalous evidence that point to such theoretical unmotivated observations, such as claims to perpetual motion, cold-fusion and more recently breaches of the speed of light by neutrinos, are only taken seriously as theoretic support for the observed phenomena becomes available. Even if scientists persist in exploring the observations, it is the ability to account for them with plausible and supported explanations that make them possible challenges to prevailing views. The notion of progressive model chain (MET, 1.2) points to the fact that empirical evidence must be reliable, that is, further investigation corroborates the claim (by duplicating it or by increasingly approximating the theoretic model in relation to the specific claim over time). If the claim strikes deeply at a warrant as in anomalies of type 2, the claim loses credence unless it too is warranted, that is, is embedded in some theoretic account (MET, 1.3). If both views are warranted by networks of generalizations (warrants) and supported by shared methodological and other aprioristic constraints (backing), the situation becomes much more complex. At the limit we have theoretic replacement that include fundamental conceptual change. The most powerful modern example is the replacement of Newtonian mechanics by relativistic and probabilistic theories after the work of Einstein and Heisenberg. An early example in astronomy is the discovery of the moons of Jupiter by Galileo. The initial evidence based on telescopic observations could be seen as mere aberrations of the telescope. It was only as the telescope was seen as reliable that even initial credence was given to the observations. And it was only with the development of the theory of optics that it was seen as productive enough to warrant serious study. The acceptance of Galileo's work had to wait upon its application to new phenomena (free-fall) and ultimately to developments in both the technology (both optical and mathematical) and the theories of Kepler and Newton (Toulmin and Goodfield [16, pp. 189ff]).

The intuition drawn from physical science is clear, We expect our descriptions and measurements to increase in reliable detail and to afford predictive success at an increasingly nuanced level, that is to say, the models of our assertions must be increasingly adequate to the reality that they attempt to describe. But increasingly adequate descriptions and measurements reflect the contrast between what we see and what our theories ask us to see. The building of adequate theories and their associated models is a confrontation with evidence pro and con. It is the dynamics of this advance that moves truth from truth in a model to truth in a chain of models, the members of which are increasing in adequacy to the phenomena as we test and probe in standard ways. Those standard ways are themselves difficult to ascertain, but determine them we do in practice, in all of those domains in which we are increasingly successful with our inquiries. We are rarely right or wrong simpliciter (at least not in areas that requires argument). Rather the picture emerges through argumentation and other methods of inquiry.

Whatever the intuitive power of the MET it is both simple and technically unsophisticated when contrasted with recent work by logicians and especially computer scientists who have attempted to capture the recent interest in the dialectics of argument by creating formal systems. In the next section I turn to a recent synoptic presentation of such efforts and give some indications of the relation of these approaches to the MET.

6 The Underlying Logic

Earlier I maintained that a generalization in mature science put forward based on empirical evidence has the following four essential properties:

1. The domain of its application is to be successively redefined in relationship to the ontological commitment of supporting theories.
2. It supports contrary to fact conditionals as a function of the generalization being reducible to a theory that explains the generalization in terms of these underlying theories.
3. Inferences (including resistance to counter examples via modes tollens) are adaptive, that is they respond to new information (monotonic if the new information is a consequence of the original set of proposition; ampliative adaptive logics are non-monotonic if they respond to additions to the set of propositions).
4. Estimates of truth likeness of the generalization can be rationally supported in light of evidence over time.

It should be clear that the MET meets the first condition. Whether I met the second condition, that is have a general solution to the problem of counterfactuals is debatable. But in the terms of the condition as specified, the solution is obvious as long as there is a reducing theory, for if a description contrary to fact is in fact reduced the falsity of the antecedent is irrelevant to the derivation of the conditional. This is the power of representing functions, recall MET, 2, above: for functions f^* in F^* for reducing theory, T^*, and reduced theory, T, f^* *reduces* T to T^* is equivalent to, for every model m^* of

T^* there exists a model of m of T such that, for every sentence s of T, s is true in m if and only if $f^*(s)$ is true in m^*. Whence the truth of a conditional of T is guaranteed in the relevant case via its truth in models of T^* and indifferent to the assignment in a model of T in which the antecedent may be false.

There is a price to pay, however, the highest-level reducers do not support counterfactuals. This is less of a problem than it might seem. If there is an alternative to a highest order theory, T^*, it is completely unconstrained by conditions other than those on the range of its reductive power. This means, among other things that highest order alternatives are not easy to find. On the other hand, if we think of the deepest physical theories, they are unconstrained by anything other than the imagination of scholars so deeply knowledgeable about the range of its concerns, even the wildest imaginings have to reflect the most profound knowledge of the field and the immediate problems that highest order theories most confront as they struggle to be adequate as reducers at the next lower levels. Nevertheless, since there is no guarantee from above, counterfactual analysis is always open. That is a replacement theory, $T\#$, if it addresses the range of reduction of T^* has comparative warrant strength of zero since arguments pro and con will be assigned the same value. The only support possible is for one of the pair to result in massive increases as the reinterpretation of the lower-order theories lead to profound differences in the breadth of reducing theories. Think of how quantum mechanics opened up vast realms of new applications even while it managed the vast yield of chemical and physical results under the older theories by offering reinterpretations of older explanations in its terms.

The intent behind the MET is to capture the underlying epistemic structure of mature physical science. But inquiry does not unfold in an instant and so the underlying logic must be adaptive, that is, sensitive to both newly realized and newly acquired knowledge. In what follows I offer a brief sketch of an attempt at amalgamating the distinctions that the MET affords with the adaptive logic, **LA**, of Straßer and Seselja [14]. Adaptive logics are concerned with defeasible reasoning, and are sometimes non-monotonic (ampliative abductive logics). Here we are concerned with the ability to formalize an account of the logic of attack and defense relevant to our concern with A/O opposition and so will assume that all of the relevant arguments are available and so we follow Straßer and Seselja in grafting adaptive rules onto classical logic.

In what follows I will not attempt more than a brief sketch of relevant aspects of Straßer and Seselja's work, which includes a wealth of logical detail that warrants careful study. Nor will I include a more recent paper that includes their attempt at a more nuanced approach to adaptive logic relevant to scientific explanation (Seselja and Straßer [13]). The more recent paper is based on a rather different image of science than the one that forms the paradigm for the MET. Their concern is with the polytheoretic arguments surrounding continental drift, rather than the unified theoretic context that physical chemistry exemplifies. I focus on the earlier paper since it includes a policy on defeating challenges that grows out of the classical concern with counter-examples. This gives me a firm background against which I can integrate my own proposals.

The MET offers a metamathematical image of explanation that supports a theory of warrant strength. Like adaptive logic it should be sensitive to the arguments that surround attempts at explanation. Such attempts are epitomized by A/O oppositions; challenges to generalizations by counter-examples. In what follows, I will attempt to amalgamate the distinctions that the MET supports into a system of adaptive logic developed by Straßer and Seselja [14] In what follows the set of arguments are stable (monotonic). We are concerned with the ability to formalize an account of the logic of attack and defense relevant to our concern with A/O opposition and can assume that all of the propositions involved are in principle available.

The basis for the construction, *AF*, is as follows, with lower case italic letters $a_1, a_2, a_3, \ldots$ for arguments and lower case letters a, b, c, $\ldots$ as meta-variables for arguments. (Dung [5] described in Straßer and Seselja [14, p. 136].)

Let $An =$ df $\{a_1, a_2, \ldots, a_n\}$.

Definition 1 An *AF* is a pair $\langle A, \rightarrow \rangle$ where $A \subseteq An$ is a finite set of arguments, and $\rightarrow \subseteq A \times A$ is a relation between arguments. The expression a $\rightarrow$ b is pronounced as 'a attacks b'.

Given an $AF \langle A, \rightarrow \rangle$, we are particularly interested in giving an account of reasonable choices of arguments in A: a minimal criterion is, for instance, that no argument in a selection S should attack another argument in S. Of course, more interesting selection types can be defined:

Definition 2 Given an *AF* $A = \langle A, \rightarrow \rangle$ we define the following notions.

(i) An argument a is *attacked* by a set of arguments $B \subseteq A$ iff there is a b $\in B$ such that b $\rightarrow$ a.

(ii) An argument a is *acceptable* with respect to a set of arguments $C \subseteq A$, iff every attacker of a is attacked by a member of C. It is said that C *defends* a.

(iii) A set of arguments $S \subseteq A$ is *conflict-free* iff no argument in S attacks an argument in S.

(iv) A conflict-free set of arguments $S \subseteq A$ is *admissible* iff each argument in S is acceptable with respect to S.

(v) A set of arguments $S \subseteq A$ is a *preferred extension* iff it is a maximal (w.r.t. $\subseteq$) admissible set.

(vi) A conflict-free set of arguments $S \subseteq A$ is a *stable extension* iff it attacks every argument in $A \backslash S$.

(vii) An admissible set of arguments $S \subseteq A$ is a *complete extension* iff $F(S) = S$, where $F(S) =$ df $\{c \mid S$ defends $\{c\}\}$.

(viii) A set of arguments $S \subseteq A$ is a *grounded extension* iff it is the minimal (w.r.t. $\subseteq$) complete extension.

(ix) A complete extension $S \subseteq A$ is a *semi-stable extension* iff $S \cup S^+$ is maximal (w.r.t. $\subseteq$), where S^+ is the set of all arguments in $A \backslash S$ which are attacked by S [14, p. 136].

The intuition should be clear; a set of arguments partitioned into arguments and attackers yields levels of acceptability to the extent that attacks on arguments are responded to by other arguments (defenses). The core ideal is (ii). An argument is acceptable is it meets the challenge from attacking argument. In terms of Johnson [8] the argumentation has met its dialectical obligation. The levels of acceptability reflect what might be seen as pragmatic consistency. This is clear in (iii) and (iv) arguments are admissible if 'conflict free' both in terms of pragmatic self-consistency and in appropriate responses to attacks.

The following criteria (v) through (ix) build up to admissibility in terms of extensions. Preferred extensions are maximal admissible sets, the largest admissible subset of A. Stable extension attacks all arguments in A other than those in S. If in addition, S defends against attacks it is a complete extension (vii). The smallest subset of S that is complete is a grounded extension (presumably, it grounds the argument by defending the attacking arguments). Finally, a semi-stable extension takes the largest complete extension that includes the arguments attacked and defended.

This leads to two key epistemological definitions:

(x) A set of arguments $S \subseteq A$ is *credulously accepted* according to preferred [(semi)-stable, complete or grounded] semantics (w.r.t. A) iff it is contained in at least one preferred [(semi)-stable, complete or grounded] extension of A.

(xi) A set of arguments $S \subseteq A$ is *sceptically accepted* according to preferred [(semi)-stable, complete or grounded] semantics (w.r.t. A) iff it is contained in every preferred [(semi)-stable, complete or grounded] extension of A' [14, p. 136].

The last remark leads to a footnote: "Our notion of defeat differs from the way defeat is defined in various preference or value-based enhancements of Dung's abstract *AF*. Defeat is there usually defined as a binary relation between arguments, which is a subset of the attack relation: a_1 defeats a_2 iff a_1 attacks a_2 and a_2 is not 'preferable' to a_1. The preferability of one argument over another is modeled in different ways" [14, p. 137]. The authors offer as examples, Amgoud and Cayrol [1] and Bench-Capon [3].

My contribution via the theory of warrants can be seen as an alternative version of preference, or better, epistemological warrant. What my view shares with many policies on preference, is that it is dependent on pragmatic considerations. This will become clearer as we turn to my amalgamation of the theory or warrant strength and the adaptive logic of Straßer and Seselja [14].

In what follows, Vn is the set of propositions that constitute an argument and Wn the set of arguments in attack and counter (defense). The amalgamation is straightforward, requiring no more than a redefinition of the valuation space.

Substituting directly in the definition of **LA** (Straßer and Seselja [14, p. 138]) we generate **LA*** with the following essential change, for valuations v and **L**: v^* and **L***

$$v^* : Vn \cup Wn \to \{0 \leq i \leq 1\}$$

L*-valuation v^***LM**: $Wn \to \{0 \leq i \leq 1\}$, where i is a selection from the set of one-place decimals $\{0, .1, .2, .3, .4, .5, .6, 1\}$. Vn is a set of proposition letters, representing the substance of an argument (in my construction conditionals composed of a conjunction of irrelevance free premises, including a nomic generalization and a consequent). Wn is the set of formula representing arguments composed of arguments and arrows, '$\to$' (attacks and defenses). The assignment to a proposition letters (arguments), a_i, is fairly straightforward:

$$V^* : a_i \in Vn \to \{0, .1, .2, .3, .4, .5, .6, 1\}$$

For propositions, 0 is assigned to unsupported empirical claims, whether singular statements or nomic generalization, 1 is assigned to the theory at the end of the Peirceian rainbow, including both its generalizations and specific implications. When 0 or 1 are assigned to arguments it is an argumentational dead-heat, the first for lack of warrant and the latter for power beyond resolution.

The assignment to the elements of Wn is more complex. Recall the formulas in Wn are of the form $a \to b$, for arguments a and b, where in the first instance a is a counter example to b.

$$V^* : a \to b \in Wn = \{v^*b - v^*a\}$$

That is, as a first approximation, the power of an attack is the strength of the argument attacked minus the strength of the attacker.

And so we have, based on **L** (Straßer and Seselja [14, pp. 138ff]). **LA*** is well-managed classical propositional logic (minimally) enriched by rules $(R \twoheadrightarrow)$, (Rad). The rule $(R \twoheadrightarrow)$ characterizes the core of the logic for abstract argumentation: 'if α is valid and it attacks $\beta, \alpha \twoheadrightarrow \beta$, then β should not be a consequence of our logic.' (Rad) is the generalization across arguments α and $\alpha \twoheadrightarrow \beta$ in Vn.

$$\frac{\alpha \; \alpha \twoheadrightarrow \beta}{\text{not-}\beta} \quad (R \twoheadrightarrow)$$

$$\frac{a \; \beta \twoheadrightarrow \alpha}{\text{def } \beta} \quad (Rad)$$

Where 'def β' is the disjunction of arguments $a \in Vn$ such that

$$\left(a \& (a \twoheadrightarrow \beta)\right)$$

My inclusion of the phrase 'well-managed,' for **LA*** satisfies the constraints on the explanation relation in the MET. Since this is a constraint on propositional elements it doesn't alter the logic of the situation, only its management in the name of relevance. The inclusion of 'minimally' is required since the full system of **LA** has additional constraints to guard against inconsistency among other things that a fuller elaboration of **LA*** would require.

We define the semantics for logics $\mathbf{L}^* \in \{\mathbf{LA}^*\}$ via assignment functions:

$$v^* : a_i \in Vn \to \{0, .1, .2, .3, .4, .5, .6, 1\}$$
$$v^* : a \twoheadrightarrow b \in Wn = \{v^*b - v^*a\}$$
$$v : Vn \cup Wn \to \{0 \le i \le 1\}$$

where i is a one place decimal and an $\mathbf{L}^*$-valuation $v\mathbf{L}^*M \colon Wn \to \{0 \le i \le 1\}$ determined by the assignment.

We follow Straßer and Seselja's structure, but expand the range of the assignment function as before; the assignment to rules remains the same.

$$V^* : Vn \cup Wn \to \{0 \le i \le 1\}$$

that assigns truth values to both, propositional letters and 'attacks', i.e. formulas in Wn. A model M is defined by an assignment function v. The following definitions are used in order to define the **LA**-valuation under, v, for the rules as follows:

$$v R \twoheadrightarrow =\text{df } 1 - \max\left(\min(v(\alpha), v(\alpha \twoheadrightarrow \beta), v(\beta))\right) \quad \text{for } \alpha, \beta \in Vn$$
$$v Rad =\text{df } 1 - \max\left(\min(v(\alpha), v(\beta \twoheadrightarrow \alpha))\right) \quad \text{for } \alpha, \beta \in Vn$$
$$1 - \max\left(\min(v(\gamma), v(\gamma \twoheadrightarrow \beta))\right) \quad \text{for } \gamma \in Vn$$

(following, Straßer and Seselja [14, p. 140]).

The next step is too modify the extensions by increasing the levels of acceptance and in light of the power of an attack based on the strength of the warrants involved. The problem for me is that the requirement to address all challenges is excessive. Some consideration must be given to the appropriateness of attack given the argumentational surround. The issue is usefully framed in terms of Ralph Johnson's [8] distinction between the illative core, the argumentation brought forward in support of a claim, the dialectical tier, the arguments attacking the illative core and the dialectical obligation to respond. The issue within the discussion of Johnson's distinction is the extent of the dialectical obligation, that is, which arguments against must be countered if the dialectical obligation is to be met. I recall a paper at an argumentation conference focused on Johnson's account, the presenter was straightforward: 'If I have to respond to every counter-argument I wouldn't bother to argue.' A sound policy, given the effort it takes to support a position and the indefinite availability of countervailing considerations. Still even if impractical in practice concern with the dialectical tier is essential to the theory of argument. But since the tension is between complementary dialogical virtues a solution needs to be fine-tuned to the circumstances. There are two approaches, a policy approach and a computational one. First as to policy:

We modify the notion of attack (and defense) to arguments worth of consideration (we will use the word 'appropriate' and its variants). This leads to a modification of the notion of valid argument in relation to rules $(R \rightarrow)$ and (Rad) where valid arguments are those certified by the rules but restricted to appropriate arguments. Appropriate arguments, in turn, meet the level of stringency determined by the practice in the discourse community. This has been at the core of policy decision in the MET and continues to govern our constructions. It essentially connects logic to argument and argument to well-functioning discourse communities, of which, the community of physical chemists remains an exemplary example.

(i*) An argument a is *appropriately attacked* by a set of arguments $B \subseteq A$ iff there is a $b \in B$ such that $b \rightarrow a$ and either $v^*b \geq v^*a$ or $v^*b \geq i \leq v^*a$, where i is the difference of the values of v^*b and v^*a, for some value, i, set by the discourse community.

The first condition is relatively trivial. The second is the more serious contender for A/O oppositions of the sort we envision, a counter-example to a generalization.

We may think of i by analogy to a significance level; it sets the threshold for appropriate attacks. Although, unlike a significance level, its value is not constant, but the difference between the values assigned to the arguments as a function of the strength of the anomaly needed and the warrant for the anomaly itself (P1 and P2 above). That is, i, is an additive function of the values assigned to the arguments (P*). It is a ratio rather than a fixed quantity. This, like so much of the MET, cannot be defined a priori, the argumentational practice of experts in a field determine the level of warrant a attack must sustain to be taken seriously. Although the apparent relativization of (i*) to actual practice may seem too arbitrary for a logical system, the MET permits a number of internal metrics based on the available information in a field (for example, MET, 1.2.1). Since we enumerate models and have chains of particular lengths it is obvious that we not only order model chains by the subset relation, but we can count their length. The problem is that if we construe models (at least at the lowest levels) as models of data, replication beyond a certain point quickly reaches a point of diminishing returns for explanatory power. Of more interest is the internal structure of approximations. Instead of counting models we can give numerical estimates in terms of approximations and so index model chains in relation to their distance from the intended model. Reduction relations offer similar metrical possibilities (for example, MET, 1.3). This if elaborated gives us the possibility of a computational approach.

The rest of the modifications of follow the same pattern (based on the formulation in [14, p. 136], see above).

(ii*) An argument a is *acceptable* with respect to a set of arguments $C \subseteq A$, iff every appropriate attacker of a is attacked by a member of C. It is said that C *defends* a.

(vi*) A conflict-free set of arguments $S \subseteq A$ is a *stable extension* iff it appropriately attacks every argument in $A \backslash S$.

(vii*) An admissible set of arguments $S \subseteq A$ is a *complete extension* iff $F(S) = S$, where $F(S)$ =df $\{c \mid S$ appropriately defends $\{c\}\}$. (Note: an appropriate defense is parasitic upon an appropriate attack in that if b appropriately attacks a and c appropriately defends a, then c appropriately attacks b.)

(ix*) A complete extension $S \subseteq A$ is a *semi-stable extension* iff $S \cup S^+$ is maximal (w.r.t. $\subseteq$), where S^+ is the set of all arguments in $A \backslash S$ which are appropriately attacked by S.

Without attempting the specifics of the logical process a clarification of the underlying intuition gives some sense of what the give and take of attack and defense looks like (for plausible definitions of connectives and propositional rules that seem modifiable to meet our needs in terms of a many-valued logic, see Straßer and Seselja [14, pp. 140ff]). If we assume, as in AF, that the set of arguments is given (AL is not ampliative) the strength of the warrants indicate the available resources for meeting challenges.

We will turn to the language of the square of opposition to make the intuition clear, using integers to reflect the levels of strength in supporting warrants. So, for example, an empirical counter-example, O, is offered to an empirical generalization A. The first question is whether the counter example is appropriate, that is, it is supported by a warrant available in the set of arguments. That is, it is at least at the first level of strength, O is a type-1 anomaly. If so, we now have a problem of A/E opposition, competing generalizations. But since an A proposition implies a corresponding I there is also E/I opposition, I is (at least) a type-1 anomaly for the corresponding E. This requires us to see if either A or E have additional strength, that is if they are type-2 warrants or stronger. This in turn generates the possibility of further evidence (I or O propositions) implied by warrants of greater strength, resulting in crucial experiments to challenge strong or even super strong warrants (anomalies of type-3 through type-6).

How to resolve these arguments where the levels for competing generalizations are at the same or adjacent levels is, again, a pragmatic decision based on the confidence of the proponents in the total set of arguments as they become available through theoretical dialog. In the history of science this, more often than not, includes new evidence and new supporting theoretical hypotheses (new generalizations). Capturing such a complex historical phenomena requires a non-monotonic ampliative adaptive logic that permits the set of arguments to increase over time. The MET sets a framework for further work in logic by indicating what the epistemic stakes are in terms of the depth and breadth of warrants. My hope is that by making this intuitive construction available to logicians and computer scientists further technical elaboration will result and ultimately students of the relevant sciences will collaborate with computer scientists to develop a computational account based on the actual facts of relevant inquiry.

References

1. Amgoud, L., Cayrol, C.: A reasoning model based on the production of acceptable arguments. Ann. Math. Artif. Intell. **34**, 197–215 (2002)

2. Apostel, L.: Toward the formal study of models in the formal and non-formal sciences. In: Freudenthal, H. (ed.) The Concept and the Role of the Model in the Mathematical and Natural Sciences, pp. 1–37. Reidel, Dordrecht (1961)
3. Bench-Capon, T.J.M.: Persuasion in practical argument using value-based argumentation frameworks. J. Log. Comput. **13**, 429–448 (2003)
4. Cartwright, N.: How the Laws of Physics Lie. Oxford University Press, Oxford (1983)
5. Dung, P.M.: On the acceptability of arguments and its fundamental role in nonmonotonic reasoning, logic programming and n-person games. Artif. Intell. **77**, 321–358 (1995)
6. Eberle, R.: Replacing one theory by another under preservation of a given feature. Philos. Sci. **38**, 486–500 (1971)
7. Goldstein, I., Goldstein, F.: The Experience of Science. Plenum, New York (1984)
8. Johnson, R.: Manifest Rationality. Lawrence Erlbaum, Mahwah (2000)
9. Nagel, E.: The Structure of Science. Harcourt, Brace and World, New York (1961)
10. Popper, K.: Conjectures and Refutations. Routledge and Kegan Paul, London (1963)
11. Riordan, M., Hoddeson, L.: Crystal Fire. Norton, New York (1997)
12. Scerri, E.: The Periodic Table: Its Story and Its Significance. Oxford University Press, New York (2007)
13. Seselja, D., Straßer, C.: Abstract argumentation and explanation applied to scientific debates. Synthese (2011, in print). doi:10.1007/s11229-011-9964-y
14. Straßer, C., Seselja, D.: Towards the proof-theoretic unification of Dung's argumentation framework: an adaptive logic approach. J. Log. Comput. **21**(2), 133–156 (2011)
15. Toulmin, S.: The Uses of Argument. Cambridge University Press, Cambridge (1969)
16. Toulmin, S., Goodfield, J.: The Fabric of the Heavens. Harper, New York (1961)
17. Weinstein, M.: Exemplifying an internal realist theory of truth. Philosophica **69**(1), 11–40 (2002)
18. Weinstein, M.: A metamathematical extension of the Toulmin agenda. In: Hitchcock, D., Verheif, J. (eds.) Arguing on the Toulmin Model: New Essays on Argument Analysis and Evaluation, pp. 49–69. Springer, Dordrecht (2006)
19. Weinstein, M.: A metamathematical model of emerging truth. In: Béziau, J.-Y., Costa-Leite, A. (eds.) Dimensions of Logical Concepts, Campinas, Brazil. Coleção CLE, vol. 55, pp. 49–64 (2009)
20. Weinstein, M.: Arguing towards truth: the case of the periodic table. Argumentation **25**(2), 185–197 (2011)

M. Weinstein (✉)
Department of Educational Foundations, Montclair State University, Upper Montclair, NJ 07044, USA
e-mail: weinsteinm@mail.montclair.edu